CROP PRODUCTION
Principles and Practices

Under the Editorship of James J. Vorst

CROP PRODUCTION
Principles AND Practices
FOURTH EDITION

DARREL S. METCALFE
University of Arizona

DONALD M. ELKINS
Southern Illinois University

MACMILLAN PUBLISHING CO., INC.
New York
COLLIER MACMILLAN PUBLISHERS
London

Dedicated to
aspiring young agronomists
in our colleges and universities,
and to
those devoted teachers and researchers
who have touched our lives,
and inspired and influenced us

COPYRIGHT © 1980, MACMILLAN PUBLISHING CO., INC.

PRINTED IN THE UNITED STATES OF AMERICA

Earlier edition entitled *Crop Production: Principles and Practices* by Harold D. Hughes and Edwin R. Henson, copyright 1930 by Macmillan Publishing Co., Inc. Revised editions entitled *Crop Production: Principles and Practices* by Harold D. Hughes, Darrel S. Metcalfe, Edwin R. Henson, and Iver J. Johnson, © 1957 by Macmillan Publishing Co., Inc., and *Crop Production* by Harold D. Hughes and Darrel S. Metcalfe copyright © 1972 by Macmillan Publishing Co., Inc.

MACMILLAN PUBLISHING CO., INC.
866 Third Avenue, New York, New York 10022

COLLIER MACMILLAN CANADA, LTD.

Library of Congress Cataloging in Publication Data

Metcalfe, Darrel S
Crop production
First-2d ed. by H. D. Hughes and E. R. Henson;
3d ed. by H. D. Hughes and D. S. Metcalfe.
Bibliography: p.
Includes index.
1. Field crops. I. Elkins, Donald M., joint author. II. Hughes, Harold De Mott, 1882–1969.
Crop production. III. Title.
SB185.H75 1980   633   79–18404
ISBN 0–02–380710–5

Printing: 4 5 6 7 8   Year: 3 4 5 6

ISBN   0-02-380710-5

PREFACE

FOR 25 years the first edition of *Crop Production* was used extensively in many universities and colleges. The second edition was an entirely reorganized and rewritten text. In it the revising authors preferred to include only such representative information as they believed would stimulate thinking, in establishing certain basic principles in relation to crop production, rather than to be encyclopedic, as in the first edition.

In the third edition entirely new chapters were added, including "The World Food-Population Problem"; "Climate, Weather, and Crops"; and "Crop Production and Air, Water, and Soil Pollution." New approaches and new materials were used in chapters on world production and distribution, United States production and distribution, agriculture production adjustments, and science and agriculture research. Other chapters were combined. For example, the material on weeds, insects, and disease, which extends for three chapters in the second edition, was combined into one chapter titled "Crop Enemies." The total number of chapters was reduced from 55 to 42.

Most of the basic chapters from the third edition have been retained in the fourth edition, but materials have been completely reorganized and rearranged under the areas of history and current status, environmental factors, botany, crop production practices, field crops, forage crops, and crop research and improvement. In addition, the field crop chapters are organized into the sections grain crops, oil and fiber crops, sugar crops, and drug and miscellaneous crops.

Many chapters have been expanded considerably. The most notable example is the retitled chapter "Tillage and Cultivation," which has been revised extensively and now includes a more complete coverage of minimum tillage systems.

The chapters "The Corn Crop" and "Corn Production Practices" from the third edition have been combined into the chapter "Corn or Maize." Selected material from the chapters "Cotton and Minor Oil and Fiber Crops" and "Tobacco and Other Specialty Crops" has been removed from these chapters, revised, and expanded into the new chapter "Minor Oil and Fiber Crops." The chapter, "Peanuts and Other Large-Seeded Legumes," has been revised into "Peanuts and Other Edible Legumes." Large-Seeded forage legumes are discussed in appropriate forage chapters.

The chapters on "Grasses for the Humid North," "Grasses for the Humid South," and "Dryland Grasses" have been reorganized into "Cool-

v

Season Perennial Grasses," "Warm-Season Grasses," and "Native and Related Grasses." The chapters "Hybrid Corn" and "Improvement of Field Crops" have been combined into the new chapter "Crop Breeding and Improvement." This chapter and a chapter on "Science and Agricultural Research" are included as an appropriate last section on looking toward the future in crop production.

In certain areas, where comparatively little research has been done in more recent years, the authors have not hesitated to include these earlier results. But wherever new research data were available, they were added to the fourth edition.

The authors realize that greater emphasis will be given to certain production phases in one part of the country than in some others and that there will be great differences in the organization of the course this text will serve. For this reason this new edition lends itself to chapter selection based on regional and geographic crop variations.

The illustrative material for figures has been carefully selected, with the interest of the student in mind. Many more photographs, charts, and other figures have been included in the fourth edition.

The authors acknowledge with thanks the assistance of their colleagues who gave generously of their time and thought to chapter material having to do with their fields of respective specialized knowledge. Our appreciation is extended to all agencies, companies, or individuals who contributed photographs related to specialized topics. We express our special thanks to Vicky Hagemann and Wendy George for their tireless efforts in typing the manuscript.

D. S. M.
D. M. E.

CONTENTS

Part I
HISTORY AND CURRENT STATUS OF CROP PRODUCTION

CHAPTER

1 Agricultural Beginnings   3
2 World Production-Distribution   13
3 U.S. Production-Distribution   30
4 The World Food-Population Problem   49
5 Agricultural Production Adjustments   70

Part II
ENVIRONMENTAL FACTORS AFFECTING CROP PRODUCTION

CHAPTER

6 Climate, Weather, and Crops   99
7 Crop Production and Air, Water, and Soil Pollution   117
8 Water Use by Crops   129
9 The Soil and Cropping Practices   143

Part III
BOTANY OF CROP PLANTS

CHAPTER

10 Crop Groups and Classification   167
11 Structure and Growth of Plants   176
12 Roots of Field Crops   188

Part IV
CROP PRODUCTION PRACTICES

CHAPTER

13 Soil and Water Conserving Practices   205
14 Rotations and Cropping Systems   220
15 Crop Seed   235
16 Tillage and Cultivation Practices   254

17 Irrigation, Drainage, and Dryland Farming   279
18 Crop Enemies   303

Part V
FIELD CROPS

A. GRAIN CROPS

CHAPTER
19 Corn or Maize   333
20 The Sorghums   366
21 Wheat and Rye   388
22 Oats, Barley, and Rice   416

B. OIL AND FIBER CROPS

CHAPTER
23 Soybeans   444
24 Cotton   468
25 Minor Oil and Fiber Crops   492

C. SUGAR CROPS

CHAPTER
26 The Sugar Crops   511

D. DRUG AND MISCELLANEOUS CROPS

CHAPTER
27 Tobacco   529
28 Potatoes   546
29 Peanuts and Other Edible Legumes   563

Part VI
FORAGE CROPS

CHAPTER
30 Forages   581
31 Pastures and Pasture Improvement   596
32 Hay and Haymaking   609
33 Silage, Green Chop, and Succulage   625
34 Alfalfa   636
35 The True Clovers   648
36 Birdsfoot Trefoil, Lespedeza, and Other Forage
    Legumes   663
37 Cool-Season Perennial Grasses   684
38 Warm-Season Grasses   699
39 Native and Related Introduced Grasses   707

Part VII
CROP RESEARCH AND IMPROVEMENT

CHAPTER
40 Science and Agricultural Research   721
41 Crop Breeding and Improvement   733

Crop Terms   758
Index   769

Part I
History and Current Status
of Crop Production

CHAPTER 1
AGRICULTURAL BEGINNINGS

Agriculture, which had its beginning some eight or ten thousand years ago, changed man from a nomadic hunting and food-gathering animal who had spent more than half a million years searching for his next meal, to a sedentary, food-growing husbandman, exercising at least some degree of control over his means of subsistence and enjoying at least a certain amount of leisure. . . . Man during his history in all parts of the world has used for food more than 3000 species of plants. Of these only some 150 have ever been extensively cultivated and only about an even dozen are important from the standpoint of the energy which they contribute.

PAUL C. MANGELSDORF

AGRICULTURE is said to have had its beginning when man found that he could gather seed of certain wild grasses that had relatively large and numerous seed, which he could plant in land that he controlled, and later gather food for which formerly he had to search far and wide.

This beginning of agriculture is believed to have occurred about 8,000 to 10,000 years ago, probably developing in both the Old World and the New World at about the same time. For many years it was the general belief that agriculture had its beginning in Southern Asia; some had it in Southwest Asia, and others in Southeast Asia. More recently the belief has come to prevail that agriculture probably developed independently in several widely separated parts of the world.

An American Geographic Society report by Sauer "recognized three centers of seed domestication" in the Old World and one in the New World: China, western India extending to the eastern Mediterranean, Ethiopia, Mexico, and Central America.

Centers of Crop Origin

The most thorough and extensive investigation of cultivated plant origins was made by Vavilov, a distinguished Russian geneticist recognized generally as the world's best authority on the wheats and several other important crop plants. The centers of origin of some of our more important crop plants as listed by Vavilov are:

1. China, central and western—soybeans, barley, naked oats, millets, buckwheat, sugarcane (total of 136 species originated in this center).

3

2. India, including Burma and Assam—rice, cotton, cowpeas (117 total species).
3. Central Asia—common wheat, cotton, sesame, hemp, peas, lentils (42 species).
4. Near East, including Asia Minor and Iran—some wheats, two-row barley, rye, common oats, alfalfa, vetch (83 species).
5. Mediterranean area—durum, emmer, and spelt wheats, some oats and barley, hops, flax (84 species).
6. Ethiopia—common six-row barley, durum wheat, grain sorghum, millet, castorbeans, chickpeas, lentils, coffee (38 species).
7. South Mexico and Central America—corn or maize, upland cotton, beans, sweet potato (49 species).
8. South America, especially Peru, Bolivia, and parts of Ecuador—white potato, tobacco, peanuts, tomato, some cottons, lupines (45 species).

Scientists believe there are only four currently grown crops that originated in what is now the United States: sunflowers, pecans, strawberries, and cranberries.

Food Production Beginnings

We are accustomed to think of man's course in the world as one of steady progress from savagery to our present civilization. But a review of world history indicates periods of thousands of years of "dead-level" existence. Historians divide man's existence into three periods:

1. Three million years of dead-level savagery.
2. Ten thousand years of hand-labor cultivation, with no significant change in the relationship of man to his environment.
3. One hundred and fifty years of scientific progress.

It is likely that man has been on the earth for three million years. He originally was expected to subsist on the land by gathering plant and animal foods. When man ate of the Tree of Knowledge, he was banished from the Garden of Eden with God's warning: ". . . In toil you shall eat of it [ground] all the days of your life. Both thorns and thistles it shall grow for you; and you shall eat the plants of the field; by the sweat of your face you shall eat bread, till you return to the ground." . . . (*Genesis* 3:17–19, NAS). Moreover, we know that Cain, the son of Adam and Eve, was a farmer. We are told that he presented God an unacceptable sacrifice—crops from the field rather than an animal.

More than 99 percent of man's existence on earth has been as a hunter-gatherer. Only during the last 10,000 years or so has he begun to domesticate plants and animals. At first man had to supplement the food he produced with hunting and gathering, but gradually he became less dependent on wild food sources. More than 90 percent of men have lived as hunters and gatherers, and only about 6 percent as agricultural producers.

It has been suggested that the first successful domestication of plants was in Thailand. Remnants of soybeans and rice from 10,000 years ago have been discovered. Most cultures were likely collectors of grain prior to ca. 8,000 B.C. The evidence from archaeological studies, including sickles and grinding stones, indicates that about 1,000 years later man had begun to cultivate grain. One of the earliest sites to furnish adequate evidence of grain production was Jarmo in Iraq, where seeds of wheat and barley were found dating to 6750 B.C.

The Egyptians have had a well-advanced agriculture for at least 6,000 years. Most of the written documents cover only part of the last 5,000 years. Many of the records of early agriculture come from writings of Greek and Roman scholars such as Herodotus and Pliny. The Greeks have fewer records of history than have the Romans, who gave us extensive authentic information on various aspects of crop production.

Crop production began somewhat later in the New World than in the Old World, and probably had its origins in Mexico. The time of the first cultivated plants in the New World was about 5000 B.C. in the Tehuacan area of South Central Mexico, where corn, squash, chili peppers, avocado, and amaranth were found. A second important agricultural center in the Americas was Peru. About 3000 B.C. it is probable that cotton, squash, gourds, lima beans, and chili peppers were produced there. The earliest evidence of corn in coastal Peru is dated at about 2000 B.C.

Crops in Worship

Food crops, and especially the cereals, appear to have been identified from the very beginning with religion and worship. This is not surprising when we consider that down through the ages man's very existence depended on successful harvests. The great concern of ancient peoples for food took the form of worshiping and propitiating deities. Ceres was the Roman goddess of agriculture, with the first temple for her worship dedicated in 496 B.C. So closely was she identified with the small grains then grown that these have since come to be known as "cereals."

"Give us this day our daily bread," pray Christians everywhere today. In Biblical times this was a grim petition, when bread saved, man from hunger and starvation. Many Christian families daily give thanks to God for food at the table, even though they themselves have not experienced a crucial struggle for life-sustaining food. In the United States each year Thanksgiving Day is observed. It is difficult for many who are far removed from agriculture in their living and work to realize, as did our Pilgrim Fathers, that the survival of all depends on the harvest from our fields.

The Cereals

Every great advance in civilization has paralleled man's ability to produce and use the crops of the field, especially food crops. A productive cereal has supported every great civilization. Rice has been the staff of life for nearly half the world's population, including India, China, and Japan. Certain sorghums and millets have been the cereals of Africa. Wheat

A

B

has furnished "the bread of life" for the great civilizations of the Mediterranean area, including Syria, Egypt, Carthage, Greece, and Rome. Corn has been outstanding in the development of the Western Hemisphere. No other crops are equal to the cereals in making it possible for man to provide his food with a minimum of labor, land, and equipment.

The Legume Rotation

Theophrastus, the Greek, in his botanical writings about 300 B.C., reported that a bean crop was beneficial to enriching the soil. Some of the first great literary efforts of the Romans included information on legumes: Cato, the Censor (234–149 B.C.); Varro, a gentleman farmer (50 B.C.); Virgil, a vigorous advocate of alfalfa in the rotation, in one of the greatest treatises in agriculture in history, his *Georgics* (30 B.C.); and Columella, with his 12 books on agriculture in the first century.

It is quite remarkable that Greece produced so many outstanding philosophers with a keen appreciation of agriculture and agricultural practices, their writings and culture and agricultural observation to be copied by

FIGURE 1–1. *(A) There are many areas of the world where crude methods of agricultural production continue as in earlier centuries.* [SOURCE: Foreign Agriculture Service, USDA.] *(B) Indian farmers use oxen to winnow their bumper crop of wheat.* [SOURCE: A.I.D.] *(C) Settlers in an African village harvest rice by hand methods.* [SOURCE: FAO.]

C

the Romans. Some of the best minds of Rome, at the very beginning of the Christian Era and before, gave themselves to a study of plant growth and the soil that supports it; they left behind voluminous written records of their thinking and conclusions. Following the Greco-Roman period there appear to have been no new thoughts or observations on agricultural production for approximately 1,000 years.

The Rise of Scientific Research

Modern agriculture had its beginning in England early in the eighteenth century. The Englishman Jethro Tull (1674–1741), an Oxford University graduate turned farmer, invented the grain drill in 1701. In 1733 he published *Horse Hoeing Husbandry,* which established the principles of row-crop cultivation and was long considered an authoritative text on English agricultural practices. Arthur Young (1741–1820) was an agricultural experimentalist and prolific writer. His *Annals of Agriculture,* in 46 volumes, was much quoted. England remained a leader in improving crop production from 1700 to 1850.

The rise of science with fundamental studies of botanical matters began in the seventeenth century. The following are some of the milestones. First, the microscope was invented in 1590. With the aid of the microscope, Robert Hooke (1635–1703) described plant cells. Antoni Van Leewenhoek (1642–1723) observed and described protozoa and bacteria. The Dutch botanist Rudolph Jacob Camerarius (1661–1721) established the nature of sexual reproduction in plants. The Austrian Abbot Gregor Mendel (1822–1884) illustrated inheritance with the garden pea. The Swedish physician Carl von Linné (1708–1778) developed a system for classifying plants and animals. The Englishman Stephen Hales (1677–1761) pioneered studies of transpiration, and another Englishman, Joseph Priestley (1733–1804), first recognized the analogy between combustion and respiration.

Agricultural research began with John Bennet Lawes (1814–1901) and Joseph Henry Gilbert (1817–1901), in a private laboratory at Rothamsted, England. The research of Louis Pasteur (1822–1895) established that microorganisms are responsible for disease.

United States Agriculture

Except for what they learned from the Indians, the American colonists employed agricultural implements and methods that were very little changed from those in common use in ancient Rome.

Then suddenly, in half or three-quarters of a century, agricultural tech-

[OPPOSITE] FIGURE 1–2. *Early methods of harvesting crops in the United States involved (A) picking ("gathering") corn by hand, (B) shocking wheat and later threshing* [SOURCE: J. W. McManigal], *and (C) loading and hauling loose stacks of hay.* [SOURCE: Henry Field.]

A

B

C

nology and research were improved vastly, more than in the previous 2,000 years. This is referred to as the agricultural revolution.

In the last quarter of the eighteenth century, the very crude plows were made of wood, and metal points were in use in some places. The restless and progressive spirit of the eighteenth century had, however, discovered the inefficiency of the wooden plow, particularly as an implement for breaking sod. Many inventive men in Europe and America were considering the idea of a better plow made of iron. Finally, in the 1830s a steel plow became a reality when John Lane, in 1833, and John Deere, in 1837, began their commercial manufacture. Soon in all but a few remote places the wooden plow was replaced. Other inventions contributing to the mechanization of agriculture included the invention by Eli Whitney of the cotton gin in 1793 and the invention of the mechanical reaper by Cyrus McCormick in 1831.

The American farmers were relatively quick to see the advantages of labor-saving devices and of new methods and procedures in crop production.

The American Indian

The American Indian made a far greater contribution to agriculture than most people realize. Their high achievements in agriculture were their most important contribution to civilization. Sometimes these achievements are obscured by the depiction of the Indians as savages or violent people. It was not until the English settlers adopted the American Indian's agricultural plants, cultivation and harvesting methods, and processing or food preparation that they were assured of adequate food supplies. Indian corn, especially, brought the earliest colonists through those first hard years. Most agricultural students know the story of Squanto, the Wampanoag Indian, who saved the Pilgrims from almost certain starvation by teaching them to grow corn. Their English grain crop failed, but 20 acres (8 hectares) of corn succeeded when they used fish in each hill for fertilization. The Indians were first to practice row-cropping, planting in hills so the space between could be tilled. Before the discovery of America, most crops in Europe were broadcast, which did not permit intertillage. Ultimately, the union of the farming methods of the American Indian and of those of Europe produced the beginning of American agriculture and provided the essential basis for its development.

A complete list of the economic plants domesticated by the American Indian is extensive. More than 50 percent of America's farm products today consist of plants used by the Indian before Columbus discovered the New World. In all of North America, Indians cultivated 150 species of plants. The most important crop plants domesticated by the Indian are corn, cotton (New World species), peanuts, pumpkins, squashes, beans, potatoes, sweet potatoes, tobacco, and tomatoes. Of by far the greatest importance was the development of the many different types of corn, which were adapted to special uses and to widely different climatic conditions from the Equator north to Canada and from Montana to the far northeast Atlantic Coast. Corn yields in Indians' times averaged 20–50 bushels per acre (12.5–31.3 quintals/hectare) but sometimes yields of more than 100 bushels per acre (62.6 Q/ha) were achieved.

In growing their crops, the Indians had neither draft animals nor plowing machinery. All the work of planting, cultivating, and harvesting was done by hand. They used pointed and spadelike tools in turning the soil. The white man introduced the ox and the horse to supply power and ultimately developed field machinery that replaced the arduous manual labor. But the fundamental system of cultivation remained essentially the same as the white man found on his arrival in the New World.

The Food Problem

Available food always has been man's most important problem. Throughout most of the history of man, there has been a race between the supply of available food and the number of mouths to be fed. It would appear that as populations increased, nations made an effort to ensure a supply of cereals for their populations. Steps were taken in early Egypt (2000 B.C.) to store grain in years of plenty for the prospect of lean years to come (*Genesis* 21:46–57).

No period in recorded history has been free of famine. The year 879 is recorded as one of universal suffering from lack of food. The famine of 1125 is reported to have diminished the population of Germany by half. A famine devastated Hungary in 1505. As late as the middle of the seventeenth century famines were a common affliction in Europe. They occurred even in the eighteenth century. In 1870–1871 Persia lost 1.5 million inhabitants, one-fourth its entire population. In China, 9.5 million are said to have perished in 1877–1878. In the famine of 1891–1892 in Russia it was estimated that in 18 provinces 27 million inhabitants were affected. The last great famine in Russia occurred in the Volga Valley in 1921.

The standard of living depends directly on the efficiency with which food is produced. When it is necessary for more than half the population to devote itself to the production of food for maintenance, a relatively low standard of living results. A relative abundance of food in several areas of the world, and consequently a relatively high standard of living, was first achieved in the nineteenth and twentieth centuries.

Worldwide attention now is being directed to world population and food problems. An adequate food supply is basic to the security, freedom, and well-being of a people. World peace and stability depend on available crop products. Although there are surpluses of basic foods in favored areas, a large part of the world population is undernourished.

The food problem is made increasingly acute by the increasing number of children being born, greatly extended life expectancy due to advances in science, the fact that there are no longer unpopulated areas into which the excess population can overflow, and industry and war as extravagant users of the earth's natural resources.

THE MALTHUSIAN THEORY. The touchstone of thinking on the adequacy of the world food supply goes back to Malthus, a British economist of the eighteenth and early nineteenth centuries (1766–1834). Malthus observed that food resources increase by the process of addition—that is, by arithmetic progression—but that population increases by multiplication, or in geometric progression. He further observed that population inevitably increases up to the limits of subsistence. This led him to conclude that the pressure of overcrowding populations on the means of subsistence must ultimately bring misery and degradation. His theory was not entirely correct, and to date has been thwarted, at least in some areas of the world.

LIEBIG AND SOIL PRODUCTIVITY. Liebig, a noted German scientist, suggested the depletion of soil fertility as a limit to possible food production. He asserted that as crops are grown and removed from the land year after year, the supply of plant food in the soil eventually is exhausted, so that crops no longer can be grown. In support of this theory he cited the relatively nonproductive soils in Mesopotamia, Persia, Egypt, Greece, Spain, and Italy, all of which once were fertile.

REFERENCES AND SUGGESTED READINGS

1. ANONYMOUS. "The First American Farmers," *Farm Jour.* (Jan. 1976), p. 53.
2. BENDER, B. *Farming in Prehistory* (London: John Baker, 1975).
3. BROWN, H. *The Challenge of Man's Future* (New York: Viking, 1954).
4. CHAPMAN, S. R., and L. P. CARTER. *Crop Production Principles and Practices* (San Francisco: W. H. Freeman, 1976).
5. *Encyclopedia Americana* (New York: Americana Corp., 1953).
6. FIELD, H. *The Track of Man* (Garden City, N.Y.: Doubleday, 1953).
7. HARLAN, J. R. *Crops and Man* (Madison, Wisc.: Amer. Soc. Agron., 1975).
8. HEISER, C. B., JR. *Seed to Civilization, The Story of Man's Food* (San Francisco: W. H. Freeman, 1973).
9. JANICK, J., et al., *Plant Science* (San Francisco: W. H. Freeman, 1969).
10. LEE, R. B., and I. DEVORE (Eds.). *Man the Hunter* (Chicago: Aldine, 1968).
11. LIEBIG, J. *Principles of Agricultural Chemistry* (London: Walton and Mabery, 1855).
12. MALTHUS, T. R. *Essay on the Principles of Population* (London: Oxford University Press, 1798; New York: Macmillan, 1926).
13. MANGELSDORF, P. C. *"Science, Food, and People," Harvard Alumni Bulletin,* 1952.
14. MARTIN, J. H., W. H. LEONARD, and D. L. STAMP. *Principles of Field Crop Production,* 3rd ed. (New York: Macmillan, 1976).
15. RASMUSSEN, W. D. (Ed.). *Agriculture in the United States* (New York: Random House, 1975).
16. SAUER, C. O. "Agricultural Origins and Dispersals," *Amer. Geogr. Soc.,* 1952.
17. THEOPHRASTUS. *Enquiry into Plants* (New York: Putnam, 1916).
18. VAVILOV, N. S. 1935, 1949–1950 translated, "The Origin and the Development of Cultivated Plants," *Chronica Botanica,* Vol. 13, 1951.

CHAPTER 2
WORLD
PRODUCTION-DISTRIBUTION

EVERYONE is interested in the factors that influence the food supply of the world, and especially in the amount of land suitable for agricultural production. Production cannot be disassociated from distribution; agricultural production in the United States is definitely related to world production.

Various climatic, soil, and economic factors limit agricultural production over large parts of the earth's surface and result in concentrating production in limited areas. The effect of climatic and soil factors will be considered in some detail in Chapter 3.

The Amount of Land Suited to Crop Production

An early report by Pearson and Harper in *The World's Hunger* gives some understanding of the world situation with reference to the use of land for producing crops (Table 2–1). In considering the outlook for the future food supply of the world, the importance of available tillable acres may be overemphasized. Many other factors must be considered. Land that has not been already under cultivation may be of such a nature that to bring it into productivity would make the cost of production so high that it would be profitable only when prices paid for farm products are extremely high.

Pearson and Harper point out that there is no scarcity of land with a favorable topography, or of adequate sunlight, or of carbon dioxide, or of favorable temperature, or of adequate rainfall, or of fertile soil. The trouble is that there is a limit to the amount of land where these factors are properly combined for food production. Only 7 percent of the earth's land area has that combination of factors that makes food production economically feasible.

A more recent study has estimated the world land area in different climatic zones in percent of potentially arable, nonarable, and grazing land. (See Table 2–2.) A recent Food and Agriculture Organization (FAO) summary of land use by world geographic region is shown in Table 2–3. World land area presently considered as arable is about 1,400 million hectares, a small percentage of the total land area. About 90 million hectares is used for permanent crops, more than 3 billion hectares are in permanent pasture, and more than 4 billion in forests and woodlands.

TABLE 2-1. Effect of Various Combinations of Factors Limiting Food Production

	Land Area Adapted to Food Production	
Factors	Millions of Acres	Percent
Individual factors		
Adequate sunlight	35,700	100
Adequate carbon dioxide	35,700	100
Favorable temperature	29,500	83
Favorable topography	22,700	64
Reliable rainfall	16,600	46
Fertile soil	16,300	46
Adequate rainfall	15,500	43
Combination of factors		
No. combination		
1 Sunlight	35,700	100
2 Carbon dioxide and sunlight	35,700	100
3 Adequate rainfall and carbon dioxide, sunlight	15,500	43
4 Reliable rainfall and adequate rainfall, carbon dioxide, sunlight	12,200	34
5 Temperature and reliable rainfall, adequate rainfall, carbon dioxide, sunlight	11,400	32
6 Topography and temperature, reliable rainfall, adequate rainfall, carbon dioxide, sunlight	7,400	21
7 Soil and topography, temperature, reliable rainfall, adequate rainfall, carbon dioxide, sunlight	2,600	7

TABLE 2-2. World Land Areas in Different Climatic Zones*

Climatic Zone	Potentially Arable		Grazing		Nonarable		Total	
	Acres	Percent	Acres	Percent	Acres	Percent	Acres	Percent
1. Polar and sub-polar	0	0	0	0	1.38	4.2	1.38	4.2
2. Cold-temperate boreal	0.12	0.4	0.47	1.4	4.28	13.2	4.87	15.0
3. Cool-temperate	2.24	6.9	2.46	7.6	2.48	7.6	7.18	22.1
4. Warm-temperate Subtropical	1.37	4.2	2.08	6.4	3.38	10.3	6.83	21.0
5. Tropical	4.13	12.7	4.02	12.3	4.08	12.7	12.23	37.7
Total	7.86	24.2	9.02	27.8	15.60	48.0	32.50	100.0

* In billions of acres. Excluding ice-covered areas.

TABLE 2-3. Summary of World Land Use in 1975*

	Land Area	Arable Land	Permanent Crops	Permanent Pasture	Forest/ Woodland
Africa	2,965	197	14	798	641
North and Central America	2,141	286	7	321	729
South America	1,755	80	22	447	927
Asia	2,672	452	27	552	601
Europe	473	128	15	87	153
Oceania	843	47	1	469	186
USSR	2,227	232	5	372	920
World	13,075	1,415	91	3,046	4,156

* In million hectares.

Source: FAO, *Production Yearbook,* Vol. 30, 1976, pp. 45–55.

Cropland Distribution

The world's cropland acreage is very unevenly distributed among countries. The United States, the USSR, India, and China together have nearly half the cropland of the world. On the basis of cropland per capita, Australia, Canada, and Argentina are leading, with India and China ranking low. In addition to this imbalance between population and farm output are such factors as differences in climates, soils, patterns of production, and the level of technology employed.

World Centers of Crop Production

In his *Atlas of the World's Agriculture,* Van Royen lists eight of the more important crop production centers of the world:
  1. Central United States and south central Canada
  2. Northern Argentina and southern Brazil
  3. Areas of South Africa
  4. Parts of North Africa
  5. Central and Western Europe
  6. India
  7. Parts of Australia and New Zealand
  8. Parts of Asiatic Russia

Central United States and South Central Canada

The Cotton, Corn, and Small Grain belts of the United States, when combined with the Grain Belt of southern Canada, make this the greatest crop production area in the world, ranking near the top in the production of corn, soybeans, wheat, oats, hay, cotton, and tobacco. Based on the predominant crops, this area is divided roughly from south to north into the Cotton Belt, the Corn Belt (which also includes part of the Soybean Belt), and

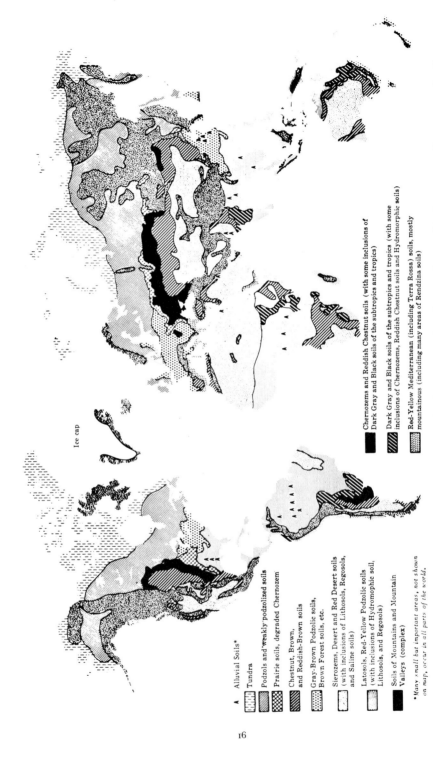

FIGURE 2–1. *Primary groups of soils in the world. There is a rather direct relationship between soils, crop production, and world population.*

▲ Alluvial Soils*

Tundra

Podzols and weakly podzolized soils

Prairie soils, degraded Chernozem

Chestnut, Brown, and Reddish-Brown soils

Gray-Brown Podzolic soils, Brown Forest soils, etc.

Sierozems, Desert and Red Desert soils (with inclusions of Lithosols, Regosols, and Saline soils)

Latosols, Red-Yellow Podzolic soils (with inclusions of Hydromorphic soil, Lithosols, and Regosols)

Soils of Mountains and Mountain Valleys (complex)

Many small but important areas, not shown on map, occur in all parts of the world.

Chernozems and Reddish Chestnut soils (with some inclusions of Dark Gray and Black soils of the subtropics and tropics)

Dark Gray and Black soils of the subtropics and tropics (with some inclusions of Chernozems, Reddish Chestnut soils and Hydromorphic soils)

Red-Yellow Mediterranean (including Terra Rossa) soils, mostly mountainous (including many areas of Rendzina soils)

Ice cap

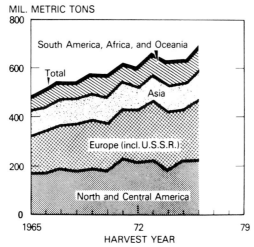

FIGURE 2–2. *World coarse grain production by world area. Production has increased steadily in most regions since the mid-1960s.* [SOURCE: USDA.]

the Small Grain Belt, with a hay and forage section in the northeastern portion of the United States.

Argentina and Brazil

In Argentina the crop area is strictly limited by lack of rainfall, the entire area being less than one-fifth that of the United States. The bulk of the cropped land corresponds more closely to that of our Great Plains area than to the Corn Belt. The more important crops are alfalfa, wheat, corn, flax, oats, barley, rye, potatoes, and sugarcane. In Brazil it is coffee, sugarcane, wheat, soybeans, cotton, cacao, and tobacco.

Areas of South Africa

The total extent of cropland in South Africa is less than one-third that of Iowa, with more than four-fifths devoted to growing cereals. The principal crops are corn, wheat, oats, grain sorghum, cotton, sugarcane, alfalfa, rye, barley, and millet. This region is more underdeveloped than are some of the other areas and is believed to have possibilities of becoming a more important crop-producing area.

North Africa

The region of North Africa is about the same latitude as that of the southern part of the United States. Were it not for the fertile Nile Valley and portions of Algeria and Morocco, crop production would be of slight importance on a world basis. Egypt has more than 5 million acres (2 million ha) in cereals out of a potential of more than 8 million acres (3.2 million ha) of arable land. Algeria has approximately 6 million acres (2.4 million ha) of arable land, the greater part of which is devoted to cereal crops. In the extent and variety of its crops, Morocco's crop production is practically identical with that of Algeria.

Central and Western Europe

A hay and pasture region extends from the northern part of Norway and Sweden across European Russia. South of this area oats and flax are grown, and still further to the south are grown sugarbeets, winter wheat, barley, rye, and white potatoes. Further to the west and south is the corn region, including northern Italy, Romania, Yugoslavia, Hungary, and parts of adjoining areas.

India

Although the total area of India is about one-third that of the United States, its crop area slightly exceeds that of the United States. Rice is of outstanding importance, particularly in the southern region; grain sorghum and grain millets are also important. Further to the north, wheat is important as well as rice, with extensive production, as well, of pulses, corn, barley, and jute. Flax and sugarcane are prominent crops in certain areas.

Australia and New Zealand

Australia and the United States have about the same total area, but the area of land with enough rainfall for the production of harvested crops in Australia is only about one-fifteenth that so used in the United States. Only a narrow coastal border receives sufficient rainfall for the production of harvested crops.

Asiatic Russia

Almost three times the area of the United States, Asiatic Russia reportedly uses less than 10 percent of this land in cultivated crops. Production apparently could be somewhat increased, although the productivity of such land is uncertain. With hay in the northern part, the chief crops in the central producing area are flax, oats, wheat, rye, and barley. In the "black earth" area of the southwestern part, spring wheat, oats, rye, barley, potatoes, and flax are the principal crops.

A summary of world production of cereals, roots and tubers, and pulses by geographic region is shown in Table 2–4. All geographic regions showed increases in production of cereals from the period 1961–1965 to 1976. Some of these increases were striking; world cereal production increased during this period by 49 percent. Significant but less dramatic increases occurred in world root and tuber/pulse production.

TABLE 2-4. Summary of World Agricultural Production*

	Total Cereals		Roots/Tubers		Total Pulses	
	1961–1965	1976	1961–1965	1976	1961–1965	1976
Africa	506	680	58	76	4	5
North and Central America	210	323	17	22	2	3
South America	40	66	35	45	3	3
Asia	394	566	158	218	22	28
Europe	159	220	139	110	4	2
Oceania	11	19	2	2	< 1	< 1
USSR	123	214	82	85	7	8
World	988	1,477	490	558	42	52

* In millions of metric tons (MT).

Exchange of Products

The United States, France, the Netherlands, Australia, Brazil, and Canada lead in agricultural exports to other countries. The value of U.S. agricultural exports increased from $3 billion in 1950 to more than $8 billion in 1978. Western Europe leads in the importation of farm products. The Federal Republic of Germany and the United Kingdom are the leading European importers. Because of high industrialization, Western Europe depends chiefly on manufacturers to pay for imports of food and raw materials, although the area exports significant amounts of certain agricultural products. Japan and the United States also are leading importers of agricultural products. Imports of goods by China and the USSR have increased markedly in recent years.

The densely populated Far East does not produce enough food, or materials that can be traded for food, to provide its inhabitants with more than a very meager diet. With more than half of the world's population, the Far East accounts for less than one-third the value of world agricultural output. In all the developing regions agriculture is the major economic activity and the major earner of foreign exchange.

Wheat, Corn, and Rice

Wheat, corn, and rice have been the leading crops worldwide, although the importance of soybeans has been increasing rapidly. The importance of the cereal grains in their relationship to world food and feed can hardly be overestimated. Used both for food and for feed, the cereals occupy nearly half of all cropland and figure prominently in the diets of most countries and regions.

WHEAT. Produced widely and extensively in the temperate zones, wheat ranks first in acreage and production of all the cereals. It is the outstanding

World Wheat Production

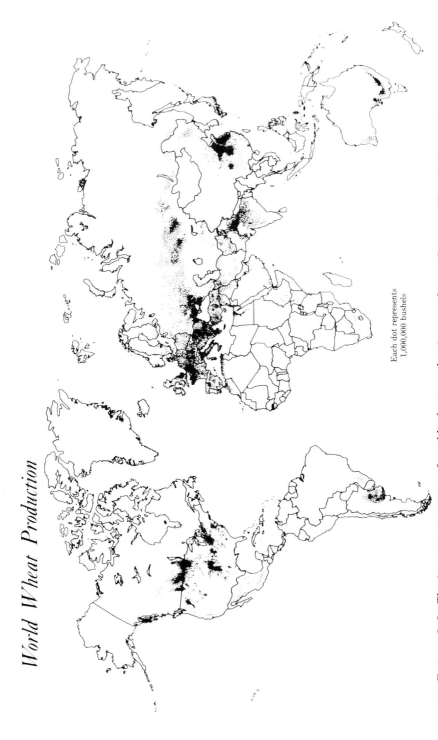

Each dot represents
1,000,000 bushels

FIGURE 2–3. *The important centers of world wheat production are shown. On a world trade basis, wheat leads all other food crops.* [SOURCE: USDA.]

food grain of the world. On a world basis, acreage has steadily increased. The Soviet Union has increased its acreage considerably in the past 25 years and now has more than one-fourth the world's wheat acreage, although acre yields have been relatively low.

For the United States, the second-largest wheat-growing country, government controls have brought a reduction in acreage, in keeping with limitations of world outlets for surpluses; but acre yields have increased. With less than half as much land in wheat as the Soviet Union, the United States has harvested almost as much total grain. Europe has the highest yields of wheat, with an average exceeding 3 metric tons per hectare (MT/ha) comparable with just over 2 MT/ha for the United States and 1.6 MT/ha for the Soviet Union. Western Europe has been the most important market for wheat, but more recently Eastern Europe and the Far East countries have been demanding a substantially larger share of the world wheat crop. Exports to Latin America and Africa also have increased substantially.

Japan had been the most important importer of wheat in the Far East, but beginning in the early 1960s China suddenly became a large cash market for wheat.

CORN. Corn ranks with wheat and rice as one of the world's leading grains. It is widely grown in the Americas, southern Europe, the Soviet Union, Africa, and the Far East. Many countries use corn largely for food, whereas others, including the United States, use most of their production as feed for livestock.

On a world basis the area devoted to corn has increased steadily, with also higher yields. In 1976 world area was 118 billion ha, as compared to 99 billion in 1961–1965. The present world yield level is more than 2,800 kg/ha, an increase of more than 30 percent since 1961–1965. For the United States, the acreage has also increased since 1961–1965, but the portion of the world crop has decreased somewhat. United States corn production still remains at nearly 47 percent of the world total, and this production comes from less than 23 percent of the world corn acreage. China ranks second and Brazil third as to world acreage and production.

RICE. Rice rivals wheat as a food grain and is the staple food of the populous Far East, though parts of this region also produce and consume substantial quantities of wheat, barley, corn, millet, and sorghum. China ranks first among the rice-growing countries, producing more than one-third the world's rice. India has a slight edge over China as to area, with the two countries combined having more than half the world's total. India's production is low, with just over one-half the production volume on China. Japan is noted for its high yields of rice, but is not among the world leaders in total production.

Rice may enter into international trade mostly in the rough or milled form. Removal of the outer husk and the bran layers reduces the weight from that of rough rice by about 28 percent.

World Corn Production

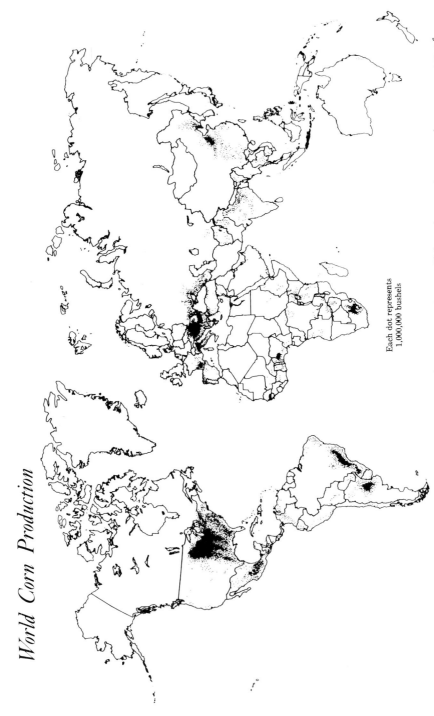

Each dot represents
1,000,000 bushels

FIGURE 2–4. *The important centers of world corn production are shown. Although corn is grown in many areas of the world, production is concentrated in certain countries.* [SOURCE: USDA.]

22

Other World Food and Feed Crops

Other important world crops, which do not enter largely into world trade but are important locally, are the sorghums and millets, oats, rye, barley, and potatoes.

SORGHUM AND MILLET. Sorghum and millet are important food and feed crops, especially in Asia and Africa. In Asia the greatest concentration of both sorghum and millet is in the west central part of India, but both are important over a much larger area. The grain sorghum durra, or Jowar, is grown very extensively as a food crop in India, and is second only to rice in acreage. It is the food of both the people and of livestock in the great Deccan Peninsula. Almost all is consumed locally. Other countries leading in sorghum production are Nigeria and Sudan in Africa, and the United States, Mexico, and Argentina in the Americas.

Millet is of greatest importance in China and India. It is grown most in drier portions of India and to the north.

OATS. The greatest concentration of oats is in the USSR, the United States, Canada, Poland, and China. In the United States the greatest concentrations of acreage have been from the Dakotas and Iowa eastward. The acreage of oats has greatly decreased in the United States since the mid-1950s with the disappearance of the large horse population.

BARLEY. In much of the subhumid and semiarid regions of North Africa and in parts of Asia, barley ranks as an important food crop. The crop has unusually wide climatic adaptation. It is grown from beyond the Arctic Circle in Scandinavia to the arid areas of the Near East. Barley is a major grain crop in parts of North Africa, where the climate is too dry for wheat. Barley is particularly important in the USSR, which has nearly 40 percent of the total world acreage, and China, with about 8 percent of the acreage.

RYE. In the past rye has been an important bread grain in central and eastern Europe. It has been important especially in the USSR, presently with more than 50 percent of the total acreage and almost half the world production. The acreage of rye has greatly decreased in recent years, with the more general acceptance of wheat as a superior bread grain. This decline in world acreage has been more than 40 percent since 1961–1965.

POTATOES. About 50 percent of the world white potato crop is planted in the USSR, with a production of nearly 40 percent of the world total. Poland is second in area planted to potatoes and production, with about 20 percent of the world production. The potato crop does best where the summers are rather humid and cool. It is reported that in many of the European countries potatoes occupy approximately 25 percent that of the small grain acreage. Potatoes are used not only for food throughout much of Europe, but also as a feed for livestock and in the manufacture of starch and alcohol.

Elsewhere potatoes are produced almost entirely for food. Countries with climates poorly suited for white potato production grow large quantities of sweet potatoes, yams, and cassava for local use. In much of tropical western Africa these crops contribute more to the calorie value of the food supply than do the grains. They are very poor sources of protein, however, and their overextensive use often is associated with protein-deficiency diseases. Because of their bulk and weight in relation to food value, potatoes enter very little into world trade.

Oilseed and Oil Nut Crops. The total world trade in oil seeds and oil nut crops increased from 5 billion MT in 1951 to more than 13 billion

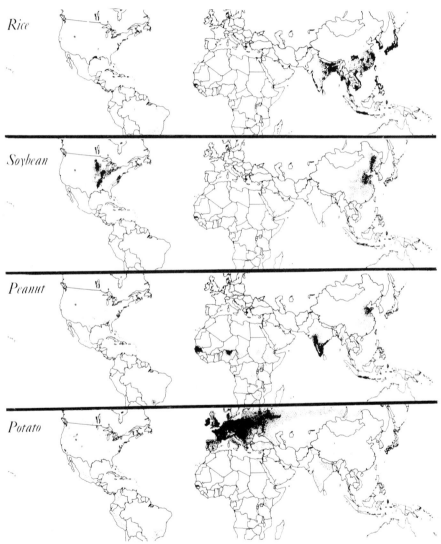

Figure 2–5. *The important centers of world production are shown for rice, soybeans, peanuts, and potatoes.* [Source: USDA.]

MT in 1975. A large share of this gain was accounted for by soybeans from the United States.

Oilseeds account for more than half the world supply of fats and oil and for an important part of its protein. In terms of quantity harvested, soybeans, cottonseed, and peanuts are the leading oilseeds. The rapidly increasing acreage of soybeans, especially in the United States, puts this crop in first place. China and Brazil are also large producers of soybeans. Additional significant world oilseed crops are sunflowers, rapeseed, sesame, and flaxseed, listed in the order of area harvested. Other oil crops of some world importance are palm, safflower, coconut, and castor.

The soybean has about the same adaptation to climate and soils as corn and has increased rapidly in acreage, especially through the U.S. Corn Belt and southern states. The soybean has become an important export crop for the United States. Its high value derives from the oil and protein content of the seed. Price has been largely dependent on the demand for soybean oil.

Peanuts are grown in many parts of the tropics and in some of the warmer regions of the temperate zones. This also is true of cotton, with cottonseed an important by-product of the cultivation for fiber.

The production of sunflower seed for oil has become very important, especially in the USSR. Sunflower oil is a competitor with soybean oil in important European markets.

Rapeseed production occurs principally in India and the USSR. These two countries account for about 60 percent of the world area devoted to this crop. Sesame is produced principally in the Far East, with India, China, and Burma the world leaders. Flax for linseed oil production is produced most extensively in India, the USSR, and Argentina. The leading countries in palm production are Nigeria, Brazil, and Malaysia; in safflower area are India and Mexico; in coconut production are Indonesia and India; and in castor area India, Brazil, and the USSR.

In the quantities of oilseed crushed locally or exported for crushing, the United States, Africa, and the Far East account for more than two-thirds of the world vegetable oil production. The soybean, peanut, cottonseed, and palm oils account for a large majority of the total.

A large share of the oilseed exports go to Western Europe. With its extensive livestock industry, Western Europe also takes most of the oilcake, which has a high protein value.

Sugar

Sugar is an important source of calories, expecially in northwest Europe, temperate North America, and parts of Latin America and South Africa. It ranks high, especially in the form of cane sugar, as an export crop for a number of the less-developed regions. Beet sugar production is confined for the most part to Europe, including the USSR, and to temperate North America. Except for Eastern Europe, the beet-sugar-producing regions are all importers of cane sugar. Cuba in the past supplied three-fifths of the sugar from all of Latin America and one-third of world sugar exports. In recent years Brazil has exceeded Cuba in sugarcane production. The United States is the leading sugar importer, having taken up to 20 percent of the world sugar trade in recent years.

Cotton

Cotton ranks first among the textile fiber crops in the value of output. It is widely produced in the warmer latitudes, chiefly as a rain-grown crop, but also under irrigation in such dry areas as the southwestern part of the Soviet Union, in the United Arab Republic, in Mexico, and in the southwestern United States. World production has increased as a result of better production methods and more irrigation, although the total world acreage has changed little.

The USSR and China are the leading world cotton producers, with the United States and India ranking third and fourth. India has the largest acreage, but its low yield puts it only fourth in production. High yields are reported for the USSR, the United Arab Republic, and Mexico, with all, or nearly all, the acreage in these countries under irrigation.

For the United States, the total cotton ouput has declined since 1961–1965, but an upward trend in acreage and production occurred during the late 1970s. The result of this overall decline has been a drop in the U.S. share of world production. Formerly the United States was the world leader in cotton production.

Tobacco

World production of tobacco has increased gradually through the years, in keeping with population growth. Per-capita use has risen in many countries, as a result of persistent advertising campaigns. Only very recently, as a result of extensive research on the effects of tobacco use on health, has there been evidence of a trend to decreased use. The United States, the world's leading producer and exporter of leaf tobacco, in 1976 produced about 2 billion pounds (0.95 billion kg) from 1.0 million acres (0.4 million ha), representing about 17 percent of the total world crop.

China, which has the largest tobacco acreage, is second in production, followed by India, the USSR, Brazil, and Bulgaria.

World Trade in Food and Feed Crops

Many of the developing countries depend heavily on their agricultural exports for exchange earnings to finance imports of capital goods. But many of these countries and areas already have had to depend heavily on imports of certain food grains, and it would appear that the need will increase. A serious attempt has been made by the United States to foresee something of what the future needs of the developing countries will be.

Agricultural Trade of Developing Countries

Agricultural exports often account for more than 50 percent of the total export earnings of a number of developing countries and territories. Several countries depend on grain exports for a large share of their earnings. Rice exports account for more than 50 percent of the export trade income of some Asian countries. With selected developing countries, more than 50 percent of the total net export earnings come from either coffee, cocoa and tea, sugar, oilseeds and vegetable oils, cotton, or fruits and vegetables.

World Pattern of Agricultural Trade

World agricultural trade volume increased by 44 percent from the period 1961–1965 to 1975. Value of world agricultural products traded increased by more than 2.3 times from 1970 to 1975. World import volume of total agricultural products increased by 44 percent from 1961–1965 to 1975. Developed countries had a 34 percent increase and developing countries a 71 percent increase during this period. World export volume of total agricultural products increased by 43 percent from 1961–65 to 1975. In contrast to imports, developed countries increased export volume by 71 percent and developing countries by only 16 percent during this period.

Importance of Grains in World Trade

About 21 percent of the value of total world agricultural trade is made up of grain exports. By far the largest part of the grain exports is wheat,

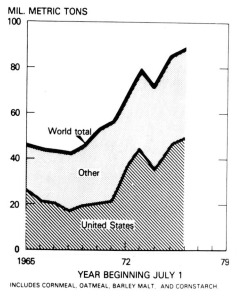

FIGURE 2–6. *World exports of coarse grains have increased dramatically since the mid-1960s. A large share of this increase can be attributed to the United States.* [SOURCE: USDA.]

FIGURE 2-7. *World trade of food and feed crops is vital to the economy of developing as well as developed countries.* [SOURCE: American Soybean Association.]

with more than 10 percent of world agricultural trade value for 1975. Corn was second in importance and rice was third. The United States alone accounted for more than 44 percent of the world grain exports in 1975; Canada accounted for nearly 10 percent; and France and Australia each accounted for almost 8 percent. In terms of trade volume of the three major export grains—wheat, rice, and corn—the United States accounted for 48 percent of the wheat, 24 percent of the rice, and nearly 65 percent of the corn in 1975. The United States' share of the value of these grains was 43, 27, and 64 percent, respectively. For Canada only wheat exports are of major importance; they accounted for almost 17 percent of the world wheat export value. Rice exports were of major importance for the Southeast Asian countries. Asia's share of the world rice exports was 64 percent of the volume and 63 percent of the value in 1975. Vietnam was the leading country, with nearly 24 percent of the Asian export volume. Much of the rice exported went to neighboring countries.

Major Grain-Importing Regions

The major grain-importing regions are Japan, the USSR, Western Europe, India, China, and North Africa. The USSR and China have become major importing regions only since 1960. In 1975 these regions accounted for the following share of the world grain import volume: Japan, 11.8 percent; the USSR, 10.4 percent; Belgium, Germany, Italy, the United Kingdom, and the Netherlands collectively, 22.2 percent; India, 4.7 percent; and China, 3.7 percent.

Historical Trends of Trade in World Grains

The volume of world grain trade increased rather steadily during the period 1951–1975, increasing from 43 million MT to 165 million MT. Food grain exports increased more than food grain exports up to the mid-1960s, the result largely of the increased demand for corn from the United States by Japan and western Europe, in support of their growing livestock production. Wheat exports have increased consistently since 1951, to 55 MT in 1970 and more than 80 million MT in 1977. Rice exports declined in the early part of this period, but increased in 1976 and 1977 to a present level of about 14 million MT. For the total period 1951–1975, exports of corn had the most rapid export increase of all the grains.

REFERENCES AND SUGGESTED READINGS

1. BLAKESLEE, L. L., E. O. HEADY, and C. F. FRAMINGHAM. *World Food Production, Demand, and Trade* (Ames: Iowa State Univ. Press, 1973).
2. FAO. *Production Yearbook,* Vol. 30, 1976.
3. FAO. *Trade Yearbook,* Vol. 29, 1975.
4. GUIDRY, N. P. *A Graphic Summary of World Agriculture,* USDA-ERS Misc. Pub. 705 (1964).
5. JILER, H. (Ed.). *Commodity Yearbook 1977* (New York: Commodity Research Bureau, Inc., 1977).
6. PEARSON, F. A., and F. A. HARPER. *The World's Hunger* (Ithaca, N.Y.: Cornell Univ. Press, 1945).
7. President's Panel on World Food Supply. *The World-Food Problem* (Washington, D.C.: The White House, May 1967).
8. USDA. *Agr. Stat.,* 1977.
9. USDA. Economics, Statistics and Cooperative Services. *Agricultural Outlook,* AO-31, April, 1978.
10. USDA. Economics, Statistics, and Cooperative Services. *U.S. Foreign Agricultural Trade Statistical Report, Fiscal year 1977,* ESCS-112, 1978.
11. USDA. Foreign Agricultural Service, *Foreign Agriculture Circular, Grain Exports by Selected Exporters,* FG-78, 1978.
12. USDA. Foreign Agricultural Service, *Foreign Agriculture Circular, World Grain Situation and Outlook for 1978/79,* FG-18-78, 1978.
13. VAN ROYEN, W. *Atlas of the World's Resources* (Englewood Cliffs, N.Y.: Prentice-Hall, 1954).

CHAPTER 3
U.S. PRODUCTION-DISTRIBUTION

The agricultural conquest of a virgin continent, mostly within a century, constitutes a pageant which may never occur again in human history. Essentially, this conquest consisted of the preparation of the soil for the production of crops. Until less than a century ago it was a slow-moving procession, through one of the largest and densest forests in the world. Then, after a pause of perplexity at the prairie margin, the pioneers brought the grassed half of the U.S. into use for crops in record time. The progress of settlement was, in general, from the poorer lands of the Atlantic coast to the better lands of the Piedmont and limestone valleys, then to the good forested soils of the Mississippi Valley, and still later to the excellent soils of the prairies. Instead of advancing onto poorer and poorer land, as the classical economists of England assumed, the movement until 1880 or 1890 was onto better and better land.

These pioneers, however, had to learn about climate and soil and the suitability of the other physical conditions to the various crops. And, as the climate varied from year to year and the soil yielded at first in accordance with its virgin fertility, there is little wonder that mistakes were made in land utilization and crop selection. Out of these trials and errors has emerged an agriculture in which the crops are suited to physical and economic conditions more closely, perhaps, than in other parts of the world. Commercial competition has compelled adjustments to be made far more quickly and accurately than in a more self-sufficing agriculture.

But the adjustments are not complete—never will be so long as civilization remains dynamic. Great shifts in crop production have been made, are being made and should be made.

O. E. BAKER
A. B. GENUNG

THE student of agriculture is interested in the amount and distribution of U.S. land that is suitable for crops and pasture and range. The latest available census figures indicate that 47 percent of the total land area is in cropland and grassland, with just over one-fifth classified as cropland. This amounts to 467 million acres (189 million ha) of 2,264 million acres (916 million ha) total land area (Figure 3–1). Geographic distribution of cropland is shown in Figure 3–2.

Not all cropland produces harvested crops each year. In 1974, crops were

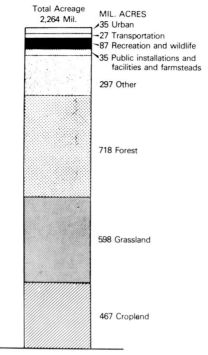

FIGURE 3–1. *Major uses of land in the United States.* [SOURCE: USDA.]

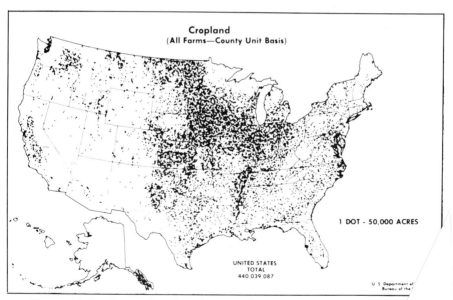

FIGURE 3–2. *Geographic distribution of cropland in the United Sta*
[SOURCE: U.S. Dept. of Commerce.]

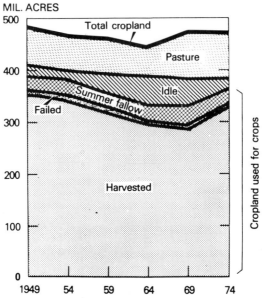

FIGURE 3-3. *Major uses of cropland in the United States.* [SOURCE: USDA.]

harvested from 303 million acres (123 million ha), with the remainder
of the 440 million acres (178 million ha) failing before harvest, being summer fallowed, idle, or in pasture (Figure 3-3). When pasture and idle land
is considered as cropland, total amount of cropland is virtually the same
as that in 1949, just under 500 million acres (202 million ha) (Figure 3-
3). When only harvested cropland is considered, an 11 percent increase
has occurred from 1969 to 1974, and another large increase occurred after
1974 (Figure 3-4). Figure 3-4 also shows that crop production per acre
has increased since 1965 by about 13 percent.

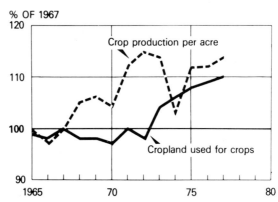

FIGURE 3-4. *Crop production per acre and cropland used for crops in
the United States.* [SOURCE: USDA.]

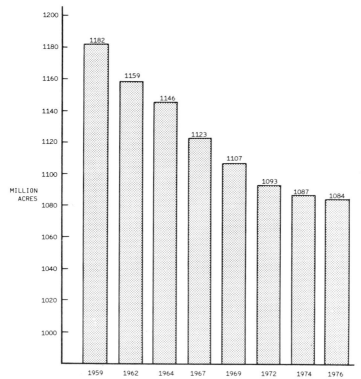

FIGURE 3-5. *Total land in United States farms.* [SOURCE: Southern Illinois University.]

Despite the increase in harvested cropland, total land in farms has declined each year since 1959. In 1976, 1084 million acres (439 million ha) were in farms, as compared to 1182 million acres (478 million ha) in 1959, a decrease or more than 8 percent during this period (Figure 3-5).

Agricultural Land Uses

Land used for agricultural purposes in the United States accounts for about 45 percent of the total U.S. land area, with 55 percent nonfarmland (Figure 3-6). In addition to agricultural land in farms, there is some additional agricultural land not in farms, including grazing lands, grasslands, brush-browse, and forested grazing lands. Figure 3-7 shows that a large percentage of the pasture and rangeland, forest, and special use acreage is owned by federal, state, or other public owners. This type of ownership amounts to 40 percent of the total U.S. land area.

After a considerable decrease in cropland acreage from 1959 to 1970, there was a marked upsurge in cropland used for crops, particularly from 1972 to 1977 (Figure 3-4). This upsurge was partly the result of a relaxation

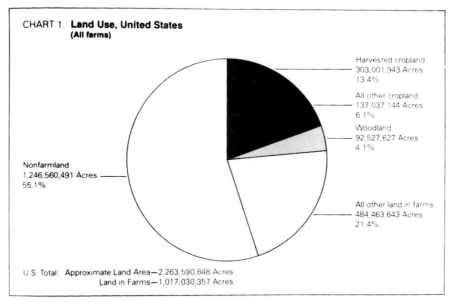

FIGURE 3–6. *Nonfarmland and land use of United States farmland.* [SOURCE: USDA.]

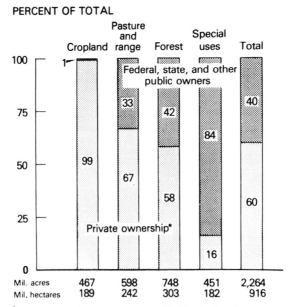

FIGURE 3–7. *Major land uses by ownership, private as compared to public owners.* [SOURCE: USDA.]

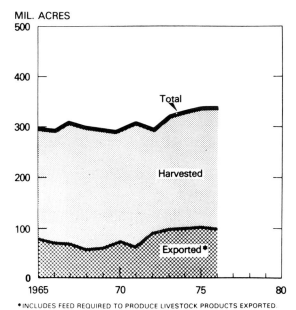

FIGURE 3–8. *United States crop acreage harvested and exported.* [SOURCE: USDA.]

of federal diversion programs, and partly the result of increased export demands. Figure 3–8 shows the increase since 1965 in total crop acreage devoted to producing crops for export, as well as the increase in total crop acreage harvested.

Major Uses of Cropland

Less than 70 percent of the land suitable for regular cultivation is currently being used for cropland. The nonharvested cropland included areas on which crops failed, areas summer-fallowed, and acres idle. Generally, crop failure is only 2 or 3 percent or less of the acreage planted. Idle land varies from year to year depending on federal set-aside farm programs.

Cultivated summer fallow is widespread in the semiarid regions where the acreage of small grains is large and grown mostly without irrigation. Under limited rainfall conditions, increases in yield result from summer-fallowing land for a year before seeding to small grains to accumulate enough soil moisture to produce a profitable crop. Some of the acreage diverted from production under federal farm programs is in cultivated summer fallow.

Some cropland is used for only soil improvement crops and is not harvested or pastured. Some cropland is put in soil-improving crops for a year or two before it is replanted to crops for harvest. This use of cropland in humid areas is similar to cultivated summer fallow in subhumid areas to improve productivity by controlling weeds, conserving moisture, and increasing the organic matter content of the soil.

Cropland as a percent of land area is about 19 percent, as a percent of land in farms is about 47 percent, and harvested cropland as a percent of land in farms is 30 percent. Pastureland as a percent of land in farms is over 50 percent. Part of the total cropland acreage is cropland used only for pasture. Most of this land is included in the rotation and regularly returns to the harvested-crop category. Some of this acreage, however, has become uneconomic for continued cropping and may remain in pasture indefinitely.

Trends in Major Uses of Cropland

Total cropland acreage experienced a downward trend from 1949 to 1964, followed by an upward trend, resulting in a present acreage almost as large as that in 1949. The earlier shifts were principally associated with the impact of federal programs designed to divert crop acres to idle but soil-conserving uses. The total of such diversions or set-aside programs was 14 million acres (5.7 million ha) in 1956, the first year, 65 million acres (26 million ha) in 1962, and 41 million acres (16.6 million ha) in 1967. A trend away from some of these acreage diversions, an increase in land devoted to pasture, and an increased export demand resulted in an increase in total cropland acres in the mid-1960s and early 1970s. It is evident that these trends can be reversed rather quickly as conditions seem to warrant. To build up depleted feed grain and wheat stocks or to meet export demands, diverted acres or idle land can be activated quickly into harvested crops to fulfill these needs.

Trends in Acreage, Production, and Value of Crops

Land in farms and number of farmers have decreased since 1965 and population and food demand has increased during the same period, but crop and livestock production has more than kept pace, increasing by 25 and 10 percent, respectively (Figure 3–9). About 31 percent of the total cropland acreage is devoted to feed-grain production, 21 percent to food grains, and 43 percent to soybeans, oil, seed, and miscellaneous crops (Figure 3–10). Acreage of feed grains has not increased greatly since 1965, but yield per acre and total production have experienced large percentage increases during this period.

In 1974 the largest acreages were planted to wheat, corn, and soybeans, all of which had increased appreciably since 1969 (Figure 3–11). There were about 60 million acres (24 million ha) devoted to wheat and corn, and about 45 million acres (18 million ha) to soybeans. Other crops had relatively small acreages in comparison to these three. The most recent census figures show that grain crops currently have a total value of nearly $25 billion, as compared to about $7.5 billion in 1969, and grains account

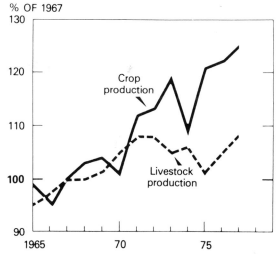

FIGURE 3–9. *Crop and livestock production in the United States.* [SOURCE: USDA.]

for over 30 percent of total farm sales (Figure 3–12). Animal enterprises hold the next four places as to value of farm sales.

The pattern of land use by major crops has shifted significantly in recent years. The most dramatic increases in acreage in recent years were shown by wheat, soybeans, and corn (Figure 3–11). Cotton, sorghums, alfalfa and other hay crops experienced slight acreage increases; peanuts, potatoes,

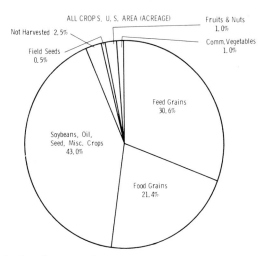

FIGURE 3–10. *Land use in the United States by crops.* [SOURCE: Southern Illinois University.]

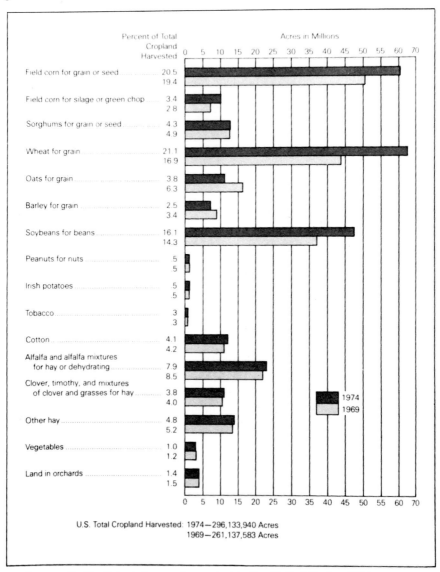

FIGURE 3-11. *Selected crops harvested in the United States, 1974 as compared to 1969.* [SOURCE: USDA.]

and tobacco remained about the same; and oats and barley acreage declined.

Table 3-1 illustrates the changes in acreages of selected crops grown in representative states for 1870–1950, 1969, and 1976. The relative importance of 15 principal field crops in acreage and value is shown in Table 3-2, with crops arranged in order of their relative value.

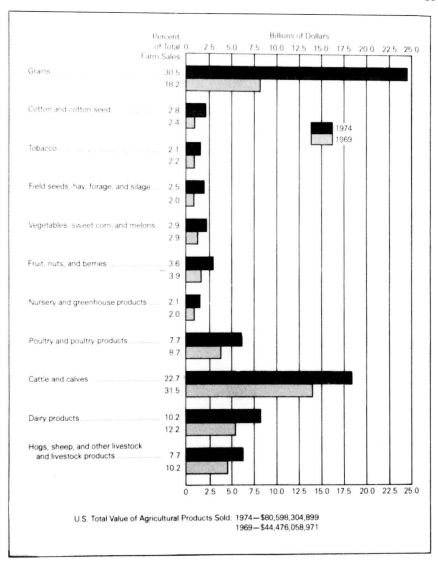

FIGURE 3–12. *Value of sales of selected agricultural commodities in the United States, 1974 as compared to 1969.* [SOURCE: USDA.]

Pasture and Range Resources

All kinds of pastures, rangelands, and grazing areas total about 0.5 billion acres (0.2 billion ha) or about 54 percent of the total farmland area. Cropland used only for pasture occupies about 13 percent of the total pasture and range area; permanent grassland pasture and range about 80 percent, and woodland grazed about 7 percent. Grassland pasture and range consists

T ABLE 3–1. Acreage of Selected Field Crops in Representative States*

	1870	1890	1910	1930	1950	1969	1976
Iowa							
Corn	2,919	8,771	9,473	11,335	11,309	10,219	13,800
Winter wheat	—	—	180	387	345	52	100
Spring wheat†	1,636	1,685	350	45	28	—	—
Barley	47	196	510	548	29	5	—
Rye	29	110	32	45	18	18	24
Oats	550	2,767	4,800	6,303	6,140	2,866	1,850
Flax	—	284	16	20	109	1	—
Timothy and clover‡	1,194	3,636	3,600	1,413	1,710	605	—
Alfalfa and mixtures	—	—	20	437	1,055	1,835	1,750
Buckwheat	9	27	8	4	—	—	—
Soybeans	—	—	—	72	1,344	5,464	6,610
New York							
Corn	571	643	680	555	624	799	1,170
Winter wheat	—	—	444	214	379	196	175
Spring wheat	662	641	—	10	6	—	—
Barley	312	343	78	168	25	14	13
Rye	172	237	170	24	16	134	123
Oats	915	1,343	1,338	872	615	410	360
Timothy and clover‡	3,651	4,933	4,811	1,696	2,305	1,165	—
Alfalfa and mixtures	—	—	—	224	327	1,075	990
Buckwheat	191	311	313	186	73	—	—
Soybeans	—	—	—	2	7	5	12
North Dakota†							
Corn				1,035	1,172	571	510
Spring wheat	—	4,209	7,221	9,896	10,178	6,828	8,080
Barley	—	257	987	2,588	1,600	2,238	2,200
Rye	—	18	15	1,223	213	268	120
Oats	—	1,183	1,628	1,827	1,672	2,660	1,320
Flax	—	—	1,605	1,677	1,775	1,547	486
Timothy and clover‡	—	935	188	28	23	—	—
Alfalfa and mixtures	—	—	—	222	285	1,256	1,620
Buckwheat	—	5	—	15	—	—	—
Soybeans	—	—	—	—	21	188	150
Alabama							
Cotton	—	2,852	3,560	3,770	1,850	566	460
Corn	2,019	2,489	3,524	2,819	2,471	738	880
Winter wheat	124	293	130	2	10	114	200
Rye	6	6	2	—	—	—	—
Oats	45	405	297	90	112	116	110
Timothy and clover‡	47	41	120	—	22	46	—
Alfalfa and mixtures	—	—	—	15	25	4	—
Soybeans	—	—	—	83	116	683	1,250

* In millions.
† Wheat not designated as spring or winter prior to 1910.
‡ 1910 and earlier reports use general term *hay*. Now includes mixtures of clover and grasses.
  Source: USDA Agr. Stat. 1977 and earlier.

TABLE 3-2. Relative Importance in the United States of 15 Principal Field Crops Arranged in Order of Value, 1976

Crop	Acreage Percent of Total	Value Percent of Total
Corn, all	24.6	29.4
Soybeans, beans	14.5	18.8
Hay, all	17.8	14.0
Wheat, all	23.4	12.4
Cotton	3.4	6.7
Tobacco	0.3	4.9
Sorghum, all	5.5	3.0
Potatoes	0.4	2.4
Oats	5.1	1.8
Barley	2.7	1.7
Rice	0.7	1.6
Peanuts	0.5	1.5
Sugarbeets	0.4	1.2
Dry Edible Beans	0.4	0.5
Flaxseed	0.3	0.1

Source: USDA Agr. Stat. (1977).

mainly of tame and native grasses and legumes, but this classification also includes shrub and brushland not included as forest. Cropland used only for pasture can be converted at any time to harvested crops and is considered to be included in the crop rotation.

Factors Affecting Production

It has been pointed out that only 7 percent of the world's lands are adapted to the production of food crops; 100 percent have sufficient carbon dioxide in the atmosphere and sufficient sunlight, 83 percent have a favorable temperature, 64 percent have a favorable topography, 46 percent have a reliable rainfall, and 46 percent have satisfactorily fertile soil. But only 7 percent of the land area has such a combination of these factors as to make the actual production of food feasible at present without additional technological advances.

Water

It is immediately evident that water is essential to plant growth. About 80–90 percent of actively growing plant tissue usually is water. Water is the medium through which the soil nutrients move into the plant. Moreover, water is essential in the photosynthetic nutrient-manufacturing processes.

Over a very large part of the earth's surface the relative scarcity or abundance of water is the most important factor in determining whether plants can grow, or what kinds of plants will survive. When we consider water in its relation to plant growth we must take into consideration not only the total rainfall, but also its distribution throughout the year and its dependability from year to year. The average relative humidity of the air influences the efficiency with which the plant uses the soil moisture available about its roots.

Rainfall

The average annual rainfall in different parts of the United States varies from less than 2 inches (5 cm) in Death Valley, California, to more than 100 inches (254 cm) in areas of the states of Washington and Oregon. On a world basis, the extremes in the average annual amount of rainfall are from less than 0.05 inch (0.1 cm) at a point in Chile to more than 450 inches (1,143 cm) in parts of India and Hawaii.

Total rainfall is important, but often the time of year in which the rain comes is much more important, as well as its dependability from one year to another. The rainfall may be excessive in certain parts of the year and deficient in others.

Of the total world land area, approximately 55 percent is semiarid to arid, averaging less than 20 inches (51 cm) of rainfall per year; 20 percent subhumid, with 20–40 inches (51–102 cm); 11 percent humid, with 40–60 inches (102–152 cm); and 14 percent with an average rainfall of 60 inches (152 cm) or more.

Dew

The cooling of the air near the soil surface during the night period may result in condensation of the air moisture and formation of dew. Opinions have differed as to the possible value of dew to plant growth. It has been suggested that the presence of dew delays the rise in temperature and transpiration as the day develops and that dew on the surface of leaves may be taken up by the plant.

Temperature

Temperature often is the factor limiting the growth and distribution of plants. Different methods have been developed for evaluating the effectiveness of temperature as related to crop distribution. Of these, the length of the growing season comes at once to mind. The number of days from the last killing frost in the spring to the first in the fall has been used to designate the growing season. In any given locality, however, the length of the growing season is known to vary as much as 30 days for different years.

The "physiological growing season" takes into account the lowest temperature at which appreciable growth can be expected to take place. The crops adapted to an area with a short growing season may be quite limited.

Potatoes, barley, and buckwheat are among the best adapted to very short growing seasons.

Of the winter small grains, it has been shown that the 10°F (−12°C) isotherm for the daily minimum temperature for January and February is about the northern limit; 20°F (−7°C) is the limit for winter barley; and 30°F (−1°C) is the limit for winter oats.

HOPKINS' BIOCLIMATIC LAW. Hopkins' bioclimatic law has been shown to be applicable in determining planting dates in the spring and for comparing growth stages and harvesting dates. This law establishes bioclimatic zones, which will generally show a lag of 4 days in degree of latitude, 5° longitude, and 10 feet (3 m) of altitudes, northward, eastward, and upward. As stated, this applies to temperate North America.

HEAT-UNIT SYSTEMS. A theory has been advanced that a given degree of development of a given plant is reached when it has received a certain amount of heat, regardless of the time elapsed, and for each successive stage of growth a definite additional heat requirement. For most heat-unit systems the total of positive degrees of temperature above a given base is used. For most crops grown early, such as the small grains and peas, the base temperature used is 40°F (4°C); for corn the base temperature is 50°F (10°C); and for cotton it is 60°F (16°C). The number of heat units for a day is the difference between the mean temperature and the base temperature for the given crop. The total of the heat units is the accumulated units for any specific period between the time of planting and maturity. Extensive use has been made of the heat-unit system by the canning industry in arranging planting and harvest schedules.

It has been shown that the number of heat units required for a given cultivar of oats to reach maturity varies little from one year to another. The heat units required are largely unaffected by the date of planting, whether early or late, in spite of the fact that a spread of 8 weeks in the date of planting may narrow to a difference of only 2 weeks in time of maturity.

From investigations in widely separated areas, good results have been obtained in predicting dates of heading and maturity of small grains by multiplying heat units by a daylength factor, rather than depending on the total of heat units alone.

Response of Plants to Temperature

Species and cultivars of crop plants differ in their physiological responses to ranges of temperature, each operating at the maximum within certain well-defined ranges from low to high. There must be at least the minimum essential to the initiation of activity, with the particular activity at its highest rate when the optimum temperature is reached. Wilsie suggests "a daily mean temperature at planting time" of 37°–40°F (3°–4°C) for spring wheat, 43°F (6°C) for oats, 45°F (7°C) for potatoes, 57°F (14°C) for corn, and 62°–64°F (17°–18°C) for cotton.

The minimum growth-initiation temperature for rye and peas is 29°–41°F (−2° to 5°C), sorghum requires temperatures of 59°–64°F (15°–18°C) to make appreciable growth. These examples give an idea of the range in temperature required by different species. The optimum temperature for most temperate crops is 75°–85°F (24°–29°C) with a maximum of 95°–105°F (35°–41°C). The minimum temperature for appreciable growth of corn is 50°F (10°C), the optimum 85°–95°F (29°–35°C), and the maximum approximately 113°F (45°C). Wide ranges are recognized in the minimum temperature necessary for the germination of different crop seed and for growth. Representative differences for germination are alfalfa, 34°F (1°C); rye, 35°F (2°C); barley, 39°F (4°C); corn, 40°F (4°C); sorghum, 48°F (9°C); rice, 52°F (11°C); and melons, 57°F (14°C).

Injury to plants by low temperatures will depend on the rapidity and the duration as well as the extent of the temperature drop, the general physiological conditions of the plant, and the inherent general growth characteristics of the species. Striking differences are recognized, not only between species in their tolerance to low temperature, but also between cultivars of a given species. Certain hybrids of corn and of sorghum, for example, are killed by temperatures considerably above the freezing point.

Studies in Missouri showed the temperature limits for appreciable growth of Canada bluegrass, Kentucky bluegrass, and orchardgrass to be 40°F (4°C); for bermudagrass the limit was 50°F (10°C). The optimum for orchardgrass was 70°F (21°C), for the bluegrasses it was 80°–90°F (26°–32°C), and for bermudagrass it was 100°F (38°C). The root temperatures for optimum growth were 50°F (10°C) for Canada bluegrass, 60°F (16°C) for Kentucky bluegrass, 70°F (21°C) for orchardgrass, and 100°F (38°C) for bermudagrass.

Light

It is recognized that green plants obtain the energy for their growth processes directly from sunlight, which is converted into energy by the photosynthetic process. It has been shown that there is a direct relationship between light intensity and photosynthetic activity, but his varies greatly with different species. Light intensity is less in humid than in arid climates; it may be reduced greatly by the presence of clouds and fog.

DURATION OF LIGHT. Light intensity and duration determine the amount of light a plant receives. The length of the day may be more important than light intensity. In the more northern latitudes a smaller angle of incidence results in a reduction in solar intensity compared to that in the southern latitudes. However, the greatest length of the day in the northern latitudes during much of the growing season may more than compensate for the smaller angle of incidence of the sun's rays. For certain species full sunlight may cause a lower photosynthetic rate and a higher respiration rate than a lower light intensity.

PHOTOPERIODISM. A light factor of great importance is the effect of the relative length of day and night as related to latitude, known as photoperiodism. Different species, and in some cases even different cultivars of the same species, respond very differently to a given day–night light ratio. Crops that flower and reproduce normally when the light period is longer than a critical minimum are known as *long-day plants,* while those that develop normally only when the photoperiod is less are called *short-day plants.* We are indebted to Garner and Allard for their pioneering work in this field of research. In their first published report, in 1920, they reported on a cultivar of tobacco that continued to make a purely vegetative growth under normal light conditions in the Washington, D.C., area, but that flowered profusely in a greenhouse in the winter under short-day conditions. The 'Biloxi' soybean cultivar behaved in a similar way. Their later research indicated a critical photoperiod of between 12 and 14 hours; this photoperiod divided species into long-day and short-day plant groups. Certain species, however, were found to be unaffected by photoperiodism; such species are called day-neutral plants.

Of our more commonly grown field crops, the small grains, potatoes, timothy, biennial sweetclover, and red clover are classed as long-day crops; most cultivars of soybeans, millet, lespedeza, and tobacco are identified as short-day crops. However, different cultivars of timothy have been shown to respond differently to daylength; certain cultivars flower under a 12-hour day, others required 13.5 hours, and still others as much as 16 hours.

Natural selection through many generations has resulted in markedly different ecotypes, which respond very differently to a given daylength. Striking examples are found in the native grasses of the United States, such as big bluestem and sideoats grama, which are distributed and flower north to the Dakotas and Canada and south to Texas.

It often is difficult to determine the specific temperature requirements because of the interactions between temperature and daylength.

Soil

The plant is dependent on the soil for water and nutrients as well as for anchorage; this makes soil an important part of the plant's environment. All the elements, or plant nutrients, needed for plant growth are obtained from the soil itself, except the carbon, hydrogen, and oxygen obtained from the air and water. The relative availability of these soil nutrients often determines whether crops can be grown economically.

Topographic features of elevation and slope have a marked effect on plant growth. High altitudes mean lower soil and air temperatures. Degree of slope, nature of the underlying rock, soil type, and time and frequency of hard rains will determine the amount of erosion.

A relatively level topography is a distinct advantage in crop production, favoring mechanical field equipment. Much of the world's grain is produced on relatively level expanses, such as those found in the Mississippi and Missouri river valleys, west central Canada, the Argentine Pampas, the area extending across France eastward into the USSR, the delta of the lower Nile, and the deltas of India and China.

The "prairie wedge" extending from Missouri into Minnesota and eastward across Illinois and Indiana into western Ohio includes some of the most productive soils. These soils developed under grasslands and resulted in the production of the dark-colored prairie soils. To the west of the humid prairie is the fertile, black Chernozem Belt, extending from Mexico into Canada. With precipitation less than evaporation and transpiration, a permanent dry layer develops in the soil profile below the depth of penetration of the water precipitation. Under these conditions, carbonates accumulate instead of being leached away. Typical chernozem soils, such as those in the Red River Valley of North Dakota, Minnesota, and Manitoba, have a deep, dark horizon, with the general condition very favorable to the production of wheat. Further to the west are the chestnut soils, occupying areas of slightly drier climate, with crops grown at some risk because of uncertain rainfall.

SOIL ACIDITY. The tendency to develop an acid reaction of soil solutions is caused mainly by the presence of CO_2. In the presence of water this gas tends to form carbonic acid. In humid regions where the rainfall is sufficient to cause leaching there is a continuing trend toward increased acidity. The various crops, and most especially the legumes, differ greatly in their adaptation to acid soil conditions. Most grow well on slightly or moderately acid soils. Small grains, corn, forage grasses, and some legumes are broadly tolerant, growing well within the pH range 5.8 to 7, or slightly above. Alfalfa and sweetclover are recognized as very intolerant of acid soil conditions.

SODIC-SALINE SOILS. Lack of rainfall in the drier regions results in an accumulation of sulfates, carbonates, and bicarbonates. When such accumulations exist the production of crops may be uncertain, or even impossible, over relatively large areas. Some crops are much more tolerant than others to sodic-saline soils.

The Atmosphere

In the photosynthetic process green plants release free oxygen during the daylight hours, whereas respiration releases carbon dioxide throughout both the day and night periods. All animals and nongreen plants use oxygen and release carbon dioxide constantly. The normal concentration of carbon dioxide in the air is sometimes considered below the optimum for the most efficient photosynthetic process to proceed, and under certain conditions is thought to be the limiting factor in production of such crops as corn and sugarcane.

SOIL AIR. Under normal growing conditions there is a high concentration of carbon dioxide in the soil and a low concentration of free oxygen.

The carbon dioxide content of the soil air remains relatively uniform, whereas the oxygen content may vary widely.

AIR POLLUTION. It is only in relatively recent years that the seriousness of air pollution has come to be fully appreciated. In highly industrialized areas one of the most common causes of plant injury is the release of fumes from industrial plants. Sulfur dioxide, a common pollutant often present in fumes from industrial plants, has been shown to be particularly injurious to plant growth. Others are ozone, hydrogen fluoride, chlorine, peroxyacetyl nitrate (PAN), nitrogen oxides, ethylene, ammonia, and particulates. Among the crop plants more sensitive to air pollution are alfalfa, oats, sugarbeets, and tomatoes. Some of the crops considered as less likely to show injury are common beans, wheat, corn, and bermudagrass. Herbicide and insecticide sprays are recognized as distinct air pollution hazards and are carried considerable distances in air currents, proving particularly damaging to many crop plants.

REFERENCES AND SUGGESTED READINGS

1. ANGUS, D. E. "Agricultural Water Use," In *Advan. Agron.*, 11 (1959). A. G. Norman, ed., pp. 19–35.
2. BAKER, O. E., and A. B. GENUNG. *A Graphic Summary of Farm Crops*, USDA Misc. Pub. (1938).
3. BROWN, E. M. *Some Effects of Temperature on Growth and Composition of Certain Pasture Grasses*, Mo. Agr. Exp. Sta. Res. Bul. 229 (1939).
4. CARDER, A. C. "Growth and Development of Some Field Crops as Influenced by Climatic Phenomena at Two Diverse Latitudes," *Can. Jour. Sci.*, 37:395–406 (1957).
5. CHAPMAN, H. W., *et al.* "The Carbon Dioxide of Field Air," *Plant Phys.*, 29:500–503 (1959).
6. EVANS, M. W., and H. A. ALLARD, "Relations of Length of Day to Growth in Timothy," *Jour. Agr. Res.*, 48:571–586 (1934).
7. GARNER, W. W., and H. A. ALLARD. "Effect of the Relative Length of Day and Night and Other Factors of the Environment on Growth and Reproduction of Plants," *Jour. Agr. Res.*, 18:553–605 (1920).
8. HOPKINS, A. D. *Bioclimatics—A Science of Life and Climate Relations*, USDA Misc. Pub. 280 (1938).
9. KATZ, Y. H. "Relation Between Heat Units Accumulated and Planting and Harvesting of Canning Peas," *Agron. Jour.*, 44:74–78 (1952).
10. LEMON, E. R. "An Aerodynamic Method for Determining the Turbulent CO_2 Exchange Between the Atmosphere and a Corn Field," *Agron. Abs. Amer. Soc. Agron.* (1959), p. 72.
11. MIDDLETON, J. T., *et al.* "Airborne Oxidants as Plant Damaging Agents," *Proc. Nat. Air Pollution Symposium* (Pasadena, Calif.: 1955), pp. 191–198.
12. OLMSTED, C. E. "Photoperiodic Response of 12 Geographic Strains of Sideoats Grama," *Bot. Gaz.*, 106:46–74 (1944).
13. SHIRLEY, H. L. "Light as an Ecological Factor and Its Measurement," *Bot. Rev.*, 11:497–532 (1945).

14. Thomas, G. W., S. E. Curl, and W. F. Bennett, Sr. *Food and Fiber for a Changing World* (Danville, Ill.: The Interstate Printers and Publishers, Inc., 1976).

15. USDA Agr. Stat. (1977).

16 U.S. Dept. of Commerce. *1974 Census of Agriculture Graphic Summary,* Vol. 4 (1978), p. xv–xxi.

17. Wiggans, S. "The Effect of Season Temperature on Maturity of Oats Planted at Different Dates," *Agron. Jour.,* 48:21–25 (1956).

18. Wilsie, Carroll P. *Crop Adaptation and Distribution* (San Francisco: Freeman, 1962).

CHAPTER 4
THE WORLD FOOD-POPULATION PROBLEM

Eight billion people are coming to dinner in the year 2000.

The Earth provides enough to provide every man's need but not enough for every man's greed.

M. K. GANDHI

History may bear out that the greatest launchings of the twentieth century were not from rocket pads, but from plots of dedicated agronomists and farmers, breaking the yield barriers to more food.

THERE are none more directly related to food production than those in the field of agronomy. It is the agronomist who works directly with the different crops. He is concerned with the increased efficiency of their production and with the soils on which the crops are grown. He also is concerned with the maintenance of these soils and the increase of their productivity potential.

With food in abundance all about us and knowledge of acreage or production limitations imposed on several crops in the United States, plus the fact that Americans spend more than $3 billion a year on food for dogs and cats, it is difficult to realize the possibility of starvation and famine on a mammoth scale and in many different countries within a few years. But this is the considered judgment of many of those who have given special and continued study to the food–population problem.

In relation to the seriousness of the food–population problem, on one extreme are the alarmists or neo-Malthusians, who have coined such phrases as "population bomb." On the other extreme are the "technocrats," who believe that technology will solve all problems and allow an acceptable level of living for the indefinite future.

Early in 1968 some 250 U.S. educators, government officials, businessmen, and editors spent two days together in a study of *The Strategy for the Conquest of Hunger.* All were concerned either directly or indirectly with international agricultural developments. Leaders from a dozen or more of the developing countries also were in attendance and brought the conference firsthand information on programs and progress in their respective countries. The seriousness of the world food–population problem was fully recognized at this time. The report of the conference indicated the general

49

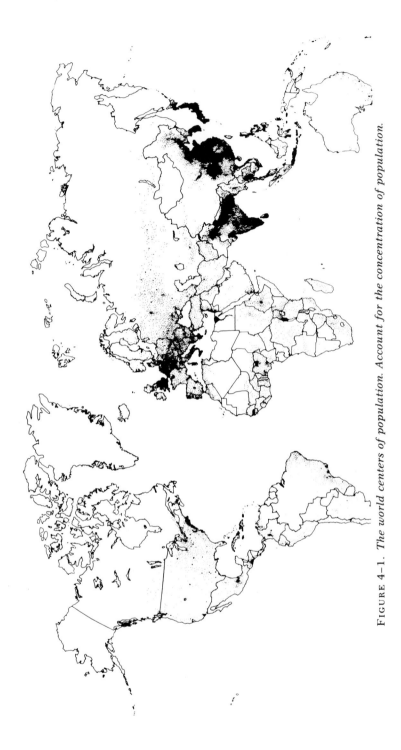

FIGURE 4–1. *The world centers of population. Account for the concentration of population.*

belief that if governments, national planners, and investors paid proper attention to the modernization of agriculture in the developing countries and if the upsurge of population increase could be halted, it would be possible during the years to come to meet the food needs in most of the developing nations.

But there are individuals of experience and ability, well informed on the agriculture of the United States, who do not foresee a serious food problem. They believe that the potential for increased food production on a worldwide basis is so great that we are not facing the probability of a worldwide starvation within the near future. Further belief is that man will find a way to supply the known, and yet to be learned, facts that will keep population and food on a reasonable balance for a long time to come.

Some of those who express such confidence recognize, however, that tools and techniques that must be applied include greatly improved crop cultivars, increased use of fertilizers, irrigation, insect and disease control, and perhaps greater profit incentives. Most recognize also that there are many problems of major magnitude that must be solved before much of the potential can be realized, including political, economic, social, developmental, and other problems. They point out that the potential production is so much greater than the present average acre production that a doubling of acre yields would seem to be within the range of possibility.

Currently, a consensus of opinion is that world food demands for the next 25 years will require crop yields and cultivated acreage to double. Food production in both the developed nations and in developing countries must be increased in order to feed the growing population.

It generally is recognized that the research of the state agricultural experiment stations and of the USDA is, in large part, responsible for the abundance of the agricultural production achieved in the United States. The results of applied research and the enterprise of U.S. farmers, in combination with natural resources such as are to be found to only a more limited extent elsewhere, have helped make the United States the bread basket of the world.

When considering the possibility of exporting the know-how of this greatly increased food production to the developing, food-deficient nations and of getting it applied there quickly enough to meet the increasing food needs, it is easy to forget that it took nearly half a century after the establishment of the agricultural research institutions before the phenomenal explosion in agricultural productivity ignited. It seems reasonable to believe that a considerable number of years will be required for significant breakthroughs to occur in most of the developing countries. The crucial difference is the urgency. Hungry mouths cannot wait.

Early Predictions

THE MALTHUSIAN THEORY. More than 150 years ago, Malthus, a British economist, observed that resources, such as food and fiber from agricultural production, increase only in an arithmetic ratio (for example, 1, 2, 3, 4, 5, 6), but that population, when unchecked, increases in a geometric ratio (2, 4, 8, 16, 32, 64).

LIEBIG AND SOIL PRODUCTIVITY. The noted German scientist, Liebig, more than 125 years ago suggested depletion of soil fertility as a limiting factor in possible future food production. He asserted that as crops are grown and removed from the land year after year, the supply of the necessary soil nutrients ultimately will become exhausted, to such an extent that crops can no longer be produced on the land.

Recent Recognition of the Food Problem

Early in the twentieth century individuals who had given special study and thought to the food–population problem gave vigorous voice to their concern over the rapidly increasing populations in different parts of the world, with no equal increases in the production of food. By mid-century many such volumes were clamoring for attention.

Individual effort in the food–population field of thought soon gave way to national concerns, with many government units and private agencies organizing to give concerted and continuing effort to obtaining accurate information on the problem and to render service in such ways as might be most effective. Beginning in the early 1960s almost every national journal, whether popular, scientific, or business and industrial, featured the subject, reporting the findings and conclusions from many comprehensive studies.

In addition to the early recognition given to the urgency of the problem by the United Nations soon after it came into being by the establishment of the Food and Agriculture Organization (FAO), nearly all the national scientific and educational organizations, as well as many business and professional groups, gave the world food problem high priority in their deliberations.

The American Society of Agronomy focused much national attention during the 1960s and 1970s on the food problem and the role of agronomists in helping solve this problem. The 1962 annual meeting gave attention to the Food for Peace Program. This was followed in 1964 by a symposium on World Population and Food Supplies: 1980. The 1967 program centered on Food for Billions. The theme of the 1974 meeting was All-Out Food Production: Strategy and Resource Implications. In 1975 it was Agronomic Research for Food, and in 1976 it was centered on Agronomists and Food: Contributions and Challenges.

The President's Science Advisory Committee

In 1966 President Johnson directed his Science Advisory Committee to make a comprehensive study of the world food supply. The panel that carried out the directive, in reporting its findings to the Food and Nutrition

Board (National Academy of Sciences, Washington, D.C.) in 1967, included the following:

Unless the situation changes markedly, food shortages and actual famine will occur. . . . The bulk of the increased food needs for the developing countries will need to come from agricultural production within these countries themselves. . . . Population control alone is not a solution. During the next 20 years food needs will more than double in the hungry countries if present population growth continues. Optimistic estimates of success in family planning in the next 20 years will only reduce food needs by 20 percent. There is still hope. . . . In the developing countries, until now, the technological revolution has had only one critically important effect, a striking reduction in death rates. . . . The problem of population–food imbalance is extremely complex and its dimensions are overwhelming. . . . The temptation to act on the basis of superficial or incomplete information has been irresistible, and has led to seizure upon panaceas and piecemeal solutions which are inappropriate, inapplicable, ineffectual, and inadequate. The cumulative delays engendered by false starts and stopgap measures have masked the broad needs for comprehensive programs. Food shortage and rapid population growth are separate but interrelated problems. . . . The choice is not to solve one or the other; to solve both is an absolute necessity. . . .

Unless farmers in a traditional subsistence agriculture can be persuaded to use fertilizers, pesticides, improved seed, and other modern inputs to increase output, all other efforts to increase food production will fail.

Journals and the Mass Media

In scientific journals and in the mass media we are reminded almost daily of an awakening by responsible authorities of two major facts: (1) there is a major worldwide food–population problem, which will be solved only by maximum contributions from the developed as well as the developing nations; and (2) the long-term solution to the food aspects of the problem will not be obtained by giving away American surplus food or by merely transferring our know-how to the developing countries. It is recognized that raising the food-production capacity in developing countries requires a combination of inputs, one of which is new knowledge and understanding, which we do not now have, and which we can obtain only through well-conceived continuous, long-range research programs under the different environments.

Mankind today faces history's greatest struggle for food. In many parts of the world human populations are increasing at rates greater than the production of basic food crops. Surprisingly, over the past decade, total food production in the developing countries has increased more rapidly each year than it has in the developed countries (about 3 percent, as compared to 2.7 percent). But rapid population growth in the developing countries means that the annual gain in per-capita output in developing countries was minimal—less than 0.5 percent—while that in developed countries was 1.7 percent each year during 1968–1977.

Decisive and vigorous action is needed to increase food availability. The need for population stabilization is becoming more widely understood and accepted, but at an agonizingly slow pace. The hard fact is that reductions in birthrates will not be achieved quickly enough to avert hunger and famine on a scale never before experienced.

If present trends continue, it seems likely that famines will reach serious proportions in India, Pakistan, and Bangladesh, and in several other countries of Asia, Africa, and Latin America in the near future. Indonesia, Iran, Turkey, Egypt, and several other countries may follow within a few years. Such famines may be of massive proportions, affecting hundreds of millions—perhaps even billions—of persons. If this happens, as appears possible, it will be the most colossal catastrophy in human history.

The deteriorating population–food situation around the world is the most important single factor affecting the outlook for American agriculture. World population is growing rapidly; food production barely is keeping pace. If the more than 8 billion people predicted by the year 2000 are to be sustained, with no improvement in their diet whatever, man will need to develop the capacity to feed another 4 billion people. This means we must duplicate within the next generation the greatest production record that man has achieved since the dawn of history. And we must do this at a time when all the easily developed lands of the world already have been brought into production and when we face increasing inroads on arable land by the urban sprawl, expanding highway systems, airports, parks, recreation sites, and the like.

Gloom in the 1960s

Two authoritative volumes of the 1960s having to do with the world food needs were *Famine: 1975* (1967) and *Toward the Year 2018* (1968). The most pessimistic of the two is the first, based on first-hand knowledge and observations by the two brother authors, William and Paul Paddock. They served or traveled in the developing countries for a combined total of some 40 years. These workers predicted the inevitability of long-term famines, which will last for years and perhaps for decades. They concluded that famine will reach catastrophic proportions and that revolutions, turmoil, and economic upheavals will sweep areas of Asia, Africa, and Latin America. Neither a new agricultural method nor a birth-control technique on the horizon can avert the inevitable. While the year 1975 has come and gone, it remains to be seen if their prediction of long-term famines will come true.

"Population," in the second work, *Toward the Year 2018,* is authored by Philip M. Hauser, Director of Population Research, University of Chicago, who also was the U.S. representative on the United Nations Population Commission. The author comments that almost a doubling of world food output is needed now to supply a nutritionally adequate diet to the population in the less-developed areas. If an adequate diet is to be achieved for a world population that will be more than doubled by 2000, production must more than quadruple by the end of the century. This will require greater annual increases in food production than have ever been attained over a prolonged period of time. The picture does have an encouraging side, however. Never before have so many nations adopted family planning programs as a matter of national policy; never before has so much been done in the fields of biomedicine and in the social sciences in the hunt for better methods of fertility control; and never before has the prospect been so good that the United States will, as it should, use its fabulous resources to support the necessary research. The question remains whether world population control is possible.

A more optimistic view was presented by the volume *Overcoming World Hunger,* under the sponsorship of The American Assembly, Columbia University. It brings together basic information by recognized authorities on the food-population program—"whether mankind is doomed to starvation and misery within a few years, due to the repeated doubling of the world's population." The authors believe that the grim consequences of hunger are not inevitable, that recent accomplishments hold out the hope that the world's population can be fed, but that before the threat of starvation subsides the nations of the world must make efforts of unprecedented magnitude to promote family planning and agricultural developments. They believe the tasks that still confront us in problem areas of world food, technical assistance, and economic development are proving more difficult, more complex, and more long-lasting than had been supposed.

Opinions in the 1970s

Authoritative views in the 1970s appear to be more optimistic than the writings of the "doomsday prophets" of the 1960s. Clifton R. Wharton, Jr., in a paper presented to a meeting of the American Society of Agronomy in 1976, called attention to the difficulty of the problem and the magnitude of the challenge ahead in the next 25 to 50 years. A puzzling argument of the neo-Malthusian prophets of doom is their contention that man somehow has reached his limits of technology. Wharton believes we have only begun to explore our capacities for increasing agricultural productivity through basic and applied research. He says that research is the key for unlocking the "Malthusian prison." Wharton further elaborates:

Just as Malthus could not foresee chemical fertilizers, the modern tractor, hybrid corn, or any of the countless innovations that have helped create modern agriculture, we cannot today predict the developments which will take place in our own future. But we can make one forecast with great confidence: innovation and new discoveries will continue to take place as long as thinking man exists.

The worldwide food crisis is real. It is serious, and it will probably worsen. Worldwide famine in all probability will be a recurring theme in our lifetime. But it is not inescapable, provided we hold fast to the following article of faith: while the world's resources may be limited, we have yet to discover the bounds of human creativity.

J. D. Ahalt, with the USDA World Food and Agricultural Outlook and Situation Board, in a paper presented in December, 1978, sees the problem as not just a production problem. In his view, the solution must address several facets. The developing world must increase its food production, reduce its rate of population growth, distribute incomes more equitably within its countries, and develop the infrastructure to move food supplies from rural and port areas to the more severely food-deficit areas. Developed countries, international institutions, and private industry can all help the developing countries achieve these goals.

World Population Increase Rates

In reviewing the conclusions arrived at from population–food studies, frequent mention has been made of the changes in rates of population increase at different times and in different world areas. What have these changes been and where have they occurred?

At the beginning of the Christian Era world population is estimated to have been about 250 million. Through the next 2,600 years the annual birthrate was but little more than the death rate, so that there was little change in the total world population. In the early 1600s, however, populations began to increase rapidly and in only 300 more years was nearly six times that of the 1600s. In the years following 1900 the rate of population increase changed from about 1 percent annually to nearly 2 percent by the late 1960s. From 1935 to 1970 the world population increased twice as rapidly as over the previous 35-year period. Even if the rate of increase for the years ahead can be materially reduced, it is predicted that by the year 2000 the world population may be as much as 8 billion.

Since 1935 the most rapid increase in population has been in Asia, Africa, and Latin America. These areas have more than 70 percent of the world population but produce only about 45 percent of the food grains.

In 1920 the combined population of the United States and Canada was very nearly the same as the total population of Latin America (the Western

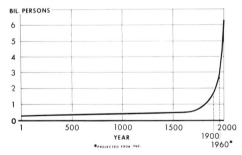

FIGURE 4–2. *Twenty centuries of population growth, projected to the year 2000. Account for the rapid population increase in later centuries.* [SOURCE: USDA.]

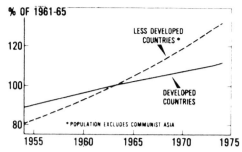

FIGURE 4-3. *Population growth in developed and less-developed countries since 1955.* [SOURCE: USDA–ESCS.]

Hemisphere south of the U.S. border). If present trends continue, however, the Latin American population will have increased to twice that of the United States and Canada by the year 2000.

Population Growth and Its Control

Hope that the world's population in the year 2000 can be fed arises from accomplishments recorded in the late 1960s and 1970s. It is still a hope, not a certainty. This optimistic goal is attainable only with the continuing, concerted efforts of unprecedented magnitude by the world's community of nations. There is no realistic possibility that the control of human fertility can be a substitute for economic development.

Expanding Food Production in Hungry Nations

The principal problems in the developing nations concern local attitudes and organizations not properly geared to modern agricultural technology.

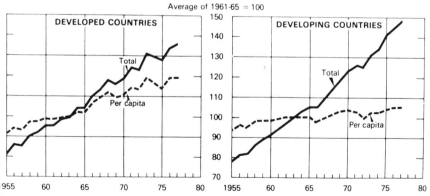

FIGURE 4-4. *Index of world total and per capita agricultural production in developing countries as compared to developed countries. Is the per capita production increase rapid enough to avert a catastrophe in developing countries?* [SOURCE: USDA.]

At the same time efforts of the developed nations to assist these countries have been hampered by conflicting views as to the seriousness of the world food problem and the approaches needed to solve it; by political and budgetary constraints; by bureaucratic intractibility; and by frequent changes in goals. Until major populations are stabilized and until most nations seriously-devote attention and resources to accelerate agricultural production, the future welfare of mankind must remain in doubt.

The United States has made some progress in recent years toward aid to developing nations. During the 1950s, the United States made major decisions and evolved programs for carrying its research and education activities to help the developing countries increase food production. This was climaxed in the 1970s by development of world cooperation through a Consultative Group in International Agricultural Research, which finances and administers 12 international institutes and centers. Participants in the Consultative Group include about 30 nations, three foundations, the World Bank, and agencies of the United Nations.

Also in the 1970s came Title XII, Famine Prevention and Freedom From Hunger, or the Findley–Humphrey Amendment to the Foreign Assistance Act of 1975. The central intent of Title XII is to promote an expanded role of U.S. agricultural colleges and universities in helping to solve the world food problem. It is supervised by the Board of International Food and Agricultural Development. It assists the Agency for International Development (AID) in planning and implementing all international food and nutrition programs.

Many Plants Used for Food

The world total of different plant species is estimated at some 35,000. Bailey's *Manual of Cultivated Plants* lists 5,347, for the United States and Canada; on a worldwide basis the number would be from two to four times this number. It has been variously reported that there are some 80,000 different plant species that produce nutritious food and that something like 800 of these have been cultivated at one time.

Some 38 food crops listed as of major importance are grown in the United States, but about 95 percent of the total food production comes from only about 20 of them. In contrast, in 87 food-deficient countries the average number of food crops reported as important is fewer than six, with the world average per country only about 11.

In recent years some of the "forgotten foods" are again receiving attention as a result of research and the race against worldwide hunger. It was reported in 1968 that plant hunters from Israel and the USSR, in search of a nourishing tuber crop that could thrive in soil too dry and hot for potatoes, found such in the American *Helianthus*. It grows in the wild in the Southwestern United States.

Some scientists believe we should return to more diversity in food production. Many of the world's peoples could increase nutrition and reduce

Protein

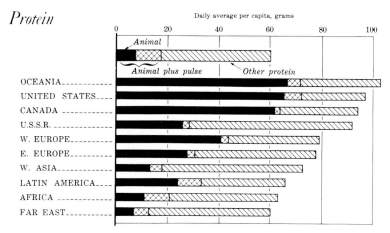

FIGURE 4–5. *Average daily grams per capita of different kinds of protein by country. It is recognized that the deficiency of available proteins is critical in less developed nations. Pulse protein, like animal protein, is effective in supplementing grain protein.*

vulnerability and the insecurity of depending on a few crops. These scientists feel that a good place to start is in the arid and semiarid regions, which constitute one-third the earth's land mass. The Sonoran Desert of Southwestern America, for example, is the home of 375 species of wild food plants. Natives presently use about 40 of them as food staples. Examples are the tepary bean, which is high in protein, and the buffalo gourd. The key point is to broaden our food base.

Research for useful plants goes on constantly. The USDA has long had its foreign plant explorers searching on a worldwide basis. Many plant species and many different cultivars of these species grown extensively in the United States were introduced by these federal plant hunters.

The more important plant species entering world trade as food crops number fewer than 20 and include wheat, rice, corn, sorghum, barley, oats, rye, millet, sugarcane and sugarbeet products, and potatoes. Other crop products entering world trade in significant amounts include cotton, sisal, tea, and cocoa.

Importance of the Cereals, Rice, and Soybeans

The importance of the cereals in the world food picture, and more recently of soybeans, can hardly be overemphasized. The countries of the Mediterranean area, including Rome, Greece, Carthage, and Egypt, had wheat as the principal basic food. In the Western Hemisphere the productivity of the new food plant maize (Indian corn) made possible the Inca, the Mayan, and the Aztec civilizations.

In the past it was said that rice was the "staff of life for half of the people of the world," including areas of India, China, and Japan. The sorghums and millets have been important food crops in much of Africa and in portions of India and Asia. More than 50 percent of the world's arable land is devoted to the grains. When the oilseed crops soybeans, peanuts, and sunflowers are added to the grain acreage, the total constitutes by far most of the world's harvested cropland area.

The 30° and 55° Parallels

World population concentrations have been greatest, and the increase in population has continued most rapidly, between the 30° north and 30° south parallels. This is also the area that has experienced the most severe food shortages through past years. And it is *not* between the 30° north and 30° south parallels that the world's most productive agricultural lands are located, where climates are best suited to the cereal grain crops that provide man's most dependable and satisfactory food.

The fact that the great world population centers of Asia, Africa, and Latin America, with about 70 percent of the total world population, are in those parts of the world where only one-fourth of the world's grain is produced is of the very greatest importance when viewed from the standpoint of food–population relationships.

The most productive agricultural areas of the world, and the areas in which most of the cereal grains are produced, are between the 30° and 55° parallels in both the Northern and the Southern Hemispheres.

Changes in Grain Export–Import Patterns

The picture of grain imports and exports has changed completely in recent years. In the period 1934–1938 Western Europe was receiving grain from all other major exporting regions. By 1966, however, the regions throughout the world that had been exporting grains had become net importing regions, with nearly all imports coming from the United States and Canada. Australia and New Zealand export relatively small amounts of grain. The remainder of the world's countries, with the exception of Argentina and the Republic of South Africa, encounter annual deficits and frequently must import grain. This fact provides the most significant evidence of the increasing seriousness of the world food problem.

Encouraging World Food Development

Future economic historians may well choose to refer to the decade of the 1960s as the decade of the green revolution. For what has started to happen to rice, wheat, corn, and soybean productivity in a number of tropical countries may be the start of a major transformation in the well-being of two-thirds the world's population whose economic fortunes are largely determined by how much their human resources and their soil and water can be made to produce. The 1970s will also be remembered as the time when the United States and the world made concerted efforts to formulate

TABLE 4-1. Summary of World Food Production
Indices, 1961–1965 = 100.

	1966	1971	1976
Africa	105	125	136
North and Central America	109	125	140
South America	109	126	158
Asia	106	125	143
Europe	107	121	130
Oceania	117	126	147
USSR	119	128	145
World	109	125	140

programs and organize groups, such as Title XII and the Consultative Group in International Agricultural Research, to help solve the world food problem.

Among the 12 international centers and institutes overseen by the Consultative Group are the Rice Research Institute (IRRI) in the Philippines, the Maize and Wheat Improvement Center (CIMMYT) in Mexico, the Institute of Tropical Agriculture (IITA) in Nigeria, the Center of Tropical Agriculture (CIAT) in Colombia, the Crops Research Institute for the Semi-arid Tropics (ICRISAT) in India, and the Potato Center (CIP) in Peru.

The IRRI and the CIMMYT are providing outstanding leadership, looking to increased food production in the developing countries. They are giving very special and thorough agricultural research and promotion training to a large number of young scientists from these developing countries, with special attention to the world's three most important food crops, rice, corn, and wheat. These institutes were established by the Rockefeller and Ford Foundations and later were taken over by the Consultative Group.

These organized efforts toward greater production have been successful, as evidenced by a summary of world food production indices (Table 4–1). Food production has risen since 1961–1965 in every major area of the world by 30–58 percent. This success has been achieved not only by improvements in cultivars available but also through increased adoption of better production techniques. A substantial expansion in irrigated areas has occurred in every major region of the world (Table 4–2).

Rice Improvement

Rice yields have been unreasonably low in the most important rice-producing regions of the world. The research with rice at the Rice Institute in the Philippines has been remarkably productive and efficient. From crosses made in 1962, one that proved to have outstanding value has been 'IR 8.' This cultivar not only matures early, but also is insensitive to photoperiodism, thus making it possible to plant at any time in the tropics, whenever year-round irrigation is available. This makes possible as many as

TABLE 4-2. Summary of World Agricultural Areas Irrigated*

	1961–1965	1970	1975
Africa	5,891	7,020	7,697
North and Central America	19,582	21,281	22,549
South America	5,403	6,100	6,616
Asia	138,617	151,891	161,098
Europe	8,954	10,794	12,697
Oceania	1,197	1,587	1,607
USSR	9,618	11,100	14,500
World	189,262	209,773	226,764

* In thousands of hectares.

three crops in a single season from the same land. The dwarf character of 'IR 8', with its great resistance to lodging, is a very important factor. It has been reported that in India more than 1 million acres (0.4 million ha) were planted to the new cultivar by 1968. One of the most important characteristics of 'IR 8' and other 'IR' cultivars is their response to heavy fertilization.

The development and spread in the 1960s of the new high-yielding cultivars of rice showed the world what agronomists can do to help alleviate the world food shortages. Evenson reported that the annual value of increased rice production attributable to new rice cultivars is nearly $3 million. Colombia is a good example of marked yield increases from new cultivars. Since 1967, when the rice improvement and introduction program was intensified in Colombia, rice yields have increased from fewer than 3 MT/ha to more than 5.4 MT/ha. Jennings reported that in 1974 alone the added production in Colombia from new rice technology was valued at $230 million.

One of the most notable improvements effected by plant breeders through manipulation of rice germplasm has been in the diversity of cultivars from which to select for variable conditions. Rice cultivars exist that will grow at soil pH values of 4.0 or less; others will tolerate values of 8.0 or more; still others will grow in soils with high salt content. Rice cultivars are available that will produce well under the conditions of cold air and water in northern Japan; still others thrive under the hot, dry climate of Pakistan and Iran. Some rices have resistance to low soil-moisture conditions and drought, while others can stand complete submergence for short periods.

Wheat and Corn Improvement

The success of the Wheat and Corn Center in Mexico in increasing the average acre yield of corn and wheat is outstanding. When this project was begun in 1943 the average wheat yield in Mexico was 11 bushels (740 kg/ha); at the end of 25 years it had increased to 39 bushels per acre (2,621 kg/ha). Mexico had been importing wheat but now exports some wheat

FIGURE 4–6. *An Iraqi farmer is pleased with a new, high-tillering rice cultivar which yields some 30 percent higher than the local cultivars. Beginning in 1963 FAO provided an expert to help improve the yields of wheat and barley in Iraq.* [SOURCE: FAO.]

and wheat products. Increases in corn yields in Mexico have been equally outstanding. Less than 25 years after the beginning of the Center, corn yields had tripled. In the mid-1970s, the CIMMYT was furnishing seed stocks of either wheat or corn to more than 90 nations.

WHEAT. Something of a revolution in wheat breeding had its beginning with the introduction from Japan of 'Norin 10.' Information about such short-straw wheats in Japan were known in the United States as early as 1874, but no use was made of this information. In 1946, S. C. Salmon of the USDA collected 16 of these types, including 'Norin 10,' and made them available to breeders in the United States. From this introduction, the first successful semi-dwarf cultivars were bred in America. These semidwarf cultivars are capable of stooling profusely, and are very resistant to lodging, even when heavily fertilized. 'Gaines', a winter wheat in its Northwest habitat, set world yield records of more than 200 bushels per acre (13,440 kg/ha). Superior lines obtained in Mexico by crossing Mexican and Columbian cultivars with 'Norin 10' proved adaptable to widely separated wheat-producing areas throughout the world. Millions of acres are seeded to the newer Mexican wheats each year. Short-strawed, photoinsensitive, and disease-resistant Mexipack wheats have become the basis for improved wheat yields around the world. Evenson has assessed the annual value of increased production attributed to newer cultivars at more than $1 billion.

Two other notable examples of wheat benefits from germplasm are:

1. A wheat from Saratov, Russia, has hairy leaves, providing a mechanism of resistance to three stages of the cereal leaf beetle. This was in the USDA collection for 40 years before it received much attention because the beetle did not reach the United States until 1962.
2. *Triticum timopheevi* became a part of the USDA collection in 1930; it was the early 1960s before it was discovered to have the mechanism now being used to produce hybrid wheat.

High-Protein Grains

The world's food problems arise not just from insufficient total food, but especially from lack of protein. Throughout most of the food-deficient areas of the world the protein content of the foods consumed in quantity is so low as to result in high infant mortality and chronic illness. It has been estimated that as a result of protein deficiency well over 25 percent of the children in certain parts of India have suffered permanent brain damage by the time they reach school age. Much effort is being expended looking to improve protein in the most used food grains: corn, rice, and wheat.

HIGH-LYSINE HYBRID CORNS. Lysine is an amino acid humans must have in their food if they are to remain healthy. The same is true for most animals other than ruminants. The high-lysine corn was developed at Purdue University, where it was discovered that a certain corn strain contained a mutant gene, called opaque-2, which had an exceptionally high lysine content. It was some 30 years after the opaque-2 gene was identified before it was discovered that it resulted in this relatively higher lysine content. Since this discovery the high-lysine gene has been introduced into many of the well-established corn hybrids. The best of the high-lysine hybrids available are reported to give yields of 90–95 percent that of hybrids without the opaque-2 gene. They are useful in human nutrition and in feeding programs of nonruminant animals such as swine.

At the Purdue Experiment Station pigs fed high-lysine corn without any

FIGURE 4–7. *A legume agronomist examines an accession of* Stylosanthes guyanensis *at the International Center of Tropical Agriculture in Colombia. This plant has the potential to improve forage production in Colombia because of insect and disease resistance.* [SOURCE: International Center of Tropical Agriculture.]

protein supplement gained weight over three times as fast as comparable animals fed commercial corn without a protein supplement. In feeding trials at the Illinois station the results indicated that corn with the opaque-2 gene can be expected to replace about 50 percent of the soybean meal ordinarily needed in feeding rations when commercial corn is fed.

In view of the extensive use made of corn as a basic food in important world areas and of the marked protein deficiency in quality protein in many diets, the availability of corn hybrids with high-quality protein, or better amino acid balance, is particularly significant. Similar work is under way on developing wheat and rice cultivars with significantly higher-quality protein than that in cultivars now commercially available.

Synthetic Proteins

The use of single-cell organisms to synthesize protein from petroleum fractions is technically feasible. Pilot production and trial feedings have been accomplished in the United States and several other countries. The first market visualized for protein from petroleum is in the animal feed industry. In this market protein from petroleum must compete on a cost basis with protein from agricultural products such as soybean meal. The impact of the production of protein from petroleum on domestic oilseed meal markets will be limited until production costs are reduced and extensive toxicity testing has been completed.

A Consensus of Judgments

—Lack of food can drive nations to desperate acts.

—There are today more hungry mouths in the world than ever before.

—The seriousness of the world food problem is fully recognized.

—The food–population problem cannot be solved without marked reductions in the rate of population increase.

—Reduced rates of population increase cannot be expected to come about quickly enough to avoid serious starvation conditions in many countries in the world.

—The potential for greatly increased food production is recognized.

—New techniques in production methods have greatly increased acre production in the United States and in a few other developed nations.

—The know-how of these new techniques must be exported to the developing nations. Some progress has been made in the last decade.

—Primarily, the food problem must be solved by increased production within the food-hungry nations.

—The food problem will not be solved by "giving away" American surplus food. Such a procedure destroys the initiative of local farmers to increase their production efforts, by taking away their markets.

—Industrial development (including the production of fertilizer and other agricultural input materials and improved transportation) must go hand in hand with agricultural development.

—Agricultural production inputs (such as fertilizers and pesticides) must be made readily available, with credit to make their purchase and use possible.

—Much long-term agricultural research will be necessary under the different environmental conditions throughout the world.

—Breakthrough increases in the production of world food has resulted when high-ability specialists have made a life career of food-increase efforts for developing nations.

—Developing high-protein grain crops, or grains with better amino acid balance (quality), such as high-lysine corn, should receive concerted research efforts in the near future.

REFERENCES AND SUGGESTED READINGS

1. AHALT, J. D. "Meeting Tomorrow's World Food Needs," Paper Presented at *IMC 14th Latin Amer. Food Prod. Conf.* (San Jose, Costa Rica, Dec. 6, 1978).

2. ALDRICH, D. G., JR. "A Challenge to American Colleges and Universities," *Food for Billions,* Spec. Pub. 11 (Madison, Wisc.: American Society of Agronomy, 1968).

3. ANONYMOUS. "Can We Broaden Our Food Base?" *The Furrow,* 82(5):8(1977).

4. *A Symposium: Strategy for the Conquest of Hunger* (New York: The Rockefeller Foundation, 1968).

5. BENNETT, I. L. JR. "Food and Population, and Overview," *Agr. Sci. Rev.,* 6:1 (1968).

6. BENNETT, I. L. JR., *et al. Food for Billions,* Spec. Pub. 11 (Madison, Wisc.: American Society of Agronomy, 1968).

7. BENNETT, M. K. *The World's Food* (New York: Harper & Row, 1954).

8. BRADY, N. C. "Agronomic Research Programs for the Future," *Challenge to Agronomy for the Future,* Spec. Pub. 10 (Madison, Wisc.: American Society of Agronomy, 1967).

9. BRADY, N. C. "The Role of Agronomists in International Agricultural Development," In *Agronomists and Food: Contributions and Challenges,* Spec. Pub. 30 (Madison, Wisc.: American Society of Agronomy, 1977).

10. BROWN, H. *The Challenge of Man's Future* (New York: Viking, 1954).

11. BROWN, L. R. *Population Growth, Food Needs, and Production Problems,* Spec. Pub. 6 (Madison, Wisc.: American Society of Agronomy, 1965).

12. BUTZ, E. L. "Agronomic Adjustments to Serve the Changing Future," *Challenge to Agronomy for the Future,* Spec. Pub. 10 (Madison, Wisc.: American Society of Agronomy, 1967).

13. COCHRANE, W. W. *The World Food Problem* (New York: Crowell, 1969).

14. EVENSON, R. E. "The Green Revolution in Recent Development Experience," *Amer. Jour. Agric. Econ.* 56(2):387 (1974).

15. EWELL, R. *World War on Hunger.* Hearings before the House Committee on Agriculture, 89th Cong., 2nd Sess., on H.R. 12152, H.R. 12704, and H.R. 12785 (Feb. 16, 1966).

16. FAO. *Trade Yearbook,* Vol. 29 (1975), pp. 3–11.

17. FAO. *Production Yearbook,* Vol. 30 (1976), pp. 57, 73–80.
18. Hardin, C. M. "For Humanity, New Hope," In *Overcoming World Hunger* (Englewood Cliffs, N.J.: Prentice-Hall, 1969). Under supervision of editor C. M. Hardin.
19. Harrar, G. J., and S. Wortman. "Expanding Food Production in Hungry Nations: The Promise, the Problems," In *Overcoming World Hunger* (Englewood Cliffs, N.J.: Prentice-Hall, 1969). ed. C. M. Hardin.
20. Hauser, P. M. "Population," In *Toward the Year 2018* (New York: Cowles, 1968). Edited by the Foreign Policy Association.
21. Heady, E. O., and L. V. Mayer. *Food Needs and U.S. Agriculture in 1980,* Natl. Adv. Commission of Food and Fiber, Tech. Paper 1 (1967).
22. Jennings, P. R. "The Amplification of Agricultural Production," *Sci. Amer.,* 235:180 (1976).
23. Liebig, J. *Principles of Agricultural Chemistry* (London: Walton and Maberly, 1855).
24. Malthus, T. R. *Essay on the Principles of Population* (London: Oxford University Press, 1798; reprinted by Macmillan, 1926).
25. Notestein, F. M. "Population Growth and Its Control," In *Overcoming World Hunger* (Englewood Cliffs, N.J.: Prentice-Hall, 1969). ed. C. M. Hardin.
26. Paddock, W., and P. Paddock. *Famine—1975. America's Decision: Who Will Survive?* (Boston: Little, Brown, 1967).
27. Pearson, F. A., and F. A. Harper. *The World's Hunger* (Ithaca, N.Y.: Cornell Univ. Press, 1945).
28. Prentice, E. P. *Food, War and the Future* (New York: Harper & Row, 1944).
29. Proc. The World Food Conference of 1976 (Ames: Iowa State Univ. Press, 1977).
30. Reitz, L. P. "Improving Germplasm Resources," In *Agronomic Research for Food,* Spec. Pub. 26 (Madison, Wisc.: American Society of Agronomy, 1976).
31. Russell, Sir John. *World Population and World Food Supply* (New York: Macmillan, 1954).
32. Stakman, E. C., *et al. Campaign Against Hunger* (Cambridge, Mass.: Harvard University Press, 1967).
33. Thompson, L. M. "Impact of World Food Needs on American Agriculture," *J. Soil Water Cons.,* 23:1 (1968).
34. Thompson, L. M. *Iowa Agriculture, World Food Needs and Educational Response* (Ames, Iowa: CAED, Rpt. 156, 1965).
35. Thorne, D. W. "Agronomists and Food—Contributions," In *Agronomists and Food: Contributions and Challenges,* Spec. Pub. 30 (Madison, Wisc.: American Society of Agonomy, 1977).
36. USDA. Econ. Res. Serv. *The World Food Situation and Prospects to 1985,* For. Agric. Econ. Rep. No. 98 (1974).
37. USDA. Economics, Statistics, and Cooperative Services, *Agricultural Outlook,* AO-31 (April, 1978).
38. Wharton, C. R., Jr. "Ecology and Agricultural Development: Striking a Human Balance," In *Agronomists and Food: Contributions and Challenges.* edited by M. D. Thorne Spec. Pub. 30 (Madison, Wisc.: American Society of Agronomy, 1977).
39. Whitney, R. S. (Ed.). *Challenge to Agronomy for the Future,* Spec. Pub. 10 Madison, Wisc.: American Society of Agronomy, 1967).

40. WILSON, C. M. "The Food Hunters," *Think,* 34:35–56 (1968).
41. WORTMAN, S. "Making Agronomy Serve Developing Countries," In *Challenge to Agronomy for the Future,* edited by R. S. Whitney, Spec. Pub. 10 (Madison, Wisc.: American Society of Agronomy, 1967).

CHAPTER 5
AGRICULTURAL PRODUCTION ADJUSTMENTS

The miracle of American agriculture is an example to all the world of the wisdom and the rewards of our democratic system. The preservation of this tradition has been the goal of all our farm policies of the last three decades. A wise and continuing partnership between American farmers and their government . . . will produce great benefits for the entire nation. Yet, all of this can be—and should be—only a beginning. New ways must be explored to keep agriculture and agricultural policy up to date, to get the full benefit of new findings and of new technology, to make sure that our bountiful land is used to the best of our ability to promote the welfare of consumers, farmers, and the entire economy.

LYNDON B. JOHNSON

THE phrase *the farm problem* is one often heard and frequently found in literature. It is an oversimplification, and may not be correct, to say that the cause of the U.S. farm problem lies in the tendency of American farms and farmers to produce more food, feed, and fiber than can be sold at a price to net a profit, either on the home market or in other parts of the world. And the production of a surplus of feed crops, with the lower prices resulting, soon brings on an oversupply of animal products—such as beef, pork, and mutton—of chicken and turkey products, and of milk and butter. Thus the "vicious circle" goes on and on.

With it all is the ever-increasing cost of the items that the farmer must purchase. The effort involves decreasing the units of production, with more fertilizer applied to the different crops, and greater use of insecticides and herbicides, with less labor employed. Ultimately there is a sharp decrease in the number of farms and farmers, as well as a change in the whole outlook of the farming community.

Our agricultural economists appear to be in agreement that it is not the overproduction of crops and livestock products, and the low net income resulting, that is responsible for the farm problem—the oversupply and low price pattern are the result rather than the cause. They emphasize that the basic cause lies in the changes in kinds and amounts of resources used in production—the changes in the relationship in the use of such inputs as capital, machinery, and labor.

Out of the Past

The attitude of the American farmer toward agricultural production, and of American businessmen and statesmen as well, underwent radical changes in the period following World War I. This attitudinal changes crystallized into action when in 1931 the nation found itself in the throes of a major depression.

Agricultural production had been stimulated to supply the strong European demands during war days and for some years later, while the European countries were making readjustments. The production of important agricultural products in some of the European countries nearly doubled, however, following the war. The unavoidable result was that the foreign market for agricultural products largely disappeared and American farmers found themselves with such surpluses that crops soon brought far less than the cost of their production.

As an average for the United States, cotton was selling in 1919 at 13.3¢ per pound; in 1932 it was 5.7¢. Corn was selling in Iowa at $1.41 per bushel in 1919, hogs at 16.71¢ per pound, and butter at 55¢. In November 1932, hogs were selling in Iowa at 3¢ per pound, beef cattle at 5¢, corn at 10¢ per bushel, oats at 7¢, and butterfat at 20¢ per pound.

Freight rates were unchanged, as were rents, insurance, payments on land, and other debts. These costs remained at close to 150 percent of the prewar prices. This situation, similar in other parts of the country, brought the nation to realize the fact that some radical changes would be necessary before conditions could be improved.

The Recovery Effort and Federal Legislation

On March 4, 1933, a financial panic struck the country. Two days later a 4-day banking holiday was declared to enable the banks to make arrangements to continue operating. Only the stronger banks were allowed to reopen.

The new Congress was convened in special session on March 5. During the short 100 days of its life, it passed in rapid succession measures that will be recorded as the most radical and revolutionary peacetime legislation in U.S. history. The President was given broad and startling powers over currency inflation. A huge program of public works was instituted. Provision was made for raising agricultural prices by limiting production. Money was made available for refinancing mortgages on homes and farms across the nation.

Federal legislation of the greatest immediate importance agriculturally was that for controlling and reducing the production of such products as corn, wheat, and pork, of which there were ruinous surpluses. This control program was inaugurated in 1933 under the Agricultural Adjustment Administration. In later years the conservation of soil resources was brought into the picture, with payments to farmers being based primarily on both the acreage of soil-conserving crops grown and on soil-improving practices and secondarily on acreage controls.

Still later programs were set up to provide for actual reductions in the acres of specified crops harvested and for the advancement of loans on stored products, as well as for payment on acreages taken entirely out of production for specified periods of years.

The Commodity Credit Corporation (CCC) was organized in 1933 as a part of a planned program to reduce surplus production and stabilize prices. The corn-acreage allotment program prior to World War II reduced the corn acreage by about 10 percent, but it did not have much of an effect on the total production of feed grains produced. Acreage diverted from wheat and cotton under the wheat and cotton programs and put into feed grains other than corn increased the total feed grains production by about 10 percent. After 1947, on the average, the CCC corn-loan program removed about 80 percent of the excess over average corn production in large-crop years and returned it to the market in small-crop years. In this way it had a substantial equalizing effect on corn consumption.

The original objective of the Federal Grain Storage Program of 1933 was to stabilize the prices of farm products in order to overcome yearly variations resulting from differences in total production. However, the programs soon began to go beyond this initial objective and were refocused from merely stabilizing prices to "stabilizing prices upward." Ultimately this stimulated production, reduced consumption, and led to accumulations of unsalable surpluses in government storage.

The storage program had some effect on farm prices and incomes. But most of the gain in prices and incomes was only temporary. Quantities of feed grains and wheat were removed from the market and held in government storage, with some of it ultimately disposed of abroad under Public Law 480. When the major share was eventually released into domestic channels, it depressed prices paid and incomes about as much as it had raised them when the surpluses were originally taken off the market.

There is increasing general recognition of the fact that U.S. agriculture can produce much more food and feed crops than domestic and export markets can absorb. Without controls of any kind on production and with support prices, the volume of grains produced would have increased far more than it did and would have resulted in markedly lower prices.

There has been little price bargaining in agriculture. Because of their inability to control output, producers have little, if any, control over the price at which agricultural products sell. As a rule, the price has been that offered by the buyer at the time of sale. To remedy this weakness at least partially, and with the majority consent of farmers, the government set up federal control programs to bring about a reduction in the quantity of grain on the market and to stabilize prices from year to year. To establish self-imposed bargaining power, producers must control the total supply. To be effective, control of supply must be on a long-term basis, not for just a few days or a few weeks. The cutoff must reduce the total supply over a relatively long period, and not simply withhold it from current marketing. Withholding farm products that have already been produced will almost certainly result in lower prices and income later, when the market supply has increased.

The national policy on export-competitive commodities has fluctuated between four general choices:

—Free markets, as largely prevailed up to 1929.

—Price supports without government control of production, which occurred from 1929 to 1931 and periodically thereafter.

—Price support and production control, as under the Agricultural Adjustment Administration in the 1930s, or land retirement and acreage-allotment programs as in the 1950s and 1960s, with direct payments and sometimes marketing quotas.

—Market subsidy, which among other things, is for keeping feed grains and other commodities competitive in commercial export markets.

In view of the high national cost of the agricultural control programs, some interests would have the federal government adopt a hands-off policy, with a resulting free market. But there seems to be no question that some kind of production control must be continued. Agricultural economists appear to agree that without any agricultural control program we would have marked overproduction and ruinously low prices. It appears that only the government has the resources to control agricultural production.

In an attempt to increase the earnings of farmers and compensate them for the tremendous gains they have made in efficiency—gains for which obviously they have not been adequately compensated in the market—society implemented government programs of various kinds. Most of these programs have worked by affecting the supply and demand of various inputs. For example, price supports and acreage allotments affect the supply of land, credit programs affect the supply of capital, and each has affected the supply, or output, of agriculture. In most cases the effect has been to increase the supply of farm commodities and to depress the price (instead of the attempted reverse effect). Consequently, government costs have increased. In response, society has required that inputs be reduced. In turn, farmers have responded by applying more capital inputs—fertilizer, insecticides, and so on—which has resulted in high production and low net returns. Thus, government programs have not been overly successful in raising farm earnings.

The present agricultural policy of the government is centered around set-aside programs, target prices, and farmer-held reserves. The Agricultural Act of 1977, which marked a return of government to prominence in the commodity markets, called for 1978 wheat and feed grain set-aside programs (taking out of production). By participating in a set-aside program, farmers are eligible for certain target prices for wheat, corn, sorghum, cotton, rice, and other commodities, and also for crop loans. Target prices refer to price levels set by the government that a farmer may expect for his commodities. If average prices received are below that level, the government will make direct payment to the farmer.

Participants in the set-aside program are also eligible for disaster payments if planting is prevented or if low yields occur. Until 1972 the government was involved in large-scale feed-grain storage. The present emphasis is on farmer-held reserves or on-farm storage. Farmers are paid for storing grain for three to five years, but the government can force selling of the grain in times of reduced supply or increased demand.

Present-Day Changing Conditions

Cropping patterns have seen important changes within recent years. Some of the more significant of these changes which relate in one way or another to our cropping patterns are changes in the (1) number of acres per farm unit, (2) farmland and farms, (3) farm population, (4) efficiency of production, (5) crop acreages, (6) export markets, (7) food preferences, and (8) costs and income.

Acres Per Farm Unit

In a period of some 57 years the number of acres in the average U.S. farm unit has more than doubled; in 1920 the average was 147 (59 ha) and in 1977 it was 393 (159 ha). Figure 5–1 shows the increase since 1960, and Table 5–1 shows the change in size of farms in selected Corn Belt states since 1850. Average of farm size is geographically in Figure 5–2.

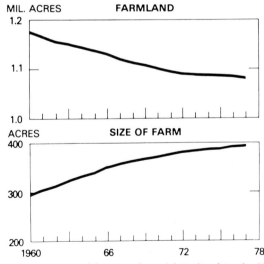

FIGURE 5–1. *Average size of farm and total farmland in the United States.*
[SOURCE: USDA.]

TABLE 5–1. Acres Per Farm in Certain Corn Belt States

Year	Ohio	Illinois	Iowa	Kansas
1850	125.0	158.0	184.8	—
1860	113.8	145.9	164.6	171.0
1870	110.8	127.6	133.6	148.0
1880	99.2	123.8	133.5	154.6
1890	92.9	126.7	151.0	181.3
1900	88.5	124.2	151.2	240.7
1910	88.6	129.1	156.3	244.0
1920	91.6	134.8	156.8	274.8
1930	98.1	143.1	158.3	282.9
1940	92.8	145.0	160.0	308.0
1950	105.0	159.0	169.0	370.0
1968	153.0	227.0	235.0	551.0
1977	149.6	241.6	261.1	636.4

Important in enlargement of farm units is the replacement of much of the farm labor with larger and more powerful equipment; other changes in techniques employed must also be made. Although total number of farms, farmland, and number of farmers continues to decrease, the trend for many years has been toward a larger farm unit, where production can be more efficient. It has been shown that the potential net income relates directly to the number of acres handled by a single farm operator.

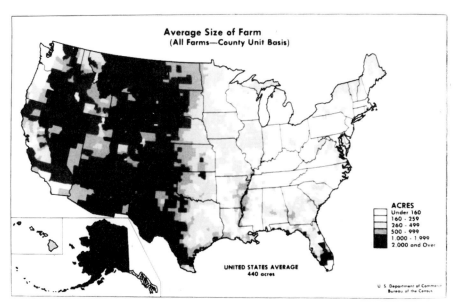

FIGURE 5–2. *Average size of farm in the United States shown geographically.* [SOURCE: U.S. Dept. of Commerce.]

Farmland, Number of Farms, and Farm Classification

Total farmland has declined steadily since 1960 as a result of conversion of farmland to other uses, such as housing developments, shopping centers, and other urbanization projects, interstate highway systems, and airports (Figure 5–1). This trend is alarming in view of the increasing population and future food requirements. Geographic distribution of land in farms is shown in Figure 5–3. Despite decreases in total farmland, cropland from which crops were harvested has increased since 1972.

Farmland value per acre has increased dramatically since the late 1960s. The per-acre value of land nearly tripled during the 10-year period of 1967 to 1977 (Figure 5–4). The percent change each year is shown in Figure 5–5. Land value has increased each year since 1955, and over the last five years, no annual increase has been less than 10 percent.

Number of farms declined from 5.9 million in 1945 to 4.1 million in 1959 to about 2.8 million presently (Figure 5–6). A sharp decrease occurred from 1959 to 1967; a more gradual decline has occurred since 1967.

Cash grain farms are the most common farm classification, making up about one-third of the total farms, with livestock farms being the second most prevalent type (Figure 5–7). Dairy farms, tobacco farms, and farms producing sugar, potatoes, or other crops are next in percent as to farm numbers. Despite the predicted decline of the family farm, individuals or families still control 90 percent of the farms and 75 percent of the farm-land (Figure 5–8). Corporations, while on the increase in the farming busi-

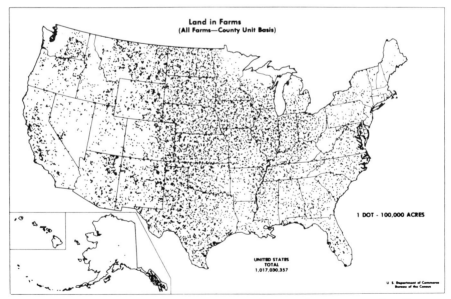

FIGURE 5–3. *Land in farms in the United States shown geographically.* [SOURCE: U.S. Dept. of Commerce.]

PERCENT OF MARCH 1, 1967

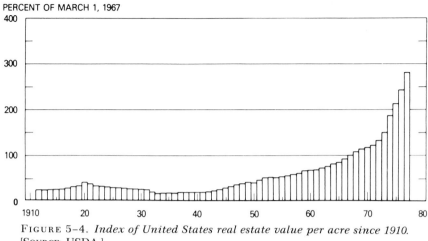

FIGURE 5–4. *Index of United States real estate value per acre since 1910.*
[SOURCE: USDA.]

ness, control less than 2 percent of the farms and only 10 percent of the
land.

Changes in Farm Population

With the larger farm units and human labor being replaced by larger
machinery, increased tractor power, and more efficient production practices, the farm population has decreased to such an extent that there are
now less than one-third as many persons on farms as there were in 1910.
The proportion of the farm population to the total U.S. population has
decreased from 35 percent in 1910 to about 4 percent in 1977. Figure 5–9
shows the decline in farm population, from more than 15 million in 1960
to a present level of just above 7 million. Net outmigration from farms

PERCENT

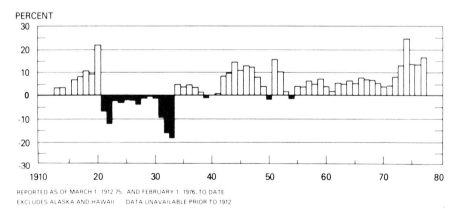

REPORTED AS OF MARCH 1, 1912-75, AND FEBRUARY 1, 1976, TO DATE
EXCLUDES ALASKA AND HAWAII   DATA UNAVAILABLE PRIOR TO 1912

FIGURE 5–5. *Percent change in per acre land value in the United States
per year since 1910.* [SOURCE: USDA.]

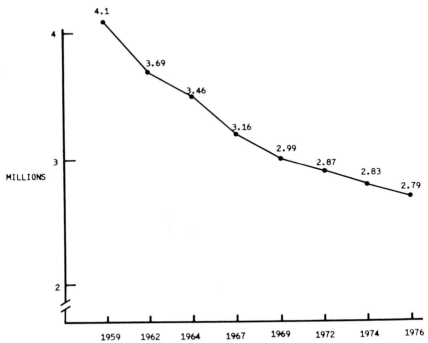

FIGURE 5–6. *Trend of number of United States farms since 1959.* [SOURCE: Southern Illinois University.]

peaked during 1940–1945, with more than 1.5 million persons leaving the farm each year during that period (Figure 5–10). While outmigration of workers from the farms has not continued at the level of 1940–1945, it still is occurring at alarming proportions.

Changes in Production Efficiency

With changes in the techniques employed in crop production, there has been a marked increase in efficiency. Although population has increased by 20 percent from 1960 to 1977, and subsequently demand for crop products has increased proportionately, cropland output increased by about 37 percent during the same period (Figure 5–11). Total farm inputs have increased only slightly—about 2 percent since 1967—but total farm output and output per unit of input have increased by 19 percent or more during 1967–1978 (Figure 5–12). Even more dramatic is the farm output per man-hour, which by 1976 was up by 60 percent over 1967 (Figure 5–13). In comparison, nonfarm business output per man-hour during the same period was up less than 15 percent.

Another view of the overall increase in production efficiency for the United States is shown by looking at production of individual crops or crop groups. With few exceptions, man-hours required to produce crops per acre (0.4 ha) has declined (Figure 5–14). This decrease from the period 1915–1919 to 1971–1975 has often been fivefold or more. But tobacco has

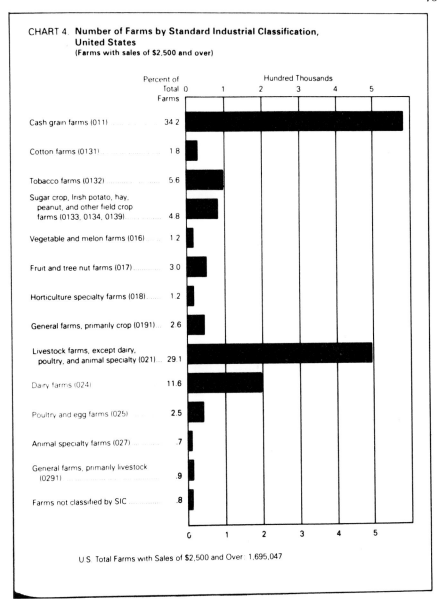

FIGURE 5-7. *Number of United States farms by classification.* [SOURCE: USDA.]

not fit the production efficiency pattern in the same way as most crops because of the large amount of hand labor required. After an increase in man-hours required from 1915–1919 to 1955–1959, a marked decrease had taken place by 1971–1975. Tobacco production is gradually becoming more mechanized, but this change has been much slower than for most other crops. Farm production per hour index is shown in Figure 5–15. Average

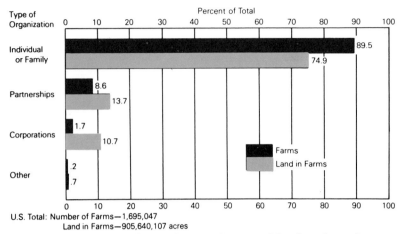

FIGURE 5-8. *Number of United States farms and land in farms by type of organization.* [SOURCE: USDA.]

production efficiency per hour for all crops has increased by 31 percent since 1967.

Changes in Crop Acreages

Marked changes in the number of acres planted to the different crops have occurred within the past 25 years. The decrease in acreage of oats, barley, rye, tobacco, and cotton has been striking, as shown in Table 5-2. Soybean acreage has increased threefold during this period and total pro-

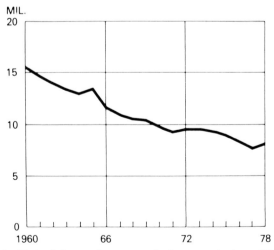

FIGURE 5-9. *United States farm population trend since 1960.* [SOURCE: USDA-ESCS.]

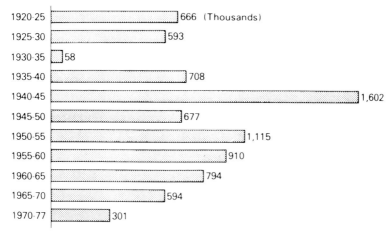

FIGURE 5–10. *Annual net outmigration from the United States farm population.* [SOURCE: USDA–ESCS.]

duction fourfold. Other crops showing significantly increased acreage include wheat and grain sorghum.

Changes in Export Markets

The United States has seen a greatly increased export demand for grain and farm products in recent years. Grains and feeds have a larger share of the export market than any other commodity group (Figure 5–16). Specifically, corn has been the feed grain with the greatest export demand, with more than 40 million metric tons exported in 1976 (Figure 5–17). The largest share of feed-grain exports are to European countries. Agricultural exports to the Soviet Union had a value of about $2.0 billion in 1978, and exports to the OPEC nations more than $2.0 billion in 1978 (Figures 5–18 and 5–

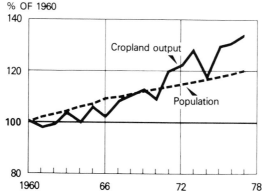

FIGURE 5–11. *Cropland output in the United States as compared to population since 1960.* [SOURCE: USDA.]

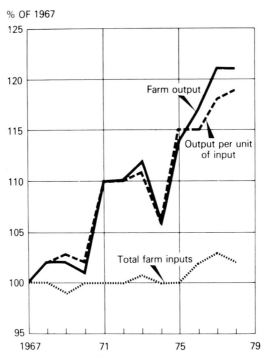

FIGURE 5–12. *United States farm inputs and outputs since 1967.* [SOURCE: USDA–ESCS.]

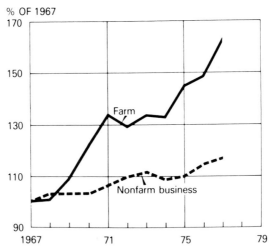

FIGURE 5–13. *United States output per man-hour for farm and nonfarm businesses since 1967.* [SOURCE: USDA–ESCS.]

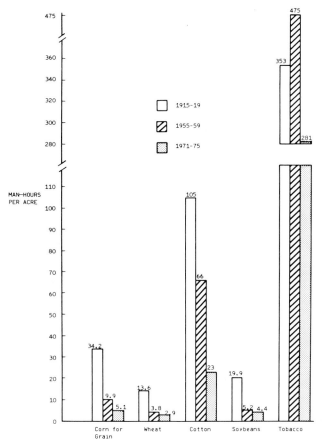

FIGURE 5–14. *Labor hours per unit of production of five crops in the United States.* [SOURCE: Southern Illinois University.]

19). Total agricultural exports increased by 51 percent from 1959 to 1975 (Figure 5–20). The sharpest increase during this period was shown by oilseeds and vegetable oils, which doubled from 1967 to 1975 and increased fourfold from 1959 to 1975. A larger percentage of soybeans and soybean products produced in the United States is exported than any other agricultural product, about 60 percent (Figure 5–21). Other crops in which a large share of the total production is exported include rice, 55 percent; cotton, 45 percent; wheat, 40 percent; grain sorghum, 31 percent; and tobacco, 30 percent.

Changes in Food Preferences

The relative demand by the average American for the different food products has changed markedly from period to period. The per-capita consumption of potatoes decreased by 50 percent and that for cereal products by 25 percent in the 40-year period from 1910 (based on the average con-

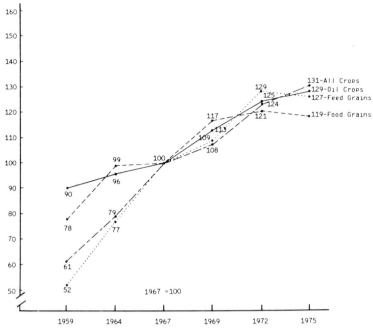

Figure 5–15. *United States farm production per hour index for oil, feed, food, and all crops since 1959.* [Source: Southern Illinois University.]

sumption for the period 1909–1913) to 1950. During the same period the consumption of eggs, fruit, and vegetables increased by 30 percent. In more recent years distinctly different trends have developed. The increase in consumption of vegetable oils and processed fruits and vegetables is most

Table 5–2. U.S. Harvested Acres and Total Production of Selected Crops, 1951–1955 and 1976

Crop	Acres Average (millions)		Production Average (millions)	
	1951–1955	1976	1951–1955	1976
Corn (bu) for grain	70.1	71.1	2,814	6,216
Wheat (bu)	60.5	70.8	1,077	2,147
Oats (bu)	37.8	12.4	1,311	562
Barley (bu)	10.8	8.4	303	377
Rye (bu)	1.7	0.8	22	17
Sorghum (bu) for grain	8.9	14.9	170	724
Soybeans (bu) for beans	15.7	49.4	313	1,265
Rice (cwt)	2.1	2.5	53	117
Potatoes (cwt)	1.4	1.4	217	353
Sweetpotatoes (cwt)	0.3	0.1	18	14
Cotton (bales) for lint	22.7	10.9	15	11
Tobacco (lb)	1.7	1.0	2,217	2,134

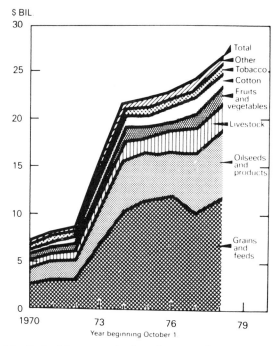

FIGURE 5–16. *United States agricultural exports by principal commodity groups since 1970.* [SOURCE: USDA–ESCS.]

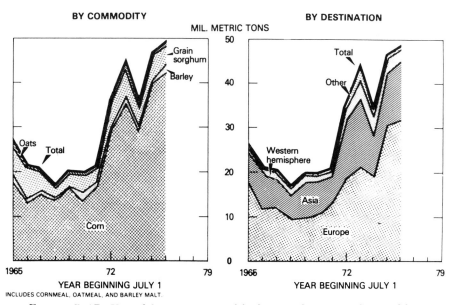

FIGURE 5–17. *United States exports of feed grains by commodity and by destination.* [SOURCE: USDA.]

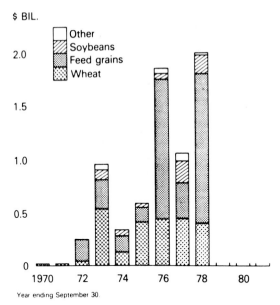

Figure 5–18. *United States agricultural exports to the Soviet Union since 1970.* [Source: USDA–ESCS.]

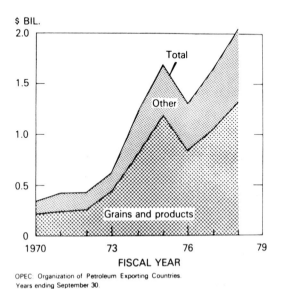

Figure 5–19. *United States agricultural exports to OPEC nations since 1970.* [Source: USDA–ESCS.]

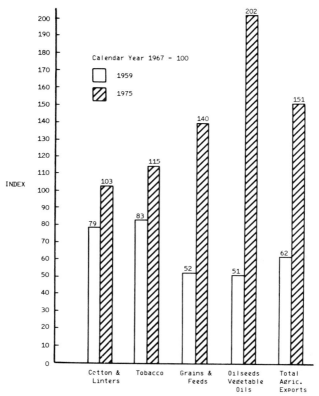

FIGURE 5–20. *United States total agricultural exports, and exports of cotton, tobacco, grains and feeds, and oilseeds and vegetable oils in 1959 vs. 1975.* [SOURCE: Southern Illinois University.]

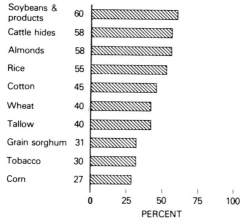

FIGURE 5–21. *Ten United States agricultural exports as a percentage of farm production.* [SOURCE: USDA.]

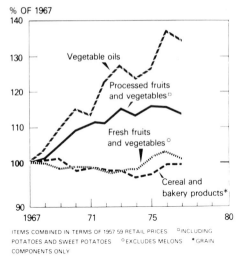

FIGURE 5–22. *United States per capita consumption of selected crop products since 1967.* [SOURCE: USDA.]

striking (Figure 5–22). Consumption of fresh fruits and vegetables and cereal and bakery products changed very little from 1967 to 1977.

Meat consumption has continued to increase. There is a direct relationship, of course, between the production of feed crops and that of animal products. In 1976 the average American consumed an average of 192.7 lb (87.4 kg) of beef, pork, veal, and mutton, in addition to 43.3 lb (19.6 kg) of chicken and turkey. Ten years earlier, the consumption of these foods was 170.9 lb (77.5 kg) of all meats except chicken, and 35.6 lb (16.1 kg) of chicken. Most of the increase in total meat consumption can be attributed to the increase in beef consumption; consumption of most other meats has not changed appreciably. Per-capita consumption of eggs, milk, and butter has decreased significantly since 1966.

Cost of Production, Prices Received, and Farm Income

Prices of selected farm inputs used in crop production increased from 80 to 200 percent during the period 1965–1977 (Figure 5–23). The sharpest rise was shown by farm real estate, three times the value of only 10 years ago. Machinery and wage rates more than doubled during this 12-year period, and fertilizer prices increased by about 80 percent.

Crop production costs continue to climb; 1977 costs per bushel were about $6 for soybeans, nearly $4 for wheat, and well over $2 for corn, soybeans, barley, and oats (Figure 5–24). Production costs for rice were over $8 per hundredweight and for cotton were over 60¢ per pound. A comparison of production costs with prices received by farmers for some of these same commodities reveals that farmers rarely have received prices for major commodities above production costs (Figure 5–25). Most years disposable

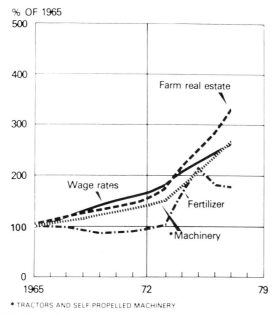

FIGURE 5–23. *United States prices of selected farm inputs since 1965.*
[SOURCE: USDA.]

personal income per capita for nonfarm population has exceeded that of the farm population (Figure 5–26). When farm income is evaluated in terms of real dollars, the net income picture has looked particularly dismal after a peak in 1973 (Figure 5–27). A specific example of the farmer's dilemma is shown by looking at the consumer's bread price and where it goes (Figure

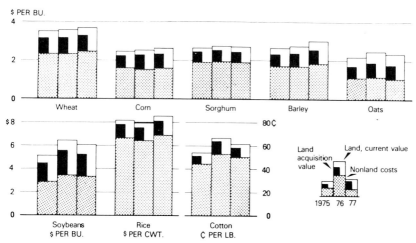

FIGURE 5–24. *United States crop production costs per bushel for eight field crops.* [SOURCE: USDA.]

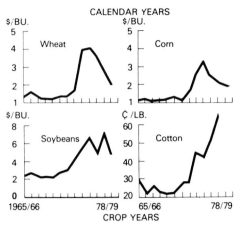

FIGURE 5–25. *Prices received by United States farmers for four major field crops since 1965/66.* [SOURCE: USDA.]

5–28). While retail price per pound of bread has almost doubled since 1960, the farm value of the wheat and other ingredients in the bread has not experienced the same proportional value increase, with the exception of a limited time period when wheat prices were high. The larger share of the increase has been due to the baker–wholesaler spread and the retail

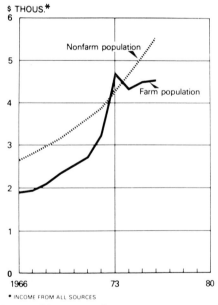

FIGURE 5–26. *Disposable personal income per capita for the farm and nonfarm population in the United States since 1966.* [SOURCE: USDA.]

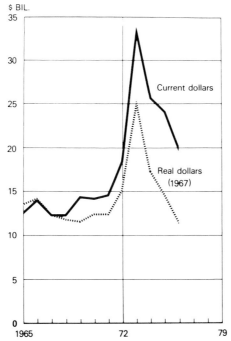

FIGURE 5-27. *Total net United States farm income in real dollars since 1965.* [SOURCE: USDA.]

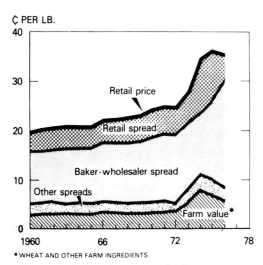

FIGURE 5-28. *Consumer's bread price and where it goes in the United States since 1960.* [SOURCE: USDA.]

spread. In 1977 farmers received 38.5¢ of each dollar spent for food in retail stores, while middlemen got 61.5¢.

The late 1970s brought farmer unrest and demands for parity or fair prices for their products. While achieving parity prices likely is not feasible or possible, some of the reasons for the farmers' distress are apparent from the preceding discussion.

Looking to the Future

Production methods and the economics of production have changed rapidly in recent years and continue to change, as is evident to all. Publications are available from federal and state agencies and from other sources in which an effort is made to predict something of the changes in agricultural production likely to develop.

A number of the more meaningful predictions and projections into the future in relationship to crop production have been condensed and grouped into related subject groups: economic operating and marketing considerations; agricultural land and farm units; farm equipment, labor and other inputs; and crop characteristics and production.

Economic Operating and Marketing Considerations

—A most serious problem will be the amount of capital required for a young farmer to get started in farming.

—For a farm operator to continue successfully it will be necessary to increase the output by about 6 percent annually, about half of which can come from increased acre production. The other half must come from farm enlargement or from increased number of livestock.

—A 6 percent annual inflation rate is expected for the next 20 years. Farmland might sell for $12,000 or more per acre (0.4 ha), but at the same time corn could sell for $8 per bushel and soybeans for $16 or more per bushel.

—The fuel problem should be solved by the year 2000, with alternative energy sources being available. Solar collectors will dry grain, warm farrowing facilities, and cool houses. Alcohol for fuel will be obtained from conversion of wastes or crop residues, such as cornstalks, and methane for fuel will be obtained from manure.

—Soybeans or feed grains might be transported from the Midwest through lubricated piplines to New Orleans or other ports.

Agricultural Land and Farm Units

—There will be a further reduction in the number of farm operators, each with a considerably larger farm unit.

—The trend to larger, more specialized farm units will continue at an even more rapid rate than in the past.

—The future will bring still fewer farm units, probably down to about 1.6 million by the year 2000.

—The total land in farms will be reduced by 10–20 percent because of increased use for highways, recreation areas, forests, and so on.

—Experts see family-managed farms still thriving by the year 2000. Although the number of corporate farms likely will increase, the vast majority of land will be controlled by family units and not by corporate giants or foreign investors.

—On a world scale, an economically feasible method of desalination of seawater could permit massive irrigation projects and agricultural production on lands now unused.

—Improved soil management techniques may allow additional agricultural land, such as the tropical rain forest, to be brought into production.

Farm Equipment

—Most farm machinery will be larger, to match the change to larger and more powerful tractors, which will make possible increased output per farm worker. Mechanization experts believe that 500-hp, 12-tire, giant tractors will be common. They will float over the field on low-pressure tires that will reduce compaction.

—Virtually every major machine will be computerized. Corn planters will not only monitor seed drop but will count seeds per acre on the go.

—Tillage tools will have sensors to manage depth control.

—Minimum tillage and no-tillage practices, first applied to corn, will be applied increasingly to other crops.

—The use of planes and helicopters will be used more widely in crop production for seeding, as well as for applying pesticides and fertilizers.

Labor and Other Inputs

—There will be a further reduction in farm labor requirements, because of larger and more efficient machinery.

—The decreased labor force will consist of 40–60 percent fewer farm operators, 70 percent fewer family workers, and 45–50 percent hired workers.

—There will be an increase in purchase inputs, in proportion to those produced on the farm.

—More than four-fifths of the production input will be capital and less than one-fifth labor within 10 years.

Crop Characteristics and Production

—There will be continued emphasis on breeding for insect and disease resistance, and for stalk or stem strength for lodging resistance.

—Breeders will give increased attention to composition of the seed or grain. Upgrading protein content of cereal grains will assume major importance.

—In developing better crop cultivars, increased attention will be given to the orientation of the leaves of such crops as corn, soybeans, and small grains, with more sturdy, erect leaves and ability for increased photosynthesis being favored.

—Additional crops, such as soybeans, will be hybridized, resulting in probable yield increases of 10–20 percent.

—Nitrogen fixation capability could be extended to new groups of plants in addition to legumes.

—Pesticide and commercial fertilizer use will continue to expand, despite some pressure from environmentalists, but with greater restrictions on kinds and application procedures.

—Integrated past management, including biological insect and disease control, will be used widely.

—Use of satellites for worldwide crop reporting, successful long-range weather predictions, and possible weather modification will be valuable aids to the farm operator in the year 2000.

—There will be an increase in all kinds of irrigation for field crop and horticultural crop production, including sprinkler, center-pivot, and trickle systems.

—Food obtained through conventional field crop-production techniques may be supplemented to some extent by the use of microbial action on waste materials.

REFERENCES AND SUGGESTED READINGS

1. ANONYMOUS. "Farm Bill Cleared, Puts Government Back in Price Support Arena," *Congressional Quarterly,* Sept.: 1977, pp. 2029–2032.

2. BROWN, L. R. "The World Outlook for Conventional Agriculture," *Science,* 158:604–611 (1967).

3. CARLSON, J. "Soybean Horoscope: Increased Acreage, Higher Yields, More Profits," *Soybean Digest* Vol. 38, pp. 32–33 (1978).

4. EICHER, C. K., and L. W. WITT. *Agriculture in Economic Development* (New York: McGraw-Hill, 1964).

5. HALCROW, H. G. *Food Policy for America* (New York: McGraw-Hill, 1977).

6. HEADY, E. O., *et al. Roots of the Farm Problem* (Ames: Iowa State Univ. Press, 1965).

7. HEADY, E. O., and L. MAYER. "Which Farm Program to Control Our Overcapacity," *Iowa Farm Science,* 22:9 (1968).

8. HEADY, E. O., and M. SKOLD. *Projections of U.S. Agricultural Capacity and Interregional Adjustments in Production and Land Use with Spatial Programming Methods,* Iowa Agr. Exp. Sta. Res. Bul. 539 (1965).

9. JOHNSON, L. B. *To the National Advisory Commission on Food and Fiber* (Washington, D.C.: The White House, January 11, 1966).

10. KELLOGG, C. E. "Interactions in Agricultural Development," *Science, Technology, and Development* (Washington, D.C.: World Food Congress, 1963).

11. MAYER, L., and E. O. HEADY. "Our Capacity to Produce," *Iowa Farm Science,* 22:8 (1968).
12. National Advisory Commission on Food and Fiber. *Food and Fiber for the Future* (Washington, D.C., 1967).
13. National Farm Institute. *Bargaining Power for Farmers* (Ames: Iowa State Univ. Press, 1968).
14. PAARLBERG, D. "Farming in the 22nd Century," *Prairie Farmer,* Sept. 1976, p. 214.
15. RASMUSSEn, W. D. (Ed.). *Readings in History of American Agriculture* (Urbana: Univ. of Illinois Press, 1960).
16. SHEPHERD, G. *Appraisal of the Federal Feed Grains Program,* Iowa Agr. Exp. Sta. Bul. 501 (1962).
17. SHEPHERD, G., and A. RICHARDS. *Effect of the Federal Program on Corn and Other Grains,* Iowa Agr. Exp. Sta. Bull. 459 (1958).
18. SHRADER, W. D., and F. F. RIECKEN. "Potentials for Increasing Production in the Corn Belt," *Dynamics of Land Use—Needed Adjustment* (Ames: Iowa State Univ. Press, 1961), pp. 61–75. Assembled and published under sponsorship of Iowa State Univ. Center for Agricultural and Economic Adjustments.
19. SOTH, L. *An Embarrassment of Plenty* (New York: Crowell, 1965).
20. THOMPSON, L. M. *Weather Variability and the Need for a Food Reserve,* CAED Rep. 26 (Iowa State Univ. Press, 1966).
21. TOMA, P. A. *The Politics of Food for Peace* (Tucson: University of Arizona Press, 1967).
22. USDA Agr. Stat. (1977).
23. USDA. *Changes in Farm Production and Efficiency,* Stat. Bul. 233 (1969).

Part II
Environmental Factors
Affecting Crop Production

CHAPTER 6
CLIMATE, WEATHER, AND CROPS

WEATHER is so important to the activities of everyone, and especially those in agricultural activities, that the old adage passed down through many centuries, "Everyone talks about the weather, but no one ever does anything about it," is most apropos.

There are three related terms in this chapter that should be defined briefly.

Meterology—the science having to do with the atmosphere and its phenomena, especially as it is related to the weather.

Climate—the composite, or generalization, of weather conditions of a region averaged over a period of years; as temperature, humidity, precipitation, sunshine, cloudiness, and winds.

Weather—the state of the atmosphere, with regard to such variables as air movements, cloudiness, moisture, and pressure. Also defined as a single occurrence, or event, in a series of conditions that in total make up the climate.

Temperature

Radiant Energy

Energy necessary to sustain life is derived from the sunlight by photosynthesis. Radiant energy from the sun is converted into chemical energy, contained in simple sugar molecules. The solar energy spectrum extends from the short wavelength of ultraviolet through the visual range and into the long wavelengths. Radiant energy within the visible band not used for photosynthesis is degraded to heat and is used in respiration and in evaporation of water from the plant in transpiration.

The photosynthetic efficiency of light utilization is very low. Plant breeders, especially of corn and small grains, are attempting to develop cultivars with leaves that will grow at such an angle as to make possible increased use of solar energy, which would result in increased acre production.

Climatic Zones

Geographers have indicated three general temperature zones: the torrid, temperate, and frigid. The north temperate zone, which includes most of the United States, generally is characterized by warm summers and cool or cold winters. Growing-season temperatures in some places favor such warm-season crops as cotton, corn, and sugarcane. In other areas such cool-season crops as oats, peas, Kentucky bluegrass, and white clover are favored. Kentucky bluegrass, for example, does not thrive and is little grown south of Kentucky. Cotton, peanuts, and many grasses are confined to the Deep South. Different cultivars of soybeans and of alfalfa are adapted to production from the Deep South all the way into Canada.

Length of Growing Season

The length of the growing season, usually stated as the frost-free days, has an important influence upon the crops that can be grown and the cultivars which are adapted to a given area. Bermudagrass, famous throughout the South, cannot be grown with success where there are fewer than 200 frost-free days. However, temperatures far above the freezing point slow or stop the growth of many crops. Although the number of frost-free days usually has been used to indicate the length of the growing season, the temperature at which plants are killed varies greatly for different species, and even for different cultivars of the same species. Some are killed at temperatures well above 32°F (0°C). A rapid drop in temperature kills more frequently than does a slow drop to the same degree.

Growing Degree Days or Heat-Unit System

"Growing degree days," or "heat-unit system," is a method for measuring at any time in the season how fast a crop is advancing toward maturity. It is calculated by adding, during the days of the growing season, the daily mean temperature degrees above a certain established base temperature. The base temperature to be used is the temperature below which the specific crop does not make any appreciable growth. The base temperature for hybrids of the Corn Belt generally is considered to be 50°F (10°C).

The method, heretofore generally applied, of designating the number of days for a given hybrid to mature has been found inadequate, because in seasons cooler than normal the development and maturity is slowed, with the result that full maturity may not be reached before a killing frost. The growing-degree-days method provides a more precise way of measuring progress toward maturity. For days with an average temperature below 50°F (10°C), corn makes no perceptible progress toward maturity. The growing-degree-days system takes this into account; days with an average below 50°F (10°C) are recorded as zero in the growing-degree-days record.

To compute growing degree days, the lowest and the highest temperatures for a given day are added and then divided by 2 to get the mean (average) temperature. Subtracting the base temperature from this mean gives the number of growing degrees for the day. Most full-season dent corn hybrids in the central Corn Belt require 2,600–2,800 degree days to reach 30 percent kernel moisture maturity. Different hybrids vary somewhat from this average. The growing degree days for a specific hybrid

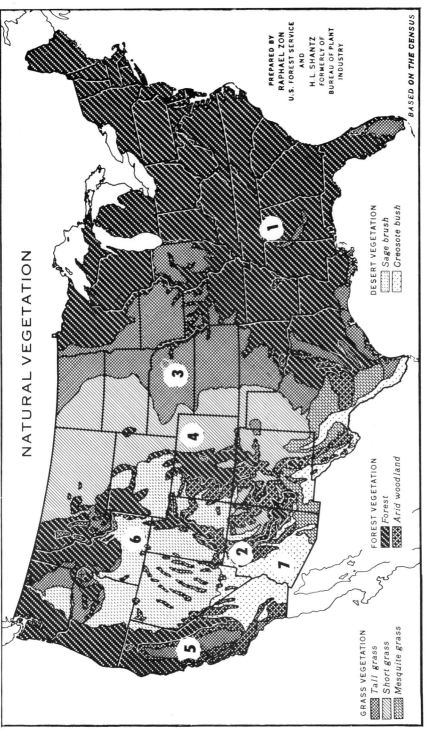

NATURAL VEGETATION

GRASS VEGETATION
  Tall grass
  Short grass
  Mesquite grass

FOREST VEGETATION
  Forest
  Arid woodland

DESERT VEGETATION
  Sage brush
  Creosote bush

FIGURE 6–1. The natural vegetation that had stabilized through the centuries in different parts of the United States reflects the prevailing climates in different parts. (1) forest, (2) arid woodland, (3) tall-grass prairie, (4) short-grass prairie, (5) mesquite grass, (6) creosote bush, (7) sage brush. [SOURCE: USDA.]

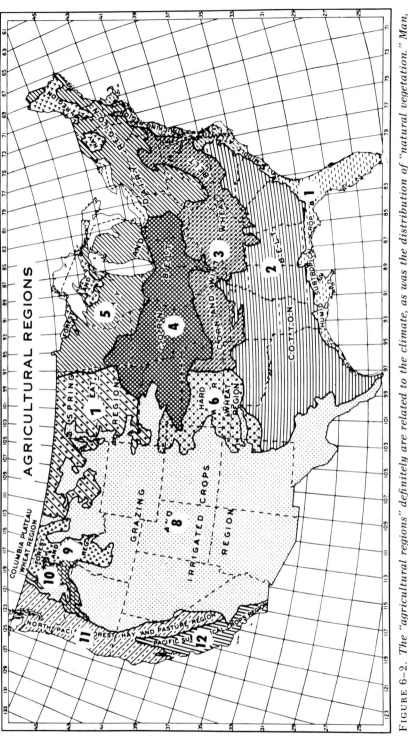

FIGURE 6–2. The "agricultural regions" definitely are related to the climate, as was the distribution of "natural vegetation." Man, through experience and experimentation, adjusts his choice of crops and farming pattern to the climatic and soil environment: (1) humid, subtropical crops; (2) Cotton Belt; (3) corn and soft red winter Wheat Belt; (4) Corn–Oats Belt; (5) hay and dairy; (6) hard red winter wheat region; (7) hard spring and durum wheat region; (8) grazing and irrigated crops; (9) forest and hay; (10) Columbia Plateau wheat region; (11) north Pacific forest, hay, and pasture; (12) Pacific subtropical crops. [SOURCE: USDA.]

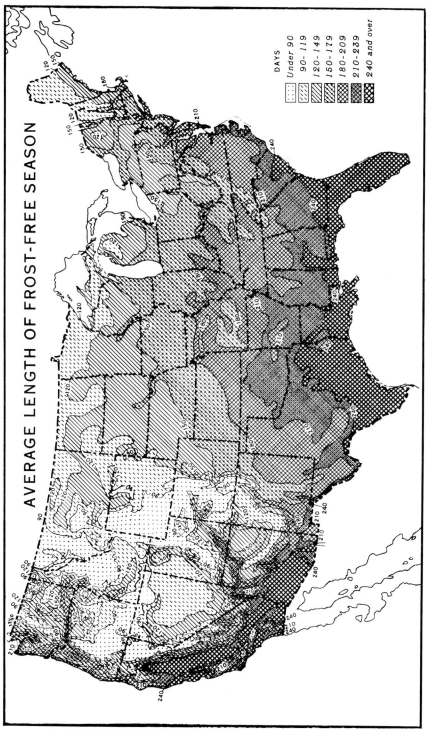

AVERAGE LENGTH OF FROST-FREE SEASON

DAYS

Under 90
90–119
120–149
150–179
180–209
210–239
240 and over

FIGURE 6–3. *The average length of the frost-free season varies from 240 days and over near the Gulf, to 210 in north central Alabama and northern Arkansas, to 180 in northern Missouri and southern Illinois, to 150 in northern Iowa and central Wisconsin, to 120 in northern Wisconsin, Minnesota, and North Dakota.* [SOURCE: USDA.]

needs to be determined, preferably by the originating seed producer. Once determined, a given hybrid can be depended upon for the same performance year after year, regardless of location.

There are certain limitations in the growing-degree-days system. Some crops, such as the soybean, are very sensitive to daylength or changes in latitude where planted. Moreover, special consideration must be given to very hot days, when it is likely that transpiration will be so high that corn is under moisture stress, and growth is reduced as a result. It usually takes a day with a maximum temperature above 90°F (32°C) and a minimum above 70°F (21°C) to produce a 30-growing-degree day. Under such conditions soil moisture is likely to be deficient for crop growth and development, with the result that growth will not be normal. It has been suggested that when the temperature goes above 90°F (32°C), the amount by which it exceeds 90°F (32°C) should be subtracted from the growing-degrees accumulation for that particular day. Moreover, any environmental conditions, which tend to depress or increase the rapidity of development or growth, such as fertility level, water availability, or weed pressure, provide another variable factor for which allowance should be made.

Canning companies have made use of the growing-degree-days system for many years to schedule planting and harvesting of peas and sweetcorn. Many hybrid corn seed companies use the system in scheduling crossing of field plantings, to obtain the desired "nicking" for pollination.

Crop Moisture

Precipitation

The seasonal distribution and total quantity of precipitation is important in determining the adaptation of a crop to a given area. Available soil moisture, in general, decreases as we proceed westward from the Atlantic Coast and northwest from the Gulf. The humid region extends west to about the 100th meridian. This area is characterized by alternating periods of warm moist air from the Gulf moving northward and cold dry air from Canada moving southward. Extremes of heating and cooling with intermediate rains occur especially toward the western part of the area.

West of the humid area is the semi-arid area, a belt 200–300 miles (322–483 km) in width. For this area the annual precipitation toward the south approaches 30 inches (76 cm); to the north the annual precipitation is only about 20 inches (51 cm), with great year-to-year variations in total precipitation for any given locality.

The arid area, extending from Mexico into Canada, is still further to the west, 400–600 miles (644–965 km) wide. Annual precipitation for this area ranges from almost nothing toward the south, to more than 30 inches (76 cm) in some of the higher mountain sections. Toward the south, rainfall occurs primarily in late and mid-summer. Further north, effective precipi-

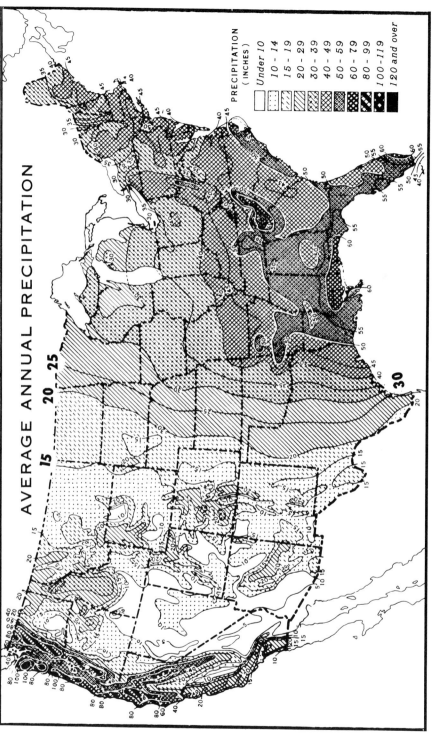

FIGURE 6–4. *The average annual precipitation in different parts of the United States varies from less than 10 inches over a considerable area in the Southwest to over 120 inches in very small areas of the Pacific Northwest. The 15-, 20-, 25-, and 30-inch isotherm lines are particularly significant in indicating limits for the production of certain crops and for different types of farming.* [SOURCE: USDA.]

tation comes in the form of winter snow and spring and early-summer rains.

Along the Pacific Coast is a summer dry belt. In this area precipitation occurs primarily during December, January, and February. Toward the extreme south the annual precipitation may be less than 5 inches (13 cm) with extremely long and dry summers.

The amount of precipitation that falls on a particular area depends to a considerable extent on its relationship to an ocean and to prevailing winds. Most of the precipitation that falls on land is the condensation of vapor from ocean surfaces.

Usually the season when rains occur is more critical than the total annual rainfall. For example, areas in eastern Washington and the Great Plains of North Dakota both have about 20 inches (51 cm) of total precipitation. For Washington, however, the precipitation is confined largely to the winter months, whereas in North Dakota the winters are dry and precipitation occurs in the spring months. As a result, in Washington annuals and winter annuals which can complete their growth early in the season before the moisture supply is depleted predominate, whereas in North Dakota corn, wheat, sorghum, and perennial grasses are the principal crops.

The relationship of temperature to moisture is also an important factor. For example, to the north, 20 inches (51 cm) of rainfall will have the same value for crop growth as about 30 inches (76 cm) toward the Gulf area.

Evapotranspiration

Growing crop plants contain large quantities of water, usually 50–90 percent. But the amount of water contained in the plant at any given time is a very small fraction of the total water that must pass up through the plant during its growth. For example, about 1.5 million pounds (0.7 million kg) of water is used to produce a 2,000-pound (907-kg) growth of alfalfa dry matter. Most of this water is evaporated from plant and soil surfaces.

The rate of water used by crop plants is a function of climate and is largely determined by solar radiation. It has been shown, however, that different plant species differ greatly in the efficiency with which water is utilized and in their tolerance to drought. The plant's capacity to get water from the soil when the supply is limited is an important species characteristic, often related to the depth and extensiveness of the root system.

Humidity and Dew Formation

Atmospheric humidity results largely from free-water surfaces. As temperatures increase a volume of air is able to hold more water vapor than at lower temperatures. *Absolute humidity* is the total amount of water in the air, whereas *relative humidity* is a percentage measurement of the approach of air to its saturation point, which changes with the temperature.

Air humidity is quite different in the same area at different heights above the soil surface. At several feet above the soil surface the degree of humidity is dependent on the air mass of a large area, whereas the air near the soil surface contains more water. Because plant parts cool more rapidly at night than does the air mass, they are soon below the temperature at which the air is saturated with water. This results in condensation of moisture on the plant as dew. Dew formation and its persistence is regarded as a primary factor in providing an environment favorable to plant diseases and insects.

Daylength

The length of the daylight period influences plant growth and the initiation and development of different parts of the plant. Daylength varies considerably with latitudes and determines the seasons. Daylengths increase after December 21 in the Northern Hemisphere, to June 21, after which time they begin to shorten. Many crop plants are very sensitive to daylength, especially in the initiation of floral and vegetative buds. The different crop species are usually quite consistent in their response to daylength, but in some cases different cultivars of a given crop respond very differently from the general trend for the species. For example, 'Louisiana S-1' white clover blooms profusely in the Deep South, whereas the 'Pilgrim' cultivar, developed in the Northeast, flowers sparsely, if at all, under the same conditions. In addition, certain of the earlier cultivars of timothy flower at a daylength of 12–12.5 hours, whereas some of the late cultivars will not flower with less than 14 hours. Although the term daylength has come to be used generally, it is the short night rather than the long day which causes many plants to flower or not to flower. It has been shown that 1 hour of artificial light at midnight, for example, effectively breaks a 14-hour winter night into two 7-hour nights.

The vegetative habit of growth also responds to daylength. In the longer summer days leaves and stems grow erect and the development of axillary buds is reduced, whereas during the shorter late-summer and fall days the growth is much more prostrate, with plants literally hugging the soil surface and more vegetative buds initiated.

Genetic inheritance is an important factor. It has been observed that the more hardy cultivars of alfalfa and winter grains make a prostrate growth in the fall whereas the less hardy cultivars continue to make the upright growth of midsummer.

Weather Forecasting

The forces that result in our day-to-day weather are far from fully understood, although meteorological observations go back as far as the fourth

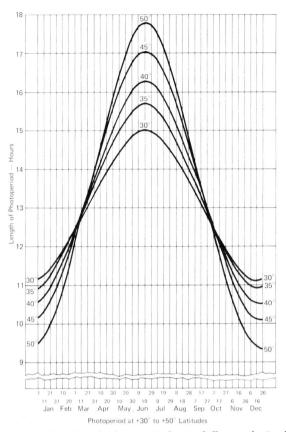

FIGURE 6–5. *The length of photoperiods at different latitudes. These curves represent the duration of sunlight from the beginning of civil twilight in the morning to the end of civil twilight in the evening. The length of day and night periods varies considerably in different parts of the United States and at different parts of the year. In northern United States higher latitudes, the daylength is as much as 18 hours. In the southern United States in June it is approximately 14 hours.*

century B.C., when Aristotle recorded the first known study of weather. But after Aristotle there were almost 2,000 years of silence.

The first barometer, invented in 1643, was a 34-foot (10.4-m) -high tube that enclosed a column of water. Later a column of mercury replaced the water, with many other modifications. The barometer measures atmospheric pressure and is very important in weather forecasting. When the barometer is falling rapidly a storm may be anticipated; a rising barometer usually is indicative of fair weather. Although the science of meteorology developed steadily through the seventeenth century and later, it was not until the invention of the telegraph, which made possible the rapid assem-

FIGURE 6-6. *The marked effect of photoperiodism on the vegetative-flow-ering relationship of the same timothy is strikingly evident when the exposure was varied from 10 hours (right) to 14.5 hours (left).*

bling of far-flung weather information, that forecasting emerged as a highly organized service.

The first U.S. government weather service was organized in 1870. Because of the great influence of the weather on all agricultural activities it operated as a unit of the Department of Agriculture. Much later, because of the importance of weather information to the rapidly developing air transportation, the Weather Bureau became a part of the Department of Commerce. In October, 1970, the United States Commerce Department's National Oceanic and Atmospheric Administration (NOAA) was formed. The NOAA brought together the functions of the Weather Bureau and at least a dozen other agencies or groups from the Department of Commerce, Department of Interior, U.S. Navy, Transportation Department's Coast Guard, National Science Foundation, and Army Corps of Engineers.

Great progress has been made in the study of weather variables and in weather forecasting. The general and extensive use of radar and of weather satellites are outstanding examples.

Although the meteorological strides thus far have been made by the United States, the study and prediction of weather is worldwide in scope. Any operative weather network will of necessity involve the widest possible international cooperation.

National Agricultural Weather Service

The extent to which weather information will be applied to crop production problems depends on (1) the extent of crop response to one or more weather factors, (2) the climatic probability of occurrence of the

FIGURE 6-7. *A typical agricultural weather station for obtaining climatological data such as precipitation, wind velocity, evaporation, and temperature at several levels.* [Photograph with permission of University of Tennessee Agricultural Experiment Station.]

several weather variables, and (3) the ability of the producer to make, and act on, alternative decisions based on timely weather information.

Weather service tailored to farming operations can make the difference between profit and loss in an agricultural enterprise. Each seed that is planted and then rots in the ground, each seedling stunted or killed by damping-off, each blossom or plant killed by frost, and each pound of pesticide applied and washed off by rain in a few hours cuts into the producer's pocketbook and hurts the entire area's economy. These losses can be minimized through an effective Agricultural Weather Service.

A federal plan for the National Agricultural Weather Service was outlined in 1971. This organization of NOAA focuses on the need for providing specialized weather services to farmers and other agribusiness interests, and includes agricultural applications of climatology. It makes maximum use of observational networks and of data processing and disseminating facilities. Features of the Service are:

1. Collection, analysis, and interpretation of weather data pertinent to optimum planning of the allocation of agricultural land, labor, and capital.
2. Technical studies in agriculture–weather relationships at one or more

Federal and State Agricultural Experiment Stations in each state aimed at future improvements in weather service.
3. Agricultural weather forecasts designed to support specific types and phases of farm operations.
4. Rapid and efficient dissemination of forecasts, warnings, outlooks, and advisories.

A pilot project to determine the best procedures for serving the interests of an important agricultural area was established in the Mississippi Delta for the 1959 crop year. The project was expanded to eight important agricultural regions by 1962 and to 12 areas by 1967. In this service all state and federal resources are coordinated to provide agriculture with both short-range and long-range planning information, provide forecasting information and interpretation and weather data collection and dissemination. A 10-phase plan is allowing for gradual expansion of the current service into presently unserviced areas of the United States.

The sophistication of current weather services is illustrated by the Agriture Weather Center located in West Lafayette, Indiana. It makes use of computers, CRT terminals, and a number of other systems to collect, process, and disseminate information for a five-state area. A daily reporting network of 25 to 30 stations are required for each state. Observers from this network report by touch tone pads directly to a computer which immediately processes the information for multiple uses by the West Lafayette Center. The stations provide air and soil temperatures, humidity, precipitation, evaporation, soil moisture, dew, and also furnish crop information.

Weather Modifications

In the last few years scientists have made headway in rain making and in certain situations can now make predictions about rainfall. Current thinking on weather modification is that cloud seeding may increase the chances of rain or snow by 5–30 percent. However, the seeding process sometimes decreases precipitation and other times has no effect at all. Cloud-seeding experiments ordinarily use silver iodide released at high altitudes in supercool clouds. This technique also is being used to prevent hail damage to crops and to disperse fog. Recently, a group of weather scientists concluded that silver iodide may be harmful to the environment. Biodegradable organic compounds, such as metaldehyde and 1,5-dihydroxynaphthalene, are being evaluated as possible replacements for silver iodide. Besides their environmental benefits, these compounds are cheaper and more readily available than silver iodide.

It is difficult to measure the effectiveness of weather modification because of the inability of scientists to predict how much rain will fall without seeding and whether rainfall that does occur is the result of the seeding. Recent evidence accumulated by NOAA indicates that dynamic cloud seeding in Florida is effective in increasing the size and rain production of convective clouds, promoting cloud merger, and increasing the rainfall from groups of conventive clouds. There are strong indications that dynamic seeding is effective in producing a net increase in rain over a fixed target area. The dynamic seeding technique uses silver iodide to seed supercooled water, which is below freezing but still in liquid form, high in developing cumulus cloud towers. Experiments in the late 1960s demonstrated that dynamic seeding stimulates greater rain production in individual clouds. Experiments in the early 1970s confirmed that seeding also promoted the merger of individual clouds. An experiment done in 1976 indicated that dynamic seeding increased total rainfall in the target area.

There have been claims of flood damage because of cloud seeding, which has a legal aspect.

Weather–Crop Yield Relationships

Almost every agricultural activity must take into consideration the expected weather conditions. Crop yields for any given locality vary greatly from year to year, and these production differences are due almost entirely to differences in the prevailing seasonal weather.

The yield-depressing effect of less than optimum growing conditions for 79 of the principal U.S. crops has been estimated to reduce income by an average of $1.6 billion per year. Another report estimates that weather-caused losses each year in crop output alone amount to between $1–$2 billion. Of necessity these losses often are passed on to consumers in higher prices for less plentiful products. Many of these losses can be prevented or reduced by weather forecast information.

Although the extreme dependence of crop growth and development on the weather factor generally is recognized, because of the complexity of the interactions among the plant, environmental, biological, and technological factors, it has been exceedingly difficult to determine the influence of any one specific weather factor on plant growth and productivity. The end-of-the-season yield data reflect the composite effect of all the environmental, biological, and technological factors on the crop for the season.

The *environmental* factors include solar radiation, soil moisture and temperature, air temperature and movement, humidity and rainfall. The *biological* factors include the possible injury effect of insects and diseases, which may be affected by the same weather variables, either favorably or unfavorably. The *technological* factors include changes in soil conditions and available nutrients, fertilizer application, plant population and distribution, cultivar or hybrid planted, mechanization, and possibly irrigation.

Employing the most modern techniques, there has been considerable research of weather–crop yield relationships. In other research the influence of weather has been separated from the influence of technology by the use of time trends for technology and multiple curvilinear regression for weather variables.

Weather and Wheat Production

There are great variations in the acre yield of wheat from year to year. It would appear that no other crop is more affected by weather variables. No single weather variable can be pinpointed as the cause; rather, there is a cumulative effect of many variables operating at different stages of growth.

As might be anticipated, for the Great Plains a high correlation has been shown between the annual precipitation and wheat production. An early study by Cole and Mathews showed that if the soil was dry to a depth of 3 feet (0.9 m) at seeding time, there was only a 30 percent likelihood that a successful crop would result. If the soil was moist to a depth of 2 feet (0.6 m) or more at planting time, the laws of probability favored the production of an average crop. Moisture to a depth of 3 feet (0.9 m) or more was recognized as the desired goal.

For the winter wheat states warmer than normal temperatures in the early stages of growth and cooler than normal in later stages favor the crop and result in the higher acre yields. Higher than normal temperatures after heading has begun usually result in injury.

In the spring wheat states of North and South Dakota, wheat responds favorably to above-normal rainfall throughout the year, to above-normal temperatures early in the growing season, and to below-normal temperatures later in the season.

Weather and Corn Yields

The effect of weather variables on the acre yield of corn has been studied by a number of individuals, with July rainfall shown to be important more often than for other variables. One of the earliest studies was done by Smith in 1914. He found that under Ohio conditions the amount of rainfall in July was of particular importance. In 1920 Wallace reported that for the central Corn Belt states July rainfall was of first importance. Hauseman, for Lincoln, Nebraska, for the years 1906–1935, also reported that rain in late July was particularly beneficial. He also showed that warmer than normal temperature was related to lower yields in each of the three months of June, July, and August, with high temperature in August more injurious than in July, and more injurious in July than in June. Runge, who used Illinois weather and yield data, found that higher than normal temperatures may not be harmful if there is an abundance of soil moisture.

Thompson made a statistical study of the relation of weather variables to the yield of corn, using weather and yield records for the five Corn Belt states of Ohio, Indiana, Illinois, Iowa, and Missouri for the years 1930–1967. Precipitation from the previous September through June provides a measure of the adequacy of the soil moisture reserve. This is of first importance. Top corn yields result when the precipitation has been normal.

It would appear that more than normal precipitation would be less desirable than a little less than normal precipitation. As in the earlier studies, however, below-normal rainfall in July results in lower acre yields. Yields are increased when the July rainfall is greater than normal. It generally is recognized that corn can do well in a relatively dry August if temperatures are normal or below. It appears that temperature is the dominant variable factor for August, when both temperature and precipitation are considered.

Normal June temperature is associated with high acre yields. For July and August temperatures, slightly below normal is associated with the highest yields. Abnormally high temperatures in August are more injurious than such termperatures in July. High temperatures in August are likely to be associated with inadequate supplies of soil moisture at the critical "filling" stage of growth and are likely to cause a more rapid rate of respiration, with relatively more of the photosynthate used in energy conversion than deposited in the filling ear.

Weather and Grain Sorghum Production

It long has been recognized that grain sorghum is more tolerant to drought and heat than is corn. The leaves of the grain sorghum tend to roll under unfavorable heat and moisture conditions, almost stopping growth during periods of moisture stress, then recovering and continuing to grow again if moisture becomes available later.

Thompson studied the effect of weather on the yield of grain sorghum in the five states of Nebraska, Kansas, Oklahoma, Missouri, and Texas for the years 1935–1961. Because of the wide range in the climate for the five states included, no consistent pattern for any one weather variable was apparent. Higher-than-average August temperature appeared to be beneficial in Oklahoma, Kansas, and Nebraska. June rainfall is more critical for Texas than for the other states, as might be expected in view of the earlier planting there. July rainfall is more critical in Oklahoma than it is in the other states, and August rainfall is more critical in Kansas than elsewhere. Higher than normal temperatures early in the season appear to be detrimental to yield, with the reverse true for temperature later in the season. High temperatures late in the season would be expected to be detrimental.

Weather and Soybean Yields

Thompson studied the effect of weather variables on soybean yields in the central Soybean Belt states, Illinois, Iowa, Indiana, Ohio, and Missouri. He found that the normal preseason precipitation, September through May, appeared to be nearly optimum for soybeans, just as it is for corn. Preseason moisture may be a limiting factor in the western part of the area (Iowa), but is rarely so further to the east. It would appear that soybeans are more sensitive to dry weather in August than is corn. Corn can tolerate low rainfall in August if subsoil moisture is adequate and if weather remains relatively cool. For soybeans August rainfall is almost as effective in increasing yields as is July rainfall. High yields of soybeans are associated with high rainfall in July and August. The two peaks for moisture are mid-July, when the vegetative growth is most rapid, and mid-August, when the seed are filling most rapidly. It has been suggested by Thompson that the dependability of August rainfall in the Soybean Belt is one of the main reasons that the United States produces a large percentage of the world's soybean crop.

REFERENCES AND SUGGESTED READINGS

1. ANONYMOUS. "Weather—Yes, Cloud Seeding Does Increase Total Rainfall," *Crops and Soils* 30(3):25 (1977).

2. ARNEY, T. J., and W. D. HANSON. *Moisture and Temperature Influence on Spring Wheat Production in the Plains area of Montana,* USDA Prod. Res. Rep. 34 (1960).

3. BATTAN, L. J. *Harvesting the Clouds: Advances in Weather Modification* (Garden City, N.Y.: Doubleday, 1969).

4. BROWN, D. M. *A "Heat Unit" System for Hybrid Corn,* Rep. 5th Natl. Conf. on Agricultural Meteorology (1963).

5. COLE, J. S. *Correlations Between Annual Precipitation and Spring Wheat in the Great Plains,* USDA Tech. Bul. 636 (1938).

6. COLE, J. S., and O. R. MATHEWS. *Use of Water by Spring Wheat in the Great Plains,* USDA Bul. 1004 (1923).

7. DALE, R. F. "Weather and Technology; A Critical Analysis of Their Importance to Corn Yields," *Agr. Sci. Rev.,* 3:4 (1965).

8. DENMEAD, O. T., and R. H. SHAW. "The Effect of Soil Moisture Stress on Different Stages of Growth on the Development and Yield of Corn, *Agron Jour.,* 52:5 (1960).

9. EGLI, D. B., *et al. Growing Degree Days for Corn in Kentucky,* Ky. Agr. Exp. Sta. Prog. Rep. 197 (1971).

10. GARNER, W. W., and H. A. ALLARD. "Effect of Day Length and Other Factors on Plant Growth and Reproduction," *Jour. Agr. Res.,* 18:11 (1920).

11. HALACY, D. S. *The Weather Changers* (New York: Harper & Row, 1968).

12. HARTLEY, J. M. "Safer, Cheaper Rainmaking," *Popular Sci.,* p. 81 (April 1975).

13. HAUSEMAN, E. E. *Methods of Computing a Regression of Yield on Weather,* Iowa Agr. Exp. Sta. Res. Bul. 302 (1942).

14. KATZ, Y. H. "Relations Between Heat Units Accumulated and Planting and Harvesting of Canning Peas," *Agron. Jour.,* 44:74–78 (1952).

15. MAGOON, C. A., and C. W. CULPEPPER. *Response of Sweet Corn to Temperature from Time of Planting to Canning Maturity,* USDA Tech. Bul. 312 (1951).

16. NEWMAN, J. E., and B. O. BLAIR. *Growing Degree Days and Dent Corn Maturity,* Crop. Prod. Proc., Top Farmer Workshop. Agron. Dept. Purdue Univ. (1968).

17. RUNGE, E. C. A. "Effects of Rainfall and Temperature Interactions During the Growing Season on Corn Yield," *Agron. Jour.,* 60:5 (1968).

18. SEWELL, W. R. D. (Ed.). *Human Dimensions of Weather Modifications* (Chicago: Univ. of Chicago Press, 1966).

19. SHAW, R. H. *Growing-Degree Units for Corn in the North Central Region,* Iowa Agr. Exp. Sta. Res. Bul. 581 (1975).

20. SHAW, L., and D. D. DUROST. *The Effect of Weather and Technology on Corn Yields in the Corn Belt,* 1929–1962, USDA Cons. Rep. 80 (1965).

21. SHAW, R. H., and L. M. THOMPSON. "Grain Yields and Weather Fluctuations," CAED Rep. 20, Iowa State Univ. (1964).

22. SMITH, J. W. "The Effect of Weather upon the Yield of Corn," *Monthly Weather Rev.,* 42:78–87 (1914).

23. THOMAS, J. R., *et al. Relationship of Soil Moisture and Precipitation to Spring Wheat Yields in the Northern Great Plains,* USDA ARS Res. Rep. 56 (1962).

24. THOMPSON, L. M. "Evaluation of Weather Factors in the Production of Grain Sorghum," *Agron. Jour.,* 55:2 (1963).

25. THOMPSON, L. M. "Evaluation of Weather Factors in the Production of Wheat," *Jour. Soil Water Cons.,* 17:4 (1962).

26. THOMPSON, L. M. "Weather and Technology in the Production of Corn and Soybeans," CAED Rep. 17, Iowa State Univ. (1963).

27. THOMPSON, L. M. "Weather and Technology in the Production of Corn in the U.S. Corn Belt," *Agron. Jour.,* 61:3 (1969).

28. THOMPSON, L. M. *Weather Influences on Corn Yield,* Proc. Agr. Res. Inst. (Washington, D.C.: Natl. Acad. Sci., 1964).

29. U.S. Dept. of Commerce. *Federal Plan for a National Agricultural Weather Service* (Jan. 1971).

30. WALLACE, H. A. "Mathematical Inquiry into the Effect of Weather on Corn Yields in the Eight Corn Belt States," *Monthly Weather Rev.,* 48:439–446 (1920).

31. WIGGANS, S. "The Effect of Seasonal Temperature on Maturity of Oats Planted at Different Dates," *Agron. Jour.,* 48:21–25 (1956).

32. WILLETT, H. C. "Evidence of Solar Climatic Relationships," CAED Rep. 20, Iowa State Univ. (1964).

33. WILSIE, C. P. *Crop Adaptation and Distribution,* (San Francisco: Freeman, 1962).

CHAPTER 7
CROP PRODUCTION AND AIR, WATER, AND SOIL POLLUTION

THE advances of our agriculture through the past 50 years have provided the United States with food at a lower real cost than ever before. Output per farm worker has increased to such an extent that less than 4 percent of the United States workforce now is required to provide the agricultural product needs of the other 96 percent.

But with changes in technology serious problems have developed. Our air, water, and soil are becoming polluted to an extent not dreamed possible a few years ago. To bring together pertinent available information on pollution problems numerous symposia have been conducted by national agriculture-related associations and societies and government agencies. Attention has been given to the effect of pollutants both on the growth of crop plants and the livestock dependent on these crops and on health problems generally. A function of the United States Environmental Protection Agency is to allow existing and new production technology to be integrated into cropping systems that will assure sustained crop production, but that will simultaneously protect or enhance the quality of our environment. According to the EPA, the cropping systems must include management elements that control soil erosion and prevent the discharge of pollutants from cropland into the nation's waters.

Injury to Crops

Extensive damage to crop plants has been noted from air pollution, especially near heavily industrialized and high-population-density centers. In addition, the continuing increase in the use of agricultural chemicals, such as pesticide sprays and dusts, is of increasing concern.

The quality of the environment as it affects agriculture, and as agriculture affects the environment, can be changed only as rapidly as society in general and agriculture in particular recognize the problems and are willing to pay the costs of such changes as may be shown to be desirable.

It has been recognized for a number of years that in certain areas the damage to crops from air pollution has been serious. Such damage has become increasingly critical. The effects of pollutants on food, fiber, and forage crops vary greatly from place to place, depending especially on the proximity of concentrations of industry and population.

Plants may be injured without visible effect, because of long-term, low-level exposure to polluted air at levels below those that cause more drastic and immediate injury. Often it is difficult to measure such injury, although the damage may be very real.

Air pollutants result principally from combustion of fuels, from vehicle exhausts, and from refining or manufacturing processes. Pollutants found deleterious to plant growth include peroxyacetyl nitrate (PAN) and ozone, photochemical pollutants (action of sunlight on nitrogen oxides and hydrocarbons from exhausts and combustion), sulfur dioxide, hydrogen fluoride, chlorine, ammonia, nitrogen oxides, ethylene, and particulates (smoke, dust, fly ash, and liquid particles).

National losses of food and fiber crops, ornamental plants, turfgrasses, and trees caused by air pollution are estimated at more than $500 million annually.

Pollution from Agriculture

Agricultural practices do not contribute pollutants in amounts comparable to those from industry, transportation, and population concentrations. But there are agricultural pollutants that must be recognized.

The major forms of pollution associated with agricultural operations include disposal of wastes from animal and poultry feeding areas, pesticides, silt and sedimentation, burning of waste material, concentration of salts and other chemicals in return flows from irrigation systems, and prevalence of ragweed pollen.

In processing agricultural products into foods, textiles, leather, and industrial chemicals, there are losses of organic matter that vary from a very small percentage to 25–50 percent of the raw material. Some of these losses are represented by washing, peeling, and slicing fruits and vegetables; cleaning dairy plant tanks and other associated equipment; slaughtering meat animals and poultry; manufacturing cornstarch and soy protein; refining sugar; wet processes in textile mills; and tanning hides.

In recent years there has been a marked increase in the number of livestock and poultry reared or fed in confined spaces. The waste materials from these produce highly undesirable odors. The problem of safe and economical disposal of animal wastes is receiving increased attention, especially in finding ways to prevent the pollution of waterways.

The use of pesticides can reduce air quality; they can drift in air currents for considerable distances from their point of application and greatly harm sensitive plants. Also, pesticides residues can accumulate in soils to dangerously high levels, with national water sources ultimately contaminated.

Air Pollution

Agricultural practices contribute relatively little to air pollution, compared to contamination caused by manufacturing processes, electric power

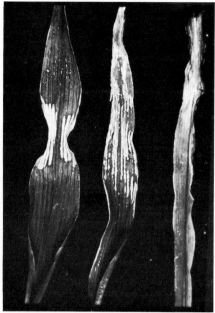

A

B

FIGURE 7–1. *Effect of air pollution on alfalfa and corn. (A) Acute marginal and intercostal necrosis of alfalfa from sulfur dioxide; (B) necrosis of corn following 2.5 hours of exposure to ozone.* [Photographs by permission from Michael Treshow, *Environmental and Plant Response,* McGraw-Hill, 1970.]

generation, combustion, and vehicle exhausts. An increasing number of our larger cities are plagued with epidemics. Records of mass illnesses and deaths resulting from smog are readily available. The injurious effect of polluted air is well illustrated by the report that sheepskin valves in New York City pipe organs need to be replaced about every five years, in contrast to an average of 25 to 30 years under more favorable air conditions.

Aside from health and life itself, the cost to the nation from air pollution has been estimated at $11–$20 billion annually. Airborne acids are reported to result in rapid deterioration of building stone, to eat holes in metal roofing materials, and to hasten house furnishing and clothing deterioration. With such damage by polluted air to so-called dead material, the injury to sensitive, living material, whether animal or plant, is immeasurably greater. The Clean Air Act of 1970 has done some good in reducing air pollution, but the program is behind schedule.

The source of air pollution varies from one community to another. In Los Angeles air pollution is estimated at 80 percent from motor vehicles and 20 percent from industry. In Chicago it is reversed, with 20 percent from motor vehicles and 80 percent from industry.

National concern is evidenced by a number of Clean Air, Solid Waste Disposal, and Air Quality Acts that have been put into effect.

Water Pollution

The philosophy that the "solution of pollution is dilution" no longer holds. The steadily increasing volume of wastes is exceeding the diluting abilities of water bodies.

So frightening are the examples of neglect of water-quality control that control has become the concern of government at all levels, as well as of business and agriculture. It is evident that much research will be necessary before the desired remedies are found. It generally is recognized that limits on the quantity and quality of water represent one of our most pressing national problems, and one that is rapidly worsening.

In the face of our ever-expanding population, with the constantly increasing water need for domestic, industrial, agricultural, and recreational purposes, the popular idea that water can be used and then thrown away will become more and more unthinkable. Water reuse is coming rapidly to be accepted by industry.

Water pollution has come to be recognized as a national problem, the result of huge volumes of untreated or inadequately treated wastes being dumped into the nation's lakes, rivers, and smaller streams. The Federal Water Quality Act of 1965 and the Clean Water Restoration Act of 1966 are evidence of this concern. For the polluters of our waters these laws mean decreased opportunity to dispose of unwanted wastes by simply dumping them into the nearest lake or stream. The Federal Water Pollution Control Act Amendments of 1972 affirm the objective of restoring and maintaining the quality of the nation's waters. They also specify that the administration of the EPA shall provide guidelines for identifying and evaluating the nature and extent of nonpoint sources of pollution. Section 208 of Public Law 92–500 refers to the Federal Water Pollution Control Act Amendments, and relates to control of nonpoint sources of pollution from agricultural land. The major mandates are development and implementation of local and regional plans to control this type of pollution from agricultural sources. Each state is to produce a plan to ensure that public waters are fishable and swimmable by 1983.

The degradation of Lake Erie, the "American Dead Sea," provides a national example of the disaster that can happen as a result of pollution. Through a 30-year period there was an average annual harvest of 25 million pounds (11 million kg) of high-quality fish from Lake Erie. Today the harvest is a few hundred pounds and the fish is of low quality. However, until recently, Americans thought that clean water resources were unlimited.

Pesticides

Records of the kinds and amounts of pesticides used on specific areas and farms for periods of 10–15 years are now available. This has made it possible to determine the amounts of pesticides from treated fields moving to water areas. The analyses of hundreds of samples of water has pro-

FIGURE 7-2. *Pesticide pollution can occur not only from runoff from treated fields but also as a result of containers that are not rinsed or disposed of properly. This study at Southern Illinois University—Carbondale is designed to evaluate and improve methods of container disposal.*

duced evidence that neither deep nor shallow wells are being contaminated by pesticides, if the well is so constructed as to provide water fit for human consumption. In most cases where pesticide contamination was found it was the result of carelessness. Far more care is now being exercised in handling and using pesticides than was the case a few years back.

But rather serious buildups of pesticide residues have been found in our larger rivers. In studies continued annually since 1964, endrin has been found in more than half the river water samples taken. Very serious fish kill by endrin occurred in the lower Mississippi River in 1963–1964. More recently, dieldrin has dominated pesticide contaminants in river basin samplings, with a slight increase in occurrence.

Pesticides have been under government regulation since 1910, but only in recent years have regulations been sufficiently stringent to assure effectiveness and safety. In 1938 accountability was given to the Food and Drug Administration, and in 1972, under the Federal Environmental Pesticide Control Act, to the Environmental Protection Agency. Among the provisions of this Act are the following:

1. All pesticides are to be registered or approved by the federal government, and are to be classified for general or restricted use. Restricted pesticides must be applied only by, or under the supervision of, certified applicators.
2. State applicator certification programs and cooperative enforcement programs are to be established.
3. Misuse of pesticides is prohibited, and the EPA is authorized to establish pesticide packaging standards and to regulate pesticide and container disposal.

The safety and effectiveness of currently used pesticides, marketed in interstate commerce, are insured by the registration requirement. Before such registration is granted, the manufacturer must provide scientific evidence that the product is effective against pests listed on the label, and that it will not injure humans, crops, livestock, and wildlife when used as directed.

Nutrients in Water Sources

The waters of our lakes and streams now contain much larger amounts of mineral nutrients than they did formerly. Ecologists are concerned about phosphates and nitrates in water, partly because they can cause excessive algae growth. Excessive algae may impart an unpleasant taste and odor to water, and fish may die when decaying residues exhaust the oxygen supply in water.

Research has shown that much of the phosphorus in surface waters comes from domestic sewage and animal waste. Phosphorus from fertilizer sources is only a secondary source. Since phosphorus applied to the soil is tightly fixed in the soil, it will enter surface waters in only small concentrations unless soil particles erode into the stream. It has been shown that the phosphorus content of the water of the western part of Lake Erie has increased more than 400 percent. Phosphorus pollution is more of a problem in water bodies of highly populated, industrial areas.

High concentrations of nitrates in water constitute a potential hazard to the health of humans and animals. Nitrate toxicity is most serious for babies under one year of age. Most cases reported are infants fed formula prepared with water from rural wells that contained excess nitrates. Numerous studies have been reported confirming earlier clinical observations that nitrates may cause acute toxicity in livestock. Most of these nitrates appear to come from sewage disposal systems or from livestock feedlots.

ALGAE CELLS PER MILLILITER

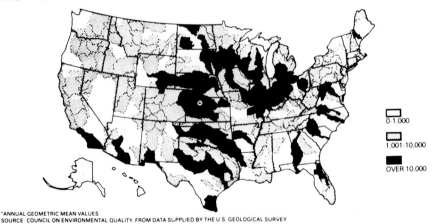

0-1,000

1,001-10,000

OVER 10,000

*ANNUAL GEOMETRIC MEAN VALUES
SOURCE: COUNCIL ON ENVIRONMENTAL QUALITY, FROM DATA SUPPLIED BY THE U.S. GEOLOGICAL SURVEY

FIGURE 7–3. *Large quantities of nutrients in water sources can result in excessive algae growth. Algae may impart an unpleasant odor and taste to water, and deplete the oxygen supply, causing fish to die.* [SOURCE: USDA.]

Many states now have livestock waste-disposal regulations. These regulations specify that existing and new livestock facilities be constructed to prevent surface water from flowing through feedlots, and runoff must be directed to disposal or storage areas. Limits are placed on the location of new facilities with regard to surface waters, flood plains, soil conditions, and population centers. Some regulations have established a permit system under which large livestock operators are required to have a National Pollutant Discharge Elimination System Permit.

Some nitrates may result from natural accumulations of organic matter in the soil, industrial sources, and commercial fertilizers under some conditions. Fertilizer applied to fields for crop production is only one of many potential sources, and likely one of the lesser sources, of nitrates in streams or lakes.

Salinity in United States Waters

It has been estimated that the total salts discharged into the oceans per year by all U.S. rivers fluctuates between 250 and 330 million tons (227–299 MT). Nearly half this salt is carried by the Mississippi River. The Ohio River, from the humid eastern part of the United States, collects more salt and more water per unit area than does the Missouri or other rivers from the more arid western areas. The indication is that the total salt removed is positively related to the amount of water that leaches through the soil of the major watersheds.

It has been estimated that 159 million acre-feet, or 3.4 percent of the total annual precipitation in the United States, is consumptively used by irrigated farm crops and pastures. This is 14 percent as much water as is consumed on all nonirrigated farms. Irrigation water may increase stream salinity whenever there is a significant salt increase in the drainage water. Factors that influence the salinity of drainage water include salt additions from soil leaching, salt concentration by evapotranspiration, salt depletion through mineral uptake and removal in harvested crops, and precipitation and absorption of salt constituents in the soil.

The growing magnitude of urban contributions to the salt burden of national water is realized more now. Salty water is one of the great national problems that must be considered.

Sediment

Erosion and sedimentation are easily seen, and their results are easily evaluated. Largely as the result of federal legislation and regulations, much has been done during the past 30 years to reduce the silt pollution of water, but much remains to be done.

It is difficult to separate erosion from sedimentation. In water pollution we are concerned especially with sedimentation. Water purification for the use of mankind is made more difficult by sediment. About 84 percent of the U.S. population is served by municipal water plants from surface

SUSPENDED SEDIMENT CONCENTRATION, MILLIGRAMS PER LITER

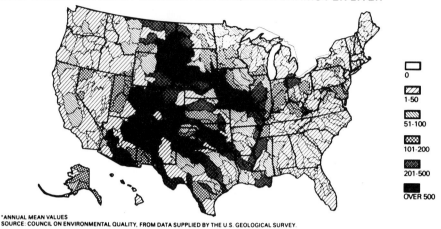

*ANNUAL MEAN VALUES
SOURCE: COUNCIL ON ENVIRONMENTAL QUALITY, FROM DATA SUPPLIED BY THE U.S. GEOLOGICAL SURVEY.

FIGURE 7–4. *Sediment in water sources means that soil erosion and nu-*
trient losses have occurred, water quality has been lowered, and reservoir
capacity has been reduced. [SOURCE: USDA.]

sources. Some 94 percent of these facilities apply some type of water treatment for the removal of sediment.

The farmer has long been aware of the consequences of erosion in terms
of future crop production potential of the land. Only in recent years has
the general public become aware of the severe economic consequences
of sedimentation in lakes, streams, and other water supplies. These "off-
farm" effects likely are more serious than the cropping losses. Probably
one of the most serious economic losses resulting from sedimentation is
the reduction in reservoir capacity. It has been estimated that the artificial
reservoirs for all purposes in the United States had an original capacity
of about 500 million acre-feet of water. About 1 million acre-feet of sediment are deposited in these artificial reservoirs each year. In many cases
the reservoirs will have been rendered useless long before their original
capacity for water storage has been reached.

Some states have appointed Agricultural Task Forces to help evaluate
non-point sources of water pollution, such as erosion, and to propose possible solutions. Increasing pressures for erosion control can be expected in
the future. The design of erosion control regulations may depend on what
the farmer does voluntarily in the meantime. It is important that soil conservation techniques and programs continue to be adopted and expanded.

Soil Pollution

A soil is said to be contaminated or polluted when foreign substances
are introduced that adversely affect the quantity or the quality of crops
produced or that result in an unfavorable environment for living. Several

recent studies have been directed to the nature, magnitude, and urgency of the problem. Limited air, water, and land not only must support a rapidly expanding population and industrialization, but also must assimilate the growing variety and volume of waste materials. Air and water no longer can be considered to provide a means of permanent disposal.

The control of agricultural pests was revolutionized with the introduction of synthetic organic pesticides. Pesticides now are used on most crops grown commercially in the United States. But the possible accumulation of some of these pesticide materials in soils has caused considerable concern.

The major concern about pesticide use and soil pollution is related to length of time pesticides stay in the environment. Some compounds such as DDT, aldrin, dieldrin, and other chlorinated hydrocarbons are very persistent, and can be concentrated in fish and the fatty reserves of land animals. Some fungicides have caused much concern because they contain toxic heavy metals such as arsenic. Herbicides have been of less concern because most are relatively nonpersistent.

Action in recent years by the EPA has helped allay fears about use of persistent and toxic pesticides. The further manufacture of DDT and most other chlorinated hydrocarbons is prohibited, and their use is greatly restricted, often on a special permit basis. The same is true with some of the more toxic fungicides. The Federal Insecticide, Fungicide, and Rodenticide Act (FIFRA), as amended, is the basis for pesticide regulation. This Act provides for a category of restricted-use pesticides, which can be used only by, or under the supervision of, certified applicators. Great progress has been made in recent years through pesticide regulation in protecting the public from soil pollution and in lessening environmental abuse.

Factors in Pollution Control

Greater care in the use of pesticides, better land practices to reduce soil erosion and sedimentation, and impoundment of wastes from feedlots are some of the measures necessary to reduce pollution caused by agricultural practices.

Disposal of vast accumulations of plant and animal wastes without affecting the quality of the air or water poses a problem requiring much research. Increasing accumulation of salts in irrigation water presents a very difficult problem.

Soil lost from cultivated agricultural lands and other earth-moving activities is a major source of the sediments that enter water courses. Losses of essential mineral nutrients have been much greater from erosion than from crop removal. Nutrient losses from soils properly fertilized can be less than those from soils to which no fertilizers have been added. A vigorously growing crop transpires more water than when growth is limited by nutrient deficiencies; in this way percolation is reduced. Erosion can be greatly reduced by a dense vegetative cover. There is little evidence to support the statements that soil fertilizer nutrients are contaminating water supplies.

It generally is recognized that our serious sedimentation problems have been accelerated by the way in which our farmlands have been handled. The rotation of crops to include meadows in the cropping sequence may reduce soil losses by as much as 75 percent. Such practices as mulching, strip cropping, and contour cultivation have been shown to be highly effective. Graded cropland terraces in combination with rotation, mulching, and minimum tillage can reduce soil loss to practically zero. Converting cropland to good grassland is recognized as most effective in eliminating erosion and consequent sedimentation.

In efforts to reduce or eliminate residue contamination by pesticide materials, we must not lose sight either of the tremendous gains in crop production of the past accredited to the use of pesticides or of the food needs of the future. There will be a continuing need for pesticides. Acreages treated with pesticides will increase. The best approach to reduce soil contamination and prevent buildup in the future is to replace persistent pesticides with readily degradable ones as they are developed and become available. This has been occurring in recent years.

The disposal of animal wastes has become an increasingly important farm problem. Although final disposal is the predominant problem, hazards to public health and neighborhood odor develop from poor management of animal wastes and create problems for both the producer and the public generally. Cattle, hogs, and poultry in the United States void more than 1 billion cubic yards of solid wastes every year, compared with 200 million cubic yards of sludge produced by the U.S. human population. The large quantity of manure voided daily, the odor and fly nuisance, the decline in value of the manure as a competitive fertilizer, and the disease-carrying potential all are factors to be considered in animal waste disposal. The problem has increased in recent years, as many farmers, farm groups, and organizations have changed from pasture to penned confinement managements. The wastes, no matter how they have been handled or treated, finally will be disposed of on land. Water and air cannot be used because of the extensive treatment of the manure required before wastes can be discharged into a public waterway or into the atmosphere. American animal producers are in search of a means for disposal with low labor requirements, reduced nuisance conditions, and improved sanitation. The problem of livestock waste disposal has been addressed in recent years, and regulations exist in some areas with regard to location of animal facilities and storage and disposal of wastes.

REFERENCES AND SUGGESTED READINGS

1. ALLISON, F. E. "The Fate of Nitrogen Applied in Crops," In *Advances in Agronomy,* Vol. 18 (New York: Academic Press, 1966). ed. A. G. Norman.

2. BARTHEL, W. F., *et. al. Pesticides and Their Effects on Soil and Water,* Soil Science Society of America, ASA Spec. Pub. 8:128–144 (1966).

3. BRADY, N. D. (Ed.). *Agriculture and the Quality of Our Environment. A Symposium,* AAAS Pub. 85 (1967).

4. BRANDT, C. S. "Effects of Air Pollution on Plants," In *Air Pollution* (New York: Academic Press, 1962). ed. A. C. Stearn.

5. BREIDENBUCK, A. W., *et al.* "Chlorinated Hydrocarbon Pesticides in Major River Basins, 1957–1965," *Public Health Rep.,* 85:2 (1967).

6. CALDWELL, R. L. Effects of Air Pollution on Vegetation," *Univ. Ariz. Prog. Agr.,* 22:2 (1970).

7. CASSELL, E. *Studies on Chicken Manure Disposal,* Res. Rep. 12, N.Y. State Dept. of Health (1966).

8. DARLEY, E. F., *et al.* "Contribution of Burning Agricultural Wastes to Photochemical Air Pollution," *Jour. Air Pollution Control Assoc.,* 16:685–690 (1966).

9. DARLEY, E. F., *et al. Identification of Air Pollution Damage to Agricultural Crops,* Calif. Dept. of Agr. Bul. 55:11–19 (1966).

10. DEEB, B. S., and K. W. SLOAN. *Nitrates, Nitrates, and Health,* Ill. Agr. Exp. Sta. Bul. 750 (1975).

11. GOTTSCHALK, L. C. "Effects of Water and Protection Measures on the Reduction of Erosion and Sediment Damages in the U.S.," *International Assoc. Sci. Hydrol.,* 59:201–213 (1962).

12. GREEN, R. S. *Pesticides in Our National Waters,* AAAS Pub. 85:137–145 (1967).

13. HOOVER, S. R., and L. B. Josewicv. *Agricultural Processing of Wastes,* AAAS Pub. 85:187–204 (1967).

14. KARDOS, L. T. *Waste Water Removal by the Land: A Living Filter,* AAAS Pub. 85:241–265 (1967).

15. KARR, J. R., and I. J. SCHLOSSER. "Water Resources and the Land–Water Interface," *Science,* 201:229 (1978).

16. LANDAU, E. *Economic Aspects of Air Pollution as It Relates to Agriculture,* AAAS Pub. 85:113–135 (1967).

17. LOVE, S. K. *Quality of Surface Water in the U.S.,* Geol. Surv. Water Supply Papers (1964).

18. MERON, A., and H. F. LUDWIG. *Salt Balance in Ground Water,* San. Eng. Div. Proc. Am. Soc. Civil Engrs., 89(S):41–61 (1963).

19. MIDDLETON, J. T., *et al.* Air Conservation for the Protection of Agricultural Production," *Proc. Agr. Res. Inst.,* 1965, pp. 61–67.

20. MIDDLETON, J. T., *et al.* The Presence, Persistence and Removal of Pesticides from the Air," In *Research in Pesticides* (New York: Academic Press, 1965). ed. C. O. Chicester.

21. MINER, J. R., *et al.* "Storm Water Runoff from Cattle Feedlots," In *Management of Farm Animal Wastes,* Am. Soc. Agr. Engrs. Pub. SP-0366:23–26 (1966).

22. SHEETS, T. J., and C. I. HARRIS. "Herbicide Residue in Soils and Their Effects on Crops Grown in Rotation," *Residue Rev. 1940–1965* (1966).

23. SMITH, E. H. "Advances, Problems, and the Future of Insect Control," In *Symposium: Scientific Aspects of Pest Control,* Natl. Acad. Sci. Natl. Res. Council Pub. 1402 (1966). A symposium conducted by Natl. Acad. Sci., Nat. Res. Council, Wash. D.C. Feb. 1–3, 1966. pp. 41–42.

24. SOBEL, A. T. "Some Physical Properties of Animal Manures Associated with

Handling," In *Proc. Natl. Symposium on Animal Waste Management* (East Lansing: Michigan State Univ., (1966).

25. SPRABERRY, J. A. *Summary of Reservoir Sediment Deposition Surveys Made in the U.S. Through 1960,* USDA Misc. Pub. 964 (1965).

26. STEWART, B. A., *et al. Control of Water Pollution From Cropland,* Vol. II, USDA, ARS, EPA (1976).

27. TAIGANIDES, E. P., and P. E. SCHLEUSENER. *Highlights of the National Symposium on Animal Waste Management,* Am. Soc. Agr. Engr. (1966).

28. TAIGANIDES, E. P., and P. E. SCHLEUSENER. *The Animal Waste Disposal Problem,* AAAS Pub. 85:385–394 (1967).

29. THORNE, W., and H. B. PETERSON. *Salinity in U.S. Waters,* AAAS Pub. 85:221–240 (1967).

30. UCHTMANN, D. L. *Pollution Laws and the Illinois Farmer,* Ill. Coop. Ext. Serv. Circ. 1130 (1976).

31. WALKER, R. D. *How Food Production Affects the Environment,* Ill. Coop. Ext. Serv. Circ. 1037 (1971).

32. WILLRICH, T. D. "Management of Agricultural Resources to Minimize Pollution of Natural Waterways," *Symposium: Water Quality Standards* (Ann Arbor: University of Michigan, 1960).

33. WILLRICH, T. D. Primary Treatment of Swine Wastes by Lagooning," *Management of Farm Animal Wastes,* Am. Soc. Agr. Engrs. Pub. SP-0366:70–74 (1966).

34. WOLOZIN, H., and E. LANDAU. "Crop Damage from Sulphur Dioxide," *Jour. Farm Econ.,* 48:394–405 (1966).

CHAPTER 8
WATER USE BY CROPS

PRODUCTION is definitely associated with precipitation, but acre yield is likely to be much more closely related to seasonal than to annual rainfall. The precipitation must penetrate the soil, and much of it must be retained for some considerable time if it is to be of any great use to the plant.

It has been shown that for a good crop of corn, about 12 acre-inches (30 cm) of water must pass up through the plants. A good crop of oats normally will require about 9 inches (23 cm), whereas a crop of alfalfa when grown under irrigation may actually pass 30 inches (76 cm) of water up through the plant tissues.

The Source of Soil Water

Rain and Plant Evaporation

Rainfall is recognized as the direct source of most of the water used by our crops. The original source of most of the water which falls as rain was credited in the past to the evaporation from the growing plants themselves. Airplanes and more recently remote sensing by satellite have made possible far more extensive studies of the upper air masses than was possible before their advent. These studies indicate that the chief source of water for precipitation on land is the ocean.

Kiesselbach estimated the evaporation from corn leaves (both surfaces) as being 32 percent of that from an equal area of free water surface. An average acre (0.4 ha) of corn has approximately 2 acres (0.8 ha) of leaves, or 4 acres (1.6 ha) of leaf surface. The evaporation from an acre (0.4 ha) of Nebraska corn therefore might be expected to be about two-thirds that from a body of water 1 acre in area. Forests are capable of transpiring far more water than would evaporate from an equal area of water surface.

Air Moisture Condensed and Absorbed

In lysimeter studies at the North Appalachian Experimental Watershed, Ohio, the moisture condensed and absorbed by the soil averaged over 6 inches (15 cm) of water annually for a six-year period. The water added to the soil by precipitation gave 81 percent of the total; condensation absorption accounted for the other 19 percent. From 80 to 85 percent of the soil-moisture depletion was due to evapotranspiration; the remainder was lost by percolation.

Data gathered by the Agricultural Research Service have shown that Ohio receives as much as 10 inches (25 cm) of dew annually.

Rainfall and Type of Crops

Rainfall may affect different crops to different degrees and in different ways, depending on such factors as the total annual precipitation, the season of the year when it falls, the rate of fall, the ability of the soil to absorb it, the air humidity, and the temperatures that prevail. In the Great Plains area seasonal precipitation was found to be much more closely related to crop yields than was annual precipitation.

It has been reported that June precipitation in most of the Corn Belt states normally exceeds that used by most of our growing crops, but July and August precipitation often is short of the requirements. With corn, moisture stress often occurs during the critical silking and tasseling stages and with soybeans in the critical pod-filling stages.

In general, our crop belt limits are the 10-inch (25-cm) and 20-inch (51-cm) rainfall lines. Areas that receive an average annual rainfall (snow or water) of less than 10 inches (25 cm) are classed as *arid,* 10–20 inches (25–51 cm) as *semi-arid,* 20–30 inches (51–76 cm) as *subhumid,* and over 30 inches (76 cm) as *humid.*

Sections receiving no more than 10 inches (25 cm) of precipitation are wholly unsuited for crop production without irrigation. Land in these sections may be used for grazing to a limited extent, although the carrying capacity may be so low as to require 50–75 acres (20–30 ha) to carry one animal unit through the grazing season.

Land in sections receiving 10–20 inches (25–51 cm) of precipitation annually can be utilized best for grazing, although some crops may be grown successfully. Wheat and sorghum probably are best suited to these conditions, although in the North corn can be grown with some degree of success. With less than 15 inches (38 cm) of precipitation, growing these crops is decidedly precarious. With more than 20 inches (51 cm) of precipitation crops are grown with ordinary farming practices. The great cereal-production areas are found in sections receiving an annual precipitation of 20–40 inches (51–102 cm). Colville states that the major corn-growing area generally ends where June, July, and August precipitation is less than 8 inches (20 cm).

Shaw has reported that a corn crop uses 25 inches (64 cm) of precipitation, mainly through evapotranspiration. Of that total, 6–7 inches (15–18 cm) are used in July.

Factors that Influence the Water Requirements of Plants

Water Requirement of Different Plants

Many experiments have been conducted in past years to determine the water requirement of plants under varying conditions of growth. *Water requirement* is defined as the ratio of the weight of water absorbed by a plant during its growth to the weight of dry matter produced. Similar terms often used synonymously are *transpiration ratio* and *water-use efficiency.* This term must not be confused with the amount of water required by an acre (or hectare) of growing crop. The water requirement of a particular plant may be relatively low, yet the crop may suffer more from drought than one in which the water requirement is greater. This is because the former crop produces more dry matter and therefore requires more water per acre (or hectare).

Crops show marked differences in water requirement. The millets, sorghums, warm-season grasses, and corn are the most efficient water users, followed by cereal grains, which are intermediate and forages (cool-season grasses and alfalfa) with a relatively high water requirement.

The relative water requirements of several crops at Akron, Colorado, as determined by Shantz and Piemeisel and cited by Wilsie, are as follows:

Crop	Relative Water Requirement
Proso millet	1.00
Common millet	1.07
Sorghum	1.14
Corn	1.31
Barley	1.94
Wheat	2.09
Oats	2.18
Rye	2.37
Legumes	2.81
Grasses	3.10

Information compiled from a number of sources about specific water requirements for crop plants and weeds is shown in Tables 8–1 and 8–2.

TABLE 8–1. Water Requirement of Certain Crop Plants and Weeds

Crop	Water Requirement	Weed	Water Requirement
Bromegrass	1,016	Cocklebur	432
Cotton	646	Lambsquarter	801
Cowpeas	576	Marigold	881
Field peas	788	Pigweed	297
Rice	710	Purslane	292
Soybeans	744	Ragweed	948
Sugarbeet	397	Russian thistle	336
Sweetclover	770	Sunflower	744
Wheatgrass	705	Tumbleweed	277
Western wheatgrass	1,076		

TABLE 8-2. Summary of the Water Requirements of Plants

Crop	Leather, India	Hellriegel, Germany	King, Wisconsin	Briggs and Shantz, Colorado	Thom and Holtz, Washington
Corn	337	—	271	368	231
Oats	469	376	503	597	313
Wheat	554	338	—	513	375
Barley	468	310	464	534	325
Rye	—	353	—	685	—
Clover	—	310	576	797	—
Buckwheat	—	363	—	578	—
Potatoes	—	—	385	636	—
Peas	563	273	477	788	385
Beans	—	282	—	736	484
Millet	—	—	—	310	339
Rape	—	—	—	743	—
Vetch	—	—	—	690	—
Sorghum	437	—	—	322	—
Alfalfa	—	—	—	831	—
Flax	807	—	—	905	—

Adaptive Plant Characters

Many factors make one plant capable of growing in sections too dry for other plants. As a striking example, the root system of the cactus plant is peculiar in that it has a shallow-feeding root system and a deep-anchorage root system. The tolerance of this plant to arid conditions appears to be due to its ability to gather water quickly and to its ability to limit transpiration and evaporation sharply.

Some plants can use the moisture available to them and then mature so quickly as to avoid the "no-water" period. Other plants, such as some of the sorghums, can stop growth and wait for more abundant water before resuming and completing their growth. Other plants, with an abundance of water may make a growth of several feet, and with less available moisture can complete a normal development, but on a greatly reduced scale.

Some of the morphological plant characters credited with modifying plant evaporation, making it possible to lower the water requirement, are:

1. Pubescent or hairy leaves
2. Rolling of the leaves
3. Bloom on leaf and stem (white powder as in sorghum)
4. Number and distribution of stomata
5. Cutinized epidermal cells, corky layer, and bark
6. Stomata sunken in pits
7. Smaller leaf area
8. Edge presentation of the leaves to the sun

FIGURE 8–1. *Although corn is relatively efficient in water use, drought stress, indicated by the leaf rolling seen here, sometimes occurs in July or August.* [SOURCE: Ohio Agricultural Research and Development Center.]

Effect of Fertility on Water Requirement

Investigations into soil factors affecting water requirement have shown, in general, a reduction of water requirement with increased soil fertility.

Studies in Utah showed that the water requirement of corn was reduced from 908 to 464 pounds (412 to 210 kg) by adding manure to the soil. At the New York station the water requirement of wheat when no fertilizer was used was 682 pounds (309 kg); with complete commercial fertilizer it was 533 pounds (242 kg); and with high-nitrogen fertilizer it was 572 pounds (259 kg).

Powers and Lewis found that commercial fertilizers along with crop rotation raised the water efficiency of plants. In 121 experimental field sites, Olson *et al.* concluded that the water requirement of several grain crops (corn, sorghum, oats, wheat, and barley) was decreased by 20 percent when optimum amounts of fertilizer were applied.

Results of more than 30 irrigation experiments in Nebraska have shown that yields of corn fertilized with adequate nitrogen averaged 38 bushels (24 quintals) more than no nitrogen plots. Only 1 extra inch (2.5 cm) of water was used. Fertilized corn was 43 percent more efficient in using water.

Black found that applications of 45–90 kg/ha of nitrogen, with or without phosphorus, increased water-use efficiency of crested wheatgrass and native range species from 1.5 to 2.5 times in Montana. Total water extracted also was increased by nitrogen fertilization. Johnston *et al.* found that water-use efficiency of native rangelands at five locations in southern Alberta was increased significantly by the addition of nitrogen. Greatest increases were observed when nitrogen and phosphorus were used together, with up to eight times greater water-use efficiency on fertilized as opposed to unfertilized rangelands.

Effect of the Amount of Soil Moisture on Water Requirement

Studies in Utah showed that for all periods of growth, the more water available to the plant, the less dry matter was produced per unit of water. In New York, wheat seedlings grown in soils with 11, 13, and 37.5 percent of moisture required 737, 696, and 854 pounds (334, 316, and 387 kg) of water respectively. The water requirement of corn grown in Nebraska in

FIGURE 8–2. *Moisture measurements are being made in this soybean water relations study at Stuttgart, Arkansas. Soybeans and most legumes have a relatively high water requirement.* [Photograph with permission of Arkansas Agricultural Experiment Station.]

a soil with 38, 31, 23, 17, and 13.4 percent moisture was 290, 262, 239, 229, and 252 pounds (131, 119, 108, 104, and 114 kg), respectively.

Briggs and Shantz, after a careful review of investigations concerning the water requirement of various crops when grown on soils varying in moisture content, concluded that there is an increase in the water requirement of most plants when the water content of the soil approaches either the extreme high or extreme low point.

Effect of Cropping on Water Requirement

In Utah, corn grown following three years of fallow had a water requirement of 512 pounds (232 kg), whereas that grown on soil continuously cropped required 593 pounds (269 kg).

In Washington state, wheat following wheat had a water requirement of 518 pounds (235 kg), whereas after fallow the water requirement was 341 pounds (155 kg). In another instance wheat after wheat had a water requirement of 487 pounds (221 kg); after oats, 400 pounds (181 kg); after alfalfa, 391 pounds (177 kg); after corn, 360 pounds (163 kg); and after clover, 310 pounds (141 kg).

Effect of Row Spacing on Water Requirement

Yao and Shaw found that narrow rows (21-inch, or 53-cm) of corn used 1–2 inches (2.5–5 cm) less water than did conventional 42-inch (107-cm) rows, and that water use was enhanced by narrow row spacings.

Effect of Humidity and Other Environmental Factors on Water Requirement

Corn grown in a greenhouse at Lincoln, Nebraska, with a humidity of 48 and 37 percent for the night and day periods, had a water requirement of 340 pounds (154 kg); with a humidity of 72 and 58 percent the water requirement was 191 pounds (87 kg).

Briggs and Shantz found that the water requirement at different places during the same period with the same crops gave quite different results, because of differences in climatic and soil conditions.

Effect of Disease on Water Requirement

Several researchers have studied the influence of rust infections on the water requirement of cereals. In most cases the water requirement was highest for infected plants. This indicates that any change in the plant's normal processes will alter its water requirement.

Relationship of Drought Resistance to Water Requirement

Drought resistance is considered to be closely associated with water requirement. However, the precise relationship between water requirement and drought resistance is not well understood. Some researchers have found a correlation between the two; others attribute little significance.

At present it is believed that many factors influence drought resistance, including any factor that delays dehydration, the leaf area and structure, stomatal behavior, and osmotic pressure. Probably the most important factor in drought resistance is the ability of a plant to endure dessication.

Plant Breeding and Water Requirement

Water requirements are not constant for a given species. Isolation of plant genotypes more efficient in water use may be obtained by selecting plants with low water requirement and by studying their breeding behavior for water requirement.

There are several examples of breeding for improvement in water-use efficiency. Certain newer cultivars of bermudagrass have proved successful because of lower transpiration ratios. Cultivars of wheat and barley have been developed for greater water efficiency.

Miller found variations in water requirement of corn and sorghum plants and concluded that these differences existed among cultivars. Kiesselbach found eight selfed lines of corn had higher water requirements than did the F_1 hybrids. He concluded that single crosses had an advantage over double crosses because variation in water requirement among double crosses was greater.

Difference in water requirement were observed in 28 species of Arizona range plants. Smaller differences existed among perennials than among summer and winter annuals. As a group, summer annuals were lowest in water requirement; the larger plants (trees and shrubs) were highest in water requirement. Variations in and between species depended on their ecological distribution.

A significant difference in both water requirement and yield among 16 selected genotypes of orchardgrass was found by Keller. Baker and Hunt, and Miller found differences in the water requirement among selected

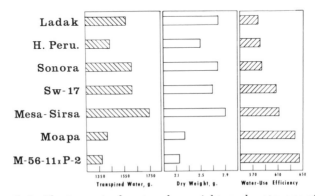

FIGURE 8–3. *The transpired water, dry weight, and water-use efficiency of five alfalfa cultivars (Ladak, Hairy Peruvian, Sonora, Mesa-Sirsa, and Moapa) and two experimental lines (SW-17 and M-56-11 × P-2).* [SOURCE: University of Arizona Agricultural Experiment Station.]

genotypes of intermediate wheatgrass. Sinha found variation among six recommended cultivars of wheat for the Great Plains with respect to water requirement on the basis of straw yield.

Absorption and Storage of Soil Water

Precipitation falling in the form of rain or snow must be absorbed and retained by the soil to be of use to plants. It is well to get an understanding of the way in which this water is held in the soil as it related to the different levels of soil water.

Forms of Soil Water

Soil water behaves differently according to the tightness with which it is held by the soil. When the water film about a soil particle gets thick enough, the attraction of the soil for the water in the outer edges of the film will be so slight that the outer water will be subject to the influence of gravity.

The surface tension effect occurring in the water–air boundary, or interface, acts like a stretched skin or rubber membrane that tends to reduce the surface to the smallest possible area. The sharper the curvature, the greater will be the pull or tension. The drier the soil, the greater the force with which the soil water is held or, in other words, the greater the tension on the soil water.

Because of the noticeable variation in the behavior of soil water at different film thicknesses, or degrees of wetness in the soil, early investigators classified soil water according to its properties at different degrees of soil wetness or dryness. On this basis four forms of soil water usually are recognized: gravitational water, capillary water, hygroscopic water, and combined water. These subdivisions are largely arbitrary, for under actual conditions there are no sharp breaks in the properties exhibited by these forms of water, or in the nature of the forces affecting them.

Attempts to set definite limits to the different kinds of moisture have proved difficult. Exactness is obtained only when all the conditions are defined. Nevertheless, such a classification is useful and convenient.

—*Gravitational water* percolates downward through the subsoil and drains away.

—*Capillary water* is held by the soil against the pull of gravity. It can move to a certain extent in any direction in response to capillary tension. This form of water can be removed by air drying, and to a certain extent by plant absorption.

—*Hygroscopic water* is that water which is retained by an air-dry soil. It can be removed only by oven drying for several hours at a temperature of 105°–110°C (221–230°F).

—*Combined water* remains after hygroscopic water has been removed;

it is held in chemical combination and is driven off only when subjected to high temperatures.

Water Available to the Plant

Because the free (or gravitational) water rapidly drains away from the root zone in well-drained soils, it is not normally available for plant use. Of the capillary water that remains, only that which is in excess of the wilting point is available to the plant. The wilting point is the percentage of water remaining in the soil at the time permanent wilting occurs. At the wilting point there is still some capillary water present in the soil, but it is held with such tension that it cannot be removed by the plant. Therefore, the capillary water in excess of the wilting point (sometimes called intercapillary water) is the main source of water to plants; the other source is the amount of gravitational water that can be utilized before it drains away.

Soil texture greatly influences the amount of water available to plants. The following are generalized data for water availability in different soil textural classes:

	Available Water	
Soil	in./ft.	(cm/m)
Coarse sand	0.5	4.2
Very fine sand	1.2	10.0
Very fine sandy loam	1.9	16.0
Silt loam	2.1	17.7
Silty clay	2.6	21.9
Clay	2.8	23.5

Crop Residues, Mulches, and Rainfall Penetration

Duley observed that the decrease in water intake by bare soils when rain continued to fall on the surface was accompanied by the formation of a thin, compact layer at the surface of the soil. Water percolated very slowly through this layer, which Duley believed resulted from changes in structure caused by the beating action of raindrops, an assorting action of the water, fitting fine particles around larger ones to form a relatively impervious seal.

Duley and Russell investigated the effect of leaving crop residues (such as straw from a small grain crop) on the surface of the land as a means of preventing loss of moisture by evaporation and runoff. The practice appeared to provide an effective method of conserving soil and moisture in the Great Plains region. They listed the following beneficial effects to be expected from leaving litter on the surface of the ground: (1) an increase in infiltration and a reduction in runoff, (2) a reduction in evaporation from the surface, (3) a reduction in water erosion, and (4) a reduction in wind erosion.

Available Water and Fertilizer Response

The economics of fertilizer use is directly related to the adequacy of soil water. Fertilizer application may show little yield response if soil water is a limiting factor.

Principles Involved in Soil Moisture Movement

Soil moisture may move either in a liquid state or in a vapor state. The movement of water in the liquid form takes place under the influence of capillary action, or of gravity, or of both.

The movement of water by capillary action is affected by anything which affects the size and continuity of the pores. Water will rise faster above the water table in a coarse-textured soil in the early stages of rise, but eventually it will rise to a greater height in a fine-textured soil. Movement of water by capillary action is faster through a wet soil than through a dry soil.

The greater the content of capillary water in a given soil, the more readily it may move in response to capillary tension. As the content of capillary water is decreased, it moves less readily. This relationship can be illustrated by comparing the rates of moisture movement at different heights above a water table. (See Table 8–3.)

Many studies have shown that a dust mulch, or a loosened surface soil, breaks the capillary contact and reduces evaporation. But such studies were made on soils where the water table was near the surface. Cultivation to produce a mulch on a soil with a water table as deep as 10 feet (3 m) would have no beneficial effect in reducing capillary rise to the surface.

It a water table does exist near the surface, a mulch would not be needed

TABLE 8–3. Water Evaporated in One Month from the Surface of Columns of Soil of Varying Heights Above Free Water (California)

Height Above Free Water		Water Evaporated			
		Berkeley		Davis	
ft.	(m)	in.	(cm)	in.	(cm)
4	(1.2)	1.57	(3.99)	3.50	(8.89)
4	(1.2)	2.10	(5.33)	3.69	(9.37)
6	(1.8)	0.54	(1.37)	1.85	(4.70)
6	(1.8)	0.70	(1.78)	1.99	(5.05)
8	(2.4)	0.00	(0.00)	0.97	(2.46)
8	(2.4)	0.15	(0.38)	1.03	(2.62)
10	(3.0)	0.00	(0.00)	0.16	(0.41)
10	(3.0)	0.00	(0.00)	0.17	(0.43)

as a water conservation measure. Quite the contrary, drainage of excess water might be the problem.

It there is no water table within 10–11 feet (3.0–3.4 m) of the surface, the water drains down in the soil for 2 or 3 days after a rain and adjusts itself to *field capacity*. When the moisture content is at field capacity, the relationship between soil and moisture is optimum for plant growth.

Depth of Water Table and Surface Evaporation

Early tests at the Wisconsin station on the loss of water by evaporation from the soil surface greatly influenced field practices through a period of years. With the water table 11 inches (28 cm) below the surface, 588 tons of water per acre (1,318 MT/ha) were lost from the soil in 100 days in the case of black marsh soil, 741.5 tons (1,662 MT/ha) in the case of a sandy loam soil, and 2,414 tons (5,410 MT/ha) for a virgin clay loam.

In Utah when the water table was maintained at 6, 12, 18, and 22 inches (15, 30, 46, and 56 cm) below the surface, the losses were 95, 70, 45, and 35 percent, respectively, of the evaporation from a free water surface. But when Alway grew plants in cylinders 6 feet (1.8 m) long, he concluded that the loss of water from the subsoil of uplands under crops was almost entirely by transpiration from the plants. In the absence of plants, the loss of water from the subsoil was negligible.

Shaw and Smith determined the capillary rise of water in tubes, 4, 6, 8, and 10 feet (1.2, 1.8, 2.4, and 3.0 m) long and 8 inches (20 cm) in diameter, that were filled with sandy loam and Yolo loams soils. These tests were duplicated at Berkeley and again at Davis, California. The tubes were filled with the soil and settled, and water was added to the top until drainage occurred below. The water that evaporated from the surface, as shown in Table 8–3, gives a measure of the amount raised by capillarity from a wet soil with free water at various distances below the surface.

Kiesselbach *et al.* reported that alfalfa grew well for four or five years when first seeded on certain Nebraska soils; then yields rapidly declined because of the depletion of available subsoil moisture. The alfalfa drew upon the subsoil moisture to a depth of 13 feet (10 m) in a six-year-old meadow and 25 feet (7.6 m) in a two-year-old meadow. After the subsoil moisture is once exhausted it is replaced very slowly under ordinary cropping. During 15 years of cropping to alfalfa, little water had entered the soil below 7 feet (2.1 m). The authors state that after six years cropping to alfalfa, 225 years would be required under crop rotation to restore the subsoil moisture.

In a Kansas study soils that had produced alfalfa for four years following 12 years of cereal crops and soils that had produced 12 cereal crops following four years of alfalfa were sampled to a depth or 25 feet (7.6 m). During the four-year period, alfalfa utilized the subsoil moisture reserves rather completely to a depth of at least 18 feet (5.5 m). Following the four years of alfalfa, the subsoil moisture supply was partially restored during the subsequent 12 years of cereal crops, when the latter period had slightly more than normal precipitation. It was concluded that under favorable rainfall conditions it should be unnecessary to use special practices, such as summer fallow, when preparing for seeding alfalfa on land previously planted in this crop, provided a sufficient number of cereal crops had been grown between the alfalfa crops.

REFERENCES AND SUGGESTED READINGS

1. ALDRICH, S. R., W. O. SCOTT, and E. R. LENG. *Modern Corn Production,* 2nd ed., (Champaign, Ill.: A & L Publications, 1975).

2. BAKER, J. N., and O. J. HUNT. "Effects of Clipping and Clonal Differences on Water Requirement of Grasses," *Jour. Range Mgt.,* 14:216–219 (1961).

3. BEVER, W. M. "Influence of Stripe Rust on Growth, Water Economy, and Yield of Wheat and Barley," *Jour. Agr. Res.,* 54:375–385 (1937).

4. BLACK, A. L. "Nitrogen and Phosphorus Fertilization for Production of Crested Wheatgrass and Native Grass in Northeastern Montana," *Agron. Jour.* 60:213 (1968).

5. BRIGGS, L. J., and H. L. SHANTZ. *The Water Requirement of Plants. II. A Review of Literature,* USDA Bur. Plant Ind. Bul. 285 (1913).

6. COLVILLE, W. L. "Environment and Maximum Yields of Corn," In *Maximum Crop Yields—The Challenge,* edited by D. A. Rohweder and S. E. Younts (Madison, Wisc.: American Society of Agronomy, 1967).

7. DULEY, F. L. "Surface Factors Affecting the Rate of Intake of Water by Soils," *Soil Sci. Soc. Amer. Proc.,* 4 (1939). pp. 60–64.

8. DULEY, F. L., and J. C. RUSSEL. "The Use of Crop Residues for Soil and Moisture Conservation," *Jour. Amer. Soc. Agron.,* 31:8 (1939).

9. EASTIN, J. D., *et al.* (Eds.). *Physiological Aspects of Crop Yield* (Madison, Wisc.: American Society of Agronomy, 1969).

10. HANKS, R. J., and C. B. TANNER. "Water Consumption of Plants as Influenced by Soil Fertility," *Agron. Jour.,* 44:98–100 (1952).

11. HARROLD, L. L., and F. R. DREIBELBIS. *Agricultural Hydrology as Evaluated by Monolith Lysimeters,* USDA Tech. Bul. 1050 (1951).

12. HOBBS, J. A. "Replenishment of Soil Moisture Supply Following the Growth of Alfalfa," *Agron. Jour.,* 45:10 (1953).

13. HUNT, O. J. "Water Requirement of Selected Genotypes of *Elymus junceus* Fisch, and *Agropyron intermedium* (Host) Beauv. and Their Parent–Progeny Relations," *Crop Sci.,* 2:97–99 (1962).

14. JOHNSTON, A., *et al.* "Seasonal Precipitation, Evaporation, Soil Moisture, and Yield of Fertilized Range Vegetation," *Can. Jour. Plant Sci.* 49:123 (1968).

15. KELLER, W. "Water Requirement of Selected Genotypes of *Dactylis glomerata* L.," *Agron. Jour.,* 45:622–625 (1953).

16. KIESSELBACH, T. A. "The Comparative Water Economy of Selfed Lines of Corn and Their Hybrids," *Jour. Amer. Soc. Agron.,* 18:335–354 (1926).

17. KIESSELBACH, T. A. *Transpiration as a Factor in Crop Production,* Neb. Agr. Exp. Sta. Res. Bul. 6 (1915).

18. KIESSELBACH, T. A., *et al.* "The Significance of Subsoil Moisture in Alfalfa Production," *Jour. Amer. Soc. Agron.,* 21:3 (1929).

19. McGINNIES, W. G., and J. F. ARNOLD. *Relative Water Requirements of Arizona Range Plants,* Ariz. Agr. Exp. Sta. Tech. Bul. 80 (1939).

20. MILLER, D. G. *Parent-Progeny Relationship for Water Requirement of Selected Genotypes of Intermediate Wheatgrass, Agropyron intermedium (Host) Beauv.* M.S. thesis, Univ. of Wyoming, 1964.

21. MILLER, D. G., and O. J. HUNT. "Water Requirement of Plants and Its Importance to Grassland Agriculture," *Wyo. Agr. Exp. Sta. Res. Jour.,* 3 (1966).

22. MILLER, E. C. *Relative Water Requirement of Corn and Sorghum,* Kan. Agr. Exp. Sta. Tech. Bul. 12 (1923).

23. MOTT, G. O. "Effectiveness of Fertilizer Use on Indiana Pastures," *Soil Sci. Soc. Amer. Proc.,* 8 (1944).

24. MURPHY, H. C. "Effect of Crown Rust-Infection on Yield and Water Requirement of Oats," *Jour. Agr. Res.,* 50:387–411 (1935).

25. OLSON, R. A., *et al.* "Water Requirement of Grain Crops as Modified by Fertilizer Use," *Agron. Jour.,* 56:427–432 (1964).

26. POWERS, W. L., and M. R. LEWIS. *Irrigation Requirements of Arable Oregon Soils,* Ore. Agr. Exp. Sta. Bul. 394 (1941).

27. ROGLER, G. A., and R. J. LORENZ. "Fertilization of Mid-Continent Range Plants," In *Forage Fertilization,* edited by D. A. Mays (Madison, Wisc.: American Society of Agronomy, 1974).

28. SCOFIELD, C. S. *The Water Requirements of Alfalfa,* USDA Circ. 735 (1945).

29. SHAW, C. F., and A. SMITH. "Maximum Height of Capillary Rise Starting with Soil at Capillary Saturation," *Hilgardia,* 2:11 (1926).

30. SHAW, R. H., and W. C. BURROWS. "Water Supply, Water Use, and Water Requirement," In *Advances in Corn Production,* edited by W. H. Pierre *et al.* (Ames: Iowa State Univ. Press, 1966).

31. SINHA, R. P. *Water Requirement and the Effect of Similar Drought on Several Winter Wheats.* M.S. thesis, Univ. of Wyoming, 1963.

32. WEISS, F. "The Effect of Rust Infection upon the Water Requirement of Wheat," *Jour. Agr. Res.,* 27:107–118 (1924).

33. WILSIE, C. P. *Crop Adaptation and Distribution* (San Francisco: Freeman, 1962).

34. YAO, A. Y. M., and R. H. SHAW. "Effect of Plant Population and Planting Pattern of Corn on Water Use and Yield," *Agron. Jour.,* 56:147 (1964).

CHAPTER 9

THE SOIL AND CROPPING PRACTICES

The social lesson of soil waste is that no man has a right to destroy soil even if he does own it in fee simple. The soil requires a duty of man which we have been slow to recognize.

H. A. WALLACE

MAN is limited in the extent to which he can modify the environment of the plant above the surface of the soil. But he can considerably influence the soil environment; soil tilth, aeration, moisture, and the nutrients available usually can be somewhat modified.

Man vs. Nature in Soil Maintenance

In different parts of the United States and of the world there are sections where it is now impossible to grow crops profitably but where at some time in the past the soil was productive.

The burden of responsibility for these ruined soils varies widely. Differences in the present productivity in different areas often are attributed to differences in the degree to which "good farming" has been practiced. It should be remembered, however, that with the same cultural practices, crop yields will continue at a higher level in sections with medium to heavy soil, moderate rainfall, relatively low winter temperatures, and a comparatively level topography than in sections having lighter soils, more rolling land, and heavier rainfall.

General Characteristics of the Soil

Soils are natural genetic bodies, a product of the environment under which they develop. An individual soil is a three-dimensional, dynamic natural body with recognizable boundaries. The earth's surface is the soil's upper boundary, the depth to which biological activity and weathering occur approximates the depth to which it extends, and laterally it is bounded by other soils or nonsoils which possess different properties. Thus, it can be seen that an individual soil occupies a definite area of the landscape.

Mineral material, organic matter, water, and air make up the soil mass. The soil that develops any place in the world is dependent on the five factors of soil formation: parent material, climate, topography, organisms, and time. The interaction of these five factors determine the kind of soil developed. Man in recent years has been mentioned as the sixth factor, for his use and abuse of the soil have often drastically altered its natural characteristics.

Soils vary in a number of characteristics. Variations in one factor may profoundly influence others, and there are important interrelationships.

Origin of Soils

Soils are classed either as residual or as transported. *Residual* soils are formed by the disintegration of rock and remain in place where the disintegration occurred. *Transported* soils have been moved from the place of origin by forces of nature and include:

1. *Glacial drift* soils, those transported by the movement of glaciers and deposited by the ice or melt water.
2. *Loessial* soils, consisting of silt, very fine sand, and a little clay, moved by the wind.
3. *Alluvial* soils, formed in river valleys and lowlands from material carried by streams from the higher lands.
4. *Marine* soils, which have been eroded from continental areas, washed into seas or oceans, and later lifted above sea level.
5. *Lacustrine* soils, those eroding into lakes from surrounding uplands and later exposed by lowering of the water level or by elevation of the land.
6. *Colluvial* soils, formed at the foot of slopes by the forces of frost and gravity.

Classification of Soils

To study any heterogeneous group in nature satisfactorily, such as plants or soils, some sort of a classification scheme is necessary. This is especially true of soils. The value of experimental work of any kind is seriously restricted and may even be misleading unless the realtionship of one soil to another is known. The crop requirements in any region depend to a marked degree on the soils in question and on their profile similarities and differences. Those differences are noted, not only from continent to continent or region to region, but from one part of a given field to another.

A basic system of soil classification has been adopted in the United States. The classification of soils is based on their properties as found in the field—properties that can be measured quantitatively and verified. Soil properties such as texture, structure, color, presence or absence of certain soil horizons and their thicknesses and depth, amount of organic matter, and chemical properties such as pH all are used in classifying soils.

Soils are classified into the following categories: order, suborder, great group, subgroup, family, and series. The *series* is the most specific and commonly used category. A series is a group of soils similar in all profile characteristics except the texture of the A horizon (or top layer). The soil *phase* is a further subdivision of soil series where differences such as degree of past erosion, amount of stones present, or the percent slope on which the soil is found also are noted. *Soil texture* refers to the relative proportions of the various soil separates—sand, silt, and clay. According to the USDA classification, sand is 2.0–0.05 mm.; silt 0.05–0.002 mm.; and clay, below 0.002 mm. There are 12 textural class names; examples are sands, sandy loam, silt loam, silty clay, and clay.

Soil Profile

A soil profile is a cross section from the surface downward. It usually shows an arrangement of layers, including the surface or topsoil and a layer, or various layers, designated as subsoil. The top or surface soil includes a layer which usually is 6 to 12 inches (15–30 cm) deep, in which is most of the nitrogen. This element is found in association with, or as a part of, the organic matter. Phosphorus is found both in the organic matter of the surface and in mineral form; it may be more or less abundant in the subsoil. Potassium is found to increase more or less regularly with depth.

Water Movement in Soil

Water moves freely in soil, either in the form of liquid or as vapor. Following rain or irrigation, water moves downward as influenced by gravity. It moves upward by capillarity to evaporate from the soil surface or into plant roots and eventually into the atmosphere through transpiration. It also moves horizontally in response to capillarity. Movement can be in any direction depending on conditions.

Water moves through the open pores between the soil particles. In an ordinary silt loam about half the soil volume is pore space. The pore space may be occupied with air or water, or a combination. For normal plant growth, it must be possible for air from the root zone to exchange with air from above the soil surface. As a result of metabolism in the roots, the air from the root zone is high in carbon dioxide content. Silty and clayey soils generally have smaller pores but many more than sandy soils. When silty and clayey soils are filled with water, these soils contain more total water than a sandy soil with all its pores filled.

A portion of the water in soils with very small pores is held so tightly that it is not available to plants. However, the amount that is available in these soils is greater than that available to plants growing in soils with larger pores.

Gravity, adhesion, and cohesion are the major forces responsible for the movement of soil water. Gravity, most important in saturated soils, causes the downward force. When a soil is saturated with water, the large pores are filled and the movement of the water is rapid. With a lower water content the large pores are occupied with air and the major force responsible for the movement of water is adhesion. When the soil moisture content is low, adhesion and cohesion cause the water particles to move on soil particle surfaces into the finer pores. These are the forces that cause water to rise in capillary tubes.

Water moves in the soil until the different forces are in balance, at which time the films on soil particles are uniform in thickness throughout any homogeneous soil, except for vertical differences because of gravity.

Soil Air

The amount of space between the soil particles varies with the type of soil. A sandy soil has approximately one-third of its volume as pore space, whereas a clay has approximately one-half. The pore space may be occupied by either air or water. Plant roots and organisms in the soil are continually taking oxygen from the air of the soil and replacing it with carbon dioxide in the respiration process. The larger the soil particles, the greater will be the circulation of air in the soil.

Soil air contains the same gases as the atmosphere, but is usually richer in carbon dioxide, which may be as high as 5 percent, as compared with 0.03 percent of the atmosphere. The living cells of roots must have oxygen to support respiration. Plant roots normally secure oxygen directly from the air in the pore spaces, and to a limited extent from the oxygen dissolved in water.

The amount of air in the soil depends not only on the pore space, but also on the water content of the soil. If water completely occupies the pore space, air is excluded. When the soil about the roots is constantly water-soaked, plants die because of their lack of sufficient oxygen.

Soil Temperature

Chemical and biological processes proceed slowly in cold soils. Processes such as biological decomposition and nitrification cease. Nitrifying bacteria begin activity at about 40°F (4°C), with the most favorable temperature being 80°–90°F (27°–32°C). Thus, low temperatures, such as occur in the spring during cold spells, check nitrate production and may result in yellow, nitrogen-deficient plants early. Absorption and transport of water and nutrients, and germination and root growth are affected adversely by low temperatures.

Plants grow best at soil temperatures ranging from 80°–85°F (27°–29°C) for corn dry matter production to 60°–70°F (16°–21°C) for potato tuber development.

The Soil Solution

The soil solution is soil water containing dissolved nutrients and other substances. It is not always continuous and it does not move freely in the soil. It is greatly changeable, and concentration of soluble salts fluctuates widely. The reaction of pH of the soil solution is an important consideration in plant growth.

Particles going into solution are ionized and taken from the solution by passing into, or becoming a part of, the plant; by being absorbed by soil particles; and by entering into combination with other ions of the solution. These processes result in reducing the concentration of a particular substance, after which more particles go into solution from the basic material. One of the most important factors influencing the concentration is the demand of the plant. Plants have the ability to absorb one ion at a greater rate than another (selective absorption).

Plants absorb nutrients in one of three ways: (1) *root interception,* the root comes in direct contact with the ion; (2) *mass flow,* nutrients move along with water to the root; (3) *diffusion,* a gradient is created between root zone and soil zones farther away by nutrient absorption by roots. A plant ordinarily does not absorb nutrients in a passive manner. Plant uptake is marked by plant reaction, such as aerobic respiration, which partly determines nutrient absorption rate. Ion or nutrient carriers help ions move across cell membranes to be released into the cell interior. In this way ions can move from a lesser concentration to a greater concentration (against a concentration gradient). A carrier is specific for ions, which enables preferential or selective absorption.

Plant species differ in the degree to which they take up the different nutrient ions from the soil solution. A notable example is that from the state of Washington, where water from the Columbia River, used to cool the Hanford nuclear reactors, is then dumped back into the river, after having become contaminated with radioactive minerals. When this water was used for irrigation further downstream, pasture grasses contained a concentration of radioactive zinc 440 times that of the irrigation water, whereas corn and beans irrigated with the same water showed little, if any, increase. The flesh and milk from cattle fed the grass showed relatively high amounts of the radioactive zinc.

Organic Matter and Soil Organisms

Organic matter influences the physical condition, moisture-holding capacity, temperature, humus content, nitrogen content, and microbial population of soils. It also accounts for much of the cation exchange capacity of a soil. The loose, friable condition of most productive soils is due to their rather high organic matter content. Organic matter increases the rate of water entry into soil, as well as the amount of water a soil can hold. In addition, organic matter is a source of energy for the soil microorganisms, which break it down into forms available to plants. Decomposition of organic matter may release available nutrients, such as nitrogen, phosphorus, and sulfur compounds. The first nitrogen compounds produced by microbial decomposition are ammonium salts. Ammonium nitrogen is available to the plant, but if conditions are favorable it may be oxidized to nitrates. Soil nitrogen may be maintained in some soils largely by the action of nitrogen-fixing bacteria, in association with legumes, which take nitrogen from the soil air and convert it to a form that can be used by plants.

From 15 to 80 percent of the phosphorus in the soil is contained in organic matter; however, decomposition of organic phosphorus compounds is so slow that plants commonly suffer from phosphorus deficiency unless fertilizer phosphorus is applied.

Soils that are fertilized and watered to the optimum will produce large quantities of top and root material, sufficient to keep the soil in optimum tilth. Many believe that the practical and economical way to get organic matter into the soil is to grow it there through proper fertilization.

THE BREAKDOWN OF ORGANIC MATTER. Many species of soil organisms are specialized and attack only one principal substance. Secondary products may in turn be attacked by other species and may be carried still further toward complete decomposition. The more resistant substances such as lignin accumulate in the soil during this decomposition process, forming humus. Connected with these changes in organic matter are the processes of ammonification, nitrification, denitrification, and cellulose decomposition.

In the process of breaking down organic matter, the soil organisms consume nitrogen. If the material on which they are working contains very little nitrogen and is high in carbohydrates, the organisms may use the available nitrates of the soil, in this way temporarily decreasing productivity. For example, a heavy application of straw may result in low available nitrates in the soil and in decreased crop yields.

THE FIXATION OF ATMOSPHERIC NITROGEN. There are two classes of bacteria that take gaseous atmospheric nitrogen and make it available to plants. First are those that live in the soil independent of growing plants; second are those that live in symbiotic relationship with plants. Examples of the nonsymbiotic or free-living nitrogen-fixers are *Azotobacter* and *Clostridium. Azotobacter* is aerobic and is affected injuriously by an acid soil condition and by low available phosphorus. *Clostridium,* an anaerobic bacterium, is more tolerant of acid conditions. Certain blue–green algae may fix nitrogen in situations of high moisture levels such as in flooded rice culture. The quantity of nitrogen that may be fixed annually by nonsymbiotic sources is 20–100 pounds per acre (22–112 kg/ha).

The relationship between the plant and the organisms which live and multiply in the root nodules is truly symbiotic; the plant gets nitrogen by means of the organism and, in turn, furnishes the organism sugar and a suitable medium in which to grow. There are several distinct strains of these *Rhizobium* bacteria. Those from the nodules of one legume will not inoculate most other legumes. The legumes that may be inoculated by the same strain of bacteria are grouped and listed in the chapter on forage crops. (See Chapter 30.)

INOCULATION. The presence in the soil of the proper strain of nodule-forming bacteria is necessary if a legume is to use atmospheric nitrogen.

When the specific kind of nodule-forming bacteria is not present, it is necessary to introduce it. It generally is recommended that all legumes be inoculated at the time of seeding if the same crop has not been grown on the land recently. There is no way to determine whether or not the proper nodule bacteria are in the soil. If present, the soil may not contain a large enough number for maximum effect.

AMOUNT OF ATMOSPHERIC NITROGEN CAPTURED BY LEGUMES. The amount of nitrogen obtained from the soil air by inoculated plants varies widely. On soils very low in nitrogen a larger proportion of the nitrogen used by the plant is from the atmosphere than on soils where this nutrient is plentiful.

Erdman has compiled in Table 9–1 estimates of the average amounts of nitrogen fixed per acre annually by the growing of different legumes. It is recognized that the range from these averages is great, depending on the environment. Soil conditions such as aeration, drainage, moisture, calcium content, and pH all have a great influence on nitrogen fixation.

Nitrogen in Soils

There are three forms of nitrogen in soils: (1) nitrogen associated with organic matter, (2) ammonium nitrogen fixed by clay materials, and (3) soluble inorganic ammonium and nitrate compounds.

Nearly all the nitrogen found in soils is an integral part of the organic matter. About one-half the organic nitrogen exists in amino compounds, and under normal conditions only 2–3 percent a year is mineralized. The nitrogen content of soils varies inversely with the mean annual tempera-

TABLE 9–1. Estimated Amount of Nitrogen Fixed from the Air by the Growth of Different Legumes

Legume	Av. Amount Nitrogen Fixed per Acre (pounds)	Legume	Av. Amount Nitrogen Fixed per Acre (pounds)
Alfalfa	194	Vetch	80
Sweetclover	119	Beans	40
Red clover	114	Peanuts	42
Ladino clover	179	Lupines	151
White clover	103	Velvetbeans	67
Alsike clover	119	Sourclover	98
Crimson clover	94	Kudzu	107
Lespedezas (annual)	85	Fenugreek	82
Soybeans	58	Bur clover	78
Peas	72	Lentils	103
Cowpeas	90	Garbanzo	66
Winter peas	50	Pastures with legumes	106

Source: Erdman, L. W. Legume Inoculation: What It Is; What It Does, USDA Farmers' Bul. 2003 (1967).

ture of a given environment, and varies from as little as 1,000 to as much as 10,000 or more pounds per acre (1,120–11,200 kg/ha).

Clay minerals fix some ammonium nitrogen between their crystal units, but this is commonly no more than 10 percent of the total soil nitrogen in surface soils. Soluble nitrogen compounds ordinarily constitute only 1–2 percent of the total nitrogen except when large applications of nitrogen fertilizer have been made.

During the ages before lands were plowed and cultivated, an equilibrium in nitrogen content of the surface soil was established between annual return of vegetation and decomposition. With cultivation the nitrogen supply in the soil decreased rather rapidly. In general, loss of nitrogen has been more rapid with intertilled crops, intermediate with cereal crops, and smallest with legumes and sod crops.

Under average conditions in the Midwest, about 25 percent of the original soil nitrogen was lost in the first 20 years of cultivation, with additional losses of about 10 percent and 7 percent during the next two 20-year periods, respectively. There is now very little farmland that supplies enough nitrogen to maintain high crop yields. The application of a supplemental nitrogen is needed.

Before organic nitrogen can be utilized by growing plants, it must be transformed into simple inorganic mineral forms, by various soil bacteria and fungi, and by other microorganisms. This overall process is known as mineralization. When crop residues are returned to the soil they are decomposed by soil microorganisms. Depending on the amount of residue and its nitrogen content, varying amounts of the mineral nitrogen released from organic forms, together with nitrogen from decomposing crop residues, will be utilized by living organisms.

The fraction of total nitrogen readily available to plants varies among soils. Only a small fraction of the total soil nitrogen is available to a crop during any one season.

In irrigated areas where large amounts of nitrogen fertilizers are being applied, nitrate pollution of ground water can pose a serious problem. Current research is under way at several experiment stations to determine whether pollution of ground water is the result of inefficient fertilizer and management practices.

Phosphorus in Soils

Phosphorus deficiency often occurs on soils containing large amounts of the element. Such a small fraction of the soil phosphorus is available to plants that phosphorus fertilizer must be applied to most soils for satisfactory yields. When soluble phosphorus is added in fertilizer, much of this may be rendered insoluble or unavailable.

Both organic and inorganic forms of phosphorus occur in soils. Sometimes the organic phosphorus makes up more than one-half the total soil supply. Organic phosphorus compounds include phytins, nucleic acids, and phospholipids. Under favorable conditions organic phosphorus can supply a significant part of what is needed.

Soil phosphorus availability often is a problem because of (1) a small total amount present in soils, less than the amount of either potassium or magnesium, (2) marked "fixation" of soluble phosphorus added from fertilization, and (3) unavailability of native phosphorus.

Availability of inorganic phosphate is determined by several factors, probably the most important of which is pH. Availability is determined to a large extent by ionic form of the phosphorus, and ionic form is determined largely by pH of the soil solution. The H_2PO_4 ion is considered more available than other forms, but its availability is complicated by the presence of soluble iron and aluminum at a low pH. This can complex the phosphorus into relatively insoluble iron and aluminum phosphate compounds highly insoluble and largely unavailable to plants. In acid soils 30–70 percent of the phosphorus is present in very insoluble iron and aluminum phosphates.

At high pH values, phosphate is precipitated by calcium compounds. Although the more strongly bound calcium phosphates are almost insoluble in water, they do become more soluble in the mildly acid environment around active plant roots. Thus, the conversion of soluble phosphorus to a less soluble form by percipitation as iron, aluminum, or calcium phosphates, is known as *fixation*. The maximum phosphorus availability occurs at a pH value of 6.0–7.0, but even within this range availability may be relatively low.

Phosphorus in acid soils becomes more available when such soils are limed. Fixation of phosphorus can be reduced by minimizing the area of contact between the fertilizer and the soil. Banding phosphate fertilizer reduces or delays fixation. Also, as one fertilizes at higher rates, fixation of the phosphorus becomes less of a concern. Moreover, at high rates of production there is less concern with fixation.

The amount of readily available phosphorus needed in soil for satisfactory yields varies considerably with the crop grown. Relatively little, often less than 1 percent, of the native soil phosphorus is available to plants in any one growing season. There is a carryover of phosphorus fertilizer for several years, more in evidence on neutral and calcareous soils than on acid soils.

Potassium in Soils

Most crops require large quantities of potassium, especially for vegetative growth. Many soils contain more than 20,000 pounds of this element per acre (22,649 kg/ha), and some soils even up to 50,000 pounds per acre (56,000 kg/ha). Yet 90–99 percent of the total potassium is held in minerals and is unavailable to plants. Examples of such minerals are the feldspars and micas. Another portion of the potassium may be fixed in relatively unavailable forms. Some soils fix large amounts of potassium, others very little, depending on the type of clay and certain other chemical factors. The available potassium, often only 1–2 percent of the total soil potassium, exists either in the soil solution or as exchangeable potassium on the soil colloids. In addition to the problem of potassium unavailability in meeting crop needs, considerable potassium is lost by leaching, and crop removal is very high. Most plants will *luxury consume* this element, taking up more than they need for optimum growth or production.

Micronutrients

Micronutrients are essential elements required only in small quantities for plant growth. These elements sometimes are called minor or trace elements. Included are iron, zinc, manganese, boron, copper, molybdenum, and chlorine. Serious deficiency of any one of these elements may cause reduced production or even crop failure.

Attention has been focused on the micronutrients only in the last 10 to 15 years. The main contributing factors to a greater concern with micronutrient fertilization have been (1) increased crop removal of these elements because of yield increases, and (2) a trend toward higher analysis fertilizers containing fewer micronutrients as impurities. In the last decade also there has been a development of more expertise in plant nutrition, which has helped diagnose problems previously not recognized.

Boron may be reduced to a deficient level where excessive leaching has occurred. Under irrigation some water contains enough boron to be toxic to many different species. Some, such as molybdenum, may be required

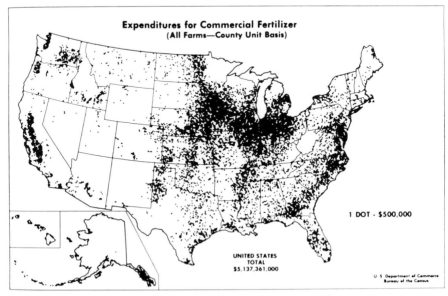

FIGURE 9–1. *United States commercial fertilizer expenditures shown geographically.* [SOURCE: U.S. Department of Commerce.]

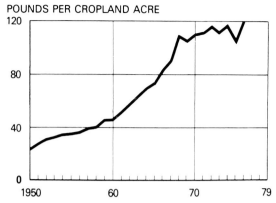

POUNDS PER CROPLAND ACRE

FIGURE 9–2. *Pounds of fertilizer nutrients used per acre in the United States. The present average is more than 120 pounds per acre (134 kg/ ha).*

at a rate as low as 1 ounce per acre (70 g/ha), while 2–3 pounds per acre (2.2–3.4 kg/ha) may be toxic.

All micronutrients cycle through plants. This is partly the reason why surface soils contain more boron, zinc, and copper than the subsoils. A number of fertilizer companies sell fertilizers containing micronutrients added. Micronutrients can be included at precise rates in liquid bulk blend fertilizers.

The availability of all micronutrients except molybdenum and chlorine

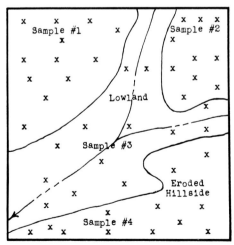

FIGURE 9–3. *Because of soil variability, four composite samples were submitted to the soil testing laboratory for the field represented here, each a composite of 10 or more subsamples, as indicated by the distribution of X's.*

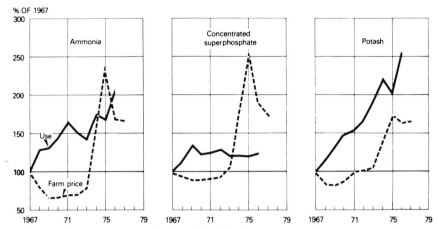

FIGURE 9–4. *Prices and uses of the fertilizers ammonia, concentrated superphosphate, and potash since 1967. Use of most fertilizers has increased despite rather sharp price increases.* [SOURCE: USDA.]

are depressed markedly by a high pH, such as would be created by overliming. Acid soil conditions lower the availability of molybdenum, while chlorine availability is not influenced greatly by pH.

Acidity in Soils

The term *acid soil* designates a soil showing an acid reaction when tested by standard chemical methods. The degree of acidity or alkalinity of a soil is expressed in pH values; the pH scale is divided into fourteen divisions or pH units. Soils with a pH value of 7.0 are neutral; those with a value below 7.0 are said to be acid and those above alkaline. A pH of 5.0 is ten times more acid than a pH of 6.0, and a pH of 4.0 is ten times as acid as a pH of 5.0, or 100 times as acid as a pH of 6.0. The pH value of most soils falls in the range between 4.0 or 8.0. A pH of about 6.5 is considered desirable for most plants.

Generally, soil acidity is attributed to the loss of calcium and other bases from the soil as a result of their removal in crops or by leaching.

CROP ADAPTATION TO ACID SOILS. A number of investigators have studied the response of different crop plants to acid soils and to liming. The results reported vary widely. The Michigan station has grouped crop plants on their tolerance to an acid soil condition as shown in Table 9–2.

THE LIMING OF ACID SOILS. The generally recommended method for treating acid soils is to apply lime. Quicklime, hydrated lime, and airslaked lime have been used, but by far the greatest use has been of finely ground limestone. The value of the limestone depends on the percentage of calcium or magnesium carbonate and the fineness of grinding. Limestone ground fine enough to pass a twenty-mesh screen is generally regarded as sufficiently fine for agricultural purposes if none of the finer particles is re-

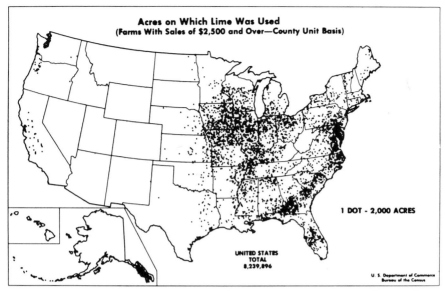

FIGURE 9–5. *Lime use in the United States shown geographically. Acid soil conditions, requiring the addition of lime, are most common in the eastern half of the country. Halomorphic soils, with soluble salt accumulations and high pH levels, are present in some of the western states.* [SOURCE: U.S. Department of Commerce.]

moved before applying. In general, the coarser the material applied, the longer it must be in the soil before it becomes effective.

The use of lime and liming material is concentrated largely in the northern states and the south Atlantic states. The use of lime and liming material increased from 8 million tons in 1939 to currently over 35 million tons.

TABLE 9–2. Relative Tolerance of Crop Plants to an Acid Soil Condition

Very Tolerant 5.0–5.5	Tolerant 5.5–6.5		Sensitive 6.0–7.0	Very Sensitive 6.5–7.0 or Above
Broom sedge	Bluegrass	Flax	Barley	Alfalfa
Cowpeas	Buckwheat	Sudangrass	Hubam clover	Sugarbeets
Meadow fescue	Alsike clover	Timothy	Tall fescue	Sweetclover
Redtop	Bur clover	Blue lupine	Red clover	
Bentgrass	Crimson clover	Oats	Bromegrass	
Korean	Kobe lespedeza	Sweetpotato	Orchardgrass	
lespedeza	White clover	Sorghum	Ryegrass	
Millet	Hop clover	Soybean	Hemp	
Potatoes	Corn	Tobacco		
Rice	Cotton	Vetch		
Rye	Chewing fescue	Wheat		
	Red fescue			

Source: Spurway, C. H. *Soil Reaction Preferences of Plants,* Mich. Agr. Exp. Sta. Bull. 306 (1941).

Halomorphic Soils

Soluble salt accumulation may occur in soils of arid regions when drainage is impeded and surface evaporation is great. Such soils are known as *halomorphic,* and may be classified as either *saline, sodic,* or *saline-sodic.* The detrimental effect saline and saline-sodic soils exert on plant growth is a result of their high soluble salt content, which creates an osmotic effect, in which plants tend to lose water into the soil. The major influence of sodic soils results from the caustic effect of high alkalinity, the toxic effect of bicarbonate, and the adverse effect of active sodium ions.

Saline soils contain sufficient soluble salts to impair productivity; sodium generally is less than 15 percent of the total with most salts of the nonsodium type. The pH is normally below 8.5. They often are referred to as white alkali soils. Excess salts may be leached out of saline soils with sodium-free water without increasing the pH greatly.

Sodic soils are sufficiently high in exchangeable sodium, which is generally more than 15 percent of total salts, to interfere with the growth of most crops. This soil condition, in the past, often has been referred to as black alkali. The detrimental effect is from the toxicity of sodium and hydroxyl ions. The pH of sodic soils may be 10 or more.

The salts in *sodic-saline* soils are of the calcium, magnesium, or sodium chlorides, sulfates, carbonates, or bicarbonates. Although sodium content is often more than 15 percent of the total exchangeable salts, the pH is likely to be below 8.5. Leaching will greatly increase soil pH unless the leaching water or soil is high in calcium or magnesium salts.

In sections where rainfall is sufficient the alkali from decomposing rock material is carried away by water; in arid and semiarid regions the salts may accumulate. Under irrigation, particularly without adequate drainage, the water table is likely to rise toward the surface and bring with it salts from the whole body of soil, these salts tending to concentrate near the surface as evaporation takes place. In more humid sections, so-called alkali spots often are found around the borders of reclaimed sloughs. Water drainage into these sloughs through many centuries carried with it salts that gradually accumulated as a result of evaporation.

RELATIVE PLANT TOLERANCE OF HALOMORPHIC SOILS. In a California trial, the total amount of alkali in pounds per acre to which various crops were subjected without showing injury was as follows:

Saltbush	156,720	Sugarbeet	59,840
Alfalfa, old	110,320	Barley	25,520
Alfalfa, young	13,120	Wheat	17,280
Sorghum	81,360	Bur clover	17,000
Vetch	69,360	Rye	12,480
Sunflower	59,840	Sweetclover	5,840

Table 9–3. Relative Tolerance of Certain Crop Plants to Salty
Soils

High Tolerance	Medium Tolerance	Low Tolerance
Bermudagrass	Alfalfa	Alsike clover
Canada Wildrye	Barley	Apples
Cotton	Birdsfoot trefoil	Ladino clover
Garden pea	Oats	Peas
Milo	Rye	Potatoes
Rape	Sweetclover	Red clover
Sugarbeets	Sudangrass	White clover
Western wheatgrass	Tomatoes	

Source: Richards, L. A. et al., Diagnosis and Improvement of Saline and Alkali
Soils (Riverside, Calif.: U.S. Regional Salinity Lab., 1947).

Richards has compiled a list of plants grouped according to their relative
salt tolerance. Selected economically important crop plants are shown in
Table 9–3.

MANAGEMENT OF SALTY SOILS. The most critical stage for most plants
is the early leaf stage. It sometimes is possible to plant after the alkali
has been either washed down into the soil or turned under by plowing.
In irrigated sections a heavy irrigation quite commonly is given before
seeding to carry the salts away from the young seedlings. Saline land often
can be freed of white alkali by flooding and draining. Use of gypsum,
calcium sulfate, may be effective on sodic soils in changing caustic alkali
carbonates to sulfates. Organic matter seems actually to remove sodium
carbonates from the soil solution in large quantities. The liberal use of
manure or other organic substances is recommended as especially beneficial to alkali soils. Lastly, the use of alkali-tolerant crops on halomorphic
soils is recommended (Table 9–3).

Plant Nutrients Removed by Crops

When crops are grown year after year and taken from the soil, plant
nutrients may be removed to such an extent that the productivity of a
soil is materially influenced. The amounts of plant nutrients contained
in a given unit of different crops is important, not only because of the
nutrients removed from the soil, but also because of the possible value
in maintaining soil productivity by means of the portion of these crops
returned to the land. The amounts of N, P_2O_5, K_2O, Ca, and Mg are shown
for certain widely grown crops in Table 9–4.

Other losses of plant nutrients from the soil may be even more important
than removal in crops. The plowing and cultivating of soils necessary in
our ordinary cropping practices encourages erosion and stimulates oxidation and other decomposition and mineralization processes, making larger
amounts of the different elements available in solution than would other-

TABLE 9–4. Pounds Per Acre of Nitrogen, Phosphorus, Potassium, Calcium, and Magnesium Removed in Selected Crops, with Acre Yields as Shown

Crop	Acre Yield	Nitrogen	Phosphate (P_2O_5)	Potash (K_2O)	Calcium	Magnesium
Corn (Ears)	150	135	53	40	2	8
Corn (Stover)	4.5 tons	100	37	145	26	20
Sorghum (Grain)	60 bu.	50	25	15	4	5
Sorghum (Stover)	3 tons	65	20	95	29	18
Oats (Grain)	80 bu.	50	20	15	2	3
Oats (Straw)	2 tons	25	15	80	8	8
Wheat (Grain)	40 bu.	50	25	15	1	6
Wheat (Straw)	1.5 tons	20	5	35	6	3
Soybeans (Beans)	40 bu.	150	35	55	7	7
Soybeans (Straw)	2 tons	90	20	50	40	18
Hay						
Alfalfa	4 tons	180	40	180	112	21
Red clover	2.5 tons	100	25	100	69	17
Timothy	2.5 tons	60	25	95	18	6

Source: National Plant Food Institute. *Our Land and Its Care,* 4th ed. (Washington, D.C., 1967).

wise be the case. These available materials may be carried away and lost by erosion, percolation, and leaching.

Animal Manure

In the past, the value of animal manure applied to the land surface as fertilizer was recognized as of considerable importance. Because of increased labor costs together with the ease of handling and applying commercial fertilizers, the disposal of animal wastes from feedlots is looked upon more and more as only a necessary expense. But under most average farm conditions the best method of disposal is its distribution on the soil surface, with distribution costs at least partially offset by the increase in acre yields resulting from its fertilizer value.

The amount of nitrogen excreted by livestock in the United States is estimated at around 10 million tons (9 million MT), contained in the 1.7 billion tons (1.5 billion MT) of animal manure, which must be disposed of annually. Most of this goes to waste by one means or another. In several states, legislation has been enacted relative to the disposal of animal wastes.

The liquid–solid excrement of farm animals is formed at an approximate 1:3 ratio. A little more than one-half the nitrogen, almost all the phosphoric acid, and about two-fifths the potash are found in the solid manure. One

FIGURE 9–6. *Animal manure being spread on a field previously cropped to corn. Despite the labor costs, manure offers considerable benefits in terms of the fertilizer value and addition of organic matter to the soil.* [SOURCE: J. I. Case.]

ton (0.9 MT) of representative farm manure is considered to contain 10 pounds (4.5 kg) of nitrogen, 5 (2.3 kg) of phosphoric acid, and 10 (4.5 kg) of potash.

As an average, the annual production of manure in tons per 1,000 pounds (454 kg) of live weight by different kinds of farm animals is about as follows: horses, 13.3 (12.1 MT); cattle, 12.6 (11.4 MT); hogs, 12.0 (10.9 MT); sheep, 12.6 (11.4 MT); poultry, 12.9 (11.7 MT). These figures are on the basis of the excrement plus the bedding, with actual output adjusted to 65 percent water content.

It is generally recognized that a large part of the plant nutrients contained in animal manure is needlessly lost, so that it does not get back to the fields. This loss is greatest when the manure is loosely piled in the open. Losses are minimized when plenty of bedding is used and the manure is compacted under cover.

Green Manure

When the soil is low in organic matter, the growing of crops to be plowed under may be desirable. When so used, the crop plowed under is referred to as *green manure,* regardless of whether it is plowed under green or after it has matured. The chief benefit from a legume green manure crop is the increased organic matter and nitrogen content of the soil with the accompanying better soil tilth. Organic matter also helps stimulate beneficial microbial activity. If the green manure is a well-inoculated legume, the increase in available soil nitrogen may be considerable.

Green manuring is practiced to some extent in all parts of the United States except under dryland farming in the Green Plains and Rocky Mountain areas, where lack of moisture is a limiting factor. For field crops, green manuring is most common in connection with the production of cotton, corn, potatoes, tobacco, and to some extent sugarbeets. Examples of good green manure crops are vetch, crimson clover, peas, and lespedeza. The combination of a legume and nonlegume is sometimes recommended, with rye and vetch a good example.

The use of green manures may not be advisable under low moisture conditions. Use of a material such as ryegrass, rye, or another nonlegume with a high C/N ratio often has resulted in a significant yield decrease. In such an instance, microorganisms tie up soil nitrogen during decomposition of the organic material.

Commercial Fertilizers

United States

Increased fertilizer use probably has contributed more than any other single practice to the phenomenal increase in crop yields. It has been estimated that 30–40 percent of the increased agricultural production in the last decades can be attributed to the increased use of fertilizer. About 49 million tons (44 MT) of fertilizer were used in 1976 compared to 25 (23 million MT) in 1960, 9 (8 million MT) in 1945, and 5 million (4 million MT) in 1940.

Because of the increasing use of fertilizers and fertilizing materials containing a higher percentage of primary plant nutrients, the increase in the quantity of the plant nutrients nitrogen, phosphorus, and potassium has been much greater than the increase in tonnage of fertilizer. From 1967 to 1976, the use of ammonia, a common nitrogen source, doubled; concentrated superphosphate use increased by about one-fourth; and potash increased by 2.5 times. A trend in recent years has been an increase in use of fertilizers containing secondary nutrient elements or micronutrients. Farmers currently apply more than 2 million tons (1.8 million MT) of such fertilizers.

Nearly 185 million acres (75 million ha) of land in the United States are fertilized annually. This area is equivalent to more than 62 percent of the total acreage of land from which crops are harvested, and to about 19 percent of the total agricultural land, including pastureland.

There are large variations in the use of fertilizer by areas and by crops. The percentage of each crop fertilized varies greatly by area. The four crops corn, wheat, cotton, and hay and cropland pasture account for approximately two-thirds the acreage of all crops fertilized in the United States.

There has been a significant increase in the use of liquid fertilizers. This includes both clear liquids and slurries, the latter of which has gained

TABLE 9–5. Percentage Content of the More Important Fertilizer Elements in Different Fertilizer Materials

Material	Percentage*				
	Nitrogen	P_2O_5	K_2O	Calcium	Magnesium
Ammonia anhydrous	82	—	—	—	—
Ammonium nitrate	33.5	—	—	—	—
Ammonium nitrate–limestone mixtures	20.5	—	—	7.3	4.4
Ammonium phosphate (mono)	11	48	0.2	1.1	0.3
Ammonium phosphate (am. phosphate–sulfate)	13–16	20–39	0.2	0.3	0.1
Ammonium phosphate (di)	16–21	48–53	—	—	—
Ammonium sulfate	20.5–21.0	—	—	0.3	—
Ammonium sulfate–nitrate	26.0	—	—	—	—
Ammoniated superphosphate	3–6	18–20	—	17.2	—
Calcium cyanamide	21	—	—	38.5	0.06
Calcium nitrate	15	—	—	19.4	1.5
Gypsum (land plaster)	—	—	0.5	22.5	0.4
Magnesium oxide	—	—	—	1.1	56.1
Nitric phosphates	14–22	10–22	0.1–16	8–10	0.1
Potassium chloride	—	—	60–62	—	0.1
Potassium–magnesium sulfate	—	—	22	—	11.2
Potassium nitrate	14	—	44–46	0.4	0.2
Potassium sulfate	—	—	50–53	0.5	0.7
Rock sulfate	—	30–36	0.2	33.2	0.2
Sewage sludge, activated	5–6	2.9	0.6	1.3	0.7
Sewage sludge, digested	2	1.4	0.8	2.1	0.5
Sodium nitrate	16	—	0.2	0.1	0.05
Superphosphate, normal	—	18–20	0.2	20.4	0.2
Superphosphate, concentrated	—	42–50	0.4	13.6	0.3
Urea	42–46	—	—	0–1.5	0.7
Urea–formaldehyde	36–40	—	—	—	—

161

* Most of the percentages larger than 1 of N, P_2O_5, and K_2O are the usual guarantees. Where more than one grade is commonly sold, the range is indicated by two numbers separated by a dash.

Source: National Plant Food Institute, Our Land and Its Care, 4th ed. (Washington, D.C., 1967).

popularity because of these advantages: higher analysis is possible, lower cost materials can be used, larger quantities of micronutrients can be added, and a wider range of pesticides can be included in suspensions.

There has also been increased interest in foliar fertilization. Foliar sprays cannot supply all macronutrient requirements, but they may be useful for temporary supplemental nutrition. Secondary elements and micronutrients have shown the most promise for this type of application. The soil will remain the principal reservoir for nutrients, but foliar fertilization likely will remain an important method for supplementing nutrient supply.

FERTILIZERS FOR THE FUTURE. Fertilizers used in quantity in the future will have to meet these criteria: safe to use, maintain quality of environment, easy to handle and apply, effective, and reasonably priced. It appears that nitrogen will continue to be supplied primarily by anhydrous ammonia, nitrogen solutions, ammonium phosphates, and solid urea. Phosphorus will be supplied mainly by concentrated or triple superphosphate and by various ammonium phosphates. Suspensions will become even more important as lower-grade raw materials are used.

World

The world is producing and consuming about 88 million tons (80 million MT) of fertilizer (N, P, and K) as compared to a 1962–1963 consumption of under 38 million tons (35 million MT). Nitrogen fertilizers accounted for more than one-half the total. The world consumption of fertilizers has increased by an average of more than 8 percent per year since 1962. Unfortunately, 80–95 percent of the world production of nitrogen, phosphate, and potash fertilizers is concentrated in the developed countries. Developing nations, where fertilizers can give the most dramatic results, are depending too heavily on fertilizer imports to meet their needs. It has been estimated that 1 kg of nutrients can increase rice yield by 10 kg and wheat yield by 8 kg in developing areas. Providing adequate fertilizers at the right place and at the right time is a high priority challenge for the world. Adequate supplies of fertilizers that are effective and reasonably priced are a principal key to avoiding widespread famine and malnutrition in developing nations. (See Table 9.5.)

REFERENCES AND SUGGESTED READINGS

1. BARBER, S. A. "Potassium in Soils," *Plant Food Rev.,* 15:1 (1969).
2. BRADY, N. C. *The Nature and Properties of Soils,* 8th ed. (New York: Macmillan, 1974).
3. BROWN, A. L. "Secondary and Micronutrients in Soils, *Plant Food Rev.,* 15:1 (1969).
4. CALDWELL, A. G. "Phosphorus in Soils," *Plant Food Rev.,* 15:1 (1969).
5. CARLSON, C. W. "Soils: Reservoir of Plant Nutrients," *Plant Food Rev.,* 15:1 (1969).

6. ENGELSTAD, O. P. *Fertilizers for the Future: Some Old, Some New,* 1979 Illinois Fertilizer Conference Proc. (Feb. 1, 1979).

7. ERDMAN, L. W. *Legume Inoculation: What It Is; What It Does,* USDA Farmers' Bull. 2003 (1967).

8. JOFFE, J. S. "Green Manuring Viewed by a Pedologist," *Adv. Agron.,* 7 (1955).

9. National Plant Food Institute. *Our Land and Its Care,* 4th ed. (Washington, D.C.: 1967).

10. NELSON, L. B. "Fertilizers for All-Out Food Production," In *All-Out Food Production; Strategy and Resource Implications,* Spec. Pub. 23 (Madison, Wisc: American Society of Agronomy, 1975).

11. NORTHERN, H. T. *Introductory Plant Science,* 3rd ed. (New York: Ronald, 1968).

12. RICHARDS, L. A. (ed.), *Diagnosis and Improvement of Saline and Alkali Soils* (Riverside, Calif.: U.S. Regional Salinity Lab, 1947).

13. SCHWARTZ, J. W. *Foliar Feeding of Essential Nutrients to Plants,* USDA-ARS NE-72, (1976).

14. SPURWAY, C. H. *Soil Reaction Preferences of Plants,* Mich. Agr. Exp. Sta. Bul. 306 (1941).

15. STANFORD, G. "Nitrogen in Soils," *Plant Food Rev.,* 15:1 (1969).

16. USDA-ERS. *The World Food Situation and Prospects to 1985,* Foreign Agr. Econ. Rep. 98, 1974.

17. USDA-SCS. *Soil Taxonomy, A Basic System of Soil Classification for Making and Interpreting Soil Surveys.* Agr. Handbook No. 436 (1975).

18. WALLACE, H. A. U.S. Secretary of Agriculture, 1933–1940.

Part III
Botany of Crop Plants

CHAPTER 10
CROP GROUPS AND CLASSIFICATION

CROPS are variously grouped and classified. For example, they are roughly classified as (1) cultivated or row crops, such as corn, soybeans, and cotton; (2) noncultivated crops, such as wheat and barley; and (3) sod, hay, or pasture crops, such as the clovers, alfalfa, bromegrass, and many other small-seeded legumes and grasses.

Crops also are grouped according to their duration. *Annuals* are those that complete their life cycles in one year—mature their seed and die. (Winter annuals established in the fall complete their growth in the following season.) *Biennials* start their growth in one season and mature their seed and die by the end of the second season of growth. *Perennials* persist through a varying number of years, from few to many.

Perhaps the most important and most used classification is the botanical classification.

Botanical Classification

Well over 300,000 different plant species have been identified and classified. Each of these species may have from one to many different subspecies or cultivars. As an aid in classification, the plant kingdom has been divided into four main divisions: *thallophytes, bryophytes, pteridophytes,* and *spermatophytes.* The *spermatophytes* include all the seed-producing plants. This division is divided into two subdivisions, *gymnosperms* and *angiosperms.* The *gymnosperms* include all coniferous trees and others. To the *angiosperm* belong all broad leaved plants, cereals, and other common crop plants. The angiosperms are divided into two classes, *monocotyledons* and *dicotyledons.* The seed of the first group have only one cotyledon; those of the second group have two. All the grasses, such as corn, wheat, rice, oats, bromegrass, and timothy belong to the first group. All the legumes, such as alfalfa, clovers, soybeans, and peas, belong to the second. Each of these classes is divided into orders, the orders into families, the families into genera, the genera into species, and the species into subspecies and cultivars.

The botanical classification of two important crop plants, corn and soybeans, is as follows:

Classification Unit	Dent Corn	Soybeans
Kingdom	Plantae	Plantae
Division	Spermatophyta	Spermatophyta
Subdivision	Angiospermae	Angiospermae
Class	Monocotyledonae	Dicotyledonae
Order	Graminales	Rosales
Family	Gramineae	Leguminosae
Genus	*Zea*	*Glycine*
Species	*mays*	*max*
Subspecies or cultivar	*indentata* (subspecies)	'Williams' (cultivar)

If two botanists unknown to each other describe the same plant and assign different names, the earliest published name (after 1753, the publication date of Linnaeus' *Species Plantarum*) will prevail. The name of the botanist who was first to identify and name a given species is indicated after the species name, usually by an initial or an abbreviation. For example, the L. that follows many plant species indicates that the identification and naming was by Linnaeus, the world-renowned plant taxonomist, the originator of our present system of identification and nomenclature. It sometimes is very difficult to determine the correct relationship of a given plant. The reproductive parts—the flowers and seed—are least affected by the environment and can be identified most dependably.

With modern microscopes and techniques for staining cell nuclei, the number of chromosomes has been counted and catalogued for a great number of different species. In very difficult cases serums from plant materials and other kinds of chemical analyses have been used to determine the grouping to which a given plant is properly related. This is somewhat in keeping with the use of blood characteristics to determine certain human relationships and relationships in other animal groups.

It will be seen from Table 10–1 that most of our field crops belong to one of two families: the *Gramineae* (grasses) and the *Leguminosae* (legumes). The relationship of the different species is shown by their classification in the genus groups.

The importance of the grass family is at once recognized, encompassing as it does all the cereal grains. In addition to the grain-bearing grasses, forage grasses are so necessary to the economical production of livestock products. Even sugarcane and bamboo are included in the grass family. The *Leguminosae* includes such crops as peanuts, field beans and peas, soybeans, and such nutritious forages as alfalfa, vetch, and the clovers. The plants of the *Leguminosae* family are the only ones which grow in a symbiotic relationship with the rhizobia bacteria, which, multiplying in the nodules on the roots, are able to utilize the gaseous nitrogen of the soil air. All the legumes are high in protein content.

Among the other plant families which contain at least one important

TABLE 10–1. Botanical Classification of Some of the More Important Crop Plants
a-Monocotyledons

Common Name	Order	Family	Genus	Species	Subspecies
Barley, two-row	Graminales	Gramineae	*Hordeum*	*distichum*	—
six-row	Graminales	Gramineae	*Hordeum*	*vulgare*	—
Corn, Pod	Graminales	Gramineae	*Zea*	*mays*	*tunicata*
Flint	Graminales	Gramineae	*Zea*	*mays*	*indurata*
Dent	Graminales	Gramineae	*Zea*	*mays*	*indentata*
Sweet	Graminales	Gramineae	*Zea*	*mays*	*saccharata*
Flour	Graminales	Gramineae	*Zea*	*mays*	*amylacea*
Pop	Graminales	Gramineae	*Zea*	*mays*	*everta*
Grasses, Timothy	Graminales	Gramineae	*Phleum*	*pratense*	—
Redtop	Graminales	Gramineae	*Agrostis*	*alba*	—
Orchardgrass	Graminales	Gramineae	*Dactylis*	*glomerata*	—
Kentucky bluegrass	Graminales	Gramineae	*Poa*	*pratensis*	—
Bermudagrass	Graminales	Gramineae	*Cynodon*	*dactylon*	—
Crested wheatgrass	Graminales	Gramineae	*Agropyron*	*desertorum*	—
Buffalograss	Graminales	Gramineae	*Buchloe*	*dactyloides*	—
Smooth bromegrass	Graminales	Gramineae	*Bromus*	*inermis*	—
Reed canarygrass	Graminales	Gramineae	*Phalaris*	*arundinacea*	—
Millets, Foxtail	Graminales	Gramineae	*Setaria*	*italica*	—
Broomcorn	Graminales	Gramineae	*Sorghum*	*vulgare*	—
Japanese	Graminales	Gramineae	*Echinochloa*	*crusgalli*	*frumentacea*
Pearl	Graminales	Gramineae	*Pennisetum*	*typhoides*	—

TABLE 10–1. Botanical Classification of Some of the More Important Crop Plants (continued)
a-Monocotyledons *(continued)*

Common Name	Order	Family	Genus	Species	Subspecies
Oats, Common	Graminales	Gramineae	*Avena*	*sativa*	—
Side	Graminales	Gramineae	*Avena*	*orientalis*	—
Hull-less	Graminales	Gramineae	*Avena*	*nuda*	—
Wild	Graminales	Gramineae	*Avena*	*fatua*	—
Red	Graminales	Gramineae	*Avena*	*byzantina*	—
Rice	Graminales	Gramineae	*Oryza*	*sativa*	—
Rye	Graminales	Gramineae	*Secale*	*cereale*	—
Sorghum, Grain	Graminales	Gramineae	*Sorghum*	*bicolor*	—
Sweet	Graminales	Gramineae	*Sorghum*	*bicolor*	—
Sudangrass	Graminales	Gramineae	*Sorghum*	*bicolor*	*sudanense*
Johnsongrass	Graminales	Gramineae	*Sorghum*	*halepense*	—
Sugarcane	Graminales	Gramineae	*Saccharum*	*officinarum*	—
Wheat, Common	Graminales	Gramineae	*Triticum*	*aestivum*	—
Durum	Graminales	Gramineae	*Triticum*	*durum*	—
Club	Graminales	Gramineae	*Triticum*	*compactum*	—
Poulard	Graminales	Gramineae	*Triticum*	*turgidum*	—
Spelt	Graminales	Gramineae	*Triticum*	*spelta*	—
Emmer	Graminales	Gramineae	*Triticum*	*dicoccum*	—
Polish	Graminales	Gramineae	*Triticum*	*polonicum*	—
Einkorn	Graminales	Gramineae	*Triticum*	*monococcum*	—

TABLE 10–1. Botanical Classification of Some of the More Important Crop Plants (continued)

b-Dicotyledons

Common Name	Order	Family	Genus	Species	Subspecies
Alfalfa	Rosales	Leguminosae	Medicago	sativa	—
Birdsfoot trefoil (broad)	Rosales	Leguminosae	Lotus	corniculatus	—
Buckwheat	Polygonales	Polygonaceae	Fagopyrum	sagittatum	esculentum
Clover, Red	Rosales	Leguminosae	Trifolium	pratense	—
White	Rosales	Leguminosae	Trifolium	repens	—
Crimson	Rosales	Leguminosae	Trifolium	incarnatum	—
Alsike	Rosales	Leguminosae	Trifolium	hybridum	—
Cotton, Upland	Malvales	Malvaceae	Gossypium	hirsutum	—
Sea Island	Malvales	Malvaceae	Gossypium	barbadense	—
Cowpeas	Rosales	Leguminosae	Vigna	sinensis	—
Field beans	Rosales	Leguminosae	Phaseolus	vulgaris	—
Field peas	Rosales	Leguminosae	Pisum	arvense	—
Flax	Geraniales	Linaceae	Linum	usitatissimum	—
Lespedeza, Common and Kobe	Rosales	Leguminosae	Lespedeza	striata	—
Korean	Rosales	Leguminosae	Lespedeza	stipulacaea	—
Peanuts	Rosales	Leguminosae	Arachis	hypogaea	—
Potatoes	Polemoniales	Solanaceae	Solanum	tuberosum	—
Rape	Papaverales	Cruciferae	Brassica	napus	—
Soybeans	Rosales	Leguminosae	Glycine	max	—

171

TABLE 10–1. Botanical Classification of Some of the More Important Crop Plants (continued)
b-Dictyledons (continued)

Common Name	Order	Family	Genus	Species	Subspecies
Sugarbeet	Chenopodiales	Chenopodiaceae	*Beta*	*vulgaris*	—
Sunflower	Campanulales	Compositae	*Helianthus*	*annuus*	—
Sweetclover, Biennial white	Rosales	Leguminosae	*Melilotus*	*alba*	—
Annual white	Rosales	Leguminosae	*Melilotus*	*alba*	*annua*
Biennial yellow	Rosales	Leguminosae	*Melilotus*	*officinalis*	—
Annual yellow (sourclover)	Rosales	Leguminosae	*Melilotus*	*indica*	—
Sweetpotatoes	Polemoniales	Convolvulaceae	*Ipomoea*	*batatas*	—
Tobacco	Polemoniales	Solanaceae	*Nicotiana*	*tobacum*	—
Velvetbean	Rosales	Leguminosae	*Stizolobium*	*deeringianum*	—
Vetch, Hairy	Rosales	Leguminosae	*Vicia*	*villosa*	—
Common	Rosales	Leguminosae	*Vicia*	*sativa*	—

crop plant are *Malvaceae, Linaceae, Solanaceae, Cruciferae, Chenopodiaceae,* and *Polygonaceae.* (See Table 10–1.)

Binomial System

Each plant must have an identifying name, recognizable on a worldwide basis. In the botanical classification of a specific crop plant, only the genus and species are used in identifying it. For example, alfalfa is *Medicago sativa* L. The genus name corresponds roughly to a person's last name, the species to the first name; that is, *Medicago sativa* would be equivalent to *Brown, John.* Scientific names are basically Latin.

Agronomic Use Classification

Agronomically, crop plants most often are grouped according to the way or ways in which they are used. Some of our crop plants have several alternative uses; for example, corn is most often grown as a *cereal grain* crop, but it is also recognized as one of our most productive *forage* crops. In the same way soybeans most often are grown as an *oil* crop, but can be grown also as a *forage* crop to be harvested as hay.

The more important crops may be grouped according to agronomic use as follows:

—*Cereals.* Corn, wheat, rice, barley, oats, rye, and grain sorghum are some of the more important crops in the cereal group. A cereal is defined as a grass grown for its edible seed. Cereals often are referred to as the *grain crops.* Flax is often thought of as a grain, but is neither a cereal nor a grass; the primary use for its seed derives from the oil extracted from it. In the same way, buckwheat is thought of as a grain crop and is grown for its edible seed. It is not a cereal, however, because it is not a member of the grass family.

—*Seed legumes.* Some of the more important field crops in the seed legume group are soybeans, field peas, field beans, cowpeas, and peanuts.

—*Forage.* Included in the forage group are most of our grasses and legumes; all crops grown as feed for animals, whether used in the form of harvested hay, soilage, silage, or pasture. Corn and other cereals are, therefore, forage crops as well as grain crops.

—*Roots.* Rutabagas, mangels, beets, turnips, and sweet potatoes are the primary members of the roots group.

—*Tubers.* The tubers group includes white or Irish potatoes and Jerusalem artichokes.

—*Fiber.* Cotton, flax, and ramie are some of the more important crops grown for their fiber, which is used in making textiles, rope, and twine.

—*Sugar.* The sugar group includes sugarcane, sugarbeet, and sweet sorghum grown for syrup.

—*Drug.* Tobacco, tea, coffee, peppermint, mustard, and hops are members of the drug group.

—*Oil.* Soybeans, flax, peanuts, sunflower, safflower, and castorbeans are the primary members of the oil group.

Special Purpose Classification

—*Green manure.* Vetch, rye, buckwheat, sweetclover, soybeans, cowpeas, and other crops, especially legumes, when grown and plowed under either in the green or mature state for soil-improving purposes, are included in the green manure group.

—*Silage.* Corn and sorghum are the crops most extensively grown to be cut and preserved in a succulent condition for silage. Many legumes and grasses are now cut and preserved as "grass silage" or haylage.

—*Soilage* or *Green Chop.* Corn, sundangrass, soybeans, and many other crops are cut when green and succulent and are fed directly to livestock without curing.

—*Catch.* Sudangrass, buckwheat, millet, and other similar short-season crops, used to fill in when regular crops have failed or when planting is for some reason delayed, are grown as catch crops (also known as *emergency* crops).

—*Cover.* Crops seeded on land needing to be protected from wind and water erosion and from nutrient losses by leaching are referred to as cover crops. Such crops as rye, buckwheat, vetches, and soybeans are examples. They often serve at the same time as green manure crops.

—*Supplement.* A crop that may be grown as a secondary crop in with a primary crop (for example, soybeans in corn) is spoken of as a supplementary crop. A supplementary crop is also a crop grown alone to be used in connection with another crop, such as sudangrass when used to provide grazing at a time when other pastures cannot be used or are not sufficiently productive.

—*Companion.* Included in companion crops are the small grain crops with which clover, alfalfa, and other legumes and grasses may be seeded. The term *companion crop* is now used instead of the former term *nurse crop,* in recognition of the fact that instead of nursing the new seedlings such crops really compete with them.

REFERENCES AND SUGGESTED READINGS

1. CARTER, G. S. *A Hundred Years of Evolution* (New York: Macmillan, 1958).
2. CHAPMAN, S. R., and L. P. CARTER. *Crop Production, Principles and Practices* (San Francisco: Freeman, 1976).
3. CORE, E. L. *Plant Taxonomy* (Englewood Cliffs, N.J.: Prentice-Hall, 1955).
4. DAVIS, P. H., and V. H. HEYWOOD. *Principles of Angiosperm Taxonomy* (Princeton, N.J.: Van Nostrand-Reinhold, 1963).
5. EHRLICH, P. R., and R. W. HOLM. *The Process of Evolution* (New York: McGraw-Hill, 1963).

6. ESAU, K. *Anatomy of Seed Plants* (New York: Wiley, 1962).
7. FERNALD, M. L. *Gray's Manual of Botany* (New York: American Book Co., 1950).
8. JANICK, J., *et al. Plant Science* (San Francisco: Freeman, 1969).
9. JAQUES, H. E. *How to Know the Economic Plants* (Dubuque, Iowa: William C. Brown, 1958).
10. MARTIN, J. H., W. H. LEONARD, and D. L. STAMP. *Principles of Field Crop Production,* 3rd ed. (New York: Macmillan, 1976).
11. STEWART, W. D. P. *Nitrogen Fixation in Plants* (London: Athone, 1966).
12. USDA. Grasses: *Yearbook of Agriculture* (Washington, D.C.: USDA, 1948).
13. WILBUR, R. L. *Leguminous Plants of North Carolina,* N.C. Tech. Bul. 151 (1963).

CHAPTER 11
STRUCTURE AND GROWTH OF PLANTS

THE student of crop production will wish to go thoroughly into the structure, the functioning, and the response of crop plants to the environment. He will do this in various botany courses: *plant morphology,* form and structure of plants, without regard to function; *plant physiology,* functioning of living plants, or their parts; *plant taxonomy,* plant identification and classification; *plant ecology;* relationship of specific plants to the various factors that constitute their environment; *genetics,* science that seeks to explain resemblances and differences in plants related to descent.

Plant Structure

Roots

The plant root is so important that the next chapter will be devoted to roots and their functions. The bulk and weight of the root portion of many plants is as great as, or greater than, the aboveground portion and usually constitutes at least a third that of the dry weight of the aboveground portion. In a study of a single rye plant four months after planting it was found to have a total of more than 15 billion living root hairs with a total surface area in excess of 4,000 square feet (372 m²).

Stems and Leaves

The stem of the plant, also known in grasses as the culm, supports and displays the leaves in such a way as to make readily possible the absorption of carbon dioxide and sunlight. The stem and its appendages constitute the shoot. The stem serves to conduct water and dissolved substances from the soil to the leaf and to transfer plant nutrients from the leaves to all parts of the plant. It has been suggested that the leaf may be viewed as

FIGURE 11–1. *Characteristics of grass parts: A, several spikelets on a central axis, enclosed in two outer glumes; B, the parts of the grass flower; C, the developed fruit, or seed, with the caryopsis shown, successively enclosed in the outer glumes, with the lemma and palea both closely adhering and free; D, spikelets arranged in a terminal spike; E, spikelets arranged in a raceme; G, the junction of the leaf blade and sheath; H, I, J, means of propagation: stolon, rhizome, and bulb, respectively.* [SOURCE: *USDA Yearbook,* 1948.]

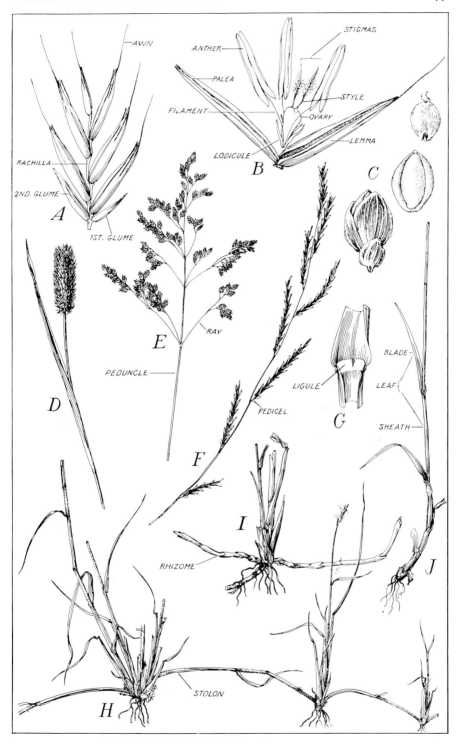

a flattened or expanded portion of the stem. The leaf contains many of the same kinds of cells and tissues as are found in the stem. It differs, however, in that the stem shows a plan of unlimited growth, whereas the growth of the leaf is limited. The leaf soon ceases to grow, matures, functions for a period, and finally falls away.

Most stems are sturdy and erect, but others are vinelike or even prostrate. White clover, for example, produces long, slender side branches called *stolons,* or *runners.* These modified stems grow above ground, but develop horizontally, producing aerial shoots and adventitious buds. Many plants have *rhizomes;* perennial, usually horizontal underground stems. Rhizomes often are confused with roots, but differ in that they have nodes and internodes, which are never found on roots. Buds commonly are found at the nodes. Quackgrass, johnsongrass, and Kentucky bluegrass have rhizomes. The tuber, as in the potato, is an enlarged portion of a modified stem, a rhizome.

The structure of the leaf is such that it is peculiarly adapted for photosynthesis and transpiration. The exposed leaf blade provides the maximum surface for the absorption of light energy. Thin as it is, all the cells lie very near to the surface, which facilitates absorption and the diffusion of gases, including water vapor, oxygen, and carbon dioxide, from the cells inside the leaf. Veins give strength to the leaf and provide the channels for the movement of water and dissolved substances. In addition, they transfer a part of the food manufactured in the leaf to other parts of the plant.

The epidermis in most plants is a single layer of cells that covers the entire leaf surface. Scattered over both sides of the leaf are modified cells,

FIGURE 11-2. *Characteristic growth of white clover, with leaves and flower heads borne on long, unbranched stems that arise directly from the nodes of the stolon.* [After Isely.]

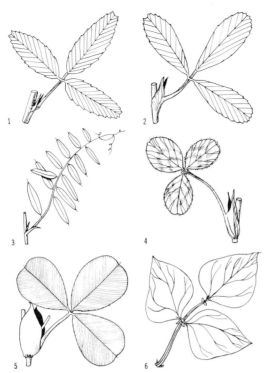

FIGURE 11-3. *The leaf arrangement of six legumes: (1) sweetclover, (2) alfalfa, (3) vetch, (4) red clover, (5) Korean lespedeza, (6) cowpea.* [(1) and (2) after Isely.]

known as the guard cells of the *stomates.* Each stomate consists of two guard cells so shaped that when turgid, or full of water, the stomate is open. When these guard cells lose water faster than it can be obtained from neighboring cells or vascular tissue the opening practically closes. This closing checks the loss of water. The closing of the stomates does not regulate transpiration but simply serves to prevent excessive drying of the tender leaf cells. The stomates are connected with the intercellular space between the cells and permit gases—particularly water vapor, carbon dioxide, and oxygen—to diffuse into the atmosphere or to be moved from the atmosphere into the plant. Beneath this outer epidermal layer of cells lies a layer of cells richly packed with *chloroplasts.* The irregular loose arrangement of cells results in a spongelike region, *spongy mesophyll,* that provides the air space necessary for the gaseous exchange involved in photosynthesis and transpiration.

Flowers, Fruits, and Seed

Flowers, fruits, and seed serve primarily for purposes of reproduction. The flower may be showy, if it is dependent on insects for pollination, or it may be inconspicuous, as are those of most of the grasses, which either are self-pollinated or are cross-pollinated by the wind. The *pistil* is the female organ, and the pistillate flower consists of *stigma, style,* and *ovary.* The ovary contains the *ovule* or ovules. Within each ovule of most crop plants are two polar nuclei, which fuse with one of the two pollen sperm cells to produce the *endosperm,* the food-storage tissues. There is also an egg cell, which unites with the other sperm cell to produce the new plant, or *embryo,* of the seed. The *stamens* are the male organs. They consist of *anthers,* which produce pollen grains, and *filaments,* the stalks or stems on which the anthers are borne.

Flowers composed of petals, sepals, stamens, and pistil are known as *complete,* while those lacking one or more of these parts are *incomplete.* The *perfect,* or bisexual, flower contains both stamens or pistils. The *imperfect* or unisexual flower contains either stamens or pistil, but not both. When the separate staminate and pistillate flowers are borne on the same plant, the plants are called *monoecious.* Examples are castorbean

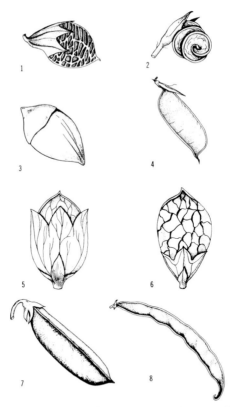

FIGURE 11–4. *Different types of legume seed pods: (1) yellow sweetclover, (2) alfalfa, (3) red clover, (4) vetch, (5) common lespedeza, (6) Korean lespedeza, (7) field pea, (8) cowpea.*

and corn. When the staminate and pistillate flowers are borne on separate plants, the plants are called *dioecious*. Examples are buffalograss and hemp.

Cell Structure

Each part of the plant is made up of minute cells, each cell consisting essentially of a mass of cytoplasm, the living matter of the plant. The cytoplasm has special structures for definite functions. The *nucleus* consists of a compact spheroidal mass separated from the cytoplasm by a membrane. *Chromatin,* found in the nucleus, is in threadlike structures that thicken into bodies called *chromosomes* when cell division is about to take place. For each type of plant the chromatin is grouped into a definite number of chromosomes. The inheritance of the plant is determined by materials in these chromosomes. Deoxyribonucleic acid (DNA) is contained in chromosomes and is a major hereditary bridge between generations. It contains genetic information, which affects the machinery of the cell through control of protein synthesis. This synthesis occurs in small particles called *ribosomes,* which are located in the cytoplasm.

The cytoplasm has cavities or vacuoles filled with cell sap and dense disc-shaped bodies called *plastids*. These plastids often carry pigments which give characteristic colors. The green pigments are *chlorophyll.* Plastids sometimes are colorless or carry other colors, as are found in the petals of flowers or in the stems or roots.

Small, dense granules called *mitochondria* are the sites of cell respiration. They contain the enzymes necessary for the oxidative breakdown of organic molecules. This releases energy-carrying substances such as adenosine triphosphate (ATP), which becomes available for essential cell processes.

The Cell Wall

The cell wall is made of cellulose, with secondary deposits of hemicellulose or lignin in certain cases, which strengthen the wall. Cell walls permit water and dissolved substances to pass through into the cell. Most substances that dissolve in water can readily pass through the cell walls. Harmful salts may get into the cell in this way. A narrower selection is made by the cytoplasmic membrane which surrounds the cytoplasm and lies next to the cell wall. Most substances are diffused through the cytoplasmic membrane as ions rather than as compounds. When an electrolytic substance goes into solution it becomes ionized. For example, sodium chloride becomes positive sodium ions and negative chloride ions, which may be absorbed in unequal proportions; this leaves either one or the other ion free to combine with ions from other solutes. In all cases there must be a balance of positive and negative charges as ions. Different ions are absorbed by the plant at different rates. One kind of ion influences the absorption of another kind. The root hair zone is the principal point of entry.

Growth Structures

The growth of a plant occurs by means of multiplication and enlargement of its individual cells. Cell growth has several phases. The first is cell formation by division. The second phase is one of enlargement, during which the cell attains its mature size. The third phase is maturation, during which the cell walls are thickened and other changes take place. For enlargement to take place, the cell wall must be stretched and additional cellulose must be added.

The rate of plant growth depends internally on many factors, among which are the rate at which food materials are supplied and the energy obtained from the oxidation of sugars. A proper temperature is essential, and light is necessary for photosynthesis, which results in sugar formation, used indirectly (as energy source through respiration) or directly for growth. Plant nutrients and raw materials necessary for photosynthesis include carbon dioxide, water, nitrogen, phosphorus, and other minerals.

Photosynthesis

The chlorophyll in green plants has the special capacity of manufacturing, with the aid of sunlight, carbohydrates from carbon dioxide and water. Energy is stored in these manufactured products; the release of this energy is our greatest source of power. Carbohydrates made by plants are the ultimate and entire source of food for all animals and plants. More than 90 percent of the dry weight of most plants consists of carbohydrates derived by photosynthesis from carbon dioxide and water. The carbon dioxide enters the plant through the stomates and diffuses through the cytoplasm to the chloroplasts. These chloroplasts are, as has been stated, the "food laboratory of the world." With the aid of the chlorophyll and light, by a series of reactions, carbon dioxide and water are transformed into energy-rich carbohydrates. Photosynthesis is a complex process involving "light reactions," which convert light energy to chemical energy, the latter resulting from the conversion of ADP to ATP and NADP to $NADPH_2$. The "dark reaction" of photosynthesis can occur in the absence of light and refers to the formation of intermediate carbohydrate compounds and eventually sugar units. This process requires use of the chemical energy resulting from previous reactions occurring in the presence of light.

The following chemical equation, although an oversimplification, summarizes the complex process of photosynthesis:

$$6CO_2 + 12H_2O \; (+ \text{ light energy}) \rightarrow C_6H_{12}O_6 \text{ (glucose)} + 6H_2O + 6O_2$$

More complete information on photosynthesis is available to the student in most botany textbooks.

The amount of chlorophyll in the growing plant varies from none to about 0.25 percent of the weight of the green plant. The green color usually is a fair indication of the conditions for growth and the supply of necessary substances. The red color of certain normal leaves, caused by anthocyanin, does not prevent photosynthetic activity if the normal chlorophyll also is present.

Respiration

Respiration in plants is very much like that in animals and is essentially the reverse process of photosynthesis. Substances are oxidized, producing carbon dioxide and water and releasing energy. Whenever growth is taking place or life is present, respiration goes on; even seeds while in a resting or dormant stage respire slowly. Respiration can be represented by the simplified equation:

$$C_6H_{12}O_6 \text{ (glucose)} + 6O_2 \rightarrow 6CO_2 + 6H_2O + \text{release of energy}$$

The principal substance oxidized in plant respiration is recently manufactured or stored carbohydrates, although other products may be used under certain conditions. Respiratory processes are controlled by enzymes formed in the cytoplasm.

Respiration proceeds both day and night. During the day photosynthesis normally occurs at a faster rate than does respiration, so that the carbon dioxide liberated in respiration is more than balanced by the oxygen liberated by photosynthesis. At night the plant gives off carbon dioxide, because respiration continues during the night while photosynthesis ceases.

Transpiration

Water enters the plant by diffusion through the walls of the root hairs and other thin-walled cells and passes deeper into the root to the fibrovascular bundles, into the *xylem*, and up through the stems to the leaves. Part of the water is used in photosynthesis and part in other cell processes, but most of it passes into the air spaces between the cells and diffuses into the atmosphere through the stomatal openings. This water loss is called *transpiration*.

Diffusion

The ultimate particles of gases, liquids, and substances in solution are in motion, with a constant tendency to spread out away from each other, distributing themselves uniformly throughout all the available space. This tendency is called *diffusion*.

Any substance in solution (solute) tends to spread from a region of high concentration of that solute to a region of less concentration. Likewise, any solvent tends to go from the region where there is more to the region where there is less of the solvent.

The cell wall usually allows the free passage of solute ions and molecules, just as it does molecules of water. However, a membrane may prevent or restrict the passage of some molecules or ions of solutes. Such differentially permeable membranes play an important role in cell physiology. *Osmosis* is the term applied to the diffusion of water through a differentially permeable membrane.

In the cells of the roots in contact with the soil solution, the cell membranes divide two solutions, each containing many solutes. Normally there are more solutes, and therefore there is less activity of water molecules, inside the cell than in the soil solution; therefore, water tends to diffuse through the membrane into the plant cell.

Movement of Materials into the Plant

The solutes in the soil solutions may be divided roughly into three classes: (1) solutes used by the plant; (2) solutes capable of entering the cell wall, but not used by the plant; and (3) solutes not capable of entering the cell wall because of exclusion by the cell membrane.

Solutes used by the plant pass through the cell wall by diffusion until they are equally distributed in the cell sap and in the soil solution; but as they are being used by the plant and taken out of solution within the cell there tends to be a continuous flow into the plant. These solutes, which typically are in an ionized state, upon passing through the cell wall may be immediately combined, precipitated, or changed, so that they are rendered inactive insofar as their preventing the entrance of other ions of the same kind is concerned.

When solutes not used by the plant pass into the cell, there soon is as much of the solute per unit volume inside the plant as in the soil solution. When this point of equilibrium is reached, this solute does not change in concentration on either side of the membrane until there is a change in concentration, either in the plant, by an increase in volume, or in the soil solution. Water may be moving freely from the soil solution into the plant, but it would not carry these solutes in with it. Certain amounts of this type of solute may be precipitated in the plant and permit the entrance of a larger amount.

Certain solutes are mechanically kept out of the plant by the impermeability of the cell membrane to them. Many elements in organic forms are of this type. They must be changed to some other form before they can be used. Many of the elements necessary for plant growth are combined in such a way as to render them unavailable. So long as the physical, chemical, and biological conditions are right, the necessary elements are rendered available, but under unfavorable conditions corrective measures must be taken to bring about the desired release of the essential substances.

Water may enter the plant, leaving behind any of these three types of substances in solution, if there already is an equal concentration of these substances inside and outside the plant. Likewise, ions of potassium or of other required substances may move into the plant, although the water itself is not moving in at the time. Any substance to which the cell wall is permeable may be found in the plant in the same concentration as in the soil solution, but only those substances that are used, precipitated, or otherwise held by the plant, so that the activity of the element within the plant continues to be lower than that without, will be accumulated.

TURGIDITY. Plants do not grow unless they are turgid. A cell so filled with water that the wall is stretched or is under strain, is said to be *turgid,* or to show turgor pressure. Turgor is really an expression of osmotic pressure in the cell by the strain on the cell wall.

If tissues are surrounded by solutions containing a greater concentration of solutes than is contained within the tissue, water will diffuse out of the tissue, with the result that the cytoplasmic membrane pulls away from the cell wall and the cell becomes placid or wilted. This loss of water is called *plasmolysis.*

Growth Differentiation in Plants

Differential growth within and among cells occurs in an orderly and systematic manner in higher plants. The controlling mechanisms of differentiation of genetically identical cells is not well understood. Apparently, this is accomplished by the cell's genetically controlled processes in combination with their external environment. Endogenous growth substances or hormones, along with genetic and environmental factors, likely play an important role in differentiation.

Essential Plant Nutrients

It is generally agreed that there are 16 elements essential to the growth of plants. Without their presence, the plant cannot complete its life cycle. Carbon, hydrogen, and oxygen are obtained from water and air. The 13 elements obtained from the soil are nitrogen, phosphorus, potassium, calcium, magnesium, sulfur, copper, boron, zinc, molybdenum, manganese, iron, and chlorine. Some elements are almost always present in higher plants but have never been proved essential. These include cobalt, vanadium, sodium, iodine, fluorine, silicon, and aluminum. Arsenic, selenium, lead, and lithium have stimulated growth of certain plants under some conditions.

The essential elements are often categorized as (1) macronutrients, required in relatively large quantities, such as nitrogen, phosphorus, and potassium; and (2) micronutrients, required in small quantities, such as copper, boron, zinc, and iron. Sometimes calcium, magnesium, and sulfur are designated as secondary elements, while nitrogen, phosphorus, and potassium are termed primary elements. Nutrient deficiencies, in general, are easily recognizable.

Factors Affecting Growth

Productivity often is influenced adversely by unfavorable environmental conditions, including temperature, available moisture, soil reaction, available mineral nutrients, radiant energy, and other biotic factors, such as insects, diseases, and weeds.

Temperature affects such plant functions as photosynthesis, respiration, cell-wall permeability, absorption of water and its solutes, and transpiration. In general, respiration takes place slowly at low temperatures. Transpiration rates are relatively low at the lower temperatures. Absorption of water and solutes increases with a rise in soil temperature. Temperature also influences plant growth by its effect on the microbial population of the soil, which generally increases with a rise in soil temperature. The relationship between yield and temperature varies for different species and for different cultivars of the same species. The so-called *cool-season* plants react differently to temperature changes than do *warm-season* plants.

The growth of many plants is in proportion to the amount of water available to them. Growth is restricted at both very low and very high levels of soil moisture. Water is required for the maintenance of carbohydrates and as a vehicle for the transportation of nutrients and raw materials. In addition, internal moisture stress causes a reduction both in cell division and in cell elongation, thereby influencing growth. Soil moisture level has a direct relationship to the uptake of plant nutrients. When moisture supplies are adequate, an increase in nutrient uptake increases the water efficiency of plants. Low moisture levels reduce the activity of microorganisms and result in lower available nitrogen supply.

Radiant energy is a significant factor depending on its quality, intensity, and duration. It is believed that, in general, a full spectrum of sunlight is most satisfactory for plant growth. Various crops respond differently to some of these factors.

Liebig, generally regarded as the father of agricultural chemistry, first advanced the idea that the growth of plants is proportional to the amount of mineral substances available and that the growth of plants is limited by the plant nutrient elements present in the smallest quantity, all of the others being present in adequate amounts. Very important is the work of Lawes and Gilbert, at the Rothamsted Station in England, who established proof that soil fertility can be maintained by the use of chemical fertilizers alone. Tremendous progress has been made in recent years in the development of crop cultivars and hybrids with much greater yield potential. This relates directly to the nutritional needs. Hybrid corn producing 200 bushels per acre requires proportionately more plant nutrients than does corn producing only half that yield. Under low fertility a high-yielding cultivar has no opportunity to produce at its potential level. High-yielding cultivars deplete soil nutrients more rapidly and heavy soil fertilization is necessary.

REFERENCES AND SUGGESTED READINGS

1. BARBER, S. "A Diffusion and Mass-Flow Concept of Soil Nutrient Availability," *Soil Sci.,* 93:1 (1962).

2. BROOK, A. J. *The Living Plant* (Chicago: Aldine, 1964).

3. DANBENMIRE, R. F. *Plants and Environment* (New York: Wiley, 1969).

4. FRANCK, J., and W. E. LOOMIS. *Photosynthesis in Plants* (Ames: Iowa State Univ. Press, 1949).

5. JANICK, J., *et al. Plant Science, An Introduction to World Crops,* 2nd ed. (San Francisco: Freeman, 1974).

6. GREULACH, V. A. *Plant Function and Structure* (New York: Macmillan, 1973).

7. LOOMIS, W. E. *Growth Differentiation in Plants* (Ames: Iowa State Univ. Press, 1953).

8. MARTIN, J. H., W. H. LEONARD, and D. L. STAMP. *Principles of Field Crop Production,* 3rd ed. (New York: Macmillan, 1976).

9. MEYER, B. S., *et al. Introduction to Plant Physiology* (Princeton, N.Y.: Van Nostrand, 1960).

10. MOUNT, M. C. N., and T. W. WALKER. "Competition for Nutrients Between Grasses and White Clover," *Plant Soil,* 11:1 (1959).

11. ROBBINS, W., *et al. Botany,* 8th ed. (New York: Wiley, 1964).

12. SAUCHELLI, V. *Trace Elements in Agriculture* (New York: Van Nostrand-Reinhold, 1969).

13. SCHUFFELEN, A. C. "Growth Substances and Ion Absorption," *Plant Soil,* 1:2 (1949).

14. SLATYER, R. O. *Plant–Water Relationships* (New York: Academic Press, 1967).

15. SPRAGUE, H. B. (Ed.). *Hunger Signs in Crops,* 3rd ed. (New York: McKay, 1964).

16. WAGGONER, P. E., *et al.* "Radiation in Plant Environment and Photosynthesis," *Agron. Jour.,* 55:1 (1963).

17. WIERSUM, L. K. "Utilization of Soil by the Plant Root System," *Plant Soil,* 15:2 (1961).

CHAPTER 12
ROOTS OF FIELD CROPS

THE roots of plants have not received the attention their functions warrant. The plant is dependent on the root system for its very life, for water and nutrients, as well as for anchorage in the soil. Plant roots often also perform an important function in storing food for future use. In extent, the root system of plants can be as great as, or greater than, the aboveground portion.

Types of Roots

Roots do not possess the symmetry noted in the top growth of most plants, because of the necessity of obtaining water, air, and nutrients wherever they are present. Within a species the tendency is to produce a certain type of root system.

The root system of all cereals and other grasses is known as *fibrous*. All the roots are slender and fiberlike, with no one root greatly more prominent than the others. In contrast to the fibrous root system are plants that have one main root—a *taproot*—which grows directly downward, and from which branch roots arise. Alfalfa, red clover, sweetclover, cotton, and flax are good examples of such plants. Taproots are likely to penetrate more deeply than fibrous roots.

Kinds of Roots

When a seed germinates, the radical portion becomes the first root and is called the *seminal* or *primary* root; its branches and subbranches are *secondary* roots. In some plants the primary root or roots remains the principal root system throughout the life of the plant, as is the case with most taproot plants. In the development of a fibrous root system, such as that of the cereals, the primary roots arising from the seed may give way to permanent roots that arise from nodes in that portion of the young plant which extends from the germinating seed to the soil surface. These are known as *adventitious* roots. Primary, or seminal, roots often are referred to as temporary, but in some plants they may remain active until maturity. In corn, the primary root system, for practical purposes, is important for 3 to 6 weeks. In general, all roots that arise from organs other than the primary root are known as adventitious. An example is the *brace roots* or *prop roots*, which may arise from several of the lower nodes of corn.

FIGURE 12-1. *Lateral roots developing from a secondary root. Note the length and abundance of root hairs. This is common for most grass and grain crops.* [SOURCE: USDA, National Tillage Machinery Laboratory, Auburn, Alabama.]

When these reach and enter the soil they branch out and cannot be distinguished from other roots.

Structure of Roots

At the tip of a growing root is a root cap, then the apical meristem, the region of elongation, and the region of root hairs. The cells of the root cap are constantly sloughed off and replaced by apical meristematic cells that are actively dividing, adding new cells not only to the root cap, but also to the region of elongation. Elongation of cells, resulting in increased lengths, does not occur in the root hair zone. Otherwise the root hairs that surround and adhere to the soil particles would be torn loose.

The root hairs are the principal absorbing structures of the plant. They are most abundant on very young roots, only a short distance back from the growing root tip. The root hair and the epidermal cell from which it grows constitute a single cell. For most plants the life of any one root hair is only a few days. New root hairs constantly are formed as others die. As the root advances through the soil, new actively growing root hairs come into contact with newly reached soil particles.

Root Functions

The functions of roots may be stated as absorption, anchorage, conduction, and storage. The roots absorb water and nutrients from the soil solution and oxygen from the soil air, anchor the plant in place, conduct water and nutrients to the aboveground parts of the plant, and distribute materials to all belowground parts.

Large amounts of water are absorbed by the roots, and this water is conducted upward throughout the plant. Much of the water is lost to the atmosphere from the leaves as water vapor (transpiration). Eighty to 95 percent of the total weight of most plants is water. But the amount retained is only a small part of the amount absorbed and transpired. For example, from about 400–1,000 pounds (181–454 kg) of water is passed through the plant during its growth for each pound (0.45 kg) of dry matter produced. The specific amount of water required for the production of 1 pound (0.45 kg) of dry matter varies greatly according to species of plant and environment.

Secretions from Roots

When roots grow along the surface of a polished marble slab, they etch the surface, making a slight depression wherever the root comes in contact with the marble. This power of roots to render minerals soluble led many to credit roots with exuding acids or other substances that dissolved nutrients from soil minerals. These secretions were supposed to accumulate in the soil when a given crop was grown on land for several years in succession, supposedly making the soil toxic to this particular crop and perhaps to other species.

But the idea developed later that it is unnecessary to assume the secretion of a permanent acid by the roots. The growing portions of the plant root release carbon dioxide, and this—especially in the concentrated solutions which must be momentarily formed in the cell walls of the root hairs—has an appreciable solvent effect on most minerals in the soil. Repeated studies have seemed to confirm this position.

It is recognized, however, that some crops do have a depressing effect on the yield of succeeding crops. Sorghum has been shown to have a depressing effect on the yield of wheat. This is attributed to a temporary tie-up of nitrogen with decomposition products, rather than to root excretions. It also may be related to the efficiency of sorghum to extract water from a soil, leaving poor moisture conditions for a succeeding crop. The yield of corn grown on land after a crop of sudangrass often will be reduced for the same reason.

Factors That Influence Root Development

The roots of a number of plants have shown great similarity when grown in uniform soil, but they have shown striking variations when the soil moisture content, fertility, soil aeration, or soil structure differ.

Available Soil Moisture

Water is essential for root growth; it will not proceed in or through a dry soil layer.

Weaver found that corn grown in a rich, loose soil with an available water content of 9 percent had a root area 2.1 times that of the transpiring surface of stems and leaves. In contrast, with an available soil-water content of 19 percent, the root area was only 1.2 times that of the top growth.

Wright found that the root weight of blue panicgrass was lowered linearly as degree of soil-moisture stress was increased. Hurd found that spring wheat cultivars showed root-pattern differences at different moisture levels, which partially explained yield differences.

Soil Fertility and pH

Fertilizer may either stimulate or depress root growth, depending on the concentration. It was demonstrated as early as 1865 that root growth became more restricted as concentration of a nutrient solution increased.

Fox *et al.* observed great variations in the root habits of the same species of grass growing in different soils in Nebraska. Extensive branching and development occurred in the lower profiles when the soil nutrients, especially phosphorus, were high. Kentucky bluegrass roots grew to a depth of 22 inches (56 cm) in a Carrington soil and to a depth of 48 inches (122 cm) in Judson soil. Roots of blue grama showed a similar response in these two soils.

Duncan and Ohlrogge found there is commonly an increased proliferation of corn roots in a fertilizer band, but it does not occur with all species. Apparently, higher-order roots, not seminal or branch roots, proliferate if the concentration in the band is relatively low.

An increased supply of nitrogen to roots generally increases the shoot–root ratio. Phosphorus is the most important element in root development. Wiersma found that a 1:5 ratio of nitrogen to phosphorus appears to favor the greatest corn–root proliferation in a fertilizer band. Pearson found that phosphorus did not have to be at the site of growth to provide normal root development; phosphorus in surface layers of soil is sufficient to allow adequate root development at greater depths. Potassium appears to have no direct effect on extent of root growth or degree of branching, but is involved in internal functions of the root.

A pH value outside the range of 5.0–8.0 generally restricts or limits root growth. At a low pH, 5.0 or below, the soil solution contains increased quantities of aluminum and manganese, which are detrimental to root development. Breeders are selecting lines less sensitive to high concentrations of these elements.

Aeration

It has been shown that the absence of oxygen retards root growth and stops the production of root hairs. The reduction of the oxygen pressure to zero can stop the development of root hairs in corn, even when all carbon dioxide is removed. With wheat, the root hairs quickly die in the absence of oxygen. Plant roots, like all other living parts, require oxygen for respiration. For most crop plants, oxygen must be present in soil air. An exception is rice, which can tranport oxygen very efficiently from the leaves through the vascular system and thus can grow well under flooded conditions.

Elliott found that in well-drained soils corn roots frequently extended to a depth of 6 feet (1.8 m) or more, whereas in peat marshes, where the water table was almost stationary at about 2.5 feet (0.8 m), the roots did not penetrate to a depth of more than 18 inches (46 cm). The downward penetration of the corn roots was bounded by a zone sharply defined by the flattening out of the roots within it. This zone, not more than 3 inches (7.6 cm) in thickness, lay 18 inches (46 cm) above and parallel to the water table.

Carbon dioxide can build up to toxic levels in compacted or wet soils. Geisler found that low CO_2 concentrations can benefit root growth, but that high concentrations can have a deleterious effect.

Roots of Selected Crops

Each plant species has a particular type of root growth, but the root development of individual plants within each species often varies considerably because of differences in environment.

One marvels at the complexity of the root system. Weaver found that a single corn plant in the eight-leaf stage has as many as 8,000–10,000 lateral roots arising from its 15–23 main roots. It is estimated that a rye plant develops an average of 3 miles (4.8 km) of new roots with 55 miles (88 km) of root hairs each day over a four-month period.

Grasses

The practice of including a grass in alfalfa seedings in order to improve soil structure and reduce soil erosion has become general in some regions. Grass growing with alfalfa has been shown to increase the organic matter of the soil, improve the soil tilth, and decrease the density of soils that have grown only grain crops for years. Grass roots also increase the aggregation of soils, in this way improving soil permeability and resistance to erosion. Woods *et al.* reported that the average acre root production of a number of grasses grown with alfalfa was 4.34 tons (9.7 MT/ha) as compared with 2.59 tons (5.8 MT/ha) when the alfalfa was grown alone. Of the 4.34 tons, 2.81 comprised grass roots and 1.53 were alfalfa roots. The root yields were determined to a depth of 8 inches (20 cm) after five growing seasons. Similar previous trials had indicated that for this soil approximately 80 percent of the roots was in the upper 8 inches (20 cm).

Roots of several southern grasses growing on a deep sand were studied at Tifton, Georgia. The relative drought tolerance is indicated by the per-acre dry-matter production in pounds in the dry summer of 1952: 'Suwanee' bermuda, 9,893 (11,080 kg/ha); 'Coastal' bermuda, 9,574 (10,723 kg/ha); Pangola, 5,480 (6,138 kg/ha); 'Pensacola' bahia, 5,176 (5,797 kg/ha); 'common' bermuda, 4,347 (4,869 kg/ha); sand lovegrass, 3,512 (3,933 kg/ha); carpetgrass, 805 (902 kg/ha).

Striking differences between the different grasses in the distribution of roots at various depths were noticed. The drought-susceptible carpetgrass had 93.6 percent of its roots in the upper 2 feet (0.6 m) of soil, whereas only 65.1 and 68.8 percent of the roots of the highly drought-tolerant 'Coastal' and 'Suwanee' bermudas were found in this layer.

The root production of five grasses was determined in 3-inch (7.6-cm)segments to a depth of 18 inches (46 cm) in a shallow Rayne silt loam at the West Virginia station, with marked differences in the relative root development for the different depths (Table 12–1).

Other Crop Plants

Pearson has found that the corn root system often extends to a depth of 5 feet (1.5 m) in a medium- to a coarse-textured glacial till, but seldom below 3 feet (0.9 m) in fine-textured subsoils.

Weaver described corn-root development in detail as early as 1926, and Foth later (1962) studied development and distribution of corn roots. Results obtained in Foth's corn root distribution study are shown in Table 12–2. Maximum root weight early, at 37 days, was in the 3- to 6-inch (7.6- to 15.2-cm) depth. Early root growth occurred largely in a downward diagonal

TABLE 12–1. Average Weight (Dried at 60°C) of Roots in Pounds per Acre-Inch at 3-Inch (7.6-cm) Depth Intervals to 18 Inches (46 cm)

Grass	Depth (inches)				
	0–3	3–6	6–9	9–12	12–18
Bromegrass	717	201	141	124	106
Orchardgrass	1,247	124	77	51	30
Bluegrass	1,126	85	53	7	3
Timothy	680	41	20	12	6
Deertongue	4,486	246	87	37	16

Source: Gist, G. R. and R. M. Smith. "Root Development of Several Common Forage Grasses," *Jour. Amer. Soc. Agron.*, 40:11 (1948).

direction followed by extensive lateral growth. Lateral growth was completed a week or two before tassel emergence. Brace root development occurred near completion of lateral growth. Extensive growth of roots below 15 inches (38 cm) occurred near tasseling time, and by early roasting ear stage root growth was completed with cessation of brace root growth.

TABLE 12–2. Vertical and Lateral Distribution of Roots from Base of Corn Plant at Four Stages After Planting

Location from Base of Plant (inches)	Weight (grams) of Roots After Planting			
	23 days	41 days	54 days	80 days
	Vertical			
0–3	0.13	0.83	2.42	5.31
3–6	0.12	0.94	1.91	2.56
6–9	0.02	1.03	1.75	1.78
9–12	—	0.86	1.55	1.11
12–15	—	0.31	0.67	0.49
15–18	—	0.13	0.17	0.30
18–21	—	0.06	0.06	0.27
21–24	—	—	0.03	0.21
24–27	—	—	0.02	0.18
27–30	—	—	—	0.09
30–36	—	—	—	0.07
	Lateral			
0–3	0.15	1.18	3.63	6.43
3–9	0.10	1.21	2.08	3.30
9–15	0.02	0.87	1.60	1.14
15–21	0.00	0.90	1.27	1.50
Total	0.27	4.16	8.58	12.37

Source: Foth, H. D. "Root and Top Growth of Corn," *Agron. Jour.*, 54:49 (1962).

FIGURE 12–2. *A corn root system can extend to a depth of nearly 2 meters in 3 months in a good environment.* [SOURCE: USDA, National Tillage Machinery Laboratory, Auburn, Alabama.]

Foth's data in Table 12–2 emphasize the heavy reliance of corn on relatively shallow roots. Most of the roots are in the top 12 inches (30 cm) of soil, with a large proportion in the upper 6 inches (15 cm).

The root habit of *sorghum* is somewhat similar to that of corn, but the

FIGURE 12–3. *Roots of biennial sweetclover early in the second year of growth, as they grew in a drilled row. These are strong, deep, fleshy taproots with many strong laterals which develop.*

roots are finer and more fibrous. The volume of soil proliferated by sorghum may be about the same as corn, but sorghum often develops twice as many roots in the same soil volume, providing more complete water-extraction capability. In sorghum, some root growth continues to occur after flowering.

Alfalfa has a strong taproot, which often reaches a depth of up to 3 feet (0.9 m) at three months, 5–6 feet (1.5–1.8 m) the first year, and 10–12 feet (3.0–3.7 m) the second year. Depths of 12–20 feet (3.7–6.1 m) are not uncommon, and depths of 100 feet (30 m) or more have been observed under very favorable conditions.

The roots of red clover frequently extend to a depth of from 4–6 feet (1.2–1.8 m). Biennial sweetclover develops a strong, deep, fleshy taproot. Many strong laterals usually are developed which penetrate deeply into the soil. The taproot often will extend to a depth of 5–8 feet (1.5–2.4 m). While crownvetch has a taproot system, this plant develops lateral roots that can send up new shoots which, in turn, establish new root systems.

Weaver has reported the depth and lateral spread of roots of numerous field crops, as shown in Table 12–3.

Top–Root Ratio

The ratio of top growth to root growth in plants may be influenced considerably by soil and climatic conditions, as well as by the stage of maturity and the characteristics of specific species.

FIGURE 12–4. *Roots of a 3-month-old soybean plant. Although the soybean, as well as other legumes, develops a tap root system, extensive secondary root development occurs.* [SOURCE: USDA, National Tillage Machinery Laboratory, Auburn, Alabama.]

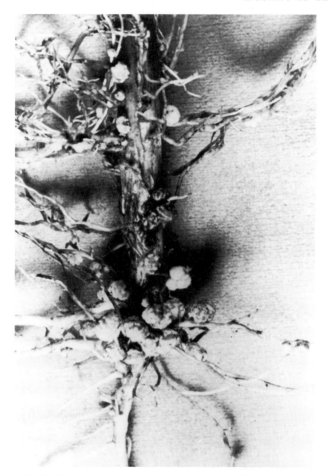

FIGURE 12–5. *Nodules on root system of soybean. Well-nodulated soybeans can meet much of their nitrogen requirement by fixation from the soil air.* [SOURCE: *Illinois Research,* University of Illinois Agricultural Experiment Station.]

In general, the increased availability of nutrients increases the weight of roots, but under these conditions the increased weight of tops shows still greater increase. Nitrogenous fertilizers especially increase the growth of tops as compared with that of the roots. For phosphorus the effect is the reverse.

Foth noted that the growth rate of corn tops exceeded that of roots, as shown by an increase, with time, in top–root ratio (Table 12–4). After 54 days, root growth and development was restricted to that below 15 inches (38 cm).

Schulze has reported the comparative weights of tops and roots and the top–root ratio for various crops, as shown in Table 12–5.

TABLE 12–3. Type of Root and Representative Depth and Spread of Root of a Number of Field Crops; Extracted from Summaries

Name of Crop	Type of Root System	Working Depth (feet)	Depth of Root System (feet)	Lateral Spread on All Sides (inches)
Spring wheat	Fibrous	3	4–5	6–9
Winter wheat	Fibrous	3.8	5–7	6–9
Rye	Fibrous	4	5	6–10
Oats	Fibrous	2.5	4–5	6–11
Barley	Fibrous	3.2	4.5–6.5	6–12
Corn	Fibrous	3.5	5–6	42
Sorghum	Fibrous	3.5	4.5–6	36
Sugarbeet	Taproot	3.5	5–6	6–18
Alfalfa	Taproot	4	15–20	24
Bromegrass	Fibrous	2.2	2–3	24
Orchardgrass	Fibrous	2.3	3.5	18
Meadow fescue	Fibrous	3	4	3–6
Bluegrass	Fibrous	2	5–7	12–18
Redtop	Fibrous	2	2	9
Reed canarygrass	Fibrous	1.2	2	15
Timothy	Fibrous	1.2	2	12
Red clover	Taproot	—	5	12–18
White clover	Taproot	—	5	12–18
White sweetclover	Taproot	—	5–8	24–60
Potatoes	Taproot	—	2–3	12–24
Sunflower	Taproot	3.0	5–9	24–60

Source: Weaver, J. E. *Root Development of Field Crops* (New York: McGraw-Hill, 1926).

TABLE 12–4. Oven-Dry Weights of Tops and Roots and Top–Root Ratios for Corn at Eight Stages

Days After Planting	Tops (grams)	Roots (grams)	Top–Root Ratio
23	1.1	0.54	2.0
37	11.6	4.36	2.6
41	26.4	8.32	3.2
47	44.7	12.26	3.6
54	87.9	17.16	5.1
67	165.3	17.68	9.3
80	189.3	24.74	7.7
100	274.4	25.56	10.7

Source: Foth, H. D. "Root and Top Growth of Corn," *Agron. Jour.,* 54:49 (1962).

TABLE 12–5. Comparative Weight of Tops and Roots, and the Top–Root Ratio of Various Crop Plants

Crop	Stage	Weight of Tops	Weight of Roots	Top–Root Ratio
Rye	Before heading	3.68	1.79	2.0
Rye	Bloom	28.80	6.08	4.7
Rye	Milk	46.17	4.66	9.9
Wheat	Before heading	3.99	1.88	2.1
Wheat	Booting	32.48	9.01	3.6
Wheat	Milk	46.50	4.90	9.5
Wheat	Ripe	31.44	2.89	10.8
Barley	Ripe	27.82	2.05	13.5
Oats	Ripe	45.40	4.09	11.1
Vetch	Bloom	14.00	2.20	6.3

Source: Miller, E. C. "The Root Systems of Agricultural Plants," *Amer. Soc. Agron.*, 8:30 (1916).

REFERENCES AND SUGGESTED READINGS

1. BURTON, G. W. "Root penetration, Distribution, and Activity of Southern Grasses," *Agron. Jour.*, 46:5 (1954).
2. DUNCAN, W. G., and A. J. OHLROGGE. "Principles of Nutrient Uptake with Fertilizer Bands. II. Root Development in the Band," *Agron. Jour.*, 50:605 (1958).
3. ELLIOTT, Q. R. B. "Relation Between the Downward Penetration of Corn Roots and Water Level in Peat Soil," *Ecology*, 5:2 (1924).
4. FOTH, H. D. "Root and Top Growth of Corn," *Agron. Jour.*, 54:49 (1962).
5. FOX, R. L., *et al.* "Influences of Soil Characteristics on Distribution of Grass Roots," *Agron. Jour.*, 45:12 (1953).
6. GEISLER, G. "Interactive Effects of CO_2 and O_2 in Soil on Root and Top Growth of Barley and Peas," *Plant Physiol.*, 42:305 (1967).
7. GIST, G. R., and R. M. SMITH. "Root Development of Several Common Forage Grasses," *Jour. Amer. Soc. Agron.*, 40:11 (1948).
8. HAYES, J. L., *et al.* "Effect of Corn Planting Pattern on Yield, Root Extension, and Interseeded Cover Crops," *Agron. Jour.*, 51:454–456 (1959).
9. HURD, E. A. "Growth of Roots of Seven Varieties of Spring Wheat at High and Low Moisture Levels," *Agron. Jour.*, 60:201 (1968).
10. LINSCOTT, D. L., R. L. FOX, and R. C. LIPPS. "Corn Root Distribution and Moisture Extraction in Relation to Nitrogen Fertilization and Soil Properties," *Agron. Jour.*, 54:185 (1962).
11. MILLER, E. C. "The Root Systems of Agricultural Plants," *Amer. Soc. Agron.*, 8:30 (1916).
12. MITCHELL, R. L. *Crop Growth and Culture* (Ames: Iowa State Univ. Press, 1970).
13. PEARSON, R. W. "Soil Environment and Root Development," In *Plant Environment and Efficient Water Use*, edited by W. H. Pierre *et al.* (Madison, Wisc.: American Society of Agronomy and Soil Science Society of America, 1966).
14. WEAVER, J. E. "Investigation on the Root Habits of Plants," *Amer. Jour. Bot.*, 12:9 (1925).

15. Weaver, J. E. *Root Development of Field Crops* (New York: McGraw-Hill, 1926).

16. Wiersma, D. "The Soil Environment and Root Development," *Advan. Agron.*, 11:43 (1959).

17. Woods, J. E., *et al.* "The Effect of Grasses on Yields of Forage and Production of Roots by Alfalfa–Grass Mixtures," *Agron. Jour.*, 45:12 (1953).

18. Wright, N. "Root Weight and Distribution of Blue Panicgrass (*Panicum antidotale* Retz.) as Affected by Fertilization, Cutting Height, and Soil-Moisture Stress," *Agron. Jour.*, 54:200 (1962).

Part IV
Crop Production Practices

CHAPTER 13
SOIL AND WATER CONSERVING PRACTICES

> In this country we have been misled by our plentiful supply of land
> into a false philosophy of inexhaustibility. We have come to regard the
> land only as a source of immediate wealth. We have forgotten that it
> is a fundamental heritage belonging as much to our children's children
> as to us in the little time we are permitted to remain here on earth.
>
> HUGH HAMMOND BENNETT

THE importance of productive land is self-evident. The land feeds and
clothes the people of a nation and provides many other necessities of life.
Many nations do not have enough productive land to make possible a good
standard of living. A large part of the United States had fertile soil when
it was first settled. On too many of these acres, however, erosion, overcrop-
ping, the oxidation of organic matter, and the leaching of soil nutrients
have greatly reduced the native fertility.

But on many soils erosion can be checked, soil fertility increased, and
the supply of water regulated so that productivity is increased. Soil- and
water-conserving practices need to be applied more generally. Conserving
practices must be fitted to the land of each field, each farm, and each
watershed.

Soil Erosion

Soil under many conditions is very unstable. When there is considerable
slope, water or wind moving across a bare surface usually carries away
some soil. Eventually, large amounts of soil may be moved from unprotected or misused areas. When land is cultivated, at times there is no vegetative cover to protect the soil, and serious erosion may result.

The National Inventory of Soil and Water Conservation Needs made by
the United States Department of Agriculture (USDA) showed that erosion
is the dominant conservation problem on 706 million acres (286 million
ha), about one-half the agricultural land in the United States. Erosion is
a problem that requires constant attention on 221 million acres (89 million
ha) of cropland. Soil erosion averages 9 tons per acre (20 MT/ha) on U.S.
farms, while topsoil is formed at a much slower rate.

Estimated average annual losses of about $800 million are attributed
to soil erosion by water or wind or both from croplands in the United
States. From the standpoint of physical deterioration of land, erosion un-

FIGURE 13-1. *Severe erosion on an Iowa farm, estimated at 200 tons per acre. The field was tilled for planting up and down the hill rather than across the slope.* [SOURCE: USDA, Soil Conservation Service.]

doubtedly causes greater damage than does any other factor. Soil erosion has forced the abandonment for cultivation of an estimated 35 million acres (14 million ha) originally suitable for crop production.

The final step in the water-erosion process is the deposition of soil particles that have been moved. Sedimentation in low-lying areas, valley lands, and reservoirs constitutes a continuing threat to land and water resources. Sedimentation attendant to soil erosion causes untold damage by covering up growing crops, reducing the productivity of good agricultural land, and reducing the effectiveness of water reservoirs.

At least 4 billion tons (3.6 billion MT) of sediment are produced annually through erosion in the United States. About 2 billion tons (1.8 billion MT) are washed into streams, and about 1 billion tons (0.9 billion MT) of this reach tidewater.

Although few reliable figures are available on the areas affected by wind erosion, it is estimated that at least 75 million acres (30 million ha) are subject to this type of loss. Perhaps 36 million acres (15 million ha) need treatment and 10 million acres (4 million ha) are seriously affected.

The effects of soil losses through wind erosion on crop yields are similar

FIGURE 13–2. *Soil drift across a county road in Iowa as a result of wind erosion. Soil blowing not only removes fertile areas of soil but also creates major problems of removal.* [SOURCE: USDA, Soil Conservation Service.]

to those caused by runoff water. The most fertile parts of the soil are re-moved, carried in the air for considerable distances, and then deposited on the land, in cities and towns, or in the sea. Much of the soil removed is redeposited on farmland, where it remains available for use. However, much of it is deposited where no use can be made of it or where it creates major problems of removal. Dust storms cause additional damage through abrasive action on crops and farm equipment.

Losses from soil blowing generally are limited to arid and semiarid areas, but are widely distributed. Losses are greatest in the Great Plains states. Soil blowing in the humid areas, although less serious, does create prob-lems, particularly on drained and cultivated organic soils and on very sandy soils.

Severe erosion, where more than 75 percent of the original surface soil has been lost, is especially common in the Piedmont Plateau and nearby upper Coastal Plain and in the Appalachian Plateau and adjacent Lime-stone Valleys and Uplands of the humid eastern United States. In the Mid-west, erosion is especially severe on the wind-laid soils along the Missis-sippi and Missouri Rivers, the claypan prairie section of Missouri and Illinois, and the Red Prairies and Cross Timbers of Oklahoma and Texas.

In the West, the Palouse section of Washington and the coastal section of California have areas of severe sheet and gully erosion.

About 30 percent of the land in the United States has slight or no erosion. Part of this land is level enough so that it does not erode easily, and part of it has been protected by vegetation. The alluvial plains, such as the Mississippi delta, swamps of the coastal plains, and level or gently undulating prairies and plains, have no major erosion problems.

Between the two extremes of severe erosion and slight or no erosion is a great deal of soil with moderate erosion.

An Infamous Dust Storm

On May 12, 1934, it was so dark at Washington, D.C., in the early afternoon that it was necessary to turn on all lights, both within buildings and on the streets. The darkness was caused by soil particles that had been picked up by high winds in the Southwest and carried clear across the country. So great was the amount of dust that it coated the decks of ships hundreds of miles at sea off the Atlantic Coast. A 1934 issue of *Time* magazine described the scene as follows:

FIGURE 13–3. *The Lincoln Memorial, at Washington, D.C., the afternoon of May 12, 1934, obscured by soil particles carried as dust from the "dust bowl" of the Middle West. This storm hastened legislation in regard to the soil conservation movement.* [SOURCE: USDA, Soil Conservation Service.]

Then came the wind; great gusty blasts. . . . It tore powdery soil from the roots of the wheat and deposited it like snowdrifts miles away. . . . The rich fertility of millions of farms took to the air; 300,000,000 tons of soil billowing through the sky. A dust storm 900 miles wide and 1,500 miles long swept out of the drought-stricken West.

Thousands of acres of native grasses in the Plains had been plowed up. The land had been stripped of its natural vegetation. When the drought came, the soil dried up and began to blow, with little vegetation cover to stop it. During this time, the nation witnessed for the first time a forced migration of some of its people.

Congress was in session. Perhaps this dust storm should be classed as famous, rather than infamous, for, although its severity was such as to cause great harm to the soil in the so-called dust bowl, it also did great good. It impressed upon the whole nation the need for concerted and vigorous action to conserve our soil resources. It helped bring speedy legislative action in Washington.

The Soil Conservation Movement

In 1929 Congress provided funds to establish 10 field stations to study erosion in cooperation with various states, but conservation of our soil resources did not receive much attention in the United States on a national basis until the 1930s. But the agricultural experiment stations had been keenly appreciative of the deterioration of soil productivity and had studied and reported on soil losses and methods of reducing such losses. Miller and Krusekopf at the Missouri Station initiated the first field plots to evaluate factors affecting runoff and erosion. A report published by them in 1932 stimulated the reporting of similar work at many other stations.

The first national movement for soil conservation was the establishment of the Soil Erosion Service, essentially an emergency agency, in 1933 under the leadership of Hugh Hammond Bennett. This agency set up erosion-control demonstrations in strategic locations. Bennett was a leader of the soil conservation movement, including development of the Soil Conservation Service. This agency was created as a successor to the Soil Erosion Service in April, 1935, as a permanent agency of the USDA through the National Soil and Water Conservation Act (Public Law 46). This was followed by a redesigned Agricultural Adjustment Act of 1938. The Soil Conservation and Domestic Allotment Act was passed, and at the same time the Agricultural Conservation Program (ACP) was created. This act provided for the continuation of the agricultural conservation program, with provision also for loans on stored products to cooperating producers and for production-marketing controls when accumulations of designated crops reached certain prescribed volumes.

The early work of the Soil Conservation Service was largely directed toward assistance in applying to individual farms the results of the research of the state agricultural experiment stations bearing on soil conservation and the maintenance of soil productivity. In addition, studies were inaugurated to determine the best methods of fitting these research results to individual farms and to specific areas. Many soil-conserving demonstration areas were set up throughout the country.

Soil Conservation Districts

The federal legislation which created the Soil Conservation Service also made possible the organization of soil conservation districts as local government units operating under state laws. The first conservation districts were formed in 1937.

The purpose in organizing districts was to enable farmers to band together on their own initiative to undertake the solution of soil and water conservation problems on a community basis, suited to local conditions. Educational programs, both on the state and district levels, have been the responsibility of the state colleges and their local representatives, i.e., the county agents and extension directors.

Technical assistance is assigned each conservation district in accordance with Soil Conservation Service agreements. Definite conservation farm plans are prepared, as agreed between the district and individual farmers.

HOW DISTRICTS ARE ORGANIZED. The organization of a soil conservation district is initiated by farmers within the area, and county committees and cooperating agencies design a program for each state and county. The programs are confined to the soil conserving practices on which federal cost sharing is most needed in order to achieve the maximum conservation benefit in state or county. In 1978 all agricultural areas of the United States were essentially covered by soil conservation districts. Conservation districts legislation has been enacted in all 50 states, with about 3,000 total districts in operation over the country.

Recent Developments

The Agricultural Conservation Program of 1975 became the successor to the Rural Environmental Assistance Program of 1971–1973 and the Rural Environmental Conservation program of 1974. The ACP provides assistance to farmers for carrying out conservation and pollution abatement practices in the United States and the Caribbean area. In 1976, nearly $150 million in assistance was provided to farmers through this program. More than 300,000 farms, covering 11.5 million acres (4.7 million ha) of land, were covered by this program. Total regular annual agreements in 1976 included the following: diversions, 130,185 acres (52,686 ha); sediment, chemical, or water runoff control measures, 765,368 acres (309,744 ha); windbreaks, 186,768 acres (75,585 ha); reorganizing irrigation systems, 1,009,297 acres (408,462 ha); establishing permanent cover, 1,104,263 acres (446,895 ha); improving permanent cover, 2,399,890 acres (971,235 ha); strip-cropping, 100,136 acres (40,525 ha); and terrace systems, 320,729 acres (129,799 ha).

More than 200,000 additional acres (80,940 ha) were included in the 1976 program year in long-term agreements on the same conservation or pollution-abatement practices listed above. The total land area affected by selected conservation measures during the period 1936–1976 is shown in Table 13–1.

Probably the most significant development relative to application of soil erosion field plots is that of the *Universal Soil-Loss Equation,* developed by Wischmeier and Smith of the USDA, located at Purdue University. This equation has become the standard reference for evaluating soil erosion problems and planning corrective practices.

In recent years, stronger actions have been taken by some states to curb excessive soil erosion from farmland. An example is the Illinois Bureau of Soil and Water Conservation, which is working under House Bill 818, adopted in 1977, to control erosion and sedimentation. The plan called for each county soil conservation district to set up a local program to fit local soil conditions. All agriculture land is subject to the *Universal Soil-Loss Equation.* The main goal is to reduce by 1988 the erosion factor for all types of soils to the existing *tolerance factor,* or permissible annual soil loss per acre, which now averages about 5 tons per acre (11 MT/ha). The average annual erosion rate on all Illinois cropland presently is about 7 tons/acre (16 MT/ha). The program has started as a voluntary effort, but stronger measures will be enforced if compliance is not sufficient to accomplish the goal.

In 1978, 43 states had some form of land-use regulation and 17 states had laws that require mandatory erosion and sediment control in certain instances. Iowa is one state that provides matching cost-sharing funds, to be combined with federal funds, for conservation needs.

Environmental concern has focused not only on keeping the soil in place, but also on runoff from the land as it affects water quality. The 1972 Amend-

TABLE 13–1. Selected Conservation and Pollution Abatement Measures Performed through the Agricultural Conservation Program, United States and Caribbean Area, 1936–1976

Practice	Acres (000)	Hectares (000)
Constructing standard terraces	34,355	13,903
Sediment or chemical runoff control	10,909	4,415
Stubble mulching	129,091	52,243
Strip cropping	114,535	46,352
Establishing permanent vegetative cover	63,373	25,647
Establishing additional vegetative cover	26,779	10,837
Interim cover crops	495,811	200,655
Contour farming	139,849	56,597
Diversion terraces, ditches, or dikes	5,657	2,289

Source: USDA. Agr. Stat. (1977).

ments to the Federal Water Pollution Control Act includes provisions for controlling runoff from agricultural lands. The goal is to eliminate all discharges of pollutants in United States waters by 1985. All states must prepare plans for identifying nonpoint sources of pollution, including agriculture, and methods of control.

The Soil Survey

The soil survey is basic to all soil conserving efforts. The different state colleges and agricultural experiment stations, in cooperation with the USDA, have had under way, over a long period of years, detailed soil surveys, mapped on a county basis. The results of these surveys are published with detailed maps in what are called *soil reports*.

Most important in the survey is to relate soil-conserving practices to the soil characteristics and climatic conditions so that the land will be used most effectively to produce at its maximum and at the same time keep it indefinitely productive.

Land-Use Classes

In land-use planning the Soil Conservation Service (SCS) has made use of a classification which places soil under eight major groups, according to the relative productivity of the land and how it can best be cropped. In making land-use maps for farm units, each class group is identified by a Roman numeral, and is indicated on the map by a different color.

Classes I, II, and III take in soils suited for cultivated crops. Soils in Class I require little conservation treatment, those in Class II require some treatment, and those in Class III require considerable treatment to protect against or overcome erosion and other hazards. The soils in these classes make up 44 percent of the private rural land in the United States, excluding Hawaii and Alaska. Class IV land also can be used for crops, but the user must choose his crops with care or manage the soil with extra care. This class accounts for 12 percent of the land. Soils in Classes V–VIII are generally not suitable for cultivation and should be kept in permanent vegetation. These soils make up 44 percent of the land.

Not all land in the same class or subclass requires the same conservation treatments. In preparing a farm conservation plan the farmer and a technician work together with careful consideration to a previously prepared, detailed, soil survey map.

Soil-Conserving Practices

Soil conservation means more than preventing erosion, checking runoff, and stopping the depletion of soil nutrients. It means using the land in such a way that it can be expected to produce indefinitely at a maximum level.

Modern soil conservation is based on the results of research and on farmer's experiences from all over the country. Conservation methods are being continually improved. The application of good soil-conserving practices results in increased acre production and decreased per-unit cost of production. Such practices not only result in an increase in farm profits, but also help lower the cost of food and clothing to people living in cities.

The tools of conservation are such practices as (1) cropping rotations suitable to the land capacity; (2) maintenance of grass waterways; (3) farming on the contour; (4) strip cropping; (5) terrace system construction and maintenance; (6) use of cover crops on otherwise bare soil surfaces; (7) mulch tillage; and (8) fertilization and liming, together with many others of a similar nature. Each of these practices helps protect or improve the land if used in the right place and in the right way. For the most part, several practices must be used together and in the right combination for best results.

Good Land Use

Good land use is a term frequently used. Essentially, it means using that rotation and producing those crops best suited to each specific soil condition or situation. The best rotation to be followed under a given soil condition will be influenced by the extent to which other soil-conserving practices are employed. Crop rotation is a systematic changing of crops to help prevent erosion and soil exhaustion. A tilled crop often is followed by a small grain and then by a grass or legume. The rotation may take three to five years to complete, depending on the soil characteristics and needs of the farmer. Table 13–2 shows the benefits of three- or four-year crop rotations in reducing runoff and minimizing soil loss. Coastal bermudagrass or tall fescue for two years followed by corn, or corn and then cotton, were effective in reducing runoff and soil erosion to very low levels during the sod crop years and in lowering the overall rotation average soil and water losses.

Grass Waterways

On rolling land—which includes most of our crop acreage—at least a portion of the water that falls as rainfall or snow must seek the lower levels through water channels. The steeper the slopes and the less permeable the soil surface, the greater the water loss and the more rapid its movement. Grass waterways are broad, gently sloping channels that carry water slowly and safely off the field. They are covered with thick grass sod, in order that water may move to lower levels, but at a decreased rate and with a minimum amount of soil loss. Although grass waterways occupy considerable space, the land is not wasted. Grass can be used for pasture or hay.

TABLE 13–2. Soil Loss and Runoff by Cropping Treatment in the Southern Piedmont

Cropping Treatment*	Slope (%)	Row† Direction	Rainfall (inches)	Runoff (inches)	Soil Loss (tons/acre)
Fallow	7	None	55.71	18.35	60.43
Continuous cotton	7	A	55.71	6.33	12.28
Continuous cotton	7	W	55.71	9.84	21.86
Continuous corn	7	A	55.71	4.54	4.02
Continuous corn	7	W	55.71	7.03	12.11
Continuous corn	11	W	55.71	2.40	20.90
Three-year rotation					
Coastal bermudagrass	7	None	54.78	2.42	1.99
Coastal bermudagrass	7	None	54.78	1.78	0.16
Corn	7	W	54.78	4.42	4.43
Rotation average	7	W	54.78	2.87	2.19
Three-year rotation					
Coastal bermudagrass	7	None	54.78	2.25	2.26
Coastal bermudagrass	7	None	54.78	2.74	0.19
Corn	7	A	54.78	2.99	1.08
Rotation average	7	A	54.78	2.66	1.17
Three-year rotation					
Tall fescue	7	None	56.73	5.42	0.88
Tall fescue	7	None	56.73	2.80	0.09
Corn	7	A	56.73	4.56	1.97
Rotation average	7	A	56.73	4.26	0.98
Four-year rotation					
Tall fescue	7	None	57.56	11.04	4.05
Tall fescue	7	None	57.56	5.07	0.05
Corn	7	A	57.56	6.04	2.93
Cotton	7	A	57.56	7.23	6.25
Rotation average	7	A	57.56	7.34	3.32

* Treatment years of record ranged from five to seven years.
† A, rows on contour across slope; W, rows uphill and downhill with slope.
  Source: Carreker, J. R., *et al.* Soil and Water Management Systems for Sloping Land, USDA-ARS-S-160 (1977).

Farming on the Contour

Farming land on the contour, including plowing, seedbed preparation, planting, and cultivation around the slope has been shown not only to reduce the per-acre cost of production, with a decrease in the power required and in machinery operation costs, but also to increase acre yields. This means a decrease in cost per unit of production. Planting row crops up and down slopes encourages serious erosion and gullying.

The effectiveness of contour farming, locating rows across the slope as compared to uphill and downhill, is shown in Table 13–2. Fields with a 7 percent slope cropped to continuous cotton lost almost twice as much soil, and fields cropped to continuous corn lost three times as much soil per acre when rows were not contoured. In addition, rainfall runoff was much greater when rows were uphill and downhill with the slope. When corn followed 'Coastal' bermudagrass in the three-year rotation, planting

FIGURE 13-4. *A permanent grass waterway in Washington. This broad (35 feet wide) gently-sloping channel can carry excess water slowly and safely off the field.* [SOURCE: USDA, Soil Conservation Service.]

on the contour reduced soil loss by four times, and the rotation average soil loss was about one-half. During the period 1936–1976, 139,849 acres (56,597 ha) were farmed on the contour through the Agricultural Conservation Program (Table (13–1).

Strip cropping

The strip-cropping system aids in maintaining and improving soil productivity by growing field crops in a systematic arrangement of strips or bands, alternated with strips of close-growing sod crops such as grass or clover. These sod strips serve as vegetative barriers to erosion and water runoff. When water spills over level furrows in a row-crop strip and starts to run downhill, it slows as it runs into the sod strip and drops the soil particles it is carrying.

Strip cropping also is used to control wind erosion. Strips are laid out at right angles to the prevailing wind. A taller-growing crop, such as sorghum, may be planted between strips of a shorter-growing crop, such as wheat, to serve as a windbreak. From 1936–1976, 114,535 acres (46,352 ha)

FIGURE 13–5. *Stripcropping on a farm in Iowa. The alternate strips of sod with a row crop serve as vegetative barriers to erosion and water runoff, and also are effective in controlling wind erosion.* [SOURCE: USDA, Soil Conservation Service.]

were strip-cropped through the Agricultural Conservation Program in this country (Table 13–1).

Terracing

In terracing, a ridge of soil is built on the contour across sloping fields to control the flow of water. The basic function of a terrace is to intercept water, which is either absorbed or conducted slowly from the field. This is the most effective mechanical practice for the control of soil and water; terraces break long slopes into short slopes. Graded terraces may reduce erosion by 75 percent and, in combination with crop rotation, mulching, and minimum tillage, may cut loss from cultivated land to practically nothing. Standard terraces that affected 34,355 acres (13,903 ha) of farmland were built during 1936–1976 in the United States through the Agricultural Conservation Program (Table 13–1).

Cover Crops

Cover crops tend to prevent soil erosion in two ways. While a crop is growing, the plant canopy tends to protect the soil from raindrop impact and splash and to retard runoff. Turning under the cover crop as green manure adds to the organic matter of the soil and increases its permeability, helping the water soak in more rapidly. Through the Agricultural Conservation Program, permanent vegetative cover was established on 63,373 acres (25,647 ha), additional vegetative cover on 26,779 acres (10,837 ha), and interim cover crops on 495,811 acres (200,655 ha) during the period 1936–1976.

Stubble-Mulch Farming

It has been shown that a mulch on the soil surface aids in soil conservation by (1) reducing water runoff; (2) reducing evaporation of water from the soil surface; and (3) increasing water intake by decreasing the puddling and surface sealing caused by raindrop splash.

The term *stubble-mulch* usually refers not only to stubble left on the soil surface, but also to small-grain straw, cornstalks, and sorghum residue, or any other material produced on the land and allowed to remain there for its mulching effect. The procedure is best suited to limited rainfall areas where moisture conservation and the prevention of wind erosion are particularly important. It usually has not given yields as high as those produced by plowing where the rainfall is more adequate and where the production of row crops such as corn predominates.

Several researchers have reported the benefits from mulch tillage. At Madison, Wisconsin, on a Miami silt loam with a 6 percent slope, shredded corn stalks on part of the surface reduced erosion 77 percent on third-year corn. At Coshocton, Ohio, on a Muskingum silt loam with 9–15 percent slope, planting corn on the contour with conventional tillage resulted in a soil loss of 7.8 tons per acre (17.5 MT/ha). Mulch-tilled plots lost only 0.03 ton per acre (0.07 MT/ha). In Indiana on a Russell silt loam, with a 5 percent slope, a chopped-hay mulch on a minimum-tilled surface reduced soil loss by 95 percent. The more complete the cover, the greater the reduction in erosion, runoff, and moisture evaporation.

The amount of residue needed to control erosion will vary with the soil and degree of slope. As a general rule, 3,000–6,000 pounds of residue per acre (3,360–6,720 kg/ha) will effectively control water erosion. Small grain, soybean, and sod residues are twice as effective as corn residues on a pound-for-pound basis. For each pound (0.45 kg) of corn produced, there is 1 pound of residue:

100 bushels (2,540 kg) of corn = 5,600 pounds (2,540 kg) of residue

For each bushel of small grain there are 100 pounds (45 kg) of residue. Soybean residue usually ranges from 1,500 to 2,500 pounds/acre (1,680–2,800 kg/ha).

Reservoirs

Many different types of storage structures may be constructed for flood control. Reservoirs also may serve for livestock water, irrigation, and fire protection. Such impoundments may be used for stocking fish as well. In the United States there are more than 2.3 million artificial ponds, reservoirs, earthen tanks, and related structures. In some cases the reservoirs will have been rendered essentially useless and may have become nuisances long before their capacities for water storage have been replaced by sediment.

Fertilizer Use and Soil Conservation

The maintenance of high fertility levels is a requisite of successful erosion control. A thick cover cushions the beating action of raindrops, offers resistance to moving water, and slows down its rate of runoff. The roots help to hold the soil in place and to improve the soil tilth, making the soil more porous and better able to absorb rainfall. The greatest protection, regardless of the kind of crop, is obtained when growth is vigorous and fast. Such growths are possible only on soils of high natural fertility or where adequate fertilizer is applied.

Conservation and Increased Earnings

The application of soil- and water-conserving practices brings an increase in net acre returns, with greater increases as the years pass. In an Illinois study, "high-conservation" farms in each of three groups of counties averaged $4.77, $6.86, and $6.41 more annual net earnings per acre for a 10-year period than comparable "low-conservation" farms.

Research at the Midwest-Claypan Soil Conservation Experiment Farm, Missouri, showed that reduction in crop yields per inch (2.5 cm) of surface soil lost through erosion was as shown in Table 13–3.

These losses indicate that in Missouri, in the places and under the conditions existing where the studies were made, on a claypan soil, a 100-acre (40-ha) farm with slight erosion became equal to only an 80-acre (32-ha) farm when it reached the stage of moderate erosion. It was equal to only 50 acres (20 ha) when it suffered severe erosion by loss of more than one-half the original surface.

It is axiomatic that any process that destroys the essential productivity of soil must ultimately destroy the industry and civilization that depends on that soil. Erosion is such a process—perhaps the most vicious. In extreme cases it leads to abandonment, rural migration, and general community disintegration.

TABLE 13–3. Crop-Yield Reduction Through Soil Erosion in Missouri

Crop	bu/acre	(kg/ha)
Oats in south Missouri	5	(179)
Corn in north Missouri	4	(251)
Soybeans on claypan soils		
Without fertilizer	1.5	(101)
With fertilizer	2.7	(181)

REFERENCES AND SUGGESTED READINGS

1. ANONYMOUS. How Close in Mandatory Soil Conservation? Successful Farm., p. 23 (March, 1978).
2. BENNETT, H. H. *Soil Conservation* (New York: McGraw-Hill, 1939).
3. CARREKER, J. R., S. R. WILKINSON, A. P. BARNETT, and J. E. BOX. *Soil and Water Management Systems for Sloping Land,* USDA-ARS-S-160 (1977).
4. CHEPIL, W. S., and N. P. WOODRUFF. "The Physics of Wind Erosion and Its Control," In *Advances in Agronomy,* vol. 15, (New York: Academic Press, 1963). ed. A. G. Norman.
5. COOK, R. L. *Soil Management for Conservation and Production,* (New York: Wiley, 1962).
6. FENSTER, C. R., H. I. OWENS, and R. H. FOLLETT. *Conservation Tillage for Wheat in The Great Plains,* USDA-Ext. Serv. PA-1190 (1977).
7. Fertilizer Institute. *Our Land and Its Care* (Washington D.C.: 1977).
8. HARLEY, A. D. *et al. Stubble Mulch for Moisture Conservation,* USDA Pro. Res. Rep. No. 6 (October, 1956).
9. HAYES, W. A. *Mulch Tillage in Modern Farming,* USDA Leaflet No. 554 (1971).
10. HOCKENSMITH, R. D., and J. G. STEELE. *Soil Erosion the Work of Uncontrolled Water,* USDA-SCS Agr. Info. Bull. 260 (1971).
11. McCALLA, T. M., and T. J. ARNY, "Stubble-Mulch Farming," In *Advances in Agronomy,* vol. 13 (New York: Academic Press, 1961). ed. A. G. Norman.
12. MILLER, M. F., and H. H. KRUSEKOPF, *Influence of Cropping Patterns on Soil Erosion,* Mo. Agr. Exp. Sta. Res. Bul. 177 (1932).
13. MOLDENHAUER, W. C., and E. R. DUNCAN. *Principles and Methods of Wind-Erosion Control in Iowa,* Iowa Agr. Exp. Sta. Spec. Rep. 62 (1969).
14. SPRABERY, J. A. *Summary of Reservoir Sediment Deposition,* USDA Misc. Pub. 964 (1965).
15. THORNE, D. W. "Agronomists and Food-Contributions," In *Agronomists and Food: Contributions and Challenges,* Spec. Pub. 30 (Madison, Wisc.: American Society of Agronomy, 1977).
16. *Time,* May 21, 1934.
17. USDA. Agr. Stat. (1977).
18. USDA-ARS, *Losses in Agriculture,* Agr. Handbook 291 (1965).
19. USDA. Office of Communication. *Fact Book of U.S. Agriculture,* Misc. Pub. 1063 (1976).
20. USDA-SCS. *Our American Land,* Agr. Info. Bul. 321 (1968).
21. USDA-SCS. *The Measure of Our Land,* Bul. PA-128 (1971).

CHAPTER 14

ROTATIONS AND CROPPING SYSTEMS

THE more or less regular changing of crops on a given area of land has been practiced by progressive farmers for many years. Growing different crops in a regular sequence has come to be called crop rotation.

In choosing a rotation for any given farm or field, its relative fertility, the erosion dangers, and the requirements of feed for livestock must all be considered. A specific rotation usually has one or more cultivated or row crops, at least one small grain crop, and a sod crop.

Crop rotation in modern times was started about 1730 in England. The well-known Norfolk four-year rotation consisted of turnips, barley, clover, and wheat. The famous Rothamsted Station in England conducted experiments including rotations that were continued for more than 100 years. In the United States, the Morrow Plots at the Illinois station, America's oldest Experiment Field, were established in 1876. Results with rotations on these plots remind us that with a good rotation and adequate soil treatment, we can reap the benefits of the soil and still keep it productive for our children and grandchildren. With proper care, some of our badly damaged soils can become productive again.

Advantages of Rotating Crops

Some of the advantages usually listed for rotating crops are:
1. Legumes in rotation can utilize atmospheric nitrogen and increase the nitrogen content of the soil.
2. Plant diseases, insects, and weeds can be more easily controlled.
3. The land can be kept more continuously in the soil-holding crop, in this way reducing soil erosion and nutrient losses by leaching.
4. Labor can be better distributed.
5. As a result of several of the preceding factors, greater acre yields may be obtained.

Soil Nitrogen and Organic Matter

Undisturbed virgin soil, when first brought under cultivation, often contains 5–6 percent organic matter. The amount of nitrogen and organic matter that a soil contains generally is regarded as one of the best indications of its relative productivity.

When soil is first cultivated there is a rapid breakdown and loss of organic matter; the rapidity of loss slows down with the years, until a relatively

FIGURE 14–1. *The Morrow Plots, established in 1876 at the University of Illinois, are America's oldest Experiment Field. Results from more than 100 years have demonstrated the beneficial effect of crop rotation on corn yield and soil improvement.* [SOURCE: *Illinois Research,* University of Illinois Agricultural Experiment Station.]

low level is reached which apparently can be maintained more or less indefinitely. With relatively constant cropping it is impossible to maintain anything like the organic matter content of virgin soils, regardless of the amount of organic matter plowed under.

Crop Rotations and Yield

The continuous or too frequent growing of row crops results in the rapid breakdown of organic matter and leaves the soil bare and exposed to the erosion action of runoff water much of the year. The loss of organic matter also reduces the water-absorbing and water-holding capacity of the soil.

The intensity of land use is governed by the physical characteristics inherent in the land. The degree of slope and porosity of the soil and the previous erosion and relative productivity are important factors governing intensity of land use. The proportion of the acreage to be planted in soil-depleting crops usually is decreased and the soil-conserving crop acres increased on the less productive soil as the depth of the topsoil decreases.

In general, a rotation with 50 percent of the land in an intertilled crop is recommended only for the more productive soils that are level and not subject to erosion. The intensity of land use may be increased somewhat when crops are contour planted, when the land is terraced, or when other soil-conserving practices are applied.

The influence of crop rotations on crop yields has been reported from

most of the experiment stations. The student of agriculture will give particular attention to the results reported from rotations under conditions comparable to those of his own area.

Continuous Corn

On certain of the more level, nonerosive soils of the Corn Belt the number
of years in continuous corn can be increased considerably without materially decreasing productivity. In continuous corn the methods of production
must include such cultural practices as maintenance of high fertility levels
by adequate fertilization, a high population rate adequate to utilize the
increased fertility available, and methods that will return the maximum
amount of crop residues to the soil. Adequate fertilization and the use of
suitable herbicides and insecticides for the control of weeds and insects
has in many areas made possible the continuous production of a given
crop year after year on the same land. Iowa research has shown that continuous corn (corn following corn each year for five or more years) appears
to be feasible on land where erosion is not a problem or where erosion
can be controlled by terracing and contouring. In South Dakota, plots
planted in continuous corn for eight years and receiving a complete fertilizer yielded the same as plots in fertilized corn following alfalfa.

Pest Control

Crop rotation aids in controlling many plant insects and diseases and
in keeping fields free of weeds.

INSECTS. Many insects are destructive to only one kind of crop. The
life cycle is broken when other crops are grown. Crop rotation often helps
reduce crop injury, particularly by insects with restricted food habits.
White grubs, the larvae of June beetles, feed on roots of crops of the grass
family and injure forage grasses and grain crops planted on land that
has been in sod. But legume crops are unfavorable to their development.
The proper use of legumes in the rotation or in combination with grasses
in pastures greatly reduces white grub injuries. The corn rootworm often
becomes abundant in fields that are planted to corn for two or three consecutive years. This insect is restricted in food-plant habits, however, and
can be eliminated as a serious factor by suitable crop rotations. Cotton
root rot and the sugarbeet nematode are controlled by rotation.

DISEASES. Certain parasites remain from season to season in the soil,
living on plant refuse from the previous crops or coming from other
sources; when susceptible crops are grown every year, the parasites tend
to accumulate to a point which makes production unprofitable. For these
parasites crop rotation is a good control measure.

Some soil-inhabiting parasites are exceedingly refractory to rotation and
sanitary measures—for example, black rot, which attacks sweet potatoes;
the flax wilt parasite; scabs of cereals; alfalfa wilt; and Texas root rot.

Rotation does not help the cereal rusts, mildews, most of the cereal smuts, and potato late blight. In all of these the spores of the parasites are carried by the wind from infested to noninfested fields. A list of plant diseases that are controlled entirely or partially by crop rotation has been compiled by Leighty.

WEEDS. Not all weeds can be controlled by crop rotation. However, weed problems are likely to be least severe on farms where crop rotation is practiced and most severe on single-crop farms.

Of about 1,200 species of plants commonly called weeds in the United States, fewer than 30 are sufficiently aggressive to be able to survive indefinitely on crop-rotated land. These are the noxious species, plants having such tenacity that no ordinary good farming measures control them. They include Canada thistle, quackgrass, bindweed or wild morning glory, johnsongrass, hoary cress, nutsedge, leafy spurge, and wild onion. All of these are perennials with spreading or creeping root systems or with underground parts, such as bulbs, that are uninjured by tillage. A few annuals, such as wild oats and crabgrass, are so difficult to control that they are classed as noxious. Most weeds are annual plants, unable to sprout from their roots and dependent upon their seeds for reproduction. Cutting them off close to the ground before the seeds ripen destroys them. In a rotation of cultivated crop, grain crop, and meadow, several opportunities occur to do this.

Among the common crops alfalfa is the one best suited to compete with noxious weeds. The frequent cutting and heavy growth of alfalfa prevent weeds from going to seed and reduce their vegetative vigor. The weeds may not be killed by the alfalfa, but they usually are subdued sufficiently so that other crops can be grown for several years without too much interference. On severely infested land the weeds may be weakened by a month or two of fallow before alfalfa is planted. The list of other good competitive crops is limited. Winter rye and winter wheat often are recommended for bindweed because they provide intervals in midsummer for fallowing and make their own growth during the season when the weed is inactive. Sorghum, soybeans, and sudangrass have been successful on infested land, although pretillage, especially in the drier areas, often is required for best results.

Crop Sequence

Several stations have observed that differences in the available soil nutrients following the growing of various crops were not sufficient alone to account for the yield differences of succeeding crops. An early study in Ohio reported the effect of previous crops on wheat yields. Whether wheat followed soybean, corn, oats, clover, or potatoes made a considerable difference in yield obtained (Table 14–1). These studies also showed that the order or sequence of crops preceding wheat influenced wheat yield.

TABLE 14-1. The Effect of the Previous Crop on Wheat Yield (Ohio)

Previous Crop	No. of Crops Averaged	Yield of Wheat (in bushels)
Soybeans	66	33.33
Corn	124	34.31
Oats	33	37.34
Clover	33	39.11
Potatoes	62	39.65

Injurious vs. Beneficial Influence of One Crop on Another

It has been observed generally through many years that soils become less productive when a given crop is grown continuously, but it also has been observed that the crop immediately preceding may have either an injurious or a beneficial effect on crops that immediately follow. At the Rhode Island station 16 different crops were grown side by side for two successive seasons, and then every third year one of these 16 crops was grown over the entire area. The yield of onions, for example, ranged from 13 to 412 bushels per acre (728 to 23,072 kg/ha) a 27-fold variation, depending on the preceding crop. The highest yields were obtained when onions followed grasses such as redtop and timothy. Among other things, soil acidity was affected by different crops; yields of crops sensitive to acid conditions could be affected greatly. A later study at Rhode Island showed that crop yields were influenced by 50 percent or more by the preceding crop. These researchers concluded that available nutrients remaining after a given crop could have an important influence on the succeeding crops.

The idea that the stubble and roots of certain plants and their breakdown are injurious to certain crops that follow has developed from the research of several workers. A notable example of such an influence is provided by winter wheat following sorghum. It has been found that sorghum roots may contain 15 times as much sugar as the roots of corn. When this is liberated into the soil after the sorghum crop is harvested, it stimulates the growth of certain microorganisms which compete with the wheat for nitrogen and probably for other elements.

Some of the explanations offered by different investigators for the effects of certain crops on succeeding ones are:

1. Removal of large quantities of nutrients tends to deplete the soil for these nutrients.
2. Removal of large amounts of elements that are chemically basic, with lesser amounts removed that are acidic, creates in this way a greater degree of soil acidity.
3. Large amounts of residues relatively high in carbohydrates and low in nitrogen (high C/N ratio), the decomposition of which by microor-

ganisms requires soil nitrogen, decreases the amount of nitrogen available to growing plants.

4. Excretion of organic toxins from the roots, or the decomposition of plant residues, may leave organic toxins in the soil temporarily or permanently.

At the Texas Station, Conrad and Holt studied a two-year grain sorghum–guar rotation. Guar yields were reduced following a grain sorghum stubble regrowth system. Available moisture for the guar crop was the main limiting factor in this rotation system.

Amount of Sod Crop in the Rotation

Leffler concluded from a study of Iowa, Indiana, and Ohio yields that increasing the proportion of sod in the rotation did not by any means give a corresponding corn yield increase. He concluded that on reasonably level land where erosion is not a problem, a corn–clover ratio somewhere around 1 : 0.5 might be expected to give satisfactory corn yields and still allow for a large total corn output for the farm. Where erosion is more of a problem, a greater proportion of sod certainly is in order, with a corn to clover ratio of perhaps 1 : 1. From its data bearing on this point the Indiana station reported:

Comparing the last period with the first, it appears that crop yields can be economically maintained regardless of the rotation, providing it contains a legume at least once in 5 years and the crop is reasonably fertilized. This allows a wide range in selection on the part of the individual farmer, permitting him to choose the rotation best suited to his soil and in keeping with the feed requirements in his livestock enterprise.

A rather markedly lower yield of second-year corn as compared with first-year corn in the same rotation generally has been observed. To reduce the proportion of the acreage in corn in an area where corn is particularly well suited reduced the total feed produced for unit of land. There is the possibility of a rearrangement in the crop sequence in some rotations, which would make possible fairly high corn acreages without having corn follow a preceding corn crop.

Considering the yield results from 16 different rotations under study at the Iowa, Indiana, and Ohio stations, in which two corn crops were grown in succession, the decreases in yield for the second corn crop in six Iowa rotations were 13.6, 14, 16.6, 21, 22.8 and 15.2 percent, respectively. In six Indiana rotations, the decreases were 10.1, 12.2, 8.9, 5.8, 10.1, and 5.7; and in four Ohio rotations, the decreases for second-year corn were 11, 15.6, 4.7, and 12.9 percent.

Turner and his co-workers studied the effects of rotations and nitrogen application on cotton production in the San Joaquin valley of California. Increased cotton yields were obtained following alfalfa and corn, as compared to continuous cotton plots (Table 14–2). No yield advantage was noted between cotton following alfalfa as compared to cotton following corn. Nitrogen fertilizer increased the yield of continuous cotton and of cotton following corn. There was an advantage of nitrogen on the second year

TABLE 14–2. Effect of Rotation and N Treatments on Yield of
Seed Cotton for an Eight-Year Period (1960–1967) in California

Treatments		Yield (8-yr. avg.) kg/ha
Continuous cotton	no N*	3,002
	N	3,434
Corn–cotton rotation	no N	3,371
	N	3,847
Alfalfa–cotton (1st-year cotton)	no N	3,726
	N	3,774
Alfalfa–cotton (2nd-year cotton)	no N	3,603
	N	3,809

* The N treatment was 89.7 kg/ha N applied each year.
Source: Turner, J. H., *et al.* Influence of Certain Rotations Upon Cotton Production in the San Joaquin Valley, *Agron. Jour.,* 64:543–546 (1972).

of cotton following alfalfa, but not in the first year following alfalfa. In addition, Verticillium wilt severity was decreased by crop rotation.

Tucker, Cox, and Eck grew winter wheat continuously and in rotation with alfalfa in north central Oklahoma. In addition, they evaluated the influence of stubble mulch tillage as compared to clean tillage, and of nitrogen fertilizer. The principal effect of alfalfa on wheat yield was to supply nitrogen (Table 14–3). Wheat following alfalfa yielded better than continuous wheat with no nitrogen, but continuous wheat fertilized with 45 kg of N/ha yielded as well as wheat in rotation (three years wheat following three years alfalfa). Clean tillage gave an advantage over stubble

TABLE 14–3. Effects of Cropping Systems, Tillage, and Nitrogen
Treatments on Wheat Yields, 1961–1966, North Central Oklahoma

Treatment			Grain Yield (kg/ha)
Cropping System	Tillage†	Nitrogen	
Continuous wheat	SM	0	1,520
	SM	45	2,130
	CT	0	1,810
	CT	45	2,230
Alfalfa–wheat rotation*			
Wheat (1 yr)	SM		1,780
Wheat (1 yr)	CT		1,760
Wheat (2 yr)	SM		2,280
Wheat (2 yr)	CT		1,980
Wheat (3 yr)	SM		2,150
Wheat (3 yr)	CT		2,180

* All wheat crops in rotation followed three years of alfalfa. Yields are six-year averages. Each treatment occurred each year.
† SM, stubble-mulch tillage; CT, clean tillage with moldboard plow.
Source: Tucker, B. B., *et al.* "Effect of Rotations, Tillage Methods, and N Fertilization on Winter Wheat Production," *Agron. Jour.,* 63:699–702 (1971).

mulching in these studies. These researchers concluded that there was little justification for an alfalfa–wheat system under their conditions as long as nitrogen fertilizer prices were reasonable enough to allow fertilization at optimum rates.

Carreker and other USDA workers recently (1977) summarized results from 20 years of research on the effect of cropping systems on productivity for the southern Piedmont regions. Among their conclusions were the following:

1. Sod crops improve soil productivity. Corn yields were higher from corn after vetch, rye, alfalfa, and grasses than corn after corn. This increase in productivity lasted two to four years after the sod was plowed under.
2. There was no advantage in applying N to corn after vetch or alfalfa. The fixed nitrogen gave yield responses comparable or superior to yields obtained from 80–120 pounds of N per acre (90–134 kg/ha) per year under the cultural practices employed.
3. Yield of corn following tall fescue was highest the first year after sod, and declined each succeeding year of corn. Yield of corn following tall fescue increased with age of the fescue through two to three years.

Soil Productivity Balance

Based on the loss of soil nitrogen and on other results from long-time rotation experiments, the Ohio station published soil productivity indices to be applied to the different crops in a rotation, as a measure of the effectiveness of a specific rotation and its management in maintaining high productivity (Table 14–4). This was followed by other indices from both Ohio and Missouri.

To apply the Ohio Productivity indices to a rotation such as C–O–W–Cl–T, with the cornstalks and the oat and wheat straw returned to the

TABLE 14–4. Soil-Productivity Indices for Crops and Crop Treatments (Ohio)

Crops and Crop Treatment	Productivity Effects
Corn, as grain or silage	−2.0
Potatoes, tobacco, and sugarbeets	−2.0
Oats, wheat, barley, and rye and buckwheat	−1.0
Credit for crop residues, as cornstalks or small grain straw	+0.25
Alfalfa, to the end of first hay year	+2.5
Alfalfa, to the end of second hay year	+0.5
Timothy and other grass sod	+0.25
Clover—timothy mixed, hay or pasture	+1.25
Clovers, hay or pasture	+2.0
Sweetclover, plowed April or May, second year	+2.5
Manure applied, per ton	+0.15
Average commercial fertilizer, each 200 pounds	+0.15

land, the result would be −1, indicating that additional soil-improving practices would need to be applied to maintain productivity.

Competitive and Complementary Relationships

The problem in attempting to determine the best rotation for a given farm is to use the equipment, livestock, labor, management, and land that will yield the greatest profit from the farm as a whole over a period of years. The most desirable rotations often are closely tied into the livestock program.

The proportionate acreage to be devoted to forages has had special attention in an Iowa study. Heady and Jensen have suggested that the economic role of forage crops is best understood when viewed from the standpoint of their *competitive* and *complementary relationships.*

Different crops are considered complementary when the production of one increases the quantity of the other from a given use of labor, land, and equipment. Forages are complementary to grain crops when an increased acreage of forage in the location has such an effect on soil productivity and other beneficial effects as to result in an increased total production of grain from a given farm or area of land. The percentage increase in *yield per acre* of the grain must be greater than the percentage decrease in *number of acres* of grain as land is shifted from grain to forage. Forage crops are considered competitive when the total output of the nonforage crop is reduced.

Sod crops are always competitive with grain crops in any single year because the acreage planted to forage reduces proportionately the acreage and total production of nonforages. Forages become complementary to other crops only over time. Any increase in production of grain results from the added nitrogen, improved soil tilth, or other contributions of grasses and legumes. Farm profits usually do not become maximum over time if the acreage of forages is not extended to the entire range in which they are complementary to other crops.

Legumes Established in Wide-Row Planted Corn

In corn rotations the legume-grass crop usually has been established in one of the small grains. The return from the year of small grain usually has been considerably less than would have been obtained had the land been in corn. For this reason there have been many attempts to work out rotation procedures by which the legume-grass might be established without including a year of small grain. One method tested widely in the 1940s and 1950s involved planting corn in wide rows, 60–72 inches (152–183 cm) apart, with the grass-legume seeded in the middles. This attempt to use corn as a companion crop for interseeded legumes and grasses proved unsatisfactory in the Corn Belt, and now is rarely seen. Some of the problems encountered included yield losses, the need for special equipment, weed strips over old corn rows, and uncertainty about obtaining stands. Yields generally were 10–20 percent lower.

Commercial Nitrogen vs. Legumes

With the prospect of an abundant supply of relatively cheap nitrogen fertilizer, the general elimination of legume crops from rotations has been suggested when the legumes are grown only for their soil-improving qualities and are not to be used for livestock feeding.

It has been pointed out that grasses give a much heavier, tougher sod and a larger tonnage of roots per acre. These grass roots result in a better crumb structure in the soil, with larger and more enduring soil aggregates, thus affording better resistance to erosion. Successful stands of grass are more readily obtained than of some legumes and often as a lower cost. In addition, grasses can be established in the late summer and fall, whereas many legumes cannot. In this way grasses make possible fall and winter cover which may greatly aid in the prevention of soil losses by erosion and by leaching. Grasses generally will grow under more adverse conditions, such as in acid soil and poorly drained soil.

In Iowa good yields were obtained without adding nitrogen fertilizer when the corn followed a legume meadow. The amount of nitrogen a legume furnishes will vary with the soil, the season, the fertilizer treatments, and the legume species grown.

Multiple Cropping Systems

Multiple cropping refers to growing and harvesting two or more crops from the same land in one year. This cropping system is practiced in many parts of the world. It is common in tropical Asia, tropical Africa, the Middle East, tropical America, and in the United States. In Southeast Asia and in the Philippines, for example, two crops of rice can be produced on the same field in one year. In the lower Midwest, southeastern, and southwestern United States, such crops as soybeans or grain sorghum can be planted after small grain (normally wheat or barley) harvest in May or early June. The principal multiple cropping systems are:

1. *Sequential Cropping:* Growing crops in sequence, one following the other. This includes double cropping, triple cropping, quadruple cropping, and ratoon cropping.
2. *Intercropping:* Producing two or more crops simultaneously on the same field. Crops used in this system may be *mixed,* with no row patterns, or in *rows* or *strips. Relay* cropping refers to a system in which part of the life cycle of one crop overlaps that of the other.

Sequential Cropping. In the Midwest, double cropping, a sequential system, may mean producing two crops in one year or else three crops in two years on the same acreage. An example is winter wheat planted in the fall, harvested in late spring or early summer, and immediately planted to soybeans, corn, or grain sorghum. The field can be planted to

FIGURE 14–2. *Double cropping, a sequential cropping system, with soybeans following winter wheat in southern Illinois. Soybeans were planted by a no-tillage method in the stubble immediately after wheat harvest. Wheat and soybeans are the two most common species for double cropping in many areas of the lower Midwest and Southeast.* [SOURCE: Southern Illinois University.]

wheat again after harvest of the previous crop, or to corn or soybeans the following year. Other types of double cropping in the Midwest include winter barley or rye followed by soybeans, corn, or grain sorghum and hay or pasture harvested early in the season prior to planting the field in a row crop.

In some regions of the United States, double cropping has been practiced for 15 years or longer. Moisture often is the most critical factor. Wheat straw or some other stubble or residue on the field preserves moisture and prevents runoff and erosion if a heavy rain does occur. Planting by zero-till or no-till methods, with a minimum of seedbed preparation and soil disturbance, has enhanced the advantages in some areas. Planting soybeans or other double crops in narrow rows, 15 inches (38 cm) or less, has proved more suitable than in wider rows. Some farmers have had success double cropping by air, seeding soybeans from an airplane following a wheat crop. After soybeans, wheat can be seeded by air just before soybeans begin to drop their leaves. Herbicide and fertilizer applications also can be made by airplane. Table 14–5 shows the yield needed, at variable prices, to break even for three second crops following wheat.

An experimental double cropping system for corn grain or silage and

FIGURE 14-3. *Double-cropped sunflowers following winter wheat. Another double cropping system may involve grain sorghum following winter wheat, barley, or rye.* [SOURCE: Southern Illinois University.]

TABLE 14-5. Estimated Breakeven Yields Per Acre to Cover Added Costs of Double Cropping

Variable Costs Per Acre	Soybeans After Wheat			Grain Sorghum After Wheat			Corn After Wheat		
Added costs									
Seed	$12.00			$ 3.00			$ 7.00		
Herbicides	21.00			15.00			15.00		
Fertilizer	7.00			18.00			18.00		
Custom spray	2.00			2.00			2.00		
Till planting	5.00			5.00			5.00		
Harvesting	10.00			15.00			15.00		
Conditioning	—			5.00			4.00		
Labor (@ $3.00)	7.50			9.00			9.00		
Total	$64.50			$72.00			$75.00		
Breakeven yields									
Prices per unit	$3.00	5.00	7.00	2.20	3.20	4.20	1.40	2.00	2.60
Yield needed to break even (bu or cwt)	22	13	9	33	23	17	54	38	29

Source: Hoeft, R. G. *et al. Double Cropping in Illinois,* Ill. Coop. Ext. Serv. Circ. 1106 (1975).

hay production using no-till methods has been reported in West Virginia and southern Illinois. Sublethal rates of herbicides or growth retardants were used to retard cool-season pasture grasses, or to kill the aboveground parts but allow regrowth, but not kill the sod completely. Several treatments produced good results, allowing a good corn grain or silage crop, as well as hay crops or pasture before corn planting and after corn harvest.

Researchers at the Georgia Station investigated several triple cropping systems, some of which were successful. The first crop was barley, second crops were short-season field corn, sweet corn, or grain sorghum, and third crops were soybeans, snapbeans, green peas, corn, field peas, or sorghum regrowth.

INTERCROPPING. This was once a common American farming method. The American Indian frequently planted beans with corn in hills, and pumpkins and squashes between hills. Intercropping is common in many areas of the world. Crops used in intercropping are corn and dry beans in Latin America, dryland rice and corn in the Philippines, sorghum and cowpeas in Africa, sorghum with lentils or chickpeas in India, and corn and soybeans in China.

Only 60–80 percent of the land is required to equal the production of monocropping systems. Crops selected for intercropping should have widely differing environmental requirements, and possibly different growth habits, such as rooting differences. Research in Canada and in South Dakota, Oklahoma, and other locations has shown increases in yield, better insect control, or other benefits from intercropping corn and soy-

FIGURE 14–4. *Intercropped cotton in a young, irrigated pecan orchard in New Mexico.* [SOURCE: Southern Illinois University.]

beans, grain sorghum and cotton, corn and sugarbeets, and various other combinations.

Relay cropping—planting a second crop in the field before the first crop is harvested—may extend double cropping to regions not previously suited because of a limited growing season. Benefits in addition to getting a head start are savings in time and in labor costs. Seed of the second crop may be flown on by airplane, or may be planted with no-till ground equipment.

Choosing a Rotation

No one final or long-time land-use system can be recommended. Any cropping system may become obsolete because of relative changes in prices and cost of producing different crops, introduction of new crops or new tillage methods, insects, diseases, weeds, or accumulative effects of any one treatment. To take the appropriate action as problems arise, one needs a thorough understanding of how various cropping treatments affect crop yield and soil properties.

REFERENCES AND SUGGESTED READINGS

1. ALDRICH, S. R., W. O. SCOTT, and E. R. LENG. *Modern Corn Production,* 2nd ed. (Champaign, Ill.: A & L Publications, 1975).
2. BAKER, W. A., and O. R. MATHEWS. "Good Farming Helps Control Insects," In *Insects: Yearbook of Agriculture* (Washington, D.C.: USDA, 1952). ed. Alfred Stefferud.
3. BENNETT, O. L., E. L. MATHIAS, and C. B. SPEROW. "Double Cropping for Hay and No-Tillage Corn Production as Affected by Sod Species with Rates of Atrazine and Nitrogen," *Agron. Jour.,* 251:250–254 (1976).
4. CARREKER, J. R., S. R. WILKINSON, A. P. BARNETT, and J. E. BOX. *Soil and Water Management Systems for Sloping Land,* USDA-ARS-S-160 (July, 1977).
5. CONRAD, B. E., and E. C. HOLT. "Influence of Post Harvest Residue Management and Fertilization on Crop Yield," *Agron. Jour.,* 62:549–551 (1970).
6. CROOKSTON, R. K. "Intercropping, a New Version of an Old Idea," *Crops and Soils,* 29(9):7–9 (1976).
7. ELKINS, D. M., J. W. VANDEVENTER, G. KAPUSTA, and M. R. ANDERSON. "No-Tillage Maize Production in Chemically Suppressed Grass Sod," *Agron. Jour.,* 71:101–105 (1979).
8. HARTWELL, B. L., and S. C. DAMON. *Influence of Crop Plants on Those Which Follow,* R.I. Agr. Exp. Sta. Bul. 210 (1917).
9. HEADY, E. O., and H. R. JENSEN. *The Economics of Crop Rotations and Land Use,* Iowa Agr. Exp. Sta. Res. Bul. 383 (1951).
10. HOEFT, R. G. *et al. Double Cropping in Illinois,* Ill. Coop. Ext. Serv. Circ. 1106 (1975).
11. LEFFLER, A. R. *Some Physical and Economic Considerations in Studies of Cropping Systems* (Ames: Iowa State Univ. mimeographed).
12. MARTIN, J. H., W. H. LEONARD, and D. L. STAMP. *Principles of Field Crop Production,* 3rd ed. (New York: Macmillan, 1976).

13. ODLAND, T. E., and J. B. SMITH. "The Effect of Certain Crops on Succeeding Crops," *Jour. Amer. Soc. Agron.,* 25:9 (1933).

14. PAPENDICK, R. I., P. A. SANCHEZ, and G. B. TRIPLETT (Eds.). *Multiple Cropping, Amer. Soc. of Agron.* Spec. Pub. No. 27 (Madison, Wisc.: American Society of Agronomy, 1976).

15. SHRADER, W. D., *et al.* "Crop Rotation—Facts and Fictions," *Iowa Farm Science,* 16:9 (1962).

16. SHUBECK, F. E., and B. E. LAWRENSEN. "Continuous Corn Is Fine But Let's Not Forget Rotations," *S. D. Research,* 20:3 (1969).

17. *The Relation of Crop Rotation to Yield of Wheat,* Ohio Agr. Exp. Sta. Bul. 402, 45th Ann. Rep. (1927).

18. TRIPLETT, G. B., JR., J. BEUERLEIN, and M. KROETZ. "Relay Cropping Not Reliable," *Crops and Soils* 29(2):8–10 (1976).

19. TUCKER, B. B., M. B. COX, and H. V. ECK. "Effect of Rotations, Tillage Methods, and N Fertilization on Winter Wheat Production," *Agron. Jour.,* 63:699–702 (1971).

20. TURNER, J. H., E. G. SMITH, R. H. GARBER, W. A. WILLIAMS, and H. YAMADA. "Influence of Certain Rotations upon Cotton Production in the San Joaquin Valley," *Agron. Jour.,* 64:543 (1972).

21. USDA-SCS. *Our American Land,* Agric. Info. Bul. 321 (1968).

Chapter 15
CROP SEED

Thou shalt not sow thy fields with mingled seed.

Leviticus 19:19

ONE kernel of wheat under favorable conditions may produce a plant with several hundred to a thousand or more seed. Because each seed is the result of self-fertilization, each in turn is capable of producing a plant with the exact characteristics of the parent. A corn plant often will produce 1,000–2,000 seeds. Because each kernel in open pollinated corn may have resulted from the fertilization of a female flower by a pollen grain from a different parent plant, the plants these seeds produce will be exceedingly variable.

Perhaps the most minute crop seeds are those of the tobacco plant—7 million of them in a pound (0.4 kg). A single tobacco plant has been known to produce 1 million seeds.

What Is A Seed?

A seed consists of a plant embryo and stored food materials surrounded by a seed coat. Seeds may vary greatly in all external characters and also in the internal structure. Seeds of the cereals and grass forages consist of the embryo, or germ, the endosperm, and the pericarp. Seeds of the legumes contain no endosperm, the bulk of the seed being made up of the two cotyledons. In these seeds the cotyledons constitute the food storage portion. The embryo of legume seed is found between and at one edge of the two cotyledons.

Each part of the seed has distinct functions to perform. The embryo of the grass seed is made up of the epicotyl, the hypocotyl, and the scutellum, the single cotyledon of monocots. The epicotyl gives rise to the shoot and the hypocotyl produces the roots. The radicle or primary, root, generally the first major structure to emerge during seed germination, may be temporary, as is generally true of the cereal grains, or may persist, as is true of the legumes. In the process of germination of the grasses the scutellum, which is located next to the endosperm, serves to dissolve the stored food of the endosperm by the secretion of enzymes; it also assists in the transfer of food to the growing points.

Most seeds become dormant on maturity. The duration of this dormancy varies greatly with different species. It ranges from a few days to months or even years. Dormant seeds have remarkable endurance and can survive conditions which would quickly kill them if they were active.

Seed Dormancy

Dormancy, an inactive condition, may be the result of either unfavorable environment or of internally imposed blocks, mechanisms that restrict germination. Some seeds must undergo a period of afterripening to remove the germination blocks.

Dormancy is an important survival mechanism with many wild species and weed seeds. These blocks can spread germination over a period of years so that one unfavorable season does not wipe out a species. Seed dormancy in domesticated species may present problems to the seedsman, but may be desirable in some instances. For example, dormancy prevents preharvest germination of winter cereals and thus maintains quality.

Several physical or physiological factors may result in seed dormancy. Blocks may be physical or chemical. Among the more important causes of dormancy are the following.

IMPERMEABLE SEED COAT. Impermeability to water (hard seed) is common among species in the legume and mallow families, as well as for many tree and shrub species, and impermeability to gases is common in the grass and composite families.

EMBRYO DORMANCY. Even though morphological growth of the embryo is complete, the embryo may be physiologically immature at harvest in some species. Such dormancy may last only a few days or for several years, and is prevalent in the woody and grass species. *Stratification,* a low-temperature pretreatment, often is effective in breaking dormancy and allowing germination.

Forward noted a marked effect from a prechilling treatment on oat germination (Table 15–1). Five days of low-temperature treatment, followed by higher-temperature germination, resulted in a germination increase of more than 35 percent.

Alternating temperatures (diurnal cycles) and certain light treatments also may be effective in overcoming embryo dormancy.

TABLE 15–1. Effect of Prechilling on Oat Grain Germination

Germination* (%)	
Tested at 20°C for 10 days	Prechilled at 10°C for 5 Days, Then 20°C for 5 Days
59.7	95.8

* Average of 10 tests.

Source: Forward, B. F. "Studies of Germination in Oats," *Proc. ISTA,* 23:20 (1958).

GERMINATION INHIBITORS. There are more than 120 known germination inhibitors present in certain fruits, seedcoats, and other membranes surrounding the seed. Any substance that interferes with metabolic succession of events may inhibit germination. Sugarbeet seed contains at least 10 inhibitory substances, and Lehmann lovegrass has 32 such compounds. In addition, inhibitors have been found in seeds of oats, rice, barley, and red and white wheats.

Grains have a dormant period following combining, but good stands can be obtained in the field when such seeds are planted. Such seeds that are dormant at harvest often lose their dormancy gradually. Nakamura observed that seeds of 15 grass species were able to germinate under a wide range of temperature conditions as seed age increased. For example, seeds of millet were dormant immediately after harvest, but the dormancy had almost disappeared a few months later.

Hard Seed

Seed of many legumes have an enforced resting period, generally attributed to an impervious seed coat which prevents the penetration of water. In nature a very large percentage of the seeds of some legumes are hard, whereas commercial seed of the same species may show a relatively small percentage of hard seed. This difference is attributed to the scarifying effect during harvesting or to the use of scarifying equipment during processing for sale.

The hard-seed character found in legume seed represents an adaptation in nature to ensure the perpetuation of these plants. For example, red clover seeds ripening in the late summer, if capable of absorbing moisture with the first rain, would germinate while still in the head, only to shrivel and die in a few days.

An impermeable seedcoat is common in alfalfa, sweetclover, true clovers, crownvetch, and sericea lezpedeza. Such impermeability may be desirable in certain *reseeding* forage legumes, which gives a greater probability that some seeds will germinate during conditions favorable for establishment.

Heavy deposits of suberin, lignin, or cutin are responsible for hard seedcoats of many legume seeds; in the mallow or cotton family, pectic substances are responsible. The hard-seeded germination block sometimes is removed naturally by alternate freezing or thawing, wetting and drying, action of soil acidity, or attack by microorganisms. Dexter observed that alfalfa, containing a large number of hard seeds, had a relatively low germination percentage in the laboratory but naturally occurring processes allowed good field emergence (Table 15–2).

DURATION OF HARD CHARACTER IN LEGUME SEED. Impermeable seed may retain their vitality for many years. Munn of the New York station kept hard legume seed submerged through a period of years in water which was changed from time to time. Alfalfa and sweetclover seed germinated gradually through a five-year period. Many hard seeds of red and alsike

TABLE 15–2. Laboratory vs. Field Germination of Hard-Seeded Alfalfa, Planted in August

Seed Lot	Laboratory Germination (%)		Field Emergence (%)			
	Hard Seed	Quick Germ.	Sept. 17	Oct. 1	Oct. 23	Total
Low in hard seed	7	85	77.1	3.4	1.5	82.0
High in hard seed	45	45	74.3	5.1	1.7	82.1

Source: Dexter, S. T. "Alfalfa Seedling Emergence from Seed Lots Varying in Origin and Hard Seed Content," *Agron. Jour.,* 47:359 (1955).

clover remained viable in the water to the end of 25 years. When the seed coats were punctured these seeds germinated with normal vigor. It is pointed out that hard seeds sometimes result in a stand after the plants from the prompt germinating seeds have failed.

SCARIFICATION. It has been found that scratching the seed coats of hard seeds makes it possible for the seed to germinate promptly. Scratching, chipping or dissolving a portion of the seed coat usually is referred to as *scarification.* Scratching the seed allows more or less free entrance of water and gases. When legume seeds are to be stored for more than one year it is best to delay scarification until a short time before using.

Mechanical scarification machines use a rotating, tumbling, or flailing action, which rubs the seed together and against an abrasive surface. Industry uses mostly mechanical methods. Excessive scarification may damage the seed.

Chemical scarification may be accomplished with sulfuric or hydrochloric acids, sodium hydroxide, acetone, and alcohols. Concentrated sulfuric acid is the most widely used chemical method. Treatment for 10 minutes to one hour is common. The acid method is used most extensively on cotton seed.

Radiofrequency radiation has been developed for scarification and is now available commercially. Nelson et al. found that two electrical treatments, radiofrequency electric fields and gas-plasma radiation, increased germination of alfalfa containing large quantities of hard seeds. The gas-plasma method, but not the radiofrequency method, increased emergence from sand in greenhouse benches.

Germination of Seeds

The necessary conditions for seed germination are a viable seed, moisture, oxygen, and proper temperature. Some need stimulation from light.

FIGURE 15–1. *Huller-scarifiers prepare seeds of small-seeded legumes by removing the pods and scratching (scarifying) the seed coat to eliminate "hard" seeds.* [Photograph with permission of Seed Technology Laboratory, Mississippi State University.]

Moisture and Oxygen

In most cases seed germination is controlled by the amount of moisture available. Water is essential for enzyme activation, which permits the breakdown, translocation, and use of stored reserves.

The chief factor considered in seed storage is moisture. Seeds in the resting stage generally are low in moisture and relatively inactive from a metabolic standpoint. The moisture content in the germinating medium must reach a certain critical level in order for a given seed to imbibe sufficient water for germination. The critical seed moisture percentages, on a fresh weight basis, to allow germination initiation is reported as: corn, 30.5 percent; soybeans, 50.0 percent; rice, 26.5 percent; sugarbeets, 31.0 percent; and wheat, 40.8 percent. Excess moisture may inhibit germination.

An adequate supply of oxygen, to allow respiration, must be available for the germination of most seeds. The concentration of oxygen in ordinary air, about 20 percent, is sufficient to allow good germination; excessive carbon dioxide has a retarding effect. Seed planted too deep or in soil that has been packed or crusted by rain or other factors may not get enough oxygen for good germination.

Rice and other aquatic plants can germinate under water where oxygen is present in only low concentrations. In such instances, anaerobic respiration generates enough energy for germination to occur.

Temperature

Seed germination is dependent on several reactions which are affected by temperature. The temperature at which certain seeds barely germinate and at which they will germinate best and the highest temperature at which they will germinate are known as minimum, optimum, and maximum germinating temperatures. The optimum temperature is that giving the greatest percentage germination during the shortest time period. For most seeds the optimum temperature is 59–68°F (15–20°C), and the maximum is 95–104°F (35–40°C). Certain flower and alpine species will germinate at temperatures approaching freezing. It has been reported that Russian pigweed has germinated in frozen soil. When several kinds of seeds were placed in a cavity in a cake of ice and surrounded with 2 feet (0.6 m) of ice, at the end of two months rye, wheat, cabbage, and mustard were germinating. The seeds of cool-season crops, such as winter wheat and rye, will germinate more readily at lower temperatures than the warm-season crops, such as soybeans and cotton.

Canadian researchers have been using plastic-coated spring wheat for a number of years to take advantage of conditions that an earlier start in the spring would provide. After the seed is provided with a three-layer coating, this normally spring-sown crop is seeded in the fall. The coating breaks down gradually over the winter and the seed germinates and starts growth in early spring. This can result in a longer growing season or an earlier harvest.

Light

The light requirement of lettuce seed, and its photoreversibility, was demonstrated by USDA researchers in 1952. Since that time the light requirement has been demonstrated in species such as tobacco, Kentucky bluegrass, peppergrass, shepherd's purse, birch, pine, and elm. Kentucky bluegrass will not germinate without light, and tobacco seeds require light stimulation as well, but often only a fraction of a second for good germination.

Both light intensity and quality affect germination. Germination of some seeds is stimulated by low-intensity light, such as moonlight, while lettuce seeds need relatively high-intensity light. From the quality standpoint, the greatest promotion of germination is by the red portion of the spectrum;

TABLE 15-3. Germination of Tobacco and Peppergrass Seeds Exposed to Red Light After 4 and 23 Hours Imbibition in Dark

Period of Exposure to Red Light (min)	Percent Germination			
	4-Hour Imbibition		23-Hour Imbibition	
	Tobacco	Peppergrass	Tobacco	Peppergrass
0	4	0	6	11
¼	6	14	22	62
1	6	46	37	65
4	14	61	58	64
16	29	66	88	76
64	69	64	95	76

Source: Toole, E. A. et al. "Physiological Studies of the Effects of Light and Temperature on Seed Germination," Proc. ISTA, 18:274 (1953).

the far-red zone is inhibitory. Germination of tobacco and peppergrass seeds exposed to variable durations of red light after imbibing water for 4 or 23 hours is shown in Table 15-3. Tobacco seed germination was increased by the longest exposure to red light, 64 minutes, while peppergrass was not stimulated appreciably by exposure for more than 4 minutes.

Seeds which normally have a light requirement for germination should not be planted deeply or they will not emerge. However, soaked seeds given an adequate light treatment retain the light stimulation and, when dried, have been found to germinate successfully when deprived of light. The relationship of light to seed germination is made complex by the influence of other environmental factors. Many seeds which require light for germination may gradually lose this requirement after storage under dry conditions. It also is possible to change the light requirement by manipulating temperature and/or by adding oxygen, acids, or nitrates.

Seed Coatings

Several seed companies are currently marketing coated seeds. A physical coating on grass seed particularly gives the seed better ballistic properties for broadcast seeding. Coated seed is particularly useful in western states for re-establishing vegetation on large acreages.

Commercially available are seed with coatings containing nutrients, fungicides, insecticides, growth regulators, and other additives. Some companies are marketing pelleted legume seeds, which have a lime or calcium carbonate coating containing the proper *Rhizobium* strain for nodulation and nitrogen fixation.

Longevity of Seed and Storage

Some seeds are inherently long-lived and others short-lived. Onions have one of the shortest viability periods. Soybeans and peanuts do not store

TABLE 15-4. Germination of High-Quality Seed Lots of Several
Crop Species at Various Storage Intervals Under Ambient
Conditions at Mississippi State University

Kind of Seed	Storage Period (Months)					
	0	6	12	18	24	30
Red clover	94	94	88	73	60	58
Corn, field	98	98	96	96	90	85
Tall fescue	95	90	85	78	37	12
Peanuts, shelled	96	93	60	5	0	0
Rice	94	92	94	93	90	88
Sorghum	96	96	93	86	82	78
Soybean	96	94	85	60	42	0
Timothy	96	96	86	76	37	0
Wheat	98	97	97	96	92	90

Source: Delouche, J. C. "Precepts of Seed Storage," Short Course Proc., Miss. State
Univ. (1973), p. 104.

and maintain viability as well as do wheat, corn, cotton, sorghum, and
rice seed.

Delouche reported on the seed germination of various crop species at
various storage intervals under ambient conditions at Mississippi State
University (Table 15-4). Corn, rice, sorghum, and wheat maintained relatively high viability after 30 months of storage, while peanuts, soybeans,
timothy, and tall fescue were largely nonviable after 30 months.

Moisture is the chief factor affecting the longevity of seeds, as commercially handled. Seeds stored under low humidity conditions retain their
vitality much longer than do those stored in a humid atmosphere.

Seeds that under average farm conditions may be unsafe to plant after
two, three, or five years may retain their vitality through a long period
of years when held in storage under very favorable conditions—dry air,
uniform temperature, and perhaps partial or total exclusion of oxygen.

It is believed generally that 150 years is the maximum lifespan of the
most durable seed. However, a Japanese botanist found some viable lotus
seeds, which tests indicated were between 830 and 1,250 years old. Many
other germination tests of old seeds have been reported. Quick reported
on germination tests with about 500 species of old seeds from a storage
room in a museum. Thirteen species of viable seeds were more than 50
years old; eleven of these species were legumes. One species had seeds
viable after 158 years; seeds of another species were viable after 115 years.
A sample of white sweetclover seed 44 years old was reported to have
germinated 52 percent and another sample of the same species 77 years
old germinated about 18 percent.

Stewart and Duncan reported on a cottonseed viability experiment initiated in 1937. Various storage moisture levels and temperatures were tested.

TABLE 15–5. Germination of Cottonseed Containing
Different Moisture Levels in Long-Term Storage at 0°C

Moisture (%)	Years in Storage					
	1	7	15	25	31	37
7	87	94	91	91	86	68
9	92	92	91	82	73	63
11	89	89	93	86	63	57
13	90	92	72	16	0	0
14	88	34	0	0	—	—

Source: Stewart, J. McD., and E. N. Duncan. "Cottonseed Viability After
Long-Term Storage," *Agron. Jour.,* 68:243–244 (1976).

Seed containing 11 percent or less moisture and stored near 32°F (0°C)
had approximately 60 percent germination after 37 years (Table 15–5).
In Arizona, some seeds produced in 1929 and stored unsealed at ambient
temperatures until 1957, then sealed and stored cold, had a maximum germination of 60 percent in 1974. This study demonstrates that it is possible
to maintain good viability with some species under the proper storage conditions.

A generally accepted practice for maintaining maximum seed viability
is to store seeds at a temperature of about 40°F (4°C) and a relative humidity
of 50 percent or less. Delouche reported on a crimson clover seed germina-

TABLE 15–6. Germination of Crimson Clover Seed After Periods of Storage Under
Various Relative Humidity and Temperature Levels

Temperature	Relative Humidity (%)	Months of Storage				
		0	3	6	9	12
50°F (10°C)	20	90	89	88	90	88
	40	90	88	87	89	88
	60	90	87	90	90	90
	80	90	86	56	8	0
	100	90	70	4	0	0
68°F (20°C)	20	90	88	87	87	88
	40	90	87	90	86	90
	60	90	87	86	90	88
	80	90	34	1	0	0
	100	90	0	0	0	0
86°F (30°C)	20	90	86	87	89	84
	40	90	87	87	88	83
	60	90	87	75	66	23
	80	90	0	0	0	0
	100	90	0	0	0	0

Source: Delouche, J. C. "Precepts of Seed Storage," *S.C. Proc., Miss. State Univ.* (1973).

tion study involving various combinations of relative humidity and temperature, and variable periods of storage (Table 15–6). High relative humidity caused rapid deterioration. Good viability was maintained at high storage temperatures, up to 86°F (30°C) if relative humidity was kept below 40 percent. A low temperature helped maintain viability at a relative humidity level of 60 percent.

NATIONAL SEED STORAGE LABORATORY. In 1957 Congress, in response to a need to preserve and rebuild the nation's supply of pure and healthy seeds, established the National Seed Storage Laboratory at Fort Collins, Colorado. This facility is operated by the USDA and SEA, who maintain stocks of seed from every type of economically important plant in America. This includes agricultural, horticultural, and forest seeds.

The National Seed Storage Laboratory holds several billion seeds representing 100,000 cultivars. In addition to preserving plant germplasm, at this location research is conducted on seed deterioration in storage, and the relationship between seed moisture, storage climate, and packaging materials.

Buried Seed and Germination

Goss buried the seeds of 107 plant species in soil at different depths in 1879. A portion of each lot was dug up 1, 3, 6, 10, 16, and 21 years later. Fifty-one species showed some viable seeds at the end of 20 years. Among the seed of common weeds, 36 different species showed some viable seed at the end of 20 years.

Beal buried 21 pint bottles of weed seeds mixed with sand. Each bottle contained 50 seeds each of 21 kinds of weedy plants. He buried the bottles, with mouths tilted downward, 18 inches (46 cm) below the soil surface. After 40 years in the soil, but not after 50 years, seeds of the following five plants were still viable: *Amaranthus retroflexus* (pigweed), *Ambrosia elatior* (ragweed), *Lepidium virginicum* (peppergrass), *Plantago major* (plantain), and *Portulaca oleracea* (purslane). After 40 and 50 years, but not after 50 years, two additional species grew: *Brassica nigra* (mustard) and *Polygonum hydropiper* (knotweed). After 40, 50, and 60 years, but not after 70 years, *Silene noctiflora* (catchfly) grew. And after 70 and 80 years, three species were still germinable: *Oenothera biennis* (evening-primrose), *Rumex crispus* (a dock), and *Verbascum blattaria* (moth mullein). At 90 years, only one species—moth mullein—germinated.

Size of Seed

The smaller, less fully developed seeds produce somewhat smaller seedlings, which make a less vigorous growth than the larger seed. But if the plant establishes itself successfully it may make nearly as great a final production as plants from the larger seed.

The Nebraska station reported a study on the size of small grain seeds. Large and small seeds, representing extreme grades of winter wheat, spring wheat, and oats, were selected by hand. Small seeds yielded 18 percent less than large ones when spaced to permit maximum individual plant development, 10 percent less when equal numbers of seeds were sown per acre at an optimum rate for the larger seeds, and 5 percent less when equal weights of seeds were sown per acre at an optimum rate for the larger seeds. Unselected seeds yielded 4 percent less than large seeds when equal numbers were sown per acre, but only 1 percent less when equal weights were sown.

With the introduction of plateless planters and lower prices of ungraded or small seed, there is interest in lower-priced grades of corn seed. Research generally has shown that there is no genetic, germination, or emergence difference in corn seed whether it is large or small, flat or round. Hicks et al. reported on a two-year Minnesota study on corn seed size and shape. Yield differences among seed grades—large round, large flat, small round, small flat, ungraded—were not significantly different. In most experiments, yields from small rounds and small flats were equal to yields from large seed.

Burris, Edje, and Wahab evaluated the laboratory and field performance of four seed sizes of four soybean cultivars. The three largest seed fractions exhibited a superior emergence percentage and greater cotyledonary and unifoliolate leaf area in the laboratory. In field studies, the three largest seed sizes showed greater overall emergence, leaf area, and height and also gave greater yields than plants from the small seed size when grown at a uniform population.

Testing Seeds

Great harm and expense may result from planting crop seeds in which the seeds of troublesome weeds are intermixed. Each state has one or more seed laboratories and the USDA maintains several in different parts of the country. In addition, there are a number of commercial laboratories that make tests for seedsmen. Many of the larger seed companies maintain their own testing laboratories. Accurate testing of seeds for purity, germination, and sometimes vigor is essential. Every state has legislation requiring the labeling of seeds offered for sale. In addition, federal seed legislation exists and is applicable to imported seed and interstate shipments.

In addition to the standard (warm) germination test, a number of other viability or vigor tests have become important. The *cold test,* used by most seed companies and offered by many state seed laboratories, measures the ability of seed to survive and emerge under adverse field conditions. It simulates cold, wet conditions that may be encountered in the field after planting and prior to emergence. Seed are planted in a sand–soil mixture and kept at a temperature of 46–50°F (8–10°C) for 5 to 10 days, commonly 7 days. After this period, seed are placed in a chamber set at 78–86°F (40–41°C) for about 4 days, at which time germination (emergence) is determined. The *accelerated-aging test* involves placing planted seed in a cham-

FIGURE 15-2. *Seed processing equipment: Air-Screen Cleaners are the basic cleaning machines for all free-flowing seeds. Good seeds are separated from all other materials through the use of air aspiration to remove light materials and screens (sieves) which are available with over 200 different size or shape perforations.* [Photograph with permission of Seed Technology Laboratory, Mississippi State University.]

ber set at 104 to 106°F (40–41°C) and 100 percent relative humidity for 36 to 96 hours. This brings the seed to the brink of germination and provides an environment where susceptibility to early fungus diseases can be measured. Seed is then placed in a chamber set at about 78°F (26°C) for 4 days, after which emergence is determined.

The *tetrazolium test,* a viability test rather than a germination test, has been recognized since 1945. It is a quick color test based on the different reactions the tetrazol solution has on living and dead tissues. When it reacts with living tissues that are actively respiring, a red substance, formazan, is formed. Dead cells do not change color. Thus the tetrazolium test is actually a test for the activity of an enzyme system. Viability can be determined by this procedure in about 2 hours.

Federal Seed Legislation

For many years the United States was the dumping ground for low-grade, foul, and adulterated seed from other countries. The Annual Appropriations Act in 1905 gave the USDA the authority to purchase seeds, test for adulteration or mislabeling, and publish the test results. The enactment of the Seed Importation Act of 1912 did much to give the American farmer

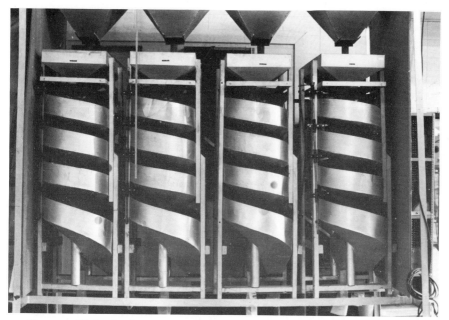

FIGURE 15-3. *Seed processing equipment: Spiral Separators are used to separate seeds differing in shape or degree of roundness, such as soybeans from split soybeans.* [Photograph with permission of Seed Technology Laboratory, Mississippi State University.]

needed protection with respect to imported forage seed. It was amended in 1916 to require minimum live seed for imported seed, and twice in 1926 to require coloration of alfalfa and red clover and to prohibit the shipment in interstate commerce of falsely or fradulently labeled seed.

The new Federal Seed Act enacted in 1939 required detailed labeling of seed in interstate commerce and extended the scope of the Act as it pertained to imported seed. It applies to all agricultural and vegetable seeds. If seeds are labeled to comply with the Federal Seed Act and are shipped in interstate commerce, normally they will comply with the laws of the state into which they are shipped. Each of the 50 states has its own seed laws that regulate the sale of seed within the state. The laws generally require that the labels attached to the containers of agricultural seeds show the percentages of pure seed, weed seeds, other crop seeds, inert matter, percentage of germination, and, if any, hard seeds. Moreover, the label must show the date of the germination test and the name and address of the shipper, seller, or person who labeled the seeds. The label also is required to show the names and rates of occurrence of seeds of noxious weeds recognized by the laws and regulations of the state in which the seed is being sold or into which the seed is shipped. Most states prohibit the sale of seeds containing seeds of certain noxious weeds or limit the number permitted in seed sold.

TABLE 15-7. Weeds Listed as Noxious in the Seed Laws of Five or More States

Common Name	Scientific Name	Number of States
Bermudagrass	*Cynodon dactylon*	15
Bindweed	*Convolvulus* spp.	5
Bindweed, field	*Convolvulus arvensis*	45
Bindweed, hedge	*Convolvulus sepium*	13
Blueweed	*Helianthus ciliaris*	12
Camel thorn	*Alhagi camelorum*	9
Carrot, wild	*Daucus carota*	7
Cheat	*Bromus secalinus*	10
Corn cockle	*Agrostemma githago*	23
Daisy, oxeye	*Chrysanthemum leucanthemum*	11
Darnel	*Lolium temulentum*	9
Dock	*Rumex* spp.	12
Dock, curled	*Rumex crispus*	11
Dodder*	*Cuscuta* spp.	48
Field cress, Austrian	*Rorippa austriaca*	7
Franseria, white-leaved	*Franseria discolor*	5
Garlic, wild	*Allium vineale*	29
Hoary cress	*Cardaria draba* and/or *Cardaria pubescens*	23
Horse nettle	*Solanum carolinense*	30
Horse nettle, white	*Solanum elaeagnifolium*	19
Johnsongrass	*Sorghum halepense*	26
Knapweed, Russian	*Centaurea repens*	26
Lettuce, blue	*Lactuca pulchella*	7
Mustard, black	*Brassica nigra*	17
Mustard, Indian	*Brassica juncea*	18
Mustard, white	*Brassica hirta*	9
Mustard, wild	*Brassica kaber*	26
Nutgrass	*Cyperus rotundus*	16
Nutgrass, yellow	*Cyperus esculentus*	6
Oats, wild	*Avena fatua*	10
Onion, wild	*Allium canadense*	21
Pennycress	*Thlaspi arvense*	12
Plantain, bracted	*Plantago aristata*	11
Plantain, buckthorn	*Plantago lanceolata*	34
Povertyweed	*Iva axillaris*	8
Puncture-vine	*Tribulus terrestris*	9
Quackgrass	*Agropyron repens*	41
Radish, wild	*Raphanus raphanistrum*	6
Ragweed, perennial	*Ambrosia psilostachya*	5
Rape, bird	*Brassica campestris*	11
Rape, turnip	*Brassica rapa*	9
Rice, red	*Oryza sativa* var.	6
Sorrel, sheep	*Rumex acetosella*	17
Sow thistle, perennial	*Sonchus arvensis*	26
Spurge, leafy	*Euphorbia esula*	21
St. Johnswort	*Hypericum perforatum*	8
Star thistle, yellow	*Centaurea solstitialis*	9
Thistle, Canada	*Cirsium arvense*	43

* Six different species are named as noxious.

The Federal Seed Act was amended in 1956 to allow civil prosecution for complaints of violation, and again in 1960 to require labeling of pesticide-treated seed.

NOXIOUS WEEDS. It is a violation of the Federal Seed Act to ship seeds, in interstate trade, that contain noxious weed seed in excess of that allowed by the receiving state. Each state seed law has established a specific list of noxious weeds for that state. In some states the list consists of primary noxious weeds and secondary noxious weeds. The sale of seed containing primary noxious weed seeds is prohibited, but sales of seed containing secondary noxious weed seed is permitted ordinarily with restrictions on the number that may be present.

Weeds named as noxious in the seed laws of five or more states are listed in Table 15–7. The number of states in which a particular weed has been so named suggests the relative extent to which it has been found troublesome. Seventy-seven additional weeds have been declared noxious in up to four states.

Certified Seed

New and improved strains and cultivars of high-yielding and adapted crop seed have been released by the different universities, by the USDA,

FIGURE 15–4. *Seed processing equipment: Width and Thickness Separators (left and center) and Disc (Length) Separator (right). After basic cleaning seeds may be separated on the basis of a difference in either length, width, or thickness.* [Photograph with permission of Seed Technology Laboratory, Mississippi State University.]

FIGURE 15–5. *Seed processing equipment: Color Sorters separate seeds which differ in the color of their seedcoats or outer coverings.* [Photograph with permission of Seed Technology Laboratory, Mississippi State University.]

and by others. To keep pure sources of these superior seed, procedures and standards for producing seed that can be certified for genuineness, or trueness to name, have been established in many states.

Certified seed is produced by outstanding farmers and seedsmen under careful quality-control standards. This involves use of pedigreed planting stock, field inspection during the growing season, and seed inspections following harvest.

Certification and registration usually are by an independent organization or association of farmer growers (in most cases a state crop improvement association) which cooperates with the state college of agriculture.

Recognizing the desirability of a degree of uniformity in the requirements for seed certification in the different states, the International Crop Improvement Association was established in 1910. Its membership represents the different state associations. This association recommends minimum requirements and standards for the different crop seeds.

A four-generation scheme is in effect for maintaining the purity of crop cultivars and subsequently producing certified seed:

1. *Breeder seed.* Produced by the originating or sponsoring breeder or institution so grown and managed as to maintain the cultivar characteristics.
2. *Foundation seed.* Produced from fields planted with breeder seed and so handled as to maintain the genetic identity and purity of the cultivar. This seed ordinarily is produced under contract by a foundation seed organization. It is the source of all certified seed, either directly or through the registered class.
3. *Registered seed.* The progeny of foundation seed so handled as to main-

TABLE 15–8. Crop Seed Characteristics and Standards

| Kind | Pounds per Bushel | Number Seeds per Pound | Length of Life in Dry Storage (years) | Standards of | |
				Purity (%)	Germination (%)
Alfalfa	60	226,720	4–7	99	90
Barley	48	13,600	5–10	97	90
Bluegrass (Kentucky)	15–28	2,150,000	2–3	88	85
Bromegrass	14	136,000	2–3	85	80
Clover, alsike	60	680,320	3–4	98	90
Clover, red	60	272,160	4–7	97	90
Clover, sweet	60	258,400	4–7	97	90
Clover, white	60	785,000	2–4	98	80
Corn	56	1,500	2–5	100	90
Flax	56	136,000	1–3	97	90
Kafir	56	17,700	3–8	98	80
Lespedeza (Korean)	25	238,000	—	97	90
Millet, Foxtail	50	213,000	—	98	85
Milo	56	13,000	3–8	98	80
Oats	32	12,640	5–10	97	90
Orchardgrass	14	586,000	2–3	85	85
Rape	50	109,000	—	—	—
Redtop	14–30	5,084,000	5–7	—	—
Reed canarygrass	45	680,320	1	98	80
Rye	56	18,000	5	97	85
Sorghum (amber)	50	25,000	3–8	97	80
Soybeans	60	3,168	2–3	97	90
Sudangrass	32	54,240	3–8	95	90
Timothy	45	1,133,900	5–8	98	90
Wheat	60	11,320	5–10	97	90

Source: Iowa State Univ. of Science and Technology. *Midwest Farm Handbook,* 6th edition (Ames: Iowa State Univ. Press, 1964, 474 pp.).

tain genetic identity and purity. Registered seed is of a quality suitable for the production of certified seed. It is intended for the purpose of increasing seed another generation before the production of certified seed. Some states bypass this generation, and go from foundation seed to certified seed.

4. *Certified seed.* The progeny of foundation or registered seed that has been handled so as to maintain satisfactory genetic purity and that has been approved and certified by the certifying agency. Certified seed of most cultivars cannot be used as planting stock for the production of certified seed.

STANDARDS OF QUALITY IN SEED. Seeds of the different crops vary greatly in the percentage of purity and of germination regarded as satisfactory for reasonably high-grade seeds. These differences usually are due to the character of the seed itself or to the methods of harvesting and curing. Standards for purity and germination generally recognized as satisfactory, the average number of seeds per pound, and the number of years that the different seeds may be expected under average conditions to retain their vitality to such an extent as to have value for planting are shown in Table 15–8.

REFERENCES AND SUGGESTED READINGS

1. ANONYMOUS. "Cold Test Predicts Bean Vigor," *Prairie Farmer,* 104a (Feb. 4, 1978).
2. ANONYMOUS. "Plastic Coated Seed," *Crops and Soils,* 19(9):8–9 (1967).
3. ANONYMOUS. "Soybeans: Seed Quality Going Up," *The Furrow,* 11–13, (May/June, 1977).
4. BARTON. L. V. *Seed Preservation and Longevity* (New York: Wiley–Interscience, 1961.
5. BURRIS, J. S., O. T. EDJE, AND H. H. WAHAB. "Effects of Seed Size on Seedling Performance in Soybeans. II. Seedling Growth and Photosynthesis and Field Performance," *Crop Sci.,* 13:207–210 (1973).
6. COPELAND, L. O. *Principles of Seed Science and Technology* (Minneapolis: Burgess, 1976).
7. DELOUCHE, J. C. "Precepts of Seed Storage," *Short Course Proc. Miss. State Univ.,* (1973), p. 104.
8. DELOUCHE, J. C. *Seed Dormancy,* mimeographed material, *Short Course Proc. Miss. State Univ.* (1964) pp. 1–12.
9. DEXTER, S. T. "Alfalfa Seeding Emergence from Seed Lots Varying in Origin and Hard Seed Content," *Agron. Jour.,* 47:359 (1955).
10. FORWARD, B. F. "Studies of Germination in Oats," *Proc. ISTA,* 23:20 (1958).
11. GALLOWAY, S. H. "The TZ Quick Test for Soybean Seed Quality," *Crops and Soils* 26(5):14–15 (1974).
12. GOSS, W. L. "The Viability of Buried Seed," *Jour. Agr. Res.,* 29:7 (1924).
13. HICKS, D. R., R. H. PETERSON, W. E. LUESCHEN, and J. H. FORD. "How Seed Size, Grade Affect Yield," *Crops and Soils,* 30(3):17 (1977).

14. HUNTER, J. R., and A. E. ERICKSON. "Relation of Seed Germination to Soil Moisture Tension," *Agron. Jour.* 44:107–109 (1952).

15. IOWA STATE UNIV. of SCIENCE and TECHNOLOGY. *Midwest Farm Handbook,* 6th edition (Ames: Iowa State Univ. Press, 1964, 474 pp.).

16. KIVILAAN, A. *Addition to the 90-Year Period for Dr. Beal's Seed Viability Experiment,* Mich. Agr. Exp. Sta. Res. Rep. 336 (1977).

17. MUNN, M. T. *Germinating Freshly Harvested Winter Barley and Wheat,* Proc. Assoc. Off. Seed Anal. (1946).

18. NAKAMURA, S. "Germination of Grass Seeds," *Proc. of the ISTA,* 27:710–729 (1962).

19. NELSON, S. O. "Radio-Frequency Electric Seed Treatment," *Seed World,* 88(12):6–7 (1961).

20. NELSON, S. O., *et al.* "Alfalfa Seed Germination Response to Electrical Treatments," *Crop Sci.,* 17:863–866 (1977).

21. POLLOCK, B. M., and V. K. TOOLE. "Afterripening, Rest Period, and Dormancy," In *Seeds: Yearbook of Agriculture* (Washington, D.C.: USDA, 1961). ed. A. Stefferud.

22. QUICK, C. R. "How Long Can a Seed Remain Alive?" In *Seeds: Yearbook of Agriculture* (Washington, D.C.: USDA, 1961).

23. RAMPTON, H. H., and T. M. CHING. "Persistence of Crop Seeds in Soils," *Agron. Jour.,* 62:2 (1970).

24. STEWART, J. McD., and E. N. DUNCAN. "Cottonseed Viability After Long-Time Storage," *Agron. Jour.,* 68:243–244 (1976).

25. TOOLE, E. H., *et al.* "Physiology of Seed Germination," *Ann. Rev. Plant Physiol.,* 7:295–324 (1956).

26. WRIGHT, L. N. "Seed Dormancy, Germination Environment, and Seed Structure of Lehmann lovegrass," *Eragrostis lehmanniana* Nees, *Crop Sci.,* 13:432–435 (1973).

CHAPTER 16

TILLAGE AND CULTIVATION PRACTICES

THE development of agriculture as an industry relates directly to the development of implements for pulverizing the soil. It has been said the "conquest of hunger began with the invention of the plow."

The earliest plows were forked sticks with which men scratched the soil surface before seeding. The first of these were used in a manner similar to a heavy hoe, or mattock. When one branch of the Y was left long it served as a beam by which it might be pulled. After several thousand years men learned to harness animal power to the plow. Unfortunately there was little or no progress in plow design for thousands of years.

History of the Plow and Tillage

The plow is the cornerstone of American history. Famous Americans such as Thomas Jefferson and Daniel Webster helped perfect the plow. Economic growth and geographic expansion during the early history of the United States followed the westward advance of the plow through the prairies to the plains.

Table 16-1 lists some of the key developments in the evolution of the plow and other tillage implements and techniques.

Something of a stir was created in 1943 with the publication of Edward Faulkner's *Plowman's Folly*, in which it was contended that the turning of the soil with the plow had resulted generally in rapid soil deterioration; that it was pure folly to continue the use of turning plows; that it was possible with the discontinuation of plowing to bring back to high productivity soils which plows had nearly ruined; that agronomists had been "asleep at the switch" in failing to bring out these facts.

Faulkner's book led to numerous reports from agronomy and agricultural engineering research workers. These reports showed a considerable amount of research through the years on soil productivity as affected by plowing. In the low-rainfall areas of the Great Plains, and especially with small grain production, leaving crop residues on the surface or mixed into the surface soil had been shown to be desirable as a means of conserving moisture and reducing soil erosion. The 1950s brought alternatives to the moldboard plow, such as chisel plowing, disk plowing, and stubble mulching. The late 1950s and early 1960s saw minimum tillage and no-tillage techniques developed, and the first commercial no-tillage planter was made

TABLE 16-1. Key Developments in Evolution of the Plow and Other Tillage Implements and Techniques

Period	Historical Development
1731	Jethro Tull's *New Horse Houghing Husbandry* published in England.
1794	Jefferson's moldboard plow of least resistance tested.
1796	Newbold patented first cast iron plow.
1819	Jethro Wood patented iron plow with interchangeable parts.
1833	John Lane began manufacture of plows faced with steel saw blades.
1837	John Deere and his co-worker began manufacture of steel plows.
1856	Two-horse straddle row cultivator patented.
1865–1875	Gang plows and sulky plows came into use.
1883–1888	Spring tooth harrows available for seedbed preparation.
1928–1933	Multiple row cultivators available.
1943	Edward Faulkner's *Plowman's Folly* published.
1950s	Conventional moldboard plow began to lose some following to chisel plow, disk plow, stubble mulcher, and other primary tillage tools.
1950–1960	Researchers studied minimum and no-tillage techniques.
1960–1965	Farmers tested no-tillage techniques.
1966	Allis-Chalmers introduced fluted coulter no-tillage planter.

Source: Phillips, S. H., and H. M. Young, Jr. *No-Tillage Farming* (Milwaukee: Reiman Associates, 1973).

available in 1966 (Table 16-1). The moldboard plow is not obsolete, but presently several primary tillage or minimum tillage alternatives are available.

Plowing and Good Tilth

A soil with a fine crumb structure is said to be in "good tilth." A soil in good tilth breaks up easily into crumbs or granules about the size of wheat grains or of soybeans; these crumbs are porous. They are made of finely pulverized bits of soil, linked together something like popcorn in a popcorn ball. They hold this structure even when wet. This allows space in the soil for air and water. Plowing and other seedbed practices, together with the cultivation of row crops, tend to increase oxidation and nitrification; organic matter is lost and there is a breakdown of the original aggregates or crumblike characteristics of the soil. This means smaller soil particles closer together, with increased difficulty of water penetration. Tillage at the same depth each year also results in a layer of compacted soil, the plow pan. Varying tillage depth, or occasionally using a chisel plow, helps keep the soil in good tilth.

When the characteristics of six virgin Iowa soils were compared with the same soils after 60 years of cropping, it was found that only 3–10 percent as much water per hour could be absorbed by the cultivated soil, that whereas the weight per cubic foot of the virgin soils ranged from 57 to 72 pounds (26–33 kg), the cultivated soils ranged from 71 to 79 pounds (32–36 kg); that the percentage of air space, which was 15–19 percent for the virgin soils, was down to a range on only 5–11 percent for the cultivated soils.

Primary Tillage

Plows and Plowing

There are several types of plows, which vary in size from a single bottom, 7–8 inches (18–46 cm) wide, to large gang plows that turn up to ten furrows at a time. In much of the Corn Belt three to seven bottom plows are common. In the Wheat Belt, very large gang plows often are used.

The moldboard plow has been used in the United States since about 1775. Despite all of the available alternatives, the moldboard plow still is used by more farmers than all other primary tillage implements in areas of medium to high rainfall. It breaks loose or shears off the furrow slice, inverts the soil, and breaks it into lumps. It is superior to other implements for breaking up tough sod and turning under green manure crops and heavy crop residues. The width of the furrow may be from 7 to 24 inches (18–61 cm).

The two-way plow is adapted to steep slopes because it throws the soil

FIGURE 16–1. *Primary tillage with a moldboard plow. It breaks loose or shears off the furrow slice, inverts the soil, and breaks it into lumps. Despite all of the available alternatives, the moldboard plow still is used by more farmers than all other primary tillage implements in areas of medium to high rainfall.* [SOURCE: J. I. Case.]

downhill when plowing in either direction. It quite often is used on irrigated fields to avoid dead furrows. Disk plows are used in loose soils, in soils too dry and hard for easy penetration of moldboard plows, or in sticky soils. They are best suited to dry regions, and work well on bare ground or small grain stubble fields. Under proper soil conditions, the disk plow gives almost complete seedbed preparation in one operation. Most of the trash and residue is left on the surface. The disks vary in size from 20 to 30 inches (51–76 cm); depth of plowing is from 4 to 10 inches (10–25 cm). In the stubble fields of the Wheat Belt, the one-way disk plow is used for seedbed preparation. It leaves the residues mixed with surface soil and leaves the surface rough to reduce wind erosion.

Listers and Listing

The lister tears open furrows in the land, throwing a furrow slice each way and leaving the ground ridged. The soil thrown out makes ridges 4–5 inches (10–13 cm) above the original surface and approximately 15 inches (38 cm) broad. The furrows between the ridges are about the same width and extend as deep below as the ridge is raised above the original surface. In single listing, as described, there is a strip of unstirred soil under each ridge. Double listing consists of single listing, as described, and of splitting the ridge by going over the field again with the lister. In planting corn and grain sorghums a lister-planter may be used, which lists the ground and plants and fertilizes the crop in one operation. Row crops may be planted in the furrow or trench in semiarid regions, but in the humid South the ridges ordinarily are dragged down and flattened somewhat and the crop is planted on top of the ridges. The lister is less popular now in semiarid regions, with a greater use of sweep-blade implements and one-way plows in these areas.

Other Methods of Seedbed Preparation

Chisel plows have rigid-tined harrows capable of penetrating to plow depth. They came into use in the 1970s in the Corn Belt as a primary tillage implement in place of the moldboard plow in some situations. The chisel plow loosens the plow layer and shatters it if the soil is fairly dry; it is ineffective in wet soils. It offers advantages over the moldboard plow of speed, requiring less draft, and leaving more trash on the surface for erosion control.

Blade or subtillage implements leave crop residues mixed in the surface soil to protect it from erosion and water runoff. Wide-sweep blades run a few inches below the surface and undercut stubble and weeds, but do not pulverize the soil. Such a stubble mulching procedure reduces erosion and runoff.

Rotary tillers are designed to prepare a seedbed in one operation. They have hooks, knives, or tines of various shapes that rotate and prepare the seedbed. This type of tillage is best suited to light-textured or organic soils

FIGURE 16-2. *Primary tillage with a chisel plow. The chisel plow loosens and shatters the plow layer. It is faster than moldboard plowing, requires less draft, and leaves more crop residue on the surface for erosion control.* [SOURCE: J. I. Case.]

where soil tilth is not a problem. On other soil types, maintaining aggregation and proper soil structure can be difficult.

Disking is a rapid method of preparing a seedbed. The disk may stir the soil 3–6 inches (8–15 cm) deep, depending on the firmness of the soil. It will kill weeds and grass that are just starting, but it does not bury most weed seed sufficiently deep to keep them from germinating during the season. Disking may leave the surface fine and loose but often packs the soil below. In recent years large, once-over, deep tillage disks have been made available. These heavier, primary tillage disks generally have cutting widths of 6–18 feet (1.8–5.5 m), and may have blades as large as 32-inch (81-cm) diameter. Tandem disk harrows are not as heavy, but models are available with a greater cutting width, up to 32 feet (9.8 m) or more.

Old-stalk fields of cotton, corn, or other cultivated crops are sometimes prepared for the next crop with a field cultivator. This stirs up the soil 3–6 inches (8–15 cm) deep, much the same as the disk, but does not cover the vegetation as much as is possible with other methods. It is a rapid way of preparing a seedbed.

The spring-tooth harrow prepares a fair seedbed on clean, mellow land that has been in a cultivated crop the preceding year. It does not incorporate much organic matter with the soil.

FIGURE 16-3. *Primary or secondary tillage with a double offset tandem disk. In some instances with certain kinds of crop residues, heavy disks are used for primary tillage. In other instances, offset or tandem disks are used for secondary tillage, breaking large clods and smoothing a rough surface.* [SOURCE: J. I. Case.]

Depth of Plowing

The cost of preparing the seedbed is greatly increased by deep plowing and by subsoiling. A report on work at the Pennsylvania station was to the effect that the draft per square foot of cross section of furrow was 724 pounds for 7.5-inch (19-cm) plowing as compared with 1,113 pounds for 12-inch (30-cm) plowing. Another report showed the pull in pounds for 4-inch (10-cm), 6-inch (15-cm), 8-inch (20-cm), and 12-inch (30-cm) plowing was 257, 332, 408, and 560 for a 12-inch (30-cm) width of furrow slice, and 417, 532, 647, and 876, respectively, for an 18-inch (46-cm) furrow.

A considerable number of studies on plowing depth were conducted in the 1920s, 1930s, and 1940s at the state experiment stations. An early study at the Illinois station revealed that increasing the depth of plowing 1 inch (2.5 cm) increased the weight of soil turned on one acre (0.4 ha) by 300,000 pounds (136,080 kg) or 150 tons (136 MT). It was estimated that subsoiling cost twice as much as ordinary plowing and that deep tilling cost three times as much.

Hume reported the yields from different depths of plowing continued through a 25-year period at the South Dakota station. There were only small increases in yield when the soil was deep-tilled or subsoiled, as compared with the shallower plowing or fitting without plowing. These increases were not nearly enough to cover the much greater cost.

Sewell concluded that plowing deeper than 7 inches (18 cm) cannot be expected to increase yields but that the best depth less than 7 inches had not been determined.

In general, experiments conducted throughout the United States indicate that a plowing depth of 8 inches (20 cm) is usually deep enough on most soils. Plowing deeper than 8 inches is feasible with the larger tractors of today, and has produced slight yield increases on sandy soils in limited instances, but very deep plowing ordinarily is not necessary on medium- or fine-textured soils.

Subsoiling

Larson and his co-workers reviewed the effect of subsoiling on corn yields on several soil types in Kansas and the Dakotas, when soil moisture was

FIGURE 16–4. *The Stoneville parabolic subsoiler to shatter compacted layers, and improve water infiltration and rooting. This was developed at the Delta Branch Experiment Station of Mississippi State University, with tines designed as a parabolic curve to decrease horsepower requirements and increase width of soil fracture.* [Photograph with permission of Mississippi State University Agriculture and Forestry Experiment Station.]

limiting. Crop yields were not improved by subsoiling or else increases were not economical. Similar results have been obtained on a number of soils in Iowa, Illinois, Indiana, and Minnesota.

Other studies regarding deep tillage to break up the claypans have been conducted with certain Planasol soils in Illinois, Iowa, Indiana, Kansas, and Missouri. Few yield increases occurred as a result of deep tillage treatments. In most experiments reported in the literature, subsoiling has had little effect on water intake, water storage, or root penetration and distribution.

Time of Plowing

Selection of a spring or fall plowing date is dependent upon a number of factors including weed control, erosion hazard, soil type and topography, and crop residue. In sections of heavy rainfall in the late fall, winter, or early spring there is a danger of erosion if the soil is loosened by plowing in the fall. Wind erosion also may be a problem. The crop and remaining residue on the land the preceding year may determine the optimum time of plowing. For example, sod crops may be plowed under in the fall on slopes or in areas where fall plowing ordinarily is not recommended. Sod roots will hold soil in place better than other kinds of vegetation or stubble.

Fall plowing may be advisable for heavy, fine-textured soils. The freezing-thawing and wetting-drying action over the winter breaks large clods into smaller granules, and makes secondary tillage for seedbed preparation easier. Fall plowing ordinarily will allow earlier planting than spring-plowed fields; in wet springs, it is difficult to prepare a seedbed early enough for planting on the optimum date.

Soils with a large silt content should not be fall plowed; soil particles tend to "seal together" rather than remain aggregated, and by spring may be as compact as before plowed. Soils that have severe erosion potential should not be fall-plowed. A 2–3 percent slope may be the safe limit for fall plowing without contouring.

Secondary Tillage

Secondary tillage refers to field operations after plowing (primary tillage) to prepare the seedbed for planting. The purpose most often is to further pulverize the soil and prepare a fine seedbed.

The disk is the most popular secondary tillage implement. Offset or tandem disks break large clods, cut some trash into the surface, and smooth a rough surface. A disk penetrates to a depth of 3–6 inches (8–15 cm) and works well on plowed ground with loose trash or on freshly plowed sod. Disking tends to pack the lower furrow slice, especially when used on wet soils.

A *field cultivator* has single- or double-pointed shovels, spikes, or small sweeps. It digs, lifts, and loosens the soil, cuts roots below the surface,

FIGURE 16-5. *Secondary tillage with a field cultivator. This implement digs, lifts, and loosens the soil and leaves some trash on the surface.* [SOURCE: J. I. Case.]

and leaves some trash on the surface. This implement is only fair for use in fields with considerable trash and in freshly plowed sod. It is excellent for controlling weeds in fallowed ground.

Harrows are widely used secondary tillage tools. The *spike-tooth harrow* is used to smooth the seedbed and break clods. It also controls small weeds if planting is delayed after primary tillage or seedbed preparation. The *spring-tooth harrow* is used widely in the Northeast and parts of the Midwest. It digs, lifts, and loosens the surface 3–4 inches (8–10 cm), breaks clods, levels a rough surface, and prepares a fine seedbed. Since this implement pulls trash to the surface, it is not suitable for use in plowed ground containing much trash, and is not suitable for freshly plowed sod. Harrows do not compact the lower part of the plow layer as much as disks.

The *cultipacker* is useful in compacting and leveling freshly plowed soil. It pulverizes clods, firms the surface to a depth of 2–4 inches (5–10 cm), and leaves the surface ridged. It is a valuable tool for making a finer, firmer seedbed, especially for establishing small-seeded forage grasses and legumes.

FIGURE 16–6. *Secondary tillage with a spring-tooth harrow. This imple-ment is used to smooth the seedbed and break clods. It can control small weeds if planting is delayed after primary tillage.* [SOURCE: Deere and Com-pany.]

Minimum Tillage

Minimum tillage has been defined as reducing tillage only to those opera-tions that are timely and essential to producing the crop and avoiding damage to the soil. It normally refers to a tillage system in which the number of field operations is reduced as compared to the number required in a conventional seedbed preparation and planting system. Minimum til-lage may range from only one less field operation on one extreme all the way to no-tillage on the other extreme.

The pioneer research on minimum tillage was done at the Ohio Station between 1938 and 1946. This work demonstrated that it was possible to produce corn on plowed ground by preparing a seedbed only in the hill. Researchers at Michigan State University started wheel-track planting in 1946. During the early 1950s workers at Cornell University planted directly after plowing by pulling a planter after the plow or attaching a planter to the plow. Various minimum tillage systems were tested by researchers and farmers during the 1950s and early 1960s. Major farm machinery man-ufacturers began marketing minimum tillage tools in the early 1960s. The first fluted coulter no-till planter was introduced in 1966.

Advantages given for minimum tillage include (1) reduced soil compaction, (2) better soil conservation because of soil left "rougher" and more residue on surface, (3) reduced energy requirements, (4) reduced capital investment for machinery, and (5) improved timeliness of farm operations and reduced labor. In addition, in some instances yield increases are obtained.

According to Voorhees in Minnesota, a tractor in the 1940s weighed less than 6,000 pounds (2,722 kg), about the same as the three- or four-horse team it replaced. Four-wheel drive tractors of today may weigh 40,000 pounds (18,144 kg), and large harvesting equipment may carry several tons of grain in addition to the weight of the machine. A six-row operation covering a width of 15 feet (4.6 m), and using 18-inch (46-cm) wide rear tractor tires, will make enough wheel tracks to cover every square inch of the field twice, if there are six total field operations during the season. Such a system can result in excessive soil compaction. A minimum tillage system could reduce the number of operations and thus reduce compaction.

The degree of water or wind erosion and greater water conservation is often related to the amount of residue left on the soil surface. The amount of surface cover remaining after several primary tillage operations is shown in Table 16-2.

Despite several potential advantages for minimum tillage systems, a recent survey of 3,000 farmers in the Midwest revealed that an average of 2.8 tillage operations is still used. Only one-third the farmers surveyed have reduced their tillage operations to one or two, and nearly one-fifth use four or more operations (Table 16-3).

Many of the minimum tillage practices are termed by some as *conservation tillage*, a practice that leaves the soil surface resistant to erosion and conserves moisture. According to Fenster *et al.,* conservation tillage may be one of these tillage systems: (1) minimum tillage, (2) no tillage, (3) stubble mulch tillage, (4) chisel planting, and (5) till-planting. Conservation tillage is designed to conserve crop residues, increase water intake, reduce wind and water erosion, and save energy. However, it may increase disease and insect problems and lower the early-season soil temperature.

TABLE 16-2. Effect of Primary Tillage Method on Surface Cover Remaining

Tillage Operation	Percent Surface Cover Remaining
Moldboard plow	1.3
Offset disk	21.0
Coulter-chisel	23.7
V-chisel	29.0
Standard chisel	39.7
Light disk	56.0

Source: "Primary Tillage," *Successful Farming,* 29, (March, 1978).

Table 16–3. Number of Tillage Operations Used
by 3,000 Midwest Farmers

No. of Tillage Operations*	Percent of Farmers
1	6
2	28
3	46
4 or more	19

* Average: 2.8 operations
Source: Reichenberger, L. *Successful Farming*, p. 30 (March, 1978).

Hayes cites several minimum tillage systems as examples of *mulch tillage,* leaving residues from the previous crop on or just beneath the soil surface throughout the cropping year. Some mulch tillage methods are (1) no-tillage, (2) slot planting, (3) chisel planting, (4) till planting, and (5) strip tillage.

Systems of Minimum Tillage

Many of the minimum tillage systems leave a large amount of the residue on the soil surface after tillage, thereby reducing wind and water erosion and improving moisture retention. In addition, energy requirements and fuel consumption may be considerably less than in conventional systems involving moldboard plowing (Table 16-4). Hinz in Arizona compiled a cost summary for tillage operations involved in farming 400 acres (162 ha) in Arizona by reduced tillage as compared to conventional tillage (Table 16-5). The reduced tillage operations of chiseling, listing, and harrowing cost only about 40 percent as much as a conventional system involving six field operations.

Among the minimum tillage systems available are (1) chisel and field cultivator planting, (2) wheeltrack planting, (3) plow-plant, (4) strip-tillage, and (5) no-tillage.

Chisel and Field Cultivator Planting. The chisel or field cultivator can be substituted for a primary tillage tool on bare ground or following crops with little residue like soybeans, field beans, sugarbeets, and potatoes. A *field cultivator* operation, coupled with planting, may eliminate soil compaction problems caused by disking and harrowing. The power requirement and machinery investments may be only slightly less than in conventional tillage.

Chisel plow planting is gaining in popularity. Chisel plowing as the primary tillage is not suitable on fields where large amounts of crop residues are present; such residues should be chopped to permit efficient operation. Chisel plowing in the fall or spring can reduce erosion and increase water infiltration. When spring operations are delayed by rain or other factors, chiseling as a substitute for moldboard plowing can speed land preparation and planting.

TABLE 16–4. The Amounts of Residue Remaining and Energy Used with Different Tillage Implements

Implement	Surface Residue Remaining After Tillage (%)	Speed (mph)	Type of Tillage	Energy Requirement PTO (Hp hr/A)	Consumption Gasoline* (Gallons/A)	Consumption Diesel† (Gallons/A)
Moldboard plow (7 in. deep)	0–5	4	Primary	23.4	2.6	1.8
Chisel plow 2 in. wide points (7 in. deep)	75	4	Primary	18.9	2.1	1.5
One-way (18–20 in. disks)	60	4	Primary / Secondary	10.0 / 13.6	1.1 / 1.5	0.8 / 1.0
One-way (24–26 in. disks)	50	4	Primary / Secondary	12.5 / 15.4	1.4 / 1.7	1.0 / 1.2
Heavy tandem or offset disks	60 / 50	4	Primary / Secondary	10.7 / 14.5	1.2 / 1.6	0.8 / 1.1
Field cultivator (12–18 in.) Sweeps	80	4	Primary / Secondary	5.3 / 7.3	0.6 / 0.8	0.4 / 0.6
V-Sweep (20–30 in. wide)	85	6	Primary / Secondary	8.0 / 10.9	0.9 / 1.2	0.6 / 0.8
V-Sweep (over 30 in. wide)	90	6	Primary / Secondary	9.3 / 12.7	1.0 / 1.4	0.7 / 1.0
Mulcher treader (spade tooth)	75–80	6	Secondary	4.0	0.4	0.3
Rodweeder (with semi-point chisel or shovel)	85	5	Secondary	8.5	0.9	0.7
Rodweeder (plain rotary rod)	90–95	5	Secondary	6.9	0.8	0.5

* 9 hp hr/gallon.
† 13 hp hr/gallon.
  Source: Williamson, E. J., et al. Conservation Tillage (South Dakota Ext. Serv., USDA, EC703, 1975).

WHEELTRACK PLANTING. This is a plow-plant method in which the soil is turned or plowed, followed by a planter aligned to plant directly behind and in the tractor tire track. Special wheels placed ahead of the corn planter may be necessary to furnish adequate tracks for all rows. The tire action firms the soil and eliminates the need for disking or other

TABLE 16–5. Tillage Cost Summary for 400 Acres (Arizona)

Operation	Annual Use (hours)	Cost Per Hour				Total Operation Cost	
		Fixed and Repair	Tractor	Labor	Total	Annual	Per Acre
Conventional Tillage							
Plowing	242	3.47	4.31	3.50	11.28	$2,730	$ 6.82
Disking	72	7.74	4.31	3.50	15.55	1,120	2.80
Disking	72	7.74	4.31	3.50	15.55	1,120	2.80
Floating	81	5.96	4.31	3.50	13.77	1,115	2.79
Listing	83	4.67	4.31	3.50	12.48	1,036	2.59
Harrowing	58	1.58	4.31	3.50	9.39	545	1.36
Total	608					$7,666	$19.16
Reduced Tillage							
Chiseling	145	2.32	4.62	3.50	10.44	$1,514	3.78
Listing	83	4.67	4.62	3.50	12.79	1,062	2.65
Harrowing	58	1.58	4.62	3.50	9.70	563	1.41
Total	286					$3,139	$ 7.84

Source: Hinz, W. W. *Reducing Tillage to Conserve Energy and Increase Profits,* Arizona Coop. Ext. Serv. Publ. Q38.

secondary tillage operations. Wheeltrack planting is an excellent system where the soil tilth is good, and on light-textured soils. It can work on heavy soils if moisture conditions are optimum. Labor and machinery costs, number of trips, and erosion may be reduced below that of a conventional system. A special planter or adaptations of existing planters are necessary. A disadvantage of wheeltrack planting is that plowing must be delayed to near planting time, thus limiting this system to small or moderate acreages.

PLOW-PLANT. Plowing and planting without additional seedbed preparation is the plow-plant method. Plow-plant and wheeltrack methods were part of the technical revolution that made the no-tillage method more acceptable to producers.

Plow-plant works best on sandy and silt loams. It is suited to about the same conditions as wheeltrack planting. Soil moisture conditions at the time of plowing and planting are critical.

Farmers have not adopted plow-plant and wheeltrack systems widely because of plowing being delayed until the planting period. Weather risks and peak labor requirements during the spring period often negate the advantages, especially on large acreages or in years with wet springs.

STRIP-TILLAGE. Strip-tillage or till-planting involves preparing a seedbed only in row strips, and not between the rows, by means of sweeps, rotary hoe blades, rotary tillers, or revolving spring teeth. This one-trip system sweeps clear a path through the row of the previous crops and the seed is planted. The width of the strip varies, but may be 10 to 14 inches (25 to 36 cm). A commercial till-planter was introduced in the 1950's, and several modified versions have been produced by a number of companies.

Strip-tillage in living grass sod generally has not been acceptable because of excessive competition from grass between the rows. It works best on plowed fields. This method has lower costs than conventional or chisel plow planting and requires less power. Operation costs are higher than for most no-tillage systems. Strip-tillage does not allow full benefits from mulch, soil moisture requirements are critical, and soil structure may be damaged by improper operation of rotary strip-tillage planters.

NO-TILLAGE. Conventional tillage systems are at least 200 years old, but the no-tillage (no-till) method is less than 30 years old. The earliest known report on no-tillage was in 1952; wheat, oats, flax, soybeans, and corn were no-till planted in Ladino clover sod killed by the herbicide 2,4,5-T.

No-tillage refers to a method of planting crops in previously unprepared soil by opening a narrow slot, trench, or band only of sufficient width and depth to obtain proper seed coverage. No other soil preparation is required, and herbicides are used for weed control.

FIGURE 16–7. *A four-row no-tillage planter. This planter is equipped with fluted coulters and can plant crops in sod or stubble, previously unprepared soil, by preparing the seedbed in a narrow band only of sufficient width and depth to obtain proper seed coverage.* [SOURCE: Southern Illinois University.]

Producers in the middle Atlantic states and the upper Southeast now use no-tillage methods on 20–30 percent of all corn acreage, and 30–50 percent of all soybean acreage.

Advantages listed for no-till include (1) reduced production costs, (2) reduced runoff, (3) less wind and water erosion, (4) better moisture retention, (5) less soil damage, such as compaction, from machinery, (6) better timing in planting and harvesting, (7) savings in labor, and (8) reduction in some weather risks. In addition, in some situations, no-till has resulted in yield increases. Significantly less total machinery investment and lower power requirements are characteristic of this method. Some land with slopes too steep for conventional row cropping may be suitable for no-tillage production without excessive erosion hazards. A summary of number of field operations in producing a crop by no-tillage as compared to minimum and conventional tillage systems is shown in Table 16–6, and a rating of some problems in crop production according to tillage system is listed in Table 16–7.

The effectiveness of no-tillage systems in minimizing runoff and reducing erosion is shown in Tables 16–8 and 16–9. In Ohio, no-tillage involving a dead sod on the surface reduced soil loss by nearly six times as compared to a tillage system with no cultivation (Table 16–8). No-tillage on the contour has been observed in some instances to reduce soil loss to only a trace, similar to a field established in meadow (Table 16–9).

A number of researchers have obtained excellent yields from no-tillage plots. Van Doren, Triplett, and Henry at the Ohio Station obtained better corn yields (six-year average) with no-till than a conventional system on a well-drained silt loam (Table 16–10). A three-year rotation involving corn, oats, and alfalfa meadow improved corn yields on both conventional and no-till plots. They concluded that reduced tillage generally has not decreased yield of corn, provided the stand of corn and weed control are adequate. However, they emphasize that more managerial skill is needed

TABLE 16–6. Number of Operations (Trips Over the Field) with Conventional, Minimum, and No-Tillage Methods

Operation	Conventional Tillage	Minimum Tillage	No-Tillage
Plowing	1	1	0
Disking	2 or more	0 or more	0
Planting	1	1 or 0	1
Spraying	0 or more	0 or 1	1
Cultivating	2 or more	1 or 2	0
Harvesting	1	1	1
Total trips	7 or more	4 or 5	3

Source: Phillips, S. H., and H. M. Young, Jr. *No-Tillage Farming* (Milwaukee: Reiman Associates, 1973).

TABLE 16-7. Rating of Problems of Crop Production According to Tillage System.

Crop-Production Problem	Conventional Tillage	Minimum Tillage	No-Tillage
1. Acceptable weed control, allowing for variations in soil types.	A*	A	A
2. Insect control.	A	A	A
3. Variety responses to tillage practices.	B	B	B
4. Proper fertilizer placement.	B	B	B
5. Best plant population and plant spacing in the row.	A	A	A
6. Best row widths.	A	A	A
7. Hedging against certain weather risks.	B	B	B
8. Crop rotations best suited to the tillage system.	B	B	B
9. Most economic machine and power balance.	B	B	B
10. Harvest stand compared to population goal at planting.	B		B
11. Lodging due to soil conditions or root development.	B	B	A
12. Soil moisture retention, in comparable soils.	C	B	A
13. Improving timeliness of operations.	B	A	A
14. Application, infiltration and retention of irrigation water on comparable soils and slopes.	C	B	A
15. Erosion control on soils subject to wind or water erosion.	C	B	A

* A, Research data and farmer experience are sufficient to offset most currently recognized problems, although continuing research and testing are needed. B, Further research, coupled with careful on-the-farm testing, is needed to help solve moderately bothersome problems. C, Severe problems exist, indicating that much research and farmer testing need to be done.
Source: Phillips, S. H., and H. M. Young, Jr. *No-Tillage Farming* (Milwaukee: Reiman Associates, 1973).

to farm a no-tillage or minimum tillage operation. The erosion control, time savings, and possible yield increase may make it worth the effort.

Carreker and his co-workers have reported excellent corn yields from no-till planting in tall fescue sod in the southern Piedmont region. Several plots in Georgia produced 191–208 bushels per acre (11,980–13,046 kg/ha), depending on herbicide, fertilizer, and manure rates (Table 16–11). These researchers concluded that the beneficial effects of mulch and sod on improving water infiltration, and reducing runoff, erosion, and nutrient losses

TABLE 16–8. Effect of Tillage Treatments on Runoff and Erosion on a Wooster Silt Loam, 5 Percent Slope (Ohio)

Tillage System	Runoff (in.)	Erosion (Tons/Acre)
No-tillage, dead sod on surface	1.35	3.55
No-tillage, bare soil surface	2.91	8.73
Conventional tillage, cultivated	1.79	16.48
Conventional tillage, not cultivated	2.19	20.92

Source: Phillips, S. H., and H. M. Young, Jr. *No-Tillage Farming* (Milwaukee: Reiman Associates, 1973).

make no-till widely adaptable on sloping lands when nutrient and water requirements are met.

Phillips has estimated that 60–80 percent of the crop acreage will be planted by no-tillage during the next 25 years. He concludes that the following factors will dictate the percentage of crops grown by no-tillage and the rate of change from traditional methods:

1. Ability to control perennial weed problems.
2. Regulation and availability of chemicals.
3. Changing economics of crop production.
4. Increasing environmental concerns for sediment control.
5. Cost and availability of energy and world need for food and fiber.

TABLE 16–9. Soil Loss Under Various Treatments (Ohio)

Treatment or Crop	Average Annual Soil Loss (Tons/Acre)
Rotation plowed cornland, sloping rows	7
Rotation plowed cornland, contour rows	2
No-tillage mulch cornland, contour	Trace
Wheatland	1
Meadow	Trace

Source: Phillips, S. H., and H. M. Young, Jr. *No-Tillage Farming* (Milwaukee: Reiman Associates, 1973).

TABLE 16–10. Effect of Cropping System (Continuous Corn versus Rotation) and Tillage (Conventional Preparation versus No-Till) on Corn Yield on a Well-Drained Silt Loam, Six-Year Average (Ohio)

	Yield (Bu/acre)	
	Conventional (Plowed)	No-Till
Continuous corn	134	150
Corn-soybeans (2-yr rotation)	139	151
Corn-oats-alfalfa meadow (3-yr rotation)	155	167

Source: Van Doren, D. M. Jr., *et al.* "No-Till Is Profitable on Many Soil Types," *Crops and Soils,* 27(9):7–9 (1975).

TABLE 16–11. Yield of No-Till Corn in Tall Fescue Sod with Fertilization Variables (Georgia)

Nitrogen (Pounds/Acre) + Poultry Litter (Tons/Acre)	Atrazine + Paraquat (Pounds/Acre)					
	0 + 0	0 + ¼	½ + ¼	1 + 0	1 + ¼	2 + ½
			Yield (Bu/Acre)			
130	4.2	12.7	42.1	37.7	77.4	154.5
230	29.1	52.8	108.6	83.0	131.7	191.6
430	36.2	89.9	105.3	112.4	143.8	184.0
130 + 2.5	55.1	—	—	—	—	181.4
130 + 5	83.3	—	—	—	—	198.4
130 + 10	82.1	—	—	—	—	173.8
130 + 20	68.3	—	—	—	—	120.8
430	26.0	—	—	—	—	208.6

Source: Carreker, J. R. *et al. Soil and Water Management Systems for Sloping Land,* USDA-ARS-S-160 (1977).

Cultivation

Row crops, or intertilled crops, are almost always included in a well-planned cropping system. One of the important advantages of including a cultivated crop in a rotation system is for control of certain perennial weeds such as quackgrass, bindweed, and thistles which can tolerate regular mowing when present as weeds in a three- or four-year meadow. The crops commonly grown in rows, and sometimes cultivated, are corn, soybeans, cotton, grain sorghum, potatoes, peanuts, sugarbeets, sugarcane, and tobacco.

Formerly, the only practical methods known to control weeds were (1) stirring the soil at frequent intervals when the weed seeds were germinating, (2) uprooting weeds when in the seedling stage, (3) cutting weeds below the soil surface, and (4) smothering the weeds. A turning point in the history of weed control was reached in 1944 when experiments showed that 2,4-D would selectively kill many broadleaf weeds without serious injury to corn. Since that time dozens of herbicides have been developed for weed control in field crops. Some herbicides will control broadleaf weeds without injury to the crop; others can be used successfully to control grass weeds without crop injury. Such chemicals are termed "selective herbicides."

Certain herbicides can be applied and incorporated before the crop is planted (preplant incorporated), others at time of planting or after planting but before the crop has emerged from the soil (pre-emergence), and others can be applied successfully after the crop and weeds are up (postemergence).

Weeds in cotton and other selected crops may be controlled by "flame cultivation." This was first used in Alabama about 1936. The flame is used to kill small weeds in the rows. With careful adjustment of the flame, proper driving speed, and correct angle of the flame direction, small weeds can be controlled in a number of crops without crop injury.

Kinds of Cultivating or Intertillage Implements

Cultivation implements vary widely in type and size. At one extreme are implements that cultivate only one side of the row, and on the other extreme are large, multirow cultivators. These implements may be equipped with shovels, sweeps, teeth, or knives.

One of the most widely used types is the *shovel* or *sweep* cultivator. It is suited to most soil types. Multirow cultivators must correspond to the row width used. Shovels or sweeps may be replaced or supplemented with

FIGURE 16–8. *Multi-row crop cultivator with sweeps.* [SOURCE: J. I. Case.]

disks when considerable soil is to be moved, or when a large amount of trash must be cut. The disk cultivator is used widely to cultivate listed crops.

The *lister cultivator* is a special tool equipped with disks, knives, or both for listed crops planted in furrows. Disks are set to cut the weeds on the sides of the furrow at the time of the first cultivation; to roll the soil into the furrow, bury the weeds, and to level the rows during the last cultivation.

The *spike-tooth harrow* can be used before corn is up and until it is 4–5 inches (10–13 cm) in height without injury to the crop. It effectively kills small weeds, those in the seedling stage, in fields relatively free of trash so that clogging is not a major problem.

The *rotary hoe* has come into widespread use for cultivating row crops or breaking a soil crust to aid in emergence. It consists of a series of hoe wheels, each of which is equipped with fingerlike teeth. The wheels rotate and teeth penetrate and stir the soil. This implement uproots small weeds and breaks the crust that may result from drying of the soil surface after an intensive rain.

Cultivation Methods and Yield

Some have concluded that the only reason for cultivation is for controlling weeds. Other researchers have suggested that cultivation aids in maintaining soil conditions favorable to the growth of plants. A number of factors are involved, some of them better understood than others. Experimental results do not always justify the expense incurred by some of the more careful farmers in their cultivation practices.

In past years the different stations did a great deal of detailed experimental work in comparing different cultivation practices. Little experimental work has been done in recent years on cultivation practices in relation to crop yields. It is believed that the student of crop production should know something of the research done in the earlier years on cultivation, and for this reason some of the earlier work and results are reported here.

Early Results from New York and Illinois

The New York station in 1886 reported a comparison of different cultivation treatments for corn. Where weeds were allowed to grow, the yield was 18.1 bushels per acre (1,135 kg/ha), where the weeds were pulled but where there was no cultivation the yield was 70.5 bushels (4,422 kg/ha), and normal cultivation and hoeing yielded 56.8 bushels (3,562 kg/ha). The New York station reported that no results that they had ever published were so widely criticized and commented on as the preceding.

The Illinois station reported work on cultivation for the years 1888 to 1893, and again in 1896. The results of this work are presented in Table 16–12.

TABLE 16–12. Bushel Yield of Corn As Influenced by Method of Cultivation (Illinois)

Nature of Cultivation	Seven-Year Average Yield
None, weeds scraped off with a hoe	68.3
Shallow, 4–5 times	70.3
Deep, 4–5 times	66.7
Shallow, 12–14 times	72.8
Deep, 12–14 times	64.5

Early USDA Report on Corn Cultivation

That the control of weeds and not the stirring of the soil is the important consideration was indicated by early USDA work in cooperation with farmers in widely scattered states, as shown in Table 16–13.

On the scraped or uncultivated plots the soil was stirred as little as possible. The weeds were controlled by cutting them off at the soil surface with the horizontal stroke of a hoe when the weeds were very small. The water runoff from surface-scraped plots was reported to be more than that from plots receiving ordinary cultivation.

Weed control experts now believe that shallow scraping with a hoe accomplished much of the effect of regular cultivation, and the conclusions that cultivation was for the sole purpose of weed control sometimes were erroneous. More recent studies which compare chemical weed control methods with cultivation have shown that cultivation may increase yields on some soils which are in less than ideal condition.

TABLE 16–13. Ordinary Cultivation Compared with No Cultivation but with the Weeds Controlled by Scraping the Surface*

State	Number of Trials	Relative Yield from Scraped Plots When Cultivated Plot = 100%
Iowa	7	102.72
Illinois	8	94.74
Indiana	9	105.36
Kentucky	9	91.28
Maryland	1	109.20
Michigan	7	116.26
Minnesota	1	100.00
Missouri	3	103.20
Nebraska	1	101.90
New Hampshire	10	112.71
South Carolina	12	99.07
Virginia	9	88.51

* Yield of scraped plots in percent of yield from ordinary cultivation (summary).

TABLE 16–14. Tons of Water Loss per Acre in 100 Days with Soil Mulches of Varying Depths

	No Mulch	1-In. Mulch	2-In. Mulch	3-In. Mulch	4-In. Mulch
Black marsh soil	588	355	270	256	253
Sandy loam	742	374	339	288	315
Virgin clay loam	2,414	1.260	980	889	884

Source: King, F. H. *Physics of Agriculture,* 6th ed. (Madison, Wisc.: Mrs. F. H. King, 1914).

Early Studies with Dust Mulch

King, at the Wisconsin station, was among the first to study the effect of mulches on the evaporation of water from the surface of the soil. He placed soil in cylinders which were approximately 4 inches (10 cm) in diameter and less than 20 inches (51 cm) in height to test mulches of different depths and their effect on the water loss by evaporation from the soil surfaces. His results, of far-reaching effect in influencing recommendations made in the past on cultivation, are given in Table 16–14.

The results shown in Table 16–14 were the main basis for the so-called dust-mulch theory with reference to cultivation. This theory called for frequent cultivation in order to maintain a loose layer of soil on the surface. The dust-mulch theory became thoroughly established, and the maintenance of a dust mulch was generally recommended through many years. It was not until considerably later that it came to be recognized generally that soil moisture can move upward in a soil by capillary attraction for only a short distance above the water table, and that under normal field conditions a dust mulch has no moisture-conserving value.

Frequency and Distribution of Cultivation

In an early six-year comparison of various frequencies of cultivation in Nebraska, plots given no cultivation, one cultivation, or two, three, and four normal cultivations yielded, respectively, 7.1, 21.6, 33.6, 35.9, and 37.2 bushels per acre (445, 1355, 2107, 2252, 2333 kg/ha). Continued late cultivation, after corn was normally laid by, reduced the yield 2 bushels per acre (125 kg/ha). The corn plots that were scraped with a hoe to prevent weed growth yielded 2.1 bushels per acre (132 kg/ha) less than corresponding corn which received four normal cultivations.

Depth of Cultivation

Different methods of cultivating corn were compared at the Arkansas station over a period of 18 years. The yield from different methods of cultivation is compared with that obtained from a standard method, corn cultivated to a medium depth three times before tasseling. The results are given in Table 16–15.

TABLE 16–15. Corn Yield from Different Methods of Cultivation Compared with That from Three Medium Cultivations (Arkansas)

Cultural Treatments	Number of Tests	Bushels (+) or (−) Yield from Three Cultivations
No cultivation	17	−31.3
Very shallow cultivation*	15	−2.0
Scraped with hoe	25	−2.3
Medium deep cultivation	15	+1.0
Deep cultivation	25	−1.87
Shallow, then deep	5	+1.4
Deep, then shallow	13	+1.4

* Very shallow cultivation was limited to one inch in depth.

Source: Nelson, M. and C. K. McClelland. *Cultivation Experiments with Corn,* Ark. Agr. Exp. Sta. Bul. 219 (1927).

Iowa and Nebraska Studies

Corn cultivation studies at the Iowa station through eight years indicated that it made little difference as to the method or the cultivating tools used if the weeds were equally well controlled. In almost any given year the yield differences were exceedingly small. The conclusion was that the important consideration is weed control with minimum expenditures for labor and power.

The Nebraska station concluded from 11 years of corn cultivation studies that it did not make much difference how corn was cultivated if weed growth was controlled without excessive injury to the corn roots. Average bushels per acre from different cultivation methods ranged only from 32.9 to 35.2 bushels (2,063–2,208 kg/ha). But when corn roots were cut to a depth of 6 inches (15 cm), 7 inches (18 cm) from the plants, yields were reduced 12 percent, and when cut 4 inches (10 cm) deep, 6 inches (15 cm) from the plants, the reduction was 20 percent.

REFERENCES AND SUGGESTED READINGS

1. ALDRICH, S. R., W. O. SCOTT, and E. R. LENG. *Modern Corn Production,* 2nd ed. (Champaign, Ill.: A&L Publications, 1975).
2. Anonymous. "Primary Tillage," *Succ. Farming,* p. 29 (March 1978).
3. BLACK, C. A. *Soil–Plant Relationships* (New York: Wiley, 1968).
4. CARREKER, J. R., S. R. WILKINSON, A. P. BARNETT, and J. E. BOX. Soil and Water Management Systems for Sloping Land. USDA-ARS-S-160 (July 1977).
5. CATES, J. S., and H. R. COX. *The Weed Factor in Corn Cultivation,* USDA Bur. Pl. Ind. Bul. 257 (1912).
6. FAULKNER, E. H. *Plowman's Folly* (Norman: Univ. of Oklahoma Press, 1943).
7. FENSTER, C. R., H. I. OWENS, and R. H. FOLLETT. *Conservation Tillage for Wheat in the Great Plains,* USDA-Ext. Serv. PA-1190 (July, 1977).
8. HAYES, W. A. Mulch Tillage in Modern Farming. USDA Leaflet 554 (1971, Reprinted 1977).

9. HINZ, W. W. "Reducing Tillage to Conserve Energy and Increase Profits," Ariz. Coop. Ext. Serv. Pub. Q38.

10. HUME, A. N. *Crop Yields as Related to Depth of Plowing.* S.D. Agr. Exp. Sta. Bul. 369 (1943).

11. KIESSELBACH, T. A., *et al. Tillage Practices in Corn Production,* Neb. Agr. Exp. Sta. Bul. 232 (1928).

12. KIESSELBACH, T. A., *et al. Cultural Practices in Corn Production,* Neb. Agr. Exp. Sta. Bul. 293 (1935).

13. KING, F. H. *Physics of Agriculture,* 6th ed. (Madison, Wisc.: Mrs. F. H. King, 1914).

14. LANE, D. E., and H. WITTMUSS. *Nebraska Till-Plant Systems.* Neb. Ext. Circ. 61–714 (1961).

15. LARSON, W. E., A. R. BERTRAND, and V. C. JAMISON. *Subsoiling in the North Central States,* Mimeographed Report (Ames: Iowa State University, 1963).

16. MARTIN, J. H., W. H. LEONARD, and D. L. STAMP. *Principles of Field Crop Production,* 3rd ed. (New York: Macmillan, 1976).

17. NELSON, M., and C. K. MCCLELLAND. *Cultivation Experiments with Corn,* Ark. Agr. Exp. Sta. Bul. 219 (1927).

18. PHILLIPS, S. H. "No-Tillage, Past and Present," *Proc. of First Annual Southeastern No-Till Systems Conf.* Ga. Spec. Pub. 5 (1978).

19. PHILLIPS, S. H., and H. M. YOUNG, JR. *No-Tillage Farming* (Milwaukee: Reiman Associates, 1973).

20. REICHENBERGER, L. "Survey of Tillage Operations," *Succ. Farming* (March 1978), p. 30, Vol. 76.

21. SEWELL, M. C. "Tillage: A Review of the Literature," *Jour. Amer. Soc. Agron.,* 11:7 (1949).

22. SHEDD, C. K., *et al. Weed Control in Corn,* Iowa Agr. Exp. Sta. Bul. (1942). 44 pp. May 1942 (no vol. no.).

23. SMITH, R. S. *Subsoiling, Deep Tilling, and Subsoil Dynamiting,* Ill. Agr. Exp. Sta. Bul. 258 (1925).

24. TOUCHTON, J. T., and D. G. CUMMINS. *Proc. of First Annual Southeastern No-Till Systems Conf.* Ga. Spec. Pub. 5 (1978).

25. VAN DOREN, D. M., JR., G. B. TRIPLETT, JR., and J. E. HENRY. "No-Till is Profitable on Many Soil Types," *Crops and Soils,* 27(9):7–9 (1975).

26. VOORHEES, W. B. "Soil Compaction, Our Newest Natural Resource", *Crops and Soils,* 29(5):13–15 (1977).

27. *Weeds in Corn,* N.Y. (Geneva) Agr. Exp. Sta. Ann. Rep. 5 (1886).

CHAPTER 17

IRRIGATION, DRAINAGE, AND DRYLAND FARMING

The vital role of water for food production is well known in both rainfed and irrigated agriculture. As in the past, today's farmers must manage water to obtain both favorable yields and profits. Less well known are the effects on water management of new societal attitudes about the environment, and resource conservation. . . . The clever mind of man has found no substitute for water in supporting all life forms and in producing our food and fiber. Forseeable demands on water must increase in both diversity and magnitude. Water demands for energy technologies alone will be substantial and necessarily compete with water for agriculture and other uses where water supplies are already over-committed, as in the Colorado River Basin. Not only are water resources physically limited in quantity, quality, and occurrence, water use is further constrained by legal and numerous institutional arrangements. The essential character of water resource management in the future will not be the development of new supplies. Instead it will be more intensive management of relatively fixed water supplies and, possibly also, reallocation of existing water supplies among competitive uses and users. Supply problems will no longer be mainly a matter of finding reservoir sites and suitable canal alignment. A mix of water uses must be developed that will support the varied goals of society.

R. M. HAGAN

LAND use is affected by physical, economic, and social conditions. Altitude, topography, climate, soil, location, and relationship between land and water, when each is considered either alone or in combination, are the principal factors that affect land use. The length of the growing season and the fertility of the soil are recognized for their importance in influencing productivity. But the amount and distribution of rainfall or the supply of water otherwise made available or controlled are the prime factors limiting the suitability of land for agricultural use.

Water and Use of Cropland

In some areas production is impossible because of lack of water; in others there is an excess of water. Based principally on precipitation, the country can be divided into two main parts, the East and the West. The dividing line approximates longitude 100° and the 20-inch (51-cm) precipitation zone. The climate of the East is mainly humid or subhumid; the West may be characterized as arid or semiarid, except along the north Pacific Coast.

Irrigation

An average annual precipitation of 20 inches (51 cm) often is considered the minimum for cropping without irrigation. With fewer than 10 inches (25 cm) or rainfall per year desert conditions prevail, and water for irrigation is necessary for agricultural production. Fifty-five percent of the world's total land area is arid or semiarid.

Early Irrigation

The Chinese are known to have been irrigating farmland for at least 4,500 years; much of China still produces crops. In both Egypt and China flooding and silting has had much to do with maintained production in certain parts of these areas. In northern Africa, in the region where Carthage once stood, extensive irrigation systems once carried large volumes of irrigation water long distances, even into parts of what is now the Sahara Desert. The Euphrates River Valley in Asia Minor contains the remains of some of the largest irrigation canals ever built. In the vicinities of Baghdad and ancient Babylon irrigation once maintained large populations; now all that remains is desolate waste.

Remains of early civilizations based on irrigation also are to be found in the American Southwest and in Mexico. The cause for the ultimate failure and abandonment of these early irrigation efforts is not always clear. In general, poor soil management and the accumulation of toxic salts are considered likely contributing factors.

Extent of Irrigation Enterprises

A survey of world agricultural efforts has shown more than 560 million acres (227 million ha) of land to be under irrigation. Countries having the largest areas of irrigated land are the following:

China: More than 209 million acres (85 million ha)
India: Nearly 80 million acres (32 million ha)
United States: More than 40 million acres (16 million ha)
USSR: More than 35 million acres (14 million ha)
Pakistan: More than 35 million acres (14 million ha)

Acreage of irrigated lands in the world is continually increasing, accounting for more than 16 percent of all arable land. An even greater proportion of the land is devoted to food production. Irrigation is very important in some countries because of climate variability and heavy reliance on certain crops. Many countries have not realized their full irrigation potential. For example, India has increased its irrigated acreage by nearly 14 million acres (5.7 million ha) since 1966, but it is estimated that India could irrigate an addition 21 million acres (8.5 million ha). Full development would enable India to increase considerably her arable land base of 162 million ha.

Irrigation in the United States

The number of acres irrigated in the United States has increased from about 8 million in 1900 to more than 40 million in 1974 (Table 17–1; Figure 17–1). Where such major export crops as cotton, wheat, and rice are grown under irrigation, export levels influence the acreage irrigated, as does public policy.

WESTERN STATES. In the 17 Western states irrigated farmland used primarily for pasture or harvested livestock feed accounts for about 50 percent of the total irrigated land. In terms of acreage, hay and pasture are the most important uses of irrigated lands. Small grains make up about 20 percent of the irrigated lands harvested. The largest wheat crop in the United States was produced recently under irrigation near Boardman, Oregon. Winter wheat averaged 104 bushels per acre (6,989 kg/ha) on 10,405 acres (4,211 ha) of semidesert land irrigated by means of a center-pivot system with water pumped from the Colombia River. The remaining irrigated land produces high-value crops. Included are tree fruits and vineyards, vegetables, potatoes, cotton, sorghum, sugarbeets, beans, and legume seed. In these states more than 35 million acres (14 million ha) are irrigated.

Many parts of the Western United States are facing present or potential water shortages. Nearly every major Western river has turned into a legal and political battleground in a struggle over the rights to dwindling water supplies. In recent years, Western states have experienced a population explosion, accompanied by an increase in industrial plants. As a result

TABLE 17–1. Acreage Irrigated in the United States by Census Periods 1890–1950, 1964, 1974

| Year | Acreage Irrigated | | |
	17 Western States 1,000 acres	31 Eastern States* 1,000 acres	Total 1,000 acres
1890	3,632	85	3,717
1900	7,543	246	7,789
1910	11,259	408	11,667
1920	13,883	599	14,482
1930	14,086	603	14,689
1940	17,243	740	17,983
1950	24,271	1,517	25,788
1964	33,208	3,704	37,056
1974	35,840	5,403	41,243

* The 1910, 1920, and 1930 acreage figures are for Arkansas and Louisiana only. Data on other Eastern states for these years are not available. The 1950 Census lists the irrigated acreage for Arkansas, Louisiana, and Florida at 422,000; 477,000; and 365,000 acres, respectively. The 1974 census period includes Alaska and Hawaii.

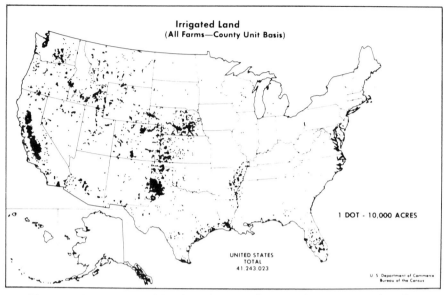

FIGURE 17-1. *Irrigated land in the United States shown geographically. More than 35 million acres are irrigated in 17 western states.* [SOURCE: U.S. Department of Commerce.]

of an increased water demand, some of the region's major underground water sources are drying up. The Colorado River has been a focal point of battles over water rights since 1922. The current claims on the Colorado River currently exceed the river's average annual flow of 14 million acre-feet. There are also excessive demands on the Yellowstone, the Missouri, the Platte, the Powder, and the Snake rivers. Efforts to expand irrigated farming in the West are meeting with strong opposition from environmentalists and others as well.

HUMID-AREA STATES. In humid areas irrigation is increasingly used to supplement natural rainfall. Almost 10 percent of the approximately 5 million irrigated acres (2 million ha) in the 31 humid-area states plus Hawaii and Alaska is in Florida, where such high-value crops as vegetables, citrus fruits, potatoes, shade-grown tobacco, berries, and nursery crops lead in importance. The other largest area of irrigated land is in Louisiana, Arkansas, and Mississippi, accounting for one-third the irrigated land outside the Western states. Three crops predominate in these Delta states— rice, cotton, and soybeans.

Basic Information

Essentially, irrigation is a method whereby water is provided for plant growth when the supply of rainfall is inadequate. It also helps control soil and air temperatures and to leach the soil of excess soluble salts.

Careful consideration must be given to a number of factors in order to determine the feasibility of irrigation under a given condition and the type of irrigation to be adopted. These factors include:

1. Adequacy and suitability of available water supply
2. Characteristics of the soil
3. Topography
4. Field sizes and shapes
5. Pump and power requirements
6. Acres that can be irrigated
7. Crops to be grown
8. Potential yields

Also to be considered, are the annual costs, among which are depreciation, interest, operation, and maintenance.

With mounting energy problems, consideration should be given to energy needs for irrigation systems. On 35 million acres (14 million ha) irrigated in the United States in 1974 with water pumped from wells, rivers, and lakes, about 260 trillion BTUs of energy costing $594 million were required to pump the water. Electricity was used for the largest acreage, followed by natural gas, diesel fuel, liquid petroleum (LP) gas, and gasoline (Figure 17–2). Irrigation costs have skyrocketed for electricity and natural gas. Solar energy may one day regularly power irrigation pumps. Several pilot projects on the use of solar energy for irrigation currently are in effect in Arizona.

Available Information on Irrigation Techniques

The different state experiment stations and several federal agencies, especially the United States Department of Agriculture (USDA) and the Bu-

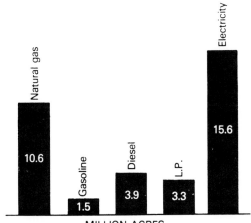

FIGURE 17–2. *Acres irrigated with pumped water by type of energy used. With mounting energy problems, and energy costs for irrigation having reached nearly 600 million, consideration is being given to alternative sources, such as solar energy.* [SOURCE: USDA.]

reau of Reclamation, Department of the Interior, have done both intensive and extensive research on irrigation problems. Numerous publications are readily available on the many factors contributing to effective irrigation procedures.

The Land Itself

Trained personnel of the Bureau of Reclamation and the Soil Conservation Service classify land according to its suitability for irrigation development, taking into consideration the topography and the characteristics of the soil itself. From the standpoint of irrigability and uses, land is graded into six classes. Class 1 represents lands best suited for development. At the other extreme, Class 6 includes lands with so low a productivity that they are permanently nonarable.

Water Relations of Soils

Probably the most important single factor in determining proper irrigation practices is the character of the soil itself. Texture, structure, and porosity relate directly to the water-retaining and water-transmitting properties of a soil.

The water-retaining capacity of a soil may be expressed as (1) the percentage of dry weight of the soil, (2) the percentage of the volume of the soil, or (3) the depth of water held in a given depth of soil. The third method is most useful in irrigation practices. A common unit of measure is the inches-depth of water per foot-depth of soil. Tensiometers, soil moisture blocks, or other sensing units, frequently are used to measure the availability of water to the plants. The tensiometer is an instrument used to determine soil moisture content in a range of 0–850 cm of surface water tension.

Heavy types of soil may have a water-holding capacity of 2 inches (5 cm) or more per foot (30 cm) of soil depth; with a sandy soil this may be less than 1 inch (2.5 cm). Soil permeability is important. The rate at which irrigation water is applied should be in accordance with the rapidity with which the particular soil will absorb it.

The infiltration capacity of a soil is of great importance. At one end of the soil spectrum the infiltration capacity may be so great that it is virtually impossible to make an irrigation stream flow down a furrow or across a field of reasonable length, such as in the very coarse sands. At the other extreme water may be absorbed by the soil so slowly that, to avoid crop damage, continuous irrigation is required during periods of rapid transpiration. In either case irrigation by ordinary surface methods is impracticable. The Soil Conservation Service has developed an accurate classification of soils according to relative intake rate.

The Crops To Be Grown

Irrigation practices must be adapted to the nature of the crops being grown, and especially to their rooting habits and soil-water requirements.

The *consumptive use* of water by plants is defined as the total quantity of water transpired by the crop, plus that evaporated directly from the soil on which the crop is growing. This usually is stated in terms of depth of water in feet, or inches, per season or year. Knowledge of consumptive-use requirements is necessary in planning irrigation and drainage systems,

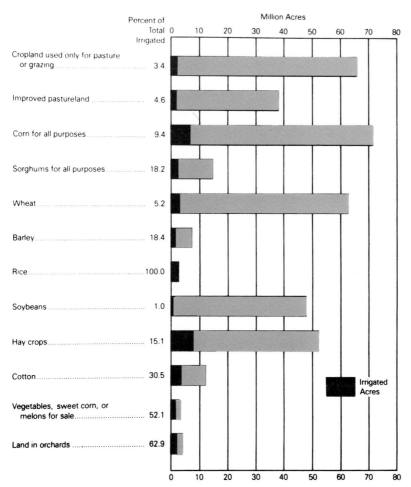

Figure 17–3. *Proportion of pastureland and selected crops irrigation in the United States. (Farms with sales of $2500 and over.) The crops with the largest percentage of acreage irrigated are rice, orchards, vegetables, and cotton. The greatest number of irrigated acres are accounted for by hay crops and corn.* [Source: USDA.]

improving irrigation practices, and aiding in scheduling irrigation. Consumptive use is affected by many factors—some artificial and others natural. The more important natural factors are climate, soils, and topography. Climatic factors include precipitation, solar radiation, temperature, humidity, wind movement, and length of growing season.

The artificial factors affecting consumptive use of water by plants usually can be controlled by man. Many are interrelated with the natural and climatic factors. They include water supply, water quality, date of planting, crop cultivar, fertility, plant spacing, water management, cultivation, and chemical sprays.

Consumptive use by various agricultural crops has been investigated at many locations for more than 60 years. The most common methods of determining the amount of water consumed by crops have been (1) to measure the amount of water necessary to maintain satisfactory plant growth in tanks or lysimeters; (2) to measure the quantity of water applied to field plots; and (3) to determine soil-moisture depletion.

Tentative irrigation schedules can be calculated directly from consumptive-use curves. Direct transfer of consumptive-use data to an area with widely divergent climatic conditions is not valid. Several methods, however, have been proposed for making such transfers.

Three practical methods are those devised by Blaney and Criddle, Penman, and Jensen and Halse. The Blaney–Criddle method determines evapotranspiration from climatological and irrigation data. The procedure is to correlate existing evapotranspiration data for different crops with the monthly temperature, percentage of daytime hours, and length of growing season. The correlation coefficients then are applied to determine the evaporation for other areas where only climatological data are available. The Penman method approaches the problem of estimating the evaporation from a free-water surface by examination of the energy balance at the surface. The solar radiation method of Jensen and Halse presents an energy-balance approach to estimating evaporation and is simpler in application than the Penman method.

The drought resistance of a crop is only slightly, if at all, related to the economy with which it uses water as expressed by the transpiration ratio. Drought resistance is chiefly the ability to slow up transpiration and other growth functions when water is not available and to start up transpiration when water can again be obtained.

Crops commonly grown under dryland farming are, in most cases, suitable for growing under irrigation. There are additional crops, however, that may be introduced and grown more profitably when the land is to be irrigated. Examples are sugarbeets, potatoes, beans, soybeans, and so on, as well as the more productive forages and forage mixtures.

To be economically successful, irrigation requires large acre yields of crops with good market value. It means that the higher acre-production cost can be met along with a surplus yield for profit. With irrigation relatively large expenditures for fertilizer usually are justified and often necessary in order that the yields made possible by the assurance of adequate water will not be reduced because of low soil-fertility levels.

Water Requirement of Crops

The water required by different crops is seasonal and varies considerably in the total required. Small grains mature relatively early and therefore require water only in the early part of the growing season. Alfalfa and pastures usually demand more water than most crops because of their long growing period. Late-maturing crops, such as corn and sugarbeets, require more water later in the growing season. In general, the amount of water to be applied per irrigation and the frequency of application are governed by the water-holding capacity of the soil in the root zone of the crops being grown.

Quality of Irrigation Water

A determination of the quality of irrigation water available is important. The total quantity of dissolved salts has most often been used as a measure of water quality. The electrical conductivity of the water is at least as good a measure as any, besides being easy to determine.

Damage to the soil is caused by the sodium that may be carried in irrigation water. Fine-textured soils absorb certain bases from the water. If the irrigation water contains more sodium than it does calcium and magnesium, there is a tendency for this sodium to replace the calcium and magnesium already in the soil. The presence of an excess of sodium makes soil less permeable. It generally is believed that the ratio of sodium ions to the sum of sodium, magnesium, and calcium ions should not exceed 50–60 percent.

Soil and water management practices must be given careful attention when saline waters are used. Proper cultural practices should be observed to avoid the accumulation of salts at or near the soil surface, particularly when the plants are small. Soil management practices in areas where salty waters are used for growing crops whenever possible should include the restriction of salty water to only the most permeable soils. Salty water can be used most effectively in the leaching of salts and in reclamation.

Optimum crop management practices should be employed to overcome the adverse effects of soluble salts in soils and to maintain an efficient permanent agriculture in arid and semiarid regions, where salts tend to accumulate under conditions of irrigated agriculture. These practices include (1) selection of crops and cultivars most tolerant to salt, (2) effective placement of seed, (3) rotation of crops, and (4) development and adoption of effective cultural practices to improve stand establishment.

Desalination and Salt-Tolerant Crops

One solution to low-quality irrigation water and salt buildup would be desalination or salt removal. One could envision massive irrigation projects with desalinated seawater in desert regions, if such a method were economically feasible.

Considerable desalination research is being conducted at several locations. The current cost of desalinating seawater is about 50 cents per 1,000 gallons (3,785 liters); the goal is to reduce the price of desalted seawater for agricultural purposes to 10–20 cents per 1,000 gallons. Current methods of desalination include distillation, freezing, reverse osmosis, electrodialysis, and ion exchange. But economical, large-scale desalination is still not a reality—although in some cases a necessity—and awaits further technological advances. Future techniques could involve the use of solar energy, tidal movements, reject steam from power plants, or geothermal power.

Recently some researchers at Purdue University tested a solar still for converting salt water into seawater directly in the field. This would allow desert lands near oceans to be irrigated with seawater. The solar still technique, which involves the principles of vaporization by the suns rays and condensation, is being adapted to field use: solar energy causes the salt water in a furrow to evaporate and condenses the vapor droplets, which run down a specially constructed plastic tent to where the crop is planted. The cost should not be much higher than what some Western growers are paying for irrigation water. This technique might be feasible now for high-value crops in desert countries forced to import most of their food.

An alternative to producing salt-free water for use on salt-susceptible crops is to develop salt-tolerant crops, which can be irrigated with seawater. At present practically all agricultural species are sensitive to salinity. Difficult salt problems currently exist in the lower Colorado basin, the San Joaquin Delta, and several other areas of the Western United States.

Salt-water irrigation of cotton and sorghum is being studied by USDA–SEA researchers from the U.S. Salinity Laboratory at Bakersfield, California. These workers are measuring the effect of saline agricultural drainage "waste" water with salt concentrations of 200–5,800 parts per million (ppm) on crop growth and yield. (Undiluted seawater is about 35,000 ppm.) Epstein and his co-workers at the California Station have set a goal of developing crops much more tolerant to salt than are present cultivars. They have made progress in selecting barley lines that can be grown in sand dunes and irrigated with undiluted seawater. Also included in the project are wheat and tomatoes.

How Water Is Applied

Water is applied to the soil in many ways. The best method for a particular case depends on such factors as the topography, the character of the soil, the water supply, the kind of crop, and the owner's experience.

Water may be applied to crops by (1) the sprinkler method, or aerial irrigation, which requires the use of pipes and a certain amount of pressure; (2) surface methods, that is, by allowing it to flow onto the land; or (3) subsurface irrigation. Subsurface irrigation may be accomplished by underground perforated pipes. Any one or a combination of methods may be best suited to a given farm unit.

SPRINKLER IRRIGATION. The availability and improvement of aluminum tubing and castings following World War II resulted in a great deal of interest in movable sprinkler irrigation systems. This is the best method to use on soils that have high intake rates, or on fields that have steep slopes or topography, and on soils too shallow to level.

A sprinkler system consists of a pumping unit, mainline pipe unit, lateral pipe unit, and sprinkler unit. Permanent pipelines are made of steel, asbestos, cement, or plastic. Portable pipelines are made of aluminum or plastics, and are equipped with quick coupling devices. Sprinkler systems may be multisprinklers, single sprinklers, boom sprinklers, all of which are portable, or may be permanent installations. Some of the multisprinklers are self-propelled. A common type is the portable rotary sprinkling system, which has revolving nozzles that operate on either low or high pressures. A multisprinkler type now popular in some regions is the center-pivot system, which has a single lateral pipe unit that rotates about a center-pivot. Rate of water application with multisprinkler systems ranges from a minimum of 0.1 inch (0.25 cm) of water per hour to about 2.0 inches (5 cm) per hour.

Sprinkler irrigation is increasing rapidly for supplemental irrigation in

FIGURE 17–4. *A self-propelled sprinkler irrigation system for alfalfa in New Mexico.* [SOURCE: Southern Illinois University.]

both the older irrigated areas and in semihumid and humid areas. It may have definite economic value in developing new land that has never been irrigated, especially if the land is rough or if the soil is porous, shallow, or highly erodible. Sprinkling is helpful in irrigation at the seedling stage when furrowing is difficult and flooding encourages crusting. Fertilizer may be applied evenly through sprinklers.

Disadvantages of sprinkler irrigation are:

The initial cost of sprinkler irrigation is rather high.

Any cost of power to provide pressure is an added charge.

Wind interferes with the distribution pattern.

Nozzles may clog.

Sprinkling may increase foliar diseases and interfere with pollination.

The stationary system has high initial but low operating costs.

The portable system costs less initially, but operating costs are more because of the labor involved.

SURFACE IRRIGATION. In surface irrigation, sometimes called gravity irrigation, water is applied on land that has sufficient slope to permit flow over the surface by gravity. Surface irrigation may be one of three main types: (1) flood irrigation, where the area is to be covered uniformly with water; (2) furrow irrigation, where water is directed down furrows between rows; and (3) trickle irrigation, where water is applied very slowly onto

FIGURE 17–5. *Basin irrigation, a type of flood irrigation, in a pecan orchard in New Mexico.* [SOURCE: Southern Illinois University.]

FIGURE 17–6. *A furrow-irrigated crop: Siphon tubes diverting water from a head ditch down furrows between rows of grain sorghum in Arizona.* [SOURCE: University of Arizona Agricultural Experiment Station.]

the surface of soil through tiny holes or valves in plastic pipe. This newer type of surface irrigation often is used for high-value agronomic or horticultural crops.

There are three types of flood irrigation: (1) Wild flooding, where water is allowed to run uncontrolled over a piece of land. This is best suited to uncultivated pastures and hay fields. (2) Border strip irrigation, where fields are divided into strips bounded by a levee or low ridge. A uniform sheet of water is allowed to move slowly across each strip. (3) Basin irrigation, which often has irregularly shaped areas to be flooded. Examples of basin irrigation include the rice paddies of Asia and fruit and nut orchards in the United States.

Basic units of a surface system are (1) water supply; (2) field supply line, which may be an open ditch or pipe; (3) measuring device, to determine the amount of water applied to a field; (4) head ditch or pipe, extending along one end or side of the field; and (5) turnouts, which may be flood gates or siphon tubes. In addition, depending on the system, one may need a dike, levee, or border to hold the water within a certain area, as well as a means of collecting excess water in a tail ditch or reservoir for re-use.

SUBSURFACE IRRIGATION. There are two types of subsurface irrigation—open ditch and underground conduit. These methods distribute water below the ground level, where it spreads by capillary action to all parts of the root zone. The basic units of a subsurface system consist of (1) water

FIGURE 17-7. *Trickle irrigation of chili peppers in New Mexico. Water is applied slowly through holes in the tubing. This system uses water efficiently and is employed mostly on high-value agronomic or horticultural crops.* [Photograph with permission of New Mexico State University Agricultural Experiment Station.]

source, (2) field supply line and disposal ditch, (3) head ditch or head pipeline, (4) open ditch laterals or pipeline laterals, and (5) check dams.

The source of water is from underground—either from surface drainage or a high groundwater table. The head pipeline may be made of concrete, galvanized steel, or plastic, while lateral pipelines are made of drain tile, jointed concrete, or perforated plastic.

Drainage

On many farms the bottomlands are the most fertile. But to be productive, excess water must be drained away before plant damage occurs.

How Drainage Helps

A wet soil is a cold soil. When soil is drained, air moves in as the water is drained away. To grow well, plants need both air and proper soil temperature in the root zones. Plant roots cannot continue to grow in a soil saturated with water. Moreover, the organisms in the soil that make plant nutrients available require air.

How Land Is Drained

Excess water can be drained away from land either in open surface ditches or through tile underdrains. Each method has its advantages and disadvantages. The subsoil in some cases is so tight and impermeable that tiling is impracticable.

The first cost of open ditches usually is considerably less than tile drains. Unless they are relatively deep, however, they remove the water only from the surface soil, and much of the subsurface soil remains waterlogged. They are difficult, and often impossible, to cross with farm equipment. They often choke up with weeds and silt and need to be cleaned out repeatedly. But for many tight soils, surface drainage is the only practical method available.

Tile drains, when once in place, waste no land and do not interfere with field machinery operation. The original cost usually is more than for open ditches. If the soil is reasonably porous, the surplus water is removed from the subsoil as well as from the surface and subsurface soils.

Surface and Subsurface Drainage

Surface drainage may be accomplished by random field ditches connecting low spots and leading to an outlet, by pumping from low spots to an outlet, or by surface inlets into tile systems. Flat areas may be surface drained by some variation of a system of shallow, parallel ditches, combined with land smoothing. Surface drainage of sloping land may be accomplished by cross-slope ditch systems similar to terraces. Subsurface drainage consists of three types of construction: open ditches, tile ditches, and levees or dikes.

Through many years ditches were dug and tiles were laid largely with hand labor—a shovelful of earth moved at a time. More recently, the digging of ditches and laying of tile has become largely mechanized, with the development of efficient ditching and tiling machines.

Extent of Drainage

Many millions of acres of the most productive land in the Corn Belt, the Lake States, and the Delta States are producing valuable crops as a result of improved drainage. Some additional fertile land in the Mississippi Valley and in the Atlantic and Gulf Coastal Plains can be made productive by drainage when justified by food and feed demands.

The amount of drained land grew from less than 7 million acres (2.8 million ha) at the turn of the century to more than 42 million acres (17 million ha) in 1974. This acreage amounts to about 5 percent of the total land in farms.

At least 80 percent of the drained land is used for harvested crops. In the Corn Belt the production of livestock feed leads, principally corn and soybeans. North of the Corn Belt in Michigan and Minnesota forage crops frequently are grown. In the Atlantic and Gulf Coast states, truck and fruit crops are important on drained land. Cotton is grown on drained land in the Mississippi Valley and in California, where drainage and irrigation are combined.

Assistance under the Agricultural Conservation Program is confined to existing cropland and improved pasture, and is not permitted for the purpose of bringing new land into production. It has been estimated that there may be as many as 60 million acres (24 million ha) of cropland on which excess water is still a problem.

Drainage may be by means of open ditches or underground tile. In recent years most of the new drainage has been by installation of tile. An exception is Florida, where open drains are most common. More than 1,200 miles of open drains were constructed in 1975 and more than 1,400 miles in 1976 in that state. In 1975 and 1976 more than 18,000 miles of new open drains were constructed, and nearly 60,000 miles of new tile drains were installed in the United States. Most of this tile installation was in Ohio, Iowa, Michigan, Minnesota, Indiana, and Illinois, the leading states in tile drainage.

One of the requirements for tile drainage is to have a suitable outlet. These ordinarily are provided by constructed open ditches. In some cases it is necessary to pump the drainage water in order to obtain an outlet.

Spacing, Size, and Depth of Tile

In humid areas tile drains should be spaced so as to lower the ground water enough within 24 hours after rain for good plant growth. The soil texture and structure affect the rapidity of drainage. In clay soils, tile drains may not work properly.

Tile is generally 12 inches (30 cm) or more in length; the diameter varies according to the amount of water to be carried. The tile is laid end to end in a trench, with sufficient grade to allow for satisfactory operation. The grade may vary from 2 to 20 inches (5–51 cm) per 100 feet (30 m), with 3–6 inches (7.6–15.2 cm) the most common. Tile often is covered with earth, straw, or topsoil to facilitate entrance of water. Drainage water enters through joints, mostly from sides and bottoms. The outlet should be carefully protected to prevent sediment clogging. The drainage tile layout may be of the *fishbone* or *gridiron* type or a modification. Each system consists of a main drain and lateral drains. The most common distance between laterals is 50–100 feet (15–30 m), but this distance in clay soils may be as low as 30 feet (9 m). When the tile must be spaced closer than 50 feet (15 m), costs may be too high to justify the undertaking.

The depth of most tile should never be less than 2.5 feet (0.8 m) because of the possibility of breakage by machinery. The most common depth is 3 feet (9 m), but a 2.5-foot depth may be used in slowly permeable clay soils and a 4-foot (1.2-m) depth or more in sandy soils.

In arid regions under irrigation, depths of 6–8 feet (1.8–2.4 m) are common for tile drains with spacings ranging from 150 to 600 feet (46–183 m) practicable.

Drainage of Irrigated Areas

Upward of 10 million acres (4 million ha) of irrigated land also are drained to prevent them from becoming waterlogged and to wash out harmful salts. Without adequate drainage many irrigated soils would soon become toxic to crops because of an accumulation of excess salts. There is a rather large acreage of land in present irrigation districts that, with proper drainage and the control of salinity, could be irrigated and made productive.

Estimating Benefits

Factors to be considered in estimating benefits of drainage include the need for drainage, thoroughness of drainage, distance to natural outlet or main drain, fertility of the soil, and increased accessibility.

Very wet land will be benefited more than land that only occasionally needs drainage. It depends on the area whether or not land can be drained adequately at a reasonable cost. The fertility of the soil is of prime importance in determining benefits. In some instances a high land area isolated by wet land may be made more accessible by drainage. Whether or not the area has rocks, trees, and shrubs also may be a deciding factor.

Legal Aspects of Irrigation and Drainage

In irrigation and drainage, technical aspects alone are not sufficient. The legal aspects also must be understood. The legal aspects for irrigation focus on two conflicting doctrines—riparian rights and appropriation.

Settlers along streams in humid regions used the water as needed under the common-law doctrine of riparian rights, each being entitled to the water flow in its natural channel, "undiminished in quantity and unpolluted in quality."

Because the doctrine of riparian rights does not contemplate or allow for consumptive use of water or pollution, changes in concept in arid regions have been inevitable. In some areas each owner may divert water for use on his adjoining lands for "domestic" and "natural purposes." In some areas diversion for these purposes is permitted even though it may consume the water previously used by a lower riparian owner. Use of water for irrigation is considered "artificial" and is not normally allowed. Modifications have been made which permit a riparian owner to divert water for irrigation only if there is a surplus to the needs of lower riparian owners, and if such diversion does not appreciably lower the water level or quality. This "reasonable use" doctrine permits a more liberal removal of water for irrigation.

The riparian doctrine or modifications of the riparian doctrine are used in the more humid areas. However, increasing use of water for irrigation in these areas has created serious water rights problems that can only be solved by a modified doctrine approaching the doctrine of appropriation.

When irrigation was begun in the arid West the doctrine of riparian rights was inadequate. Agricultural and community developments depended on a reliable supply of water. Out of this came the doctrine of appropriation, which asserts that all water rights are founded upon priority of use—that use creates the right and disuse destroys or forfeits the right.

The common-law doctrine of riparian rights is still used in most states. However, at least 15 states have some form of groundwater rights based on the doctrine of appropriation, and others apply the reasonable-use concept to the riparian doctrine.

Estimating Costs of Irrigation and Drainage

Costs for an irrigation or drainage district project include construction costs, right-of-way costs, damage to land, roads, bridges, fences, engineering expenses, and legal and other fees. On an individual farm, expenses are primarily construction and engineering. Maintenance is a continuing cost. Because costs vary from section to section and from year to year, specific data are of little value.

Dryland Farming

Dry farming, or dryland farming, is the production of crops on land that receives limited rainfall and to which no irrigation water is applied. Precipitation is so uncertain and low over much of this area as to limit its agricultural use to grazing.

About 40 percent of the total land surface in the United States is markedly deficient in precipitation for use as cropland. More than 700 million acres (283 million ha) receive fewer than 20 inches (51 cm) of rainfall. Dryland farming is practiced on the more fertile soils where the rainfall approximates 20 inches. Some of this cropland is summer-fallowed. The acre production and dependability of the different crops that might be grown will determine the type of agriculture to be followed. In some low-rainfall areas, crop yields are so uncertain without irrigation that production should be limited to feed for livestock. Other areas are particularly well suited to general mixed farming, if all dryland farming procedures are applied. The efficient conservation and utilization of rainfall determine the success or failure of the dryland operator.

Soil-Moisture Fallacies

A considerable number of theories regarding soil-moisture conservation and utilization have had temporary wide acceptance, only to be proved fallacious.

THE DUST-MULCH THEORY. The theory that moisture could be held in the soil by maintaining a dust-mulch on the surface had almost universal acceptance for a considerable period of years. This theory was based on the erroneous belief that moisture moved to the surface through capillary attraction, where it evaporated.

THE RISE OF SUBSOIL MOISTURE. This belief was associated with the dust-mulch theory. Many investigations have shown that, except as water is removed from the soil through plant roots, there is practically no upward movement unless the water table is only a very few feet from the surface.

DEEP PLOWING AND SUBSOILING. Plowing deep adds greatly to the cost of seedbed preparation, and instead of increasing the water-holding capacity of the soil, the loss by evaporation is increased, with deep tillage more harmful than beneficial in dry years and of no value in normal or wet years.

Increased Available Soil Water

A certain minimum amount of water is necessary to maintain plant life but that amount does not necessarily bring a crop, such as wheat, to the production of grain. An increase of only 2–3 inches (5.1–7.6 cm) or more than the minimum amount has in some cases more than doubled crop yields. Over the Great Plains as a whole, the value of 1 inch (2.5 cm) of conserved water is 3–4 bushels of wheat per acre (202–269 kg/ha), whenever this inch is above the plant life minimum. In parts of the Winter Wheat Belt it has been shown that the depth to which the soil is wet at seeding time can be used to predict accurately the acre yield for the following season. Normally, a small grain crop in the dry-farming area uses all the water available to its roots before harvest time, leaving the soil dry.

FALLOWING. The purpose of fallowing land is to accumulate as much as possible of the rainfall through two seasons to grow one crop. This has been one of the most widely applied practices in dryland farming. For 75 years farmers have included summer fallow as part of their crop rotation system in the Great Plains. Even with the best summer-fallow practices, however, in the Great Plains area only 20–25 percent of the rainfall of the fallow year is held in the soil. When the precipitation is in the late fall or winter, a larger proportion is held. However, fallowing often is profitable and worthwhile. Table 17–2 shows yield benefits of fallowing before wheat as compared to continuous wheat at six locations.

In the far north where corn cannot be grown, as in the great western Canada wheat area, a third of the land under cultivation is summer-fallowed. Fallow there has not been shown to increase significantly the long-time average yield but is important in helping to control weeds, spreading the summer labor load, and saving half the seed required to plant a crop each year. The primary benefit is that in years of severe drought a crop can be produced, when crops seeded on stubble land may not be worth harvesting.

Corn vs. fallow. In the northern Great Plains it often is more profitable to grow a corn crop than to fallow, since a corn crop usually leaves consider-

TABLE 17–2. Comparison of Yields From Continuous Wheat vs. Wheat–Fallow Systems at Six Locations

Cropping System	Location	Wheat Yield (Bu/Acre)	Average Annual Precipitation 1931–1960 (Inches)
Wheat–Fallow	Moccasin, MT (9 yr. av.)	32.0	13.7
Wheat–Barley–Fallow	Moccasin, MT (9 yr. av.)	39.0	—
Wheat–Fallow	Akron, CO (60 yr. av.)	21.7	16.8
Continuous wheat	Akron, CO (60 yr. av.)	7.4	—
Wheat–Fallow	Colby, KS (49 yr. av.)	19.6	17.9
Continuous wheat	Colby, KS (49 yr. av.)	9.3	—
Wheat–Fallow	North Platte, NE (56 yr. av.)	31.1	17.5
Continuous wheat	North Platte, NE (56 yr. av.)	12.4	—
Wheat–Fallow	Alliance, NE (23 yr. av.)	18.3	16.3
Continuous wheat	Alliance, NE (23 yr. av.)	7.4	—
Wheat–Fallow	Bushland, TX (29 yr. av.)	15.0	19.7
Continuous wheat	Bushland, TX (29 yr. av.)	9.6	—

Source: Haas, H. J. et al. Summer Fallow in the Western United States, USDA-ARS Cons. Research Rep. 17 (1974).

able moisture in the soil available for a following small grain. As an average for many trials in several states, the small grain yield has been approximately one-fourth larger after fallow than after corn—a 5-bushel loss of small grain does not represent a real loss, however; the net return from the corn crop, as compared with the cost of fallow with no crop, is greater than the 5-bushel small-grain decrease.

WIDE-SPACED CROPS. In the central and southern Great Plains areas both corn and sorghum have been grown in double-spaced rows, as a means of carrying over moisture to supplement the next year's rainfall. Except in drought years, some increase in wheat yields has resulted. But each bushel increase in wheat has been at the expense of a 2-bushel (125 kg/ha) decrease in the yield of corn. Sorghum yields did not suffer so much as corn. In drought years both corn and sorghum used all the moisture in the soil, so that the wheat yield was not benefited by the double-width rows of corn or sorghum. One purpose in growing the double-width row crops is to check wind erosion. This may be the chief value of this method.

REDUCED STANDS. A heavy stand will often remove all the available moisture from the soil before a crop can be brought to maturity. For this reason, it is a common practice to decrease greatly the rate of seeding under dryland conditions. By this procedure profitable returns often are obtained from crop plantings that otherwise would have failed.

REDUCED WATER LOSSES. The controllable water losses are runoff and those caused by weed growth. The more torrential the rains, the greater the losses by runoff. A rough, loose surface usually is sufficient to hold all water from ordinary rain. Much will depend on the soil type. If the soil is porous, and especially if there is a trash mulch on the surface such as stubble and straw, there will be a minimum runoff.

The moldboard plow, which buries crop residues, contributed to the dust bowl of the 1930s in the Great Plains. Those farmers who remained in the area learned to leave crop residues on the fields to protect against wind and water erosion. This stubble-mulching allows more water to soak into the soil and be used by crops.

Contour listing, contour cultivation, and level terracing are generally accepted in dry-farming areas as means for holding water in place until it can percolate into the soil, where it will be available to crop roots.

The importance of destroying weeds in fallowed fields as a means of conserving water is generally recognized. In many cases crop yields relate directly to the extent to which weeds have used water that otherwise would be available to the crops. Weeds may be controlled by herbicides or by occasional cultivation during the summer.

Dryland Crops

More acre-inches of water may be required to produce a unit of dry matter under dryland conditions than in more humid areas, because of the low humidity and high loss by transpiration in dryland areas.

Some crops are much more efficient than others in the way they use water. But the adaptation of a specific crop to dryland production may relate more directly to the particular portion of the season when maximum growth is made than to efficient water utilization. For example, small grains that make their maximum growth when it is cool and other climate conditions are not too severe, and when the best use can be made of the limited water supply, are among the best dryland crops. Winter wheat is generally preferred to all other crops throughout most dryland areas.

Sorghums are particularly adapted to southern dryland sections. Sorghums need water in June for early growth and again in August for heading. Rainfall in the southern Great Plains normally is higher in these two months than in July. The sorghums are unusual in that their growth may be temporarily suspended, thus greatly reducing transpiration during periods of low available moisture, and resumed when soil moisture again becomes available.

Some weedy plants known to flourish under extreme drought conditions have been found relatively productive, nutritious, and palatable. As an example, *Kochia scoparia* (burning bush) is reported to have proved valuable and to have been harvested extensively at times in parts of Kansas, Colorado, Utah, and South Dakota. *Kochia indica* has had similar reports from Egypt.

Rotations and Tillage Practices

ROTATIONS. The limited number of crops that can be grown profitably in many dryland areas largely determines possible rotations. The inclusion of specific crops in a rotation may depend on their effect on other crops. In the northern Great Plains, for example, a corn crop usually leaves considerable moisture in the soil, available to a following small grain. Also, the corn and small grain combination helps in the distribution of labor and works well in farming systems where livestock production is carried on in conjunction with grain farming.

In the southern Great Plains, sorghum and winter wheat do not fit well together because sorghum leaves the ground too dry for fall-seeded wheat. If sorghum is included in the rotation, it usually is necessary to summerfallow before wheat.

When a specific crop is outstandingly more productive or more valuable than any other crop, a rotation that calls for considerable acreages of other crops finds little favor. In west central Kansas, for example, winter wheat is a comparatively sure crop and is so outstandingly more valuable than other crops that every acre of cropland not devoted to wheat means a loss of revenue. Under these conditions, one year of fallow, as a weed control measure, in perhaps four years of wheat may be about the extent of the rotation.

FERTILIZER USE. The use of commercial fertilizers or of animal or green manures has little place under dry-farming conditions. Only under very unusual conditions has any economic increase resulted. The use of such materials is more likely to result in a decrease in yield than in an increase.

The soils of dryland areas developed under a light rainfall have lost little of the elements of fertility present in the parent material. They usually are rich in nitrogen and other elements needed for plant growth. The addition of nitrogen, either as commercial fertilizer or in manures, is likely to bring on a vigorous early growth, with insufficient moisture available in the soil to bring the crop to maturity.

SEEDBED PREPARATION. First essentials in seedbed preparation are (1) reduction of loss of water by runoff to a minimum, (2) destruction of competing weed growth, and (3) maintenance of a soil surface resistant to wind erosion. In the southern Great Plains, cultivation soon after the removal of a winter grain is important as a means of storing water. The use of implements that leave the residue of the preceding crop on the surface is desirable to reduce runoff and wind erosion. In the northern portion of the Great Plains, however, there is no advantage to cultivation following wheat harvest over waiting until the following spring, but there may be a disadvantage.

Under any condition, too much cultivation breaks down the cloddy structure of the soil, destroys crop residue, and leaves a surface that favors runoff and wind erosion. Also, the fact that dryland farming is on an extensive basis, with crops of relatively low acre value, makes it essential that the cost of labor and power be held to a minimum. The proper choice of implements and of the time at which needed operations will be performed contribute materially to this desired end.

CROP RESIDUE DISPOSAL. Research indicates that in the dry-farming area of the Great Plains the proportion of the crop residue left on the surface has had relatively little effect on crop yield. Such residue does reduce water erosion, but water erosion is of less importance in this area than where the rainfall is heavier. On certain soils it has important value in reducing wind erosion. Cultural practices that leave a considerable part of the crop residues on the surface are in general use over much of the dryland area of the Great Plains. They are associated with a low-cost, extensive type of farming. It would appear that the amount of effort that should be expended in keeping residues on the surface will be determined by the need for erosion control, with special regard to the long-term effect of erosion and the relative cost of tillage, rather than by the expectation of materially influencing current crop yields.

REFERENCES AND SUGGESTED READINGS

1. American Society of Agronomy. *Irrigation of Agricultural Lands,* ASA Ser. 11, Madison, Wis. (1967).
2. ANONYMOUS. "A Free-for-All Over Water in the West," *U.S. News and World Report,* 54–55 (Jan. 8, 1979).
3. BRADY, N. C. *The Nature and Properties of Soils,* 8th ed. (New York: Macmillan, 1974).
4. CHAPMAN, S. R., and L. P. CARTER. *Crop Production Principles and Practices,* (San Francisco: Freeman, 1976).
5. ERIE, L. J., *et al. Consumptive Use of Water by Crops in Arizona,* Ariz. Agr. Exp. Sta. Tech. Bul. 169 (1968).
6. FENSTER, C. T., H. I. OWENS, and R. H. FOLLET. *Conservation Tillage for Wheat in the Great Plains,* USDA Ext. Serv. PA-1190 (1977).
7. FULLER, W. H. *Water, Soil and Crop Management Principles for the Control of Salts,* Ariz. Agr. Exp. Sta. and Coop. Ext. Ser. Bul. A-43 (1967).
8. HAAS, H. J., *et al. Summer Fallow in the Western United States,* USDA-ARS Conserv. Research Rep. No. 17 (1974).
9. HAGAN, R. M. "Water Management: Some Effects of New Societal Attitudes," *Agronomic Research for Food,* Spec. Pub. 26 (Madison, Wisc.: American Society of Agronomy, 1976).
10. HALDERMAN, A. D., and K. T. FROST. *Sprinkler Irrigation in Arizona,* Ariz. Coop. Est. Ser. and Agr. Exp. Sta. Bul. A-56 (1968).
11. HUFFMAN, R. E. *Irrigation Development* (New York: Ronald Press, 1953).
12. ISRAELSEN, O. W., and V. E. HANSEN. *Irrigation Principles and Practices,* 3rd ed. (New York: Wiley, 1962).
13. LARSON, W. E. (Coordinator of Report). *Research Progress and Needs, Conservation Tillage,* USDA-ARS-NC-57 (1977).
14. National Advisory Commission on Food and Fibers. "Land Use," in *Agriculture and Foreign Economic Development,* Vol VII, Chap. III. (Washington, D.C.: U.S. Government Printing Office, 1967).

15. NELSON, A. G., and P. T. COX. *Water Priority Rights and Their Effect on Farm Planning in the San Carlos Irrigation and Drainage Districts in Central Arizona,* Ariz. Agr. Exp. Sta. Tech. Bul. 184 (1968).

16. PAIR, C. H., and A. S. HUMPHREYS. *Sprinkler Irrigation* USDA-ARS Leaflet 476 (1977).

17. PROBSTEIN, R. F. "Desalination," *Amer. Scientist,* 61:280–293 (1973).

18. REEDER, N. "Grow Crops with Salt Water?" *Farm Jour.,* F-4 (Aug. 1977).

19. RUSH, D. W., J. D. NORLYN, and E. EPSTEIN. 1976. "Salt-Resistant Crops Coming," *Crops and Soils,* 29(3):7–9 (1976).

20. SCHWAB, G. O., *et al. Soil and Water Conservation Engineering* (New York: Wiley, 1966).

21. SLOGGETT, G. *Energy and U.S. Agriculture: Irrigation Pumping, 1974.* USDA-ERS, Agr. Econ. Rep. No. 376 (1977).

22. THORNE, D. W., and H. B. PETERSON. *Irrigated Soils* (New York: Blakiston, 1954).

23. TURNER, J. H., and C. L. ANDERSON. *Planning for an Irrigation System,* (Athens, Ga.: American Association for Vocational Instructional Materials and Soil Conservation Service, 1971).

24. USDA Agr. Stat. (1977).

25. U.S. Dept. of Commerce. Bureau of The Census. *1974 Census of Agriculture,* Vol IV, Graphic Summary (1978).

CHAPTER 18
CROP ENEMIES

CROP enemies—insects, diseases, and weeds—always have confronted man in his quest for food and fiber, good health, and a pleasant environment. The increased use of chemicals for the control of these pests has created a greater public awareness of potential environmental problems.

A program by the Environmental Protection Agency was instituted in the late 1970s to educate the pesticide applicator and control the use of those pesticides that are highly toxic or that could cause environmental damage. This involves the classification of pesticides into restricted and nonrestricted categories. Most pesticides on the initial restricted list are insecticides, such as endrin, carbofuran, demeton, and methyl parathion, but included also are the herbicides 2,4,5-T and paraquat and the fungicide benomyl. This program also required that pesticide applicators be certified by October 1977, in order to use restricted pesticides, or that they apply them under the supervision of a certified person.

Production and use of synthetic organic pesticides to control crop enemies continues to expand despite public concerns about pesticide use. Production in 1975 was 1,609 million pounds (730 million kg), up from the preceding five-year average of 1,322 million pounds (600 million kg). The volume of pesticide sales by class—insecticides, herbicides, and fungicides—is shown in Figure 18–1.

Insects

Scientists estimate there are as many as 1.5 million species of insects on earth. Of this number, more than 85,000 are found in North America above Mexico, in addition to more than 2,600 kinds of ticks and mites. Some 10,000 species are classified as public enemies.

Crop losses from insect damage have been estimated at 5–10 percent of total crop values, nearly $4 billion annually. This does not include losses to humans from disease or insect-control costs. It is estimated that insect control costs another $1 billion annually.

About half of our worst insect pests have been introduced, most of them from Europe. Some of them brought in accidentally are the European corn borer, the Hessian fly, the cotton boll weevil, the pink bollworm, the sugarcane borer, the alfalfa weevil, the pea weevil, and the angoumois grain moth.

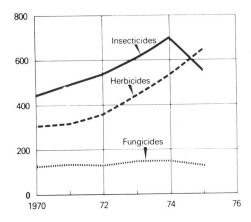

FIGURE 18-1. *Volume of sales by class of pesticide—insecticides, herbicides, and fungicides.* [SOURCE: USDA.]

FIGURE 18-2. *The boll weevil is a serious cotton pest, causing about 40 percent of the total losses attributed to cotton pests. The adult punctures the bolls and lays eggs; when the eggs hatch, the boll is damaged or destroyed.* [SOURCE: Agricultural Information Office, Oklahoma State University.]

In some cases, in addition to causing direct loss, insects may carry diseases to plants. For example, aphids carry mosaic to the potato, sugarcane, and tobacco; the leafhoppers carry curly top of sugarbeets; and the Colorado potato beetles carry spindle tuber of the potato.

The control of insect pests is partially affected by natural means. When a particular insect becomes very numerous, further increase is checked by its natural enemies, such as birds, parasitic insects, diseases, and unfavorable food or climatic conditions. This control usually does not affect noticeable reduction until after serious crop losses have been suffered.

KINDS. Insects are roughly classified as chewing or piercing–sucking. Chewing insects tear or pinch off, chew, and swallow bits of the plant. Examples include grasshoppers, caterpillars, crickets, and darkling beetles. Piercing–sucking insects pierce or rasp the plant and suck or sponge up the sap from plant body tissue. Examples include aphids, squash bugs, chinch bugs, lygus bugs, and leaf-footed bugs. Chemical control is based on this classification.

Insects that probably have caused more damage than any other to growing crops are the grasshopper, the European corn borer, the cotton boll weevil, the chinch bug, and the Hessian fly. Not all insects are destructive; many are beneficial. They are indispensable as pollinators of plants. Intense competition between insect species, predatory and parasitic, must be considered beneficial. Some insects do not feed entirely on commercial crops; they feed on weeds. Some insects improve the physical condition of the soil by helping air penetration. They hasten decomposition of plant and animal material and their return to the soil.

Control Methods

CULTURAL PRACTICES. There are a number of well-known practices and methods that can be used effectively against a large number of different insects. Some of the more important are plowing and cultivation, pasturing, rotation and selection of crops, time of planting, early harvesting, and the destruction of infested material and material affording hibernation.

In the control of insects, the time and depth of plowing may be important. For insects that pass the winter near the suface of the ground, deep plowing may crush and kill them or may bury them so deep that they are unable to get to the soil surface. Such a method may be effective for the European corn borer. Some insects in burrows, in the pupal stage, are brought to the surface by fall plowing and are subsequently killed by cold weather or perhaps are destroyed by birds and rodents.

Certain insects may be effectively controlled by introducing into the rotation a crop not attacked. All insects are checked in their development by such a practice because sites for overwintering and laying eggs are removed. Heavy infestations may occur when a susceptible crop is grown year after year. It may become necessary to discontinue the growing of that crop completely until the pest is under control. Community effort in the control of certain insects is necessary, as one field may serve as a breeding ground from which the pest will migrate and reinfest adjoining farms. Crop rotation is effective in controlling or lessening the severity of crop pests such as the corn rootworm and soybean cyst nematode.

In the case of certain insects, such as the Hessian fly, serious damage can be avoided by delaying seeding of winter grains until after the adults have emerged and the egg-laying time has passed. In the boll weevil area everything is done to get an early crop of bolls set before weevil population can be built up to destructive levels. Late-planted corn often is less seriously damaged by the European corn borer than early-planted corn. Early harvest of some crops may prevent losses. It is effective against the alfalfa weevil and the pink bollworm.

Many insects hibernate or spend a portion of their life cycle in plant residues, grassy fence rows, and weed patches. It often is advisable to destroy these sources of infestation by burning or by deep plowing. Destroying crop residues has been effective in controlling the corn earworm. Flooding and soil sterilization also have been used effectively to control insects.

Artificial barriers to prevent spread of insects are expensive and normally practical only for small-scale use on high-value crops. When there is danger of insects being brought into a new territory, quarantine often is used. Quarantine areas usually are established by the federal government.

GENETIC CONTROL. There are three types of insect resistance: preference or nonpreference, antibiosis, and tolerance. The problems in developing resistant cultivars are many and complex. Less has been done by comparison than breeding for yield, quality, and disease resistance. Corn inbred lines and hybrids differ considerably in their resistance or susceptibility to such insects as the European corn borer, corn earworm, and aphids. Corn hybrids containing a chemical substance called DIMBOA have been noted as resistant to European corn borer.

Certain strains of winter wheat combine resistance to the Hessian fly and tolerance to the wheat stem sawfly. It has been discovered recently that hairy-leaved wheat lines, from Russia and Asia, have some resistance to cereal leaf beetle. The hairs make for difficulty in egg-laying, act as a barrier to the insect, and allow drying or dessication of eggs before they hatch.

Sorghum resistant to the chinch bug, barley to the green bug, alfalfa to the pea aphid, potato leafhopper and alfalfa weevil, and sugarcane to the sugarcane borer are other examples of genetic resistance.

BIOLOGICAL CONTROL. The method of biological control is simply the use of living organisms to control pests. The introduction and breeding of parasitic and predaceous insects, mites, worms, and birds have been used to control certain insects. Also effective has been the spread and increase of fungus, bacterial, virus, and protozoal insect diseases. Interest in biological and other methods of nonchemical control has increased since Rachel Carson's book *Silent Spring*. It has stimulted more research and

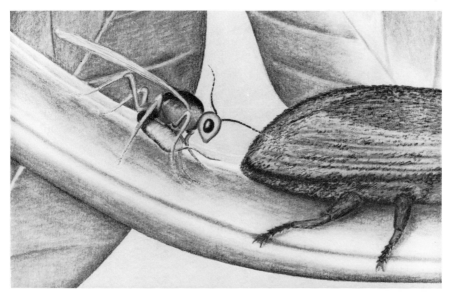

A

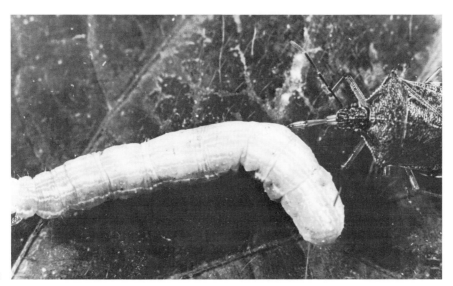

B

FIGURE 18–3. *Biological insect control is attracting interest and gaining more acceptance. (A) Tiny wasps, such as* Microctonus aethiops; *can attack adult alfalfa weevils and sterilize them so they do not reproduce.* [SOURCE: Ohio Agricultural Research and Development Center.] *(B) The* Podisus *bug may attack and kill loopers.* [SOURCE: American Soybean Association.]

development and a greater interest in achieving a balance between chemical and nonchemical control.

The alfalfa weevil, European corn borer, sugarcane borer, alfalfa caterpillar, and a host of other insects have been subjected to biological control.

FIGURE 18-4. *Bacteria such as* Bacillus thuringiensis *can control loopers and certain other insects effectively.* [Photograph at 1978 Farm Progress Show, Taylorville, Illinois.]

The following are some examples of specific biological control: (1) Tiny wasps, such as *Microctonus aethiops,* can attack adult alfalfa weevils, sterilizing them so they cannot reproduce. Other alfalfa weevil parasites are available which will attack the eggs, or the larval stage. The wasp *Bathyplectes curculionis* deposits eggs in the larvae, and parasitism results in weevil larvae feeding less and pupating early. (2) Cereal leaf beetles can be controlled with a wasp, as well as with a fungus that infects the beetle. (3) The Japanese beetle is controlled by a parasitic bacteria, *Bacillus popilliae,* which enters the body of the larvae and eventually kills them. (4) The common ladybug beetle can ingest a large number of aphids.

Other approaches involving nonchemical control procedures include the following: (1) Pedigo and Higgins at Iowa State are testing antifeeding compounds as control agents for green cloverworms, leaf-feeding soybean insects. Providing these "direct-pill" materials has reduced leaf consumption by 70–80 percent. (2) Natural hormones, ordinarily found within the insects, can be used to disrupt the life cycle. Selected hormonal chemicals can kill, prevent insects from reproducing, cause to remain a larva, or cause to remain in the dormant stage. (3) Sex attractants, or pheromones can be used to attract males, where they can be sterilized and released. Alternately, widespread distribution of pheromones could confuse the males and make it impossible for them to find females and mate. (4) Some insect growth regulators, which can prevent breeding or reproducing, are being produced commercially.

CHEMICALS. All chemicals used to control pests are referred to as pesticides. A number of terms are derived from pesticides, all self-explanatory. Some refer to the kinds of organisms that a particular insecticide kills best, such as insecticide, miticide, and nematicide. Others refer to the stages in the life cycle at which they are designed to kill, such as ovicide and

FIGURE 18–5. *A non-chemical insect control procedure involves the use of sex attractants or pheromones to attract and trap males, where they can be sterilized and released, thereby reducing the insect population.* [Photograph with permission of Mississippi State University Agriculture and Forestry Experiment Station.]

larvicide. Chemicals are either organic or inorganic compounds. These compounds have various chemical compositions, and the same formulations may each have a variety of trade names.

Chemicals with insecticidal properties have been recognized for thousands of years. Nearly 3,000 years ago, Homer, the poet and author of the *Iliad* and the *Odyssey,* mentioned sulfur as a pest control agent. By 900 A.D., the Chinese were using arsenic to control garden insects. By 1690 tobacco was employed as a contact insecticide. Dozens of insecticides were developed between 1800 and 1900. By 1920 rotenone was recognized as an insecticide, and the 1940s brought the introduction of some of the chlorinated hydrocarbons, such as DDT and chlordane. In the 1950s organic phosphates, such as parathion, and carbamates, such as carbaryl, were introduced.

Insecticides are classified by their action. Stomach poisons usually are used against chewing insects. The insect is killed when the treated plant material reaches the stomach. Piercing or sucking insects must be controlled by contact poisons which kill by penetrating the body of the insect directly or by entering breathing or sensory pores.

Chemicals are involved in other methods of insect control. Males of the insect species are made sterile or are otherwise self-destructively modified, usually, by chemosterilants or by physical sterilants, such as irradiation. A sterilizing agent may be mixed with an insect attractant. Chemicides may be used to prevent the diapause of insects, to interfere with their biological clocks. Chemical repellents include dimethyl phthalate, ethyl hexanedial, and dimethyl carbate. Systemic chemicals usually are toxic to a wide range of organisms but can be used in a highly selective manner. Systemic pesticides are absorbed by the plant or animal victim and are spread throughout its system so that organisms that feed on the victim are killed.

The use of poison baits has been effective in certain sections in the control of grasshoppers and armyworms, and sometimes of cutworms. The availability of effective new insecticides has greatly reduced the use of poisonous baits.

Stored grain. Infestations with weevil or angoumois moths in grain and other seeds are common. Several fumigants are used that produce a gas and kill the insects; these do not kill the eggs, however, so that a second treatment is needed. The use of a mixture of carbon bisulfide with carbon tetrachloride is one of the cheapest, safest, and most effective methods of treating infested bins to control weevils and other pests.

Sprays of malathion or a combination of pyrethrins with piperonzl butoxide also gives effective control. Heat often is used and is the most effective means where it can be controlled. In regions where the temperature falls as low as $-20°F$ $(-30°C)$ grain insects can be killed by the simple expedient of allowing the temperature in grain storage buildings to go as low as possible.

Use controversial. The use of such persistent chemicals as DDT, dieldrin, endrin, aldrin, and heptachlor, chlorinated hydrocarbons, has been a controversial subject since Rachel Carson's *Silent Spring.* As a result, certain of these chemicals have been banned or their use greatly restricted. Newer compounds, the organic phosphates such as malathion, parathion, methyl parathion, and the carbamates, such as carbaryl, metacil, and dimetilan, are being used instead of the chlorinated hydrocarbons. While DDT and other persistent chemicals remain in the soil for years and cause serious residue problems, newer insecticides are less persistent and break down into harmless products, generally within a few weeks after application.

Even though insecticide use has brought controversy, it continues to expand. In 1975, 666 million pounds (302 million kg) of insecticides were produced in the United States, up from the preceding five-year average of 615 million pounds (279 million kg).

An increasingly serious disadvantage from the extended use of chemical pesticides is the development of species resistant to them. Resistance cannot always be easily overcome by changing to a different kind of pesticide. Although there are thousands of kinds of chemical pesticides they fall into a few groups on the basis of their chemical structure. When a pest develops resistance to one pesticide it tends to become more or less resistant to all pesticides of the same chemical family.

It is estimated that without pesticide control food prices would increase 50–70 percent, quality would be lowered, and the general standard of living would be reduced. Nevertheless, pesticides are hazardous and the complex problems of pest control need continued research. Pesticides must be used as prescribed and according to published precautions to avoid unnecessary hazards to man, domestic animals, crops, beneficial insects, and wildlife.

A relatively new approach to insect control is a program called integrated pest management (IPM). This program was developed in response to the so-called "pesticide treadmill," where insects were developing resistance to some pesticides, rates had to be increased, and nontarget organisms were being affected. IPM is a system that involves closer monitoring of the pest population in a field, and incorporates a combination of biological, cultural, genetic, and chemical control methods. Natural enemies or predators are protected and nonchemical means of controlling pests are emphasized. Pesticides are used on an "as-needed" basis only.

Another recent development that could alter insecticide rates and application procedures is the microencapsulation process. This offers a slow or sustained release of a pesticide, extending the life of the active ingredient and decreasing the hazard to the applicator.

Plant Diseases

A published "compendium of plant diseases" runs to 1,192 pages of fine print. Some plants are subject to attack by as many as 300 different disease disorders.

It is estimated that the U.S. crop losses from diseases are 10 percent, carrying a value of $3 billion. Cost of controlling diseases annually is estimated to be over $115 million.

Diseases reduce and affect yields and crop quality. Diseased grains may be lightweight, discolored, and shriveled. The oil content of peanuts, safflower, and soybeans may be reduced. Disease-induced poor coloration, blemishes, or rots reduce the grade of potatoes and sweet potatoes. Certain plant diseases can produce illness and death of persons and animals that eat affected products. Scabby grain may be toxic to hogs. Ergot in cereals and grasses contains alkaloids that may produce convulsions and even kill animals and human beings.

Disease losses affect not only the producer, but the consumer as well. When diseases limit production, prices usually increase. Many diseases are expensive to control and the added cost is reflected in higher prices.

DISEASES CLASSIFIED. Plant diseases may be defined as deviations from normal growth of structure of the plants. They may be classified as infectious and noninfectious. Infectious diseases include those caused by fungi, bacteria, viruses, nematodes, and parasitic seed plants. Noninfectious diseases may be the result of unfavorable environmental and nutritional conditions.

Diseases may affect any part of a plant—roots, stems, leaves, flowers and fruit—at any stage in its development—seed, seedling, or mature plant. The symptoms of plant diseases may be quite variable and include changes in color; holes in leaves; wilt; necrosis; stunting or failure to develop; galls; shriveling and dying of fruit; alternations in growth habit; dropping of leaves, blossoms, and fruits; production of malformations; rots; and extrusion of liquid or oozes.

Infectious Diseases

FUNGI. Fungi are responsible for by far the greatest number and diversity of plant diseases. All economic plants apparently are attacked by one

FIGURE 18–6. *Two alfalfa plants with advanced stages of crown and root rot. This is one of the prime causes of the thinning of alfalfa stands in the irrigated Southwest.* [SOURCE: By permission from Robert B. Streets, Sr., *Diseases of the Cultivated Plants of the Southwest,* The University of Arizona Press, Tucson, Ariz., Copyright 1969.]

or another of the fungi; often a dozen or more different fungi induce disease in a single species. Fungi attack any of the plant's organs—roots, stems, leaves, flowers, fruit, or seed. They reduce vitality and germinability of seed, destroy seedlings, and damage the host at any subsequent time in life. Fungus diseases include stem rust in wheat, corn leaf blight, smut diseases of grasses and other plants, *Phytophthora* root and stem rot in soybeans, late blight in potatoes, and root rot in sugarbeets.

BACTERIA. The number of major diseases caused by bacteria, in contrast to fungi, is relatively small. Species in all major families of higher plants can be attacked by one or more bacterial plant pathogens. Bacterial diseases include bacterial wilt in alfalfa, Southern bacterial wilt of tobacco and potato, wildfire of tobacco, bacterial leaf blight of rice, angular leaf spot of cotton, potato scab, and Stewart's disease of corn.

VIRUSES. Viruses are submicroscopic infectious entities that multiply only inside living host cells. Few living organisms are immune. Most plants may be attacked by one or more viruses. Plant disease symptoms caused by viruses include changes in colors such as yellowing and mottling; malformations, such as distortion, rosetting, and proliferation; necrosis; and stunting. Diseases caused by viruses include curly top in sugarbeets, tobacco mosaic, sugarcane mosaic, soybean mosiac, common mosiac in

FIGURE 18–7. *The soybean cyst nematode is a growing concern in the Midwest as it gradually spreads from the south. This microphotograph shows a soybean nematode hatching some 400+ eggs.* [SOURCE: American Soybean Association.]

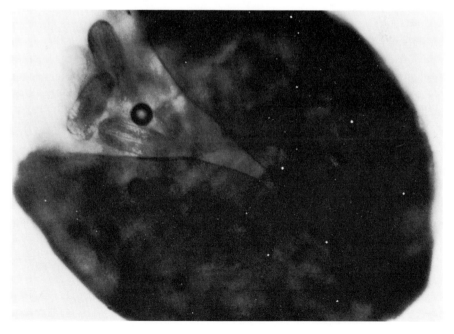

beans, dwarf maize mosiac, and streak and barley stripe mosiac of small grains.

PARASITIC SEED PLANTS. Examples of the numerous parasitic seed plants are mistletoe, dodder, and witchweed. Dodder's chief damage is to clover, alfalfa, and flax. Witchweed parasitizes roots of corn, sorghum, sugarcane, rice, rye, and oats.

NEMATODES. Nematodes are slender microscopic eelworms. One or more species attack practically all crop plants. Usually they live in the soil and attack underground parts, but occasionally they inhabit buds, stems, leaves, or inflorescences. Nematodes usually attack only small roots and kill by feeding on the cells and by toxic salivary secretions. They cause decline, stunting, or death. Although worldwide, nematodes are more prevalent in warmer climates. Examples of nematodes that can do extensive crop damage are the wheat nematode; sugarbeet nematode; rootknot nematode, which affects cotton and various forage crops; meadow nematode, which affects cotton; and the seed-borne nematode which causes the white tip disease of rice.

Noninfectious Diseases

Plants may be deficient in certain elements, resulting in a noninfectious or physiological disease. Because it is readily leached, nitrogen is the major element most frequently deficient. Micronutrient deficiencies include iron, zinc, manganese, magnesium, copper, and boron. Excessive use of fertilizers, as well as the pH condition of the soil, may cause plant injury.

In regions with an annual rainfall of 20 inches (51 cm) or less, alkali salts often remain in the root zone and cause injury. Continued use of irrigation water fairly high in soluble salts may cause a salt accumulation injurious to crops.

Environmental factors predispose plants to disease. Excessively wet soils may support root rots caused by soil-borne organisms. When water covers roots too long they may be damaged because of insufficient oxygen. Abundant moisture usually favors infection by bacteria and fungi. Lush growth produced by heavy fertilization and irrigation suffers more severely from drought than do plants grown with normal amounts of water.

The immediate effects of freezing temperatures on plants usually are recognizable, but some delayed effects are not so well known. Heat injury is more prevalent in, but not limited to, the semiarid Southwest. Lack of adequate water predisposes to heat injury, as cooling of tissues by transpiration is minimized. Plant tissues may be injured when exposed to the afternoon sun with air temperatures of 100°F (38°C) or above. Injury from high temperatures includes sunscald, sunburn, heat chlorosis, and heat canker of seedlings. Drought often accompanies high temperatures and causes stunting, wilting, and burning.

Many chemicals are used to prevent damage by insects and microorganisms or to kill weeds. If not applied properly, they may cause serious physiological disorders. Injuries produced are mostly necrotic areas, chlorosis, or russeting but are so variable that they are difficult to describe.

Most herbicides have selective activity, but only when directions are followed. However, many plants are seen with strange abnormalities caused by herbicide misuse or carryover. Each material produces distinctive symptoms, and it is often possible to tell from the plant the material used.

Smog from industry and other forms of air pollution also cause injury. Growth is retarded and leaf areas are necrotic. Injury is more serious on seedlings and young plants. Leakage of natural gas can cause wilting, browning of foliage, and quick death.

Disease Control

Plant disease control covers a broad, highly technical, and rapidly developing field of study. Disease control involves the application of one or more of the following principles:

1. *Disease resistance*—Developing resistant plants, including all techniques that alter the physiological processes, structural nature, or habits of individual plants to make them tolerate or resist disease.
2. *Protection*—Preventing infection by use of a chemical pesticide or other effective barrier between the plant and the disease.
3. *Avoidance*—Avoiding disease by planting at times when, or in areas where, disease is ineffective, rare, or absent.
4. *Exclusion*—Preventing the introduction or the establishment of a disease within an uninvaded area.
5. *Eradication*—Reducing, inactivating, eliminating, or destroying disease at the source, either from a region or from an individual plant in which it is already established.
6. *Therapy*—Reducing severity of disease in an already infected plant by chemical and nutritional treatments.

RESISTANT CULTIVARS AND STRAINS. Use of disease-resistant plants is the most effective, simple, and economical means of controlling plant diseases for the grower where acceptable resistant lines have been developed. The successful development of disease-resistant cultivars has been one of the major factors in increasing and maintaining high levels of crop productivity in the United States. Prior to 1900, only a few thousand acres were planted to resistant cultivars. For certain crops like alfalfa and small grains 95–98 percent of the acreage is planted with cultivars resistant to one or more diseases. A specific example is the development and use of soybean cultivars carrying resistance to *Phytophthora* root rot. Conservative estimates indicate a gain to the farmer from disease-resistant cultivars may exceed $1 billion annually in the United States alone.

Disease resistance may be mechanical or physiological. Mechanical disease resistance includes such external characteristics as hairiness, thick waxy layers, and small or sunken stomata. Internal characteristics include thick cell walls, production of gum, and the formation of suberized cells. Physiological resistance is associated with osmotic pressure, cell-sap acidity, tannin, nutrition, and toxic substance. Characteristics associated with resistance of susceptibility are subject to environmental variability. There also are genetic variations caused by segregation, recombination linkage, and crossing over and gene mutations.

Resistance is sought by several means, including (1) selection of resistant individuals from large heterogeneous populations; (2) crossing strains or cultivars carrying factors of resistance with plants characterized by high quality and yield, with subsequent selection and testing for desired characteristics; (3) hybridizing resistant wild species with susceptible cultivars of the cultivated species, followed by selection; (4) backcrossing to the original high-yielding parent to maintain acceptable characteristics, as a sequel to initial crossing of different cultivars or species; (5) subjecting wild host plants to radiation and other mutagenic agents to facilitate transfer of resistance to cultivated plants.

Among the special treatments for inducing gene changes is the application of such chemicals as colchicine and mustard oils. Irradiation with ultraviolet light and X-rays commonly induces deleterious genetic changes, but occasionally produces a beneficial alteration.

Not all efforts to obtain resistant cultivars have been successful. Development of new, more virulent races of rust and smut fungi has complicated the efforts on small grains. New strains may appear among diseases through importation, through development from already existing strains, or through new associations with insect vectors. Also, a new cultivar may be developed that is susceptible to a previously insignificant disease.

CULTURAL PRACTICES. Control of certain diseases or a reduction in their severity may be achieved through cultural practices, including sanitation.

Crop rotation. Many fungi and other organisms can attack only certain kinds of plants. They soon die and disappear from the soil when other crops are planted. The continuous cropping of one crop builds up disease. Green manure crops in rotation may reduce the activity of certain diseases.

Destruction of plant residues. Many diseases are carried over from season to season on the crop residue, such as on stalks, straw, and stubble. The burial of such material by deep plowing, fall spraying, or burning are methods advocated in cases of infestation with certain diseases.

Eradication as a means of control. The complete removal and destruction of diseased plants from a field, particularly in the case of diseases which spread from one plant to another, sometimes is necessary in disease control. In addition, some diseases of crop plants have alternate hosts on which the disease may carry over for a portion of its life cycle; they also must be destroyed. The destruction of the common barberry to reduce the stem rust of wheat and the destruction of buckthorn to reduce crown rust of oats are well-known examples of disease control measures taken through the eradication of alternate host plants.

Other sanitary measures. Disinfestation of machinery and other equipment and of packing and shipping containers prevents development and spread of diseases. Such sanitary measures are necessary to prevent spread of the soybean cyst nematode, for example, in areas which are quarantined. At ports of entry most countries have agricultural inspections and quarantines against injurious plant diseases and insect pests.

Repulsion of insect vectors. Control of insect vectors by sprays, dusts, or repellents has been effective in reducing certain diseases. Insects also may be excluded by enclosures.

Use of fertilizers. Fertilizers may stimulate plants to outgrow disease injury. Certain diseases result from nutrient deficiencies; other result from mineral excesses. Modification of soil pH influences the occurrence and severity of some diseases.

Irrigation practices. Excessive irrigation encourages some diseases. Overhead irrigation encourages the buildup and spread of certain fungus diseases and foliar blights. Wet-weather diseases, characteristically seed-borne, may be avoided in seed crops by maintaining the culture of the affected crop in an arid region where irrigation is by furrow only.

Soil treatment. Soil treatments often are effective where the value of the crop justifies the expense. Certain disease-producing organisms that carry over from year to year in the soil can be killed by treating the soil. Sterilizing the soil in tobacco seedbeds is a rather common practice to control seedling diseases as well as to kill weed seed. Volatile liquids that function as soil fumigants have come into extensive use since 1945.

Disease-free seed. The use of seed free of disease is an important control measure. In some cases seed is treated before planting to kill disease organisms carried on the seed. Treatment of small grain seed for the control of smut, and treatment of seed corn to control seed rots and early seedling blights are examples.

SEED TREATMENT. Many seed-borne diseases can be controlled by chemical seed treatment. Use of most of the organic mercury fungicides for seed treatment has been banned since the early 1970s because of their high toxicity to humans. They have been replaced by nonmercury organic fungicides such as thiram and captan.

Seed disinfestants vs. protectants. Fungicides may be classified either as seed disinfectants or as protectants, according to the location of the organisms to be controlled. Disinfectants destroy organisms on the surface of the seed and in many cases also may be effective against those located within the seed. Protectants help to keep the seed free of injury by organisms that are present in the soil. Practically all effective seed-treatment materials are disinfectants. Many are both disinfectants and protectants.

Fungicides may be applied (1) as solutions, in which the seed is soaked for a period and then dried; (2) as dust; (3) as a thick, souplike "slurry," which coats the seed with the fungicide; or (4) as a volatile concentrate.

The small grains. The seed-borne diseases of the small grains listed as controllable by seed treatment are the following: for barley—covered smut, black or false loose smut, stripe, the seedling-blight stages of anthracnose, bacterial blight, spot blotch, and scab; for oats—mostly loose and covered smuts, but treatment also reduces losses from seed-borne bacterial blights and some other diseases; for wheat—stinking smut, or bunt, and also seed-borne flag smut, anthracnose, and seedling blight. Hot water has been effective for loose smut in barley.

Corn and sorghum. The diseases of corn that are prevented by seed treatment are seed rot and seedling blight. Treated seeds frequently produce more vigorous plants. Treatment of sorghum seed is largely to combat seed rot and seedling blight, but kernel smuts, seed-borne anthracnose, and bacterial leaf diseases also are prevented.

Legume and grass seed. Legume and grass seed are subject to seed-rotting organisms of the soil. Seed treatment may in some cases result in better stands. The results of research thus far available are conflicting, but in general they do not justify seed treatment.

If properly done, seed treatment does not reduce the benefits of inoculation. When seed is to be inoculated with nitrogen-fixing bacteria, the seed-treating chemical should be applied before inoculation. The chemical treatment may be at any time, but the seed should not be inoculated more than 2 hours before it is to be planted.

Seed-treating equipment. Most of the seed, such as corn and sorghum, for which treatment is recommended regularly, is treated with large commercial outfits ranging in capacity from 100 to 500 bushels per hour. The latter usually are combined with seed-cleaning equipment so that both processes can be done in one continuous operation. The slurry treaters on the market are of large capacity, and their use is confined largely to seed houses and elevators.

Fungicides are poisonous and dust or fumes from them should not be inhaled. The fungicides should not touch the skin. Treated seed should not be used for foods or feed.

SYSTEMICS. Properly applied, surface applications of fungicides may effectively control fungal diseases such as downy mildew and potato late blight. However, the systemic fungicides which can penetrate and move in the plant may have an advantage. Several new chemicals have shown promise in controlling several foliar and root diseases, as well as seed rots. One such group of chemicals is readily taken up by the roots in soil containing low concentrations of the chemical, is moved in plants from foliar applications, and is systemic from seed treatment. One of the most recent foliar fungicides is benomyl which, among other uses, is applied from airplanes to control pod and stem blight of soybeans.

Weeds

The old definition of a weed as a "plant out of place" still holds. Some plants are "weeds" under certain conditions and under other conditions have considerable value. Even corn can be a weed when it volunteers in a soybean field. Weeds are taxonomically diverse and are particularly adapted to thrive in close association with man and his domesticated plants and animals.

The growth habit of weeds has been used to classify them as annuals, biennials, and perennials. The method of control usually is associated with the length of the life period or the method of reproduction.

An annual plant completes its life cycle from seed in less than one year. There are two types, summer annuals and winter annuals. Although considered easy to control, because of seed abundance and fast growth annuals are very persistent. In the aggregate, they cost more to control than do perennials.

Summer annuals germinate in the spring, make most of their growth during the summer, and usually mature and die in the fall. Summer annuals include common morning glory, pigweed, lambsquarter, common ragweed, crabgrass, and foxtail. These weeds are most troublesome in corn, sorghum, soybeans, cotton, peanuts, tobacco, and vegetable crops.

FIGURE 18-8. *A weed is a plant out of place. Even corn can be a weed when it volunteers in a soybean field.* [SOURCE: Southern Illinois University.]

Winter annuals germinate in the fall and winter and usually mature seed in the spring or early summer before dying. Winter annuals include downy brome, cheatgrass, shepherd's purse, and corn cockle. These are most troublesome in winter wheat, winter barley, and such perennials as alfalfa and pastures.

A biennial plant lives for more than one year, but not over two years. Wild carrot, bull thistle, and burdock are examples. There is some confusion between the biennials and the winter annual group, because the winter annual group normally lives during two calendar years and during at least two seasons.

Perennials live for more than two years and may live almost indefinitely. Most reproduce by seed and many are able to spread vegetatively. They are classified as simple and creeping. Simple perennials spread only by seed. They have no normal means of spreading vegetatively. However, if injured or cut, some may produce new plants. Examples include common dandelion, dock, buckhorn plantain, and broadleaf plantain. Creeping per-

FIGURE 18–9. *Weeds may be classified as grassy or broadleaf weeds, an important classification for herbicide recommendations. One of the most serious grassy weed pests, johnsongrass (in soybeans), a perennial which reproduces by underground creeping rhizomes.* [Photograph with permission of Mississippi State University Agriculture and Forestry Experiment Station.]

ennials reproduce by tubers, bulbs, creeping roots, stolons, or rhizomes. Examples are red sorrel, perennial sowthistle, field bindweed, and johnsongrass. Weeds also are classified as grassy or broadleaf weeds. This classification is referred to extensively in making herbicide recommendations.

Weeds may be beneficial in several ways. They reduce soil erosion on abandoned fields, add organic matter to the soil, provide food and cover for wildlife, yield useful drugs or delicacies, and beautify the landscape. They provide a reservoir of germplasm and biochemicals and constitute a potential source of domesticated plants.

Weed-Caused Losses

Heavy annual losses from weeds are due to a reduction in the quality and quantity of crops, depreciation of land values, dockage in crops, molding and spoilage of grain in storage, reduction in quality and quantity of livestock products through objectionable odors and flavors, livestock poisoning, reduction in value of wool, increase in labor and equipment costs for cultivation and chemicals to control weeds, increase in expenditures for cleaning small grain, increase in transportation costs, reduction in flow of irrigation and drainage of water, harbor for insects and diseases, and cause of discomfort and death to human beings and animals.

The losses caused by weeds are much greater than most realize. An estimate of total annual losses due to reduced crop yields and quality and cost of control is shown in Table 18–1.

LOWERED CROP YIELDS. As an example of the losses that may result from weed infestations, research in Washington County, North Carolina, and reported in 1974, showed a soybean yield loss to smartweed, with one smartweed per 16 feet (4.9 m), of 5.5 bushels/acre (370 kg/ha) when the field was weedy all season. Another study at Blackville, South Carolina, showed that cocklebur plants located at 1-foot (30-cm) intervals in the field reduced soybean yield to one-half that when the field was clean The yield was about 40 bu/acre (2,688 kg/ha) under weed-free conditions and only about 20 bu/acre (1,344 kg/ha) under cocklebur-infested conditions.

REDUCED LAND VALUES. The sale, rental, and loan value of land may be seriously affected if noxious weeds are prevalent. In some cases, banks

TABLE 18–1. Estimated Annual Weed-Caused Losses

Crop or Situation	Losses in Yield and Quality (millions)	Cost of Control (millions)	Total (millions)
Agronomic crops	$1,500	$1,900	$3,400
Horticultural crops	250	300	550
Grazing lands	630	370	1,000
Aquatic sites and noncropland	50	60	110
Totals	$2,430	$2,630	$5,060

and investment companies either refuse to make loans or limit loans on land badly infested with noxious weeds. Bindweed has been known to depreciate the selling value of land by more than one-half.

REDUCED UNIT VALUE OF CROPS. Weeds may greatly reduce the market value of the crop. Pastures infested with such weeds as wild garlic, wild onions, and bitterweed, when grazed by dairy animals impart a bad flavor to milk and to all dairy products. Wheat containing garlic in considerable quantity is sold only as sample grade and at a price 20–50 percent lower than No. 2 Red. The money lost from wild onions or garlic alone runs into millions of dollars yearly. Wheat growing has been abandoned in some sections because of the presence of this weed. Records for a four-year period showed that wheat arriving at terminal markets from the northern grain states averages 7 percent dockage (large weed seed); flax averaged a dockage of 16 percent.

INSECTS AND DISEASES. Some weeds may harbor insects and diseases which attack crop plants. Clubroot of cabbage is carried over on wild mustard, whereas the Colorado potato beetle also lives on buffalo bur and jimsonweed. The tomato mosaic lives on nightshade and ground cherry, whereas the corn borer is reported to carry overwinter in almost any weed growing along fence rows.

POISONOUS WEEDS. Some weeds are poisonous or otherwise injurious to livestock. The most important plants found poisonous to livestock over a large section of the country are death camas, tall larkspur, water hemlock, locoweed, whorled milkweek, salt bush, lupines, and cocklebur seedlings.

Why a Weed Problem?

In some areas a principal cause of the weed problem is the use of impure seed. Most of the worst weeds have been introduced in this way. Other factors that have made the weed problem acute are the following: (1) Weeds that retain their vitality for years often have been plowed under instead of being cut before the seed is mature. (2) Laxity in the enforcement of weed laws has permitted increased distribution of noxious weeds. (3) On many farms soil erosion and depleted fertility have given weeds the advantage over hay and pasture crops. (4) In some areas the main rotations have consisted principally of wheat, oats, corn, and other row crops, with the result that cultivation, which normally ceases in the early summer, has not prevented the summer growth of perennial weeds. (5) Herbicides have not always given complete long-season control of a wide spectrum of weeds.

NOXIOUS WEEDS. Weeds particularly difficult to control or to exterminate usually are referred to as noxious. Sometimes weeds that are noxious in one state, with the particular climatic conditions and cropping practices found there, are not so in other states. Other weeds that are particularly troublesome are not included as noxious in the seed laws because the seeds are of such a character that they are never found in commercial seed.

There are a number of weeds that are recognized as very troublesome over a wide territory. About 180 weed species are defined as noxious in one or more states. Quackgrass and Canada thistle are considered noxious in 47 and 48 states, respectively, although neither species is adapted to the warm climate of the Southern states.

No federal noxious weed law exists, nor does it appear that legislation similar to that in the states would be feasible. About half the states have weed laws which vary greatly. They require that the weeds be cut at such times as will prevent the production of seed. In most states control and eradication of weeds has largely been left up to the individual landowner. Local agencies, such as county control boards or township boards often have the authority to compel a property owner to carry out weed-sanitation measures. In some states the county or state agencies are in control. Success in controlling noxious weeds through the enforcement of weed laws ultimately depends on the availability of effective and economical control measures. In general, preventive laws have been moderately effective and weed-control laws have been relatively ineffective.

The federal government and every state in the United States have seed laws. The Federal Seed Act of 1939 regulates seed imported into the United States or shipped in interstate commerce. The Seed Act, administered by the USDA, protects purchasers from mislabeled or contaminated seed. State seed laws are relatively uniform, although there is sufficient variation to create some interstate problems.

Standards for maximum tolerance for weed seed in individual crops are determined by the International Crop Improvement Association, which is made up from seed certification agencies in the various states and Canada.

The Plant Quarantine Act and the Federal Plant Pest Act both have several aspects relating to weed control.

Control of Weeds

A number of recognized, well-known methods of weed control have been advocated and are effective for certain weeds. Some of these come under the heading of prevention, whereas others apply to the control or eradication of weeds after they have become established. One of the greatest developments in weed control was the availability of selective herbicides, those that injure or kill certain plants without injury to other kinds. Some of these are injurious to broadleaf plants but not to grasses, whereas others have the reverse activity.

USE OF CLEAN SEED. The use of high-quality, clean crop seed is important in any weed control program. Perhaps the worst offenders in the spread of weed seed are the small-seeded forage crops. It is almost impossible for the average farmer to look at seed and tell whether or not it is free of weed seed and safe to plant. It is best not to buy anything but high-quality seed, labeled as to weed content. It may be advisable to buy certified seed, grown under strict standards and inspected in the field and bin.

ROTATION OF CROPS. In some cases where a one-crop system of farming prevails, particularly of a crop such as wheat, the land becomes heavily infested with weeds adapted to growing with these crops. Crop rotation aids greatly in weed control, in that cultivated crops commonly are included and also because weeds adapted to one particular crop usually cannot meet the competition of a change of cultural methods followed in growing other crops in the rotation. Rotation also may permit the growing of crops which serve as smother crops, which tend to eradicate certain weeds.

CULTIVATION. Repeated cultivation of the soil aids in the control of weeds after the land has become infested. In certain cases, with persistent weeds, it is necessary to use the fallow system, with the land cultivated regularly to prevent any top growth. This method is particularly necessary in the case of weeds that spread by underground rootstocks, such as the Canada thistle, quackgrass, johnsongrass, perennial sowthistle, and other noxious weeds.

SMOTHERING. Rapid- and rank-growing crops, such as the sorghums, sudangrass, and the millets, are particularly valuable as smother crops. Alfalfa has been found effective in the control of Canada thistle. Bermudagrass in alfalfa fields has been eradicated by the application of fertilizer which stimulates a rapid growth of alfalfa.

PASTURING. Certain types of weeds may be controlled temporarily, and in some cases permanently, by pasturing with sheep, goats, or hogs. Where it is not practicable to graze sufficiently close to destroy perennials completely, the grazing may so weaken the root system of these plants as to make it easier to control them by cultivation or with herbicides.

MOWING. Another means of prevention or control of weeds after they have been established is to prevent reseeding by timely mowing. The mower can be used to good advantage in fence rows, pastures, roadsides, and other uncultivated places. Mowing must be done before the seed pods are formed, as many weedy plants have the ability to mature seed even after being cut.

FLAME CULTIVATION. Fire is effective in "flaming" intertilled crops such as cotton and corn. Good results have been obtained with flame cultivation following herbicides in the control of johnsongrass and other grass seedlings in sugarcane. This method is effective on small annual weeds, but less effective on perennials. For best results, the flame weeder should be used when the weeds are very small and at regular intervals. Proper adjustment and speed of the burners are important to prevent crop injury. Flame cultivation is more adapted to controlling weeds in a crop such as cotton, which has a woody stem more resistant to damage from the flame. Fire is effective in removing undesirable plants from ditch banks, roadsides, and other waste areas.

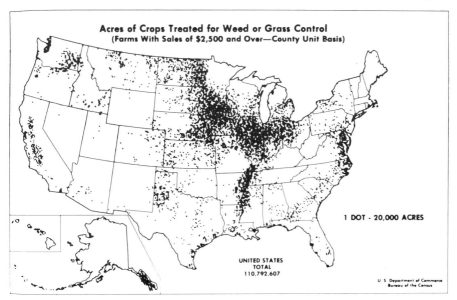

FIGURE 18-10. *Acres of crops treated with herbicides for weed or grass control, shown geographically. (Farms with sales of $2500 and over—county unit basis.)* [SOURCE: U.S. Department of Commerce.]

Chemical Weed Control

The control of weeds by chemical methods has been one of the most outstanding developments of the twentieth century. It started in 1896 with the chance discovery in France that a solution of copper sulfate killed yellow charlock without affecting oats. The discovery of 2,4-D (2,4-dichlorophenoxyacetic acid) changed the picture entirely with respect to chemical weed killers. Chemical weed control has far-reaching consequences and involves problems in chemistry, physiology, ecology, engineering, agronomy, and related sciences. Chemical weed control now is considered a vital and established part of good farming rather than only for emergency use. In 1975, 788 million pounds (357 kg) of herbicides were produced in the United States, up from 554 million pounds (251 kg) for the preceding five-year period. Acreage of crops treated with herbicides for weed or grass control is shown in Figure 18-10. Herbicides offer several advantages for weed control, such as the following:

1. Herbicides may be applied broadcast, which includes crop rows where cultivation is impossible.
2. Season-long weed control often can be achieved by careful selection of herbicides used alone or in combinations.
3. Cultivation may injure the crop root system as well as the foliage.

4. Herbicides reduce the destructive effects of tillage and excessive compaction on soil structure.
5. Erosion in perennial crops frequently is prevented by using a sod, kept thick by using herbicides to reduce weed competition.
6. Many perennial weed and brush species cannot be controlled efficiently without the use of herbicides.
7. Herbicides may eliminate the need for cultivation, thus reducing production costs.

Some chemicals for weed control have been in use for a considerable time; others have been introduced only recently. Certain herbicides are specific in their action whereas others are general. Information on how these new herbicides work to kill the plant and on how to use them effectively is voluminous. Names and structural formulas of herbicides often are long and cumbersome, so for frequent use many chemicals are known by their common or by trade names. To keep up with the latest research on these new chemicals, agricultural journals and state publications must be checked periodically. In addition, the USDA periodically makes available a list of chemical pesticides that are registered for agricultural use.

CLASSIFICATION OF HERBICIDES. Herbicides are grouped according to their effect on plants as either contact, translocated, or soil sterilant.

Contact herbicides are used when a rapid kill of foliage is required. They do not, however, prevent regeneration of perennial weeds, nor do they have any persistent effect in preventing reinfestation. Contact herbicides may be selective or nonselective. Paraquat is an extensively used contact herbicide. It is used widely in no-till cropping systems.

Herbicides commonly are referred to as being selective when treatment controls the weeds without affecting the crop. They are called nonselective when they control or kill all plant species with which they come in contact. Most herbicides do not fall rigidly into either group because at a sufficiently low rate even the so-called nonselective herbicides may be selective in certain situations; similarly, at very high rates all herbicides generally lose their selectivity. Factors other than the chemical nature of the herbicide and the rate affect selectivity. Examples are formulation of the chemical, method and time of application, stage of growth of the crop and weeds, and environmental conditions.

Translocated herbicides also are valuable for the initial control of established weeds, and, in addition, will eradicate susceptible species. They are relatively nonpersistent and will not prevent invasion and reinfestation. Translocated herbicides also are called systemic herbicides. They usually exhibit a wide range of selectivity and therefore are effective on certain plants but not on others. Translocated herbicides are those chemicals which after entering the plant are transported within it and thus may affect tissues remote from the initial point of entrance. With many herbicides, downward translocation takes place rapidly after application to the shoot and the chemical may become distributed throughout the plant, including the root system. Similarly, herbicides applied to the soil may be absorbed in the root system and translocated upward through the entire plant. Because of their ability to move within the plant, and in contrast to contact herbicides, the chemicals in this category can be effective in controlling perennial plants and do not necessarily have to be applied evenly over the whole plant for good results. Most herbicides commonly used for weed control in field crops are in the translocated category.

Any chemical that prevents the growth of green plants and is not completely inactivated in the soil is considered a soil sterilant. Soil sterilants generally are not the best control for established perennial weeds but may be useful for controlling weeds in fence rows and similar situations. They can prevent reinfestation of clean ground for considerable periods.

If a chemical sterilizes the soil for less than 48 hours, it is said to have no residual toxicity; if it sterilizes for four months or less, it is a temporary soil sterilant; if it sterilizes from four months to two years, it is semipermanent; and if it sterilizes for more than two years, it is considered permanent.

HERBICIDAL ACTION. Systemic herbicides generally are either absorbed by the roots and transported in the xylem or absorbed by leaves and translocated in the phloem. Movement of chemicals applied to leaves is through the phloem to the petioles and stems. Chemicals applied to the soil are taken up by roots and move primarily in the xylem.

With hormonelike chemicals, movement usually is correlated with food movement. Movement through shorter distances may be by normal diffusion or by a polarized movement. Finally, after entering the plant, the chemicals must kill the cells. In all modes of physiological action, whether it be a violent upset of cell reaction, a coagulation of cell proteins, or a more subtle disorganization of metabolism, the final action usually involves the denaturing of the sensitive protoplasmic system responsible for normal plant function.

Safety testing. Herbicides are safety tested before they are sold to the public. Experimental animals are used to determine the acute toxicity of a new herbicide. In addition, more detailed studies determine long-term effects. A herbicide to be applied on food and feed crops must obtain the approval of the Federal Food and Drug Administration. They help draw up specific instructions for the herbicide label. The cost to a company of obtaining a label for a single herbicide or other pesticide may cost $13–$20 million, with only one chance out of 15,000 or less for a new chemical to become a product. The Federal Food and Drug Administration has established standards and procedures to determine the safety of a material. Legal residue tolerances are established. Food that exceeds the tolerances may be condemned.

TIME OF APPLICATION. Many factors influence the effectiveness of a herbicide. The herbicide–soil–weather–plant interactions are complex. The time of herbicide application may largely determine its usefulness in various crops. Applications may be preplating, pre-emergence, or postemergence. Herbicides may be applied broadcast, in a band, as a directed spray, or as a spot treatment.

ONE APPLICATION. Foliar applications of nitrogen and minor elements combined with fungicides presently are being used in certain commercial operations. Fertilizers and insecticides are being applied together. In recent years, some herbicides have been applied with liquid fertilizers, in irrigation water (herbigation), and by other alternative methods. Application of herbicides with a foam carrier can kill more weeds at given rate and reduce spray drift. Sometimes the rate can be reduced.

Biological Weed Control

Biological methods may play an important role in overall weed control procedures in the future. The following are some specific examples where biological methods have been successful, at least on a limited scale:

1. A small white moth, *Coleophora parthenica,* from Pakistan feeds on Russian thistle and halogeton, a weed that is poisonous to some livestock. The moth already is established in parts of the West.
2. Templeton from Arkansas has found that the fungus *Colletotrichum gloeosporioides* is effective in controlling prickly sida or teaweed, and another fungus is effective against jointvetch in rice and soybeans.
3. A seed weevil, *Rhinocyllus conicus,* has been found to affect the seed development of musk thistle.
4. Other weed-eating insects, such as Altica beetle and leafy spurge hawkmoth, have controlled specific weeds.
5. Weeder geese are used in some areas to remove grassy weeds selectively from cotton fields.

Other nonherbicidal methods presently used or possibly of some future promise include:

1. Recycling sprayers, developed to control tall-growing weeds, such as johnsongrass in cotton or soybeans, and capture the excess herbicide not absorbed by the weeds.
2. The microwave Zapper kills weeds, weed seeds, nematodes, fungi, and soil insects in the top layers of soil.
3. Laser beams from planes or satellites may one day fight weeds.
4. Weed researchers in North Carolina have used ethylene gas in the soil to overcome dormancy and to stimulate premature and complete weed seed germination so that control can be effected.

Remote Sensing

Remote sensing by infrared wavelengths now offers techniques for detection and monitoring of insect infestation, crop diseases, nutrient deficiencies, and plant identification as well as soil moisture, crop yield, and the ill effect of smog on natural vegetation and cultivated crops.

Environmental analyses through remote sensing techniques include the

FIGURE 18–11. *Recycling or recovery sprayer, designed to control tall-growing weeds in shorter-growing field crops, captures the excess herbicide not absorbed by weeds. The new innovations in application equipment carry great environmental implications. This photo shows a recycling sprayer set for 38 to 42 inch (97 to 107 cm) rows.* [Photograph with permission of Mississippi State University Agricultural Experiment Station.]

spatial distribution of plant species, both native and cultivated. These techniques are sufficient to account for a large portion of the statistical variation in the statistical distribution of weed sources, nutrient and moisture deficiencies, and crop–insect reservoirs. Unmanned earth satellites can monitor and predict crop yields of wheat, barley, oats, and soybeans, along with detecting diseases of cultivated crops, forests, and rangelands.

REFERENCES AND SUGGESTED READINGS

1. AGRIOS, G. N. *Plant Pathology* (New York: Academic Press, 1969).
2. ANONYMOUS. "Diet Pill for Pests," *Crops and Soils,* 29:11 (1977).
3. ANONYMOUS. *1977 Insect Pest Management Guide—Field and Forage Crops,* Ill. Coll. of Agric. Circ. 899 (1976).
4. BOYD, J. D. "Now It's Biological Control of Weeds," *Farm Jour.,* (April 1978), pp. G2 and G4.
5. CARSON, R. *Silent Spring* (Boston: Houghton Mifflin, 1962).
6. CHAPMAN, S. R., and L. P. CARTER. *Crop Production Principles and Practices* (San Francisco: Freeman, 1976).

7. CHIPPENDALE, G. M. "Third Generation Insecticide?" *Crops and Soils,* 29:7 (1977).
8. CHRISTENSEN, C. M., and H. H. KAUFMANN. *Grain Storage* (Minneapolis: Univ. of Minnesota Press, 1969).
9. CRAFTS, A. S. *The Chemistry and Mode of Action of Herbicides* (New York: Wiley–Interscience, 1961).
10. DUGAN, R. E., *et al.* "Residue in Food and Feed," *Pesticide Monitoring Jour.,* 1:2 (1967).
11. FOWLER, D. L., and J. N. MAHAN. *The Pesticide Review 1976,* USDA-ASCS Publ. (1977).
12. GOOD, J. M. *Integrated Pest Management,* USDA-Ext. Serv. ESC 583 (1977).
13. *Insects: Yearbook of Agriculture* (Washington, D.C.: USDA, 1952).
14. JAQUES, H. E. *How to Know Insects* (Dubuque: Wm. C. Brown, 1960).
15. KLINGMAN, G. C. *Weed Control as a Science* (New York, Wiley, 1961).
16. McGLAMERY, M. D., *et al. 1978 Field Crop and Weed Control Guide,* Ill. Agr. Exp. Sta. Publ. (1978).
17. METCALF, C. L., and W. P. FLINT, rev. by R. L.METCALF. *Destructive and Useful Insects and Their Habits and Control* (New York: McGraw-Hill, 1962).
18. Michigan State Univ. "The Biological Control Story," *Mich. Science in Action* (Oct. 1970) pp. 1–19.
19. National Academy of Sciences. *Principles of Plant and Animal Pest Control,* Vol. 1 (Washington, D.C., 1968).
20. National Academy of Science. *Remote Sensing: With Special Reference to Agriculture and Forestry* (Washington, D.C.: The Academy, 1970).
21. National Academy of Sciences. "Weed Control," *Principles of Plant and Animal Pest Control,* Vol. 2, Pub. 1597 (Washington, D.C., 1968).
22. PAINTER, R. H. *Insect Resistance in Crop Plants* (New York: Macmillan, 1968).
23. PARKS, W. L. *The Use of Remote Multispectral Sensing in Agriculture,* Tenn. Agr. Exp. Sta. Bul. 505 (1973).
24. *Plant Diseases: Yearbook of Agriculture* (Washington, D.C.: USDA, 1953).
25. POEHLMAN, J. M. *Breeding Field Crops* (New York: Holt, 1959).
26. RIKER, A. J. "Plant Diseases" (Encyclopedia Britannica, 1961).
27. ROBERTS, R. S. "A Brief History of Pesticide Use," *Utah Science* (June 1971).
28. SABROSKY, C. W. "How Many Insects Are There?" *Insects: Yearbook of Agriculture* (Washington, D.C.: USDA, 1952).
29. SCHMUTZ, E. M., *et al. Livestock-Poisoning Plants* (Tucson: Univ. of Arizona Press, 1968).
30. USDA-ARS. *A Survey of Extent and Cost of Weed Control and Specific Weed Problems,* ARS 34–23–1 (1965).
31. USDA-ARS. *Losses in Agriculture,* Agr. Handbook 291 (1965).
32. USDA. *1968 Report on Pesticides and Related Activities* (1969).
33. USDA. *Suggested Guide for the Use of Insecticides to Control Insects Affecting Crops, Livestock, Households, Stored Products, Forests, and Forest Products,* Handbook 331 (1969).
34. WALKER, J. C. "The Role of Resistance in New Varieties," *Plant Breeding* (Ames: Iowa State University Press, 1966).
35. WITT, W. W., and J. W. HERRON. *Chemical Control of Weeds in Farm Crops in Kentucky—1978,* Ky. Coop. Ext. Serv. Publ. AGR-6 (1978).

Part V
Field Crops

A. Grain Crops

Chapter 19
CORN OR MAIZE

Corn is the most important crop produced in the United States. Corn is harvested for grain on more than 70 million acres (28.3 million hectares), with total production often exceeding 6 billion bushels (1.5 billion quintals) and a total value of more than $14 billion in each of the years 1973–1976. Corn for silage is harvested from another 11 million acres (4.5 million hectares), and corn for other forage purposes from 1 million acres (0.4 million hectares).

Origin and Development

The food grains wheat and barley, in the form in which we know them, were in use in southwest Asia as early as 5000–6000 B.C., and grain sorghum in Africa and India was probably in use even earlier. Corn, as we have it today, recognized to be a development of the American Indian, is believed to have developed somewhat later. A tremendous amount of research has been done through many years in an effort to determine how, where, and when corn had its beginning. Living specimens of a wild corn as a possible progenitor of modern corn have never been found. It apparently disappeared centuries ago. The oldest archaeological race of corn is a pod-popcorn; the closest living relative is teosinte.

That a wild corn existed in central Mexico at a very early time is evidenced by fossil pollen grains recovered from drill cores at a depth of 200 feet (61 m) in the erosion-filled lake bed below the present Mexico City. This material is assigned to the last interglacial period, now estimated by geologists to have occurred about 80,000 years ago. Because this period is believed to antedate the arrival of man on this continent, the pollen evidently was a wild corn. Other pollen, considered to be that of cultivated corn, occurred abundantly through a considerable range at higher levels. Just where this new plant, which produced food so much more abundantly than any other plant previously available, may first have come into use has been a subject for speculation.

The oldest and most persistent theory of corn's origin is that it was selected either directly from teosinte or from an ancestor common to both. Some evidence appears to point toward teosinte or a teosinte-type grass as the progenitor of corn. Others speculate that corn originated from a wild form of pod corn. *Tripsacum* is also mentioned frequently as a close

333

FIGURE 19-1. *Seed of teosinte, a close relative of corn. Much evidence points to teosinte or a teosinte-type grass as the progenitor of corn.* [SOURCE: Southern Illinois University.]

relative, but it probably did not contribute directly to the parentage of cultivated corn. While most archaeologists locate the center of origin in Mexico and Central America, others say that corn may have originated in the highlands of Bolivia, Peru, and Ecuador. In both areas, a diversity of wild forms still exist.

Certain investigators believe that the first Indians of the Americas to plant and cultivate corn extensively probably were those living in what is now Peru, Bolivia, and northern Chile. Rigid selection of seed from the most productive plants, continued through the long centuries, finally resulted in a new plant that produced foods so abundantly that it made possible several ancient American civilizations. There were the Incas in what is now Peru, with a culture and standard of living that rival those of ancient Egypt, Babylon, and Greece. Perhaps somewhat later, again made possible by the productivity of food by this new plant, there developed in Guatemala the Mayas and in Mexico the Aztecs and similar groups.

Perhaps the best information on the origin of corn is based on specimens unearthed as recently as 1961, when archaeologists and botanists made full-scale excavations of five major caves in the valley-exposure mountains that surround the Tehuacan Valley of southern Mexico. Over 20,000 corn specimens were unearthed in these five caves. The thousands of cobs revealed a well-defined evolution sequence for the corn plant. The earliest cobs, dated at 5200–3400 B.C., apparently were of a wild type of corn. These ears appeared to be only about an inch (2.5 cm) in length, enclosed in only two husks. These researchers are confident that a type of corn they label "early cultivated" was grown in central Mexico during the period 3400–2300 B.C.

Mangelsdorf's description of the valley and of the five caves, may well stimulate one's imagination:

Caxcatalan Cave, first found in 1960, was one of the richest in vegetal remains. Excavations revealed 28 superimposed floors, or occupation levels, covering two long unbroken periods—from 10,000 to 2300 B.C., and from 900 B.C. to A.D. 1500. Fourteen of the upper floors, those from 5200 to 2300 B.C., and from 900 B.C. to A.D. 1500 contained well preserved corn cobs.

History of Corn In America

When Columbus discovered the New World, he also discovered corn. When European colonists first came to America, they found the Indian tribes along the Atlantic seacoast growing large acreages of corn. In New England and southern Canada, flint corns prevailed, with types varying in color. Far to the south and west there were many different corns, with different types grown for different purposes. Early American civilization appears to have developed around corn as the most important food crop, probably as important as all other food crops combined. There were corns for popping as well as types grown especially for eating green on the cob, though they did not have the high sugar content of our present sweet corn.

Columbus described something of the extent and importance of corn as a food crop in his exploration reports for 1492 and 1498. Captain Miles Standish also reported his observations of a field of 500 acres (202 ha) of corn in 1620. It generally is believed that those who came from Europe to settle the ill-fated Virginia Colony would have died of starvation except for the corn obtained from friendly Indians. The Indians taught the colonists how to prepare the ground and how to plant corn. Herring or shad, which came up the streams in great numbers in the spring to spawn, were captured and used as fertilizer; the practice required usually one fish in one hill of corn.

The American Indians performed one of the greatest plant breeding jobs in history. Out of some primitive corn they developed widely-adapted types to make this crop more extensively distributed over the world than any other cereal. Using as seed the corn obtained from the Indians, following the Indian's method of mass selection of seed ears with certain desired characteristics, the colonists soon developed many new lines better suited to their needs. There were as many as 200 named and described cultivars of sweet corn alone by the late 1800s.

Dent corn is of comparatively recent origin. Early-maturing flint types were grown almost entirely in the North until well into the nineteenth century. As contrasted with the long, slender flint ears, with a small number of rows, the relatively late-maturing gourd-seed types grown farther south had very deep kernels, were white in color, and had about 20 rows per ear. The extreme variation in the number of rows of kernels and the roughness of indentation of the dent types, as well as the variations in the size and shape of ears, undoubtedly traces back to the mixing of the gourd-seed and flint types.

The Corn Plant

Vegetative Characteristics

The height of the corn plant varies greatly, from taller than 15 feet to just 3 feet (4.6–0.9 m) or less. Some of the earliest flint and popcorn types make a very small, short growth, with ears found close to the ground. Corns from Central and South America planted in the Corn Belt have been known to make a growth of nearly 25 feet (7.6 m) before the end of the season without approaching maturity. Lodging of stalks may be a problem with some corns. The degree of lodged stalks, those that break over causing harvesting difficulties or losses, varies greatly among hybrids. Modern-day hybrids have been developed with a strong stalk rind and better root systems to resist lodging. Stalk rot organisms increase lodging severity.

The stalk is make up of nodes and internodes, usually 10 to 15 per stalk for dent types. The leaves, ears, and roots are outgrowths from the nodes. The longer internodes are found toward the top of the stalks; toward the base the internodes are very short. Buds are located at most internodes. Above-ground buds may develop into ear shoots; those below the ground may develop into tillers or suckers. Modern field corn hybrids have been bred so that they seldom tiller extensively; but most sweet corn hybrids will develop numerous tillers where space, water, and nutrients are abundant.

Leaves are borne alternately. A leaf consists of a ligule, blade, and sheath, which clasps the culm or stem. The ligules, found at the junction of the leaf blade and sheath, are absent in newer liguleless or erect-leaved types. The erect-leaved characteristic allows higher plant populations and apparently makes better use of light on an acreage or unit area basis, but not on an individual plant basis.

Leaf rolling is a common phenomenon often observed during drought periods. Rolling is the result of shrinking of the large bulliform cells in the upper epidermis of the leaf.

Like other grasses, corn develops a fibrous root system. When the seed germinates, temporary seminal or seed roots develop. The permanent coronal or crown roots begin to develop a few days later at the first seven or eight, closely compacted nodes at the base of the stem, usually just below the ground. Brace roots usually make their appearance at the first two or three nodes above the ground surface.

Reproductive Development

Corn is considered monoecious, with male and female structures borne in separate flowers, with both types of flowers present in different locations on the same plant. The staminate or male flowers are borne in the tassel in the terminal portion of the plant. The pistillate or female flower is borne in the leaf axis.

The staminate tassel bears spikelets, each of which has two florets. A floret contains three stamens which produce pollen grains. As many as 10 million pollen grains are produced per plant. Only one pollen grain is required to fertilize the silk of each kernel.

The pistillate inflorescence or potential ear is a spike with a thickened, woody rachis or cob structure. Spikelets are borne in pairs in several longitudinal rows. The paired arrangement of the spikelets explains why most corns have an even number of rows of kernels. The single ovary of the fertile floret, the potential kernel, bears a long style or silk, with a forked tip and a sticky stigmatic surface, which is receptive to pollen. Each kernel has its own silk and must be fertilized separately. There are usually 800–1,000 kernels per ear.

The silks remain receptive to pollen for several days except when temperatures are extremely high. When a pollen grain falls on the silk, it germinates within a few minutes into a pollen tube which enters and grows down through the silk to the ovary. The two sperm nuclei from the pollen grain migrate down the tube into the ovary; one fuses with the egg and the other with polar nuclei. This double fertilization process, which normally is accomplished within 36 hours after pollination, results in eventual development of the embryo and the endosperm of the kernel.

Most Corn Belt hybrids have been bred to develop only one ear. In recent years, there has been renewed interest in prolific hybrids which develop two or more ears per stalk.

Types of Corn

Corn *(Zea mays)* is a grass. There are several groups or types of *Zea mays* which have some economic importance. Those of greatest interest are the following:

1. Dent corn *(Zea mays indentata)* is characterized by the presence of hard, horny endosperm at the sides and back of the kernels, with only the starchy endosperm extending to the crown. A depression or "dent" forms at the crown of the kernel as the starchy endosperm dries rapidly and shrinks.

2. Flint corn *(Zea mays indurata)* is characterized by having the starchy endosperm enclosed in a relatively thick layer of horny endosperm, with a relatively small amount of starchy endosperm. The kernels

FIGURE 19-2. *Groups or types of corn. From left to right: pod, yellow dent, white dent, flint, Indian (a flint type), sweet, yellow popcorn, white popcorn, strawberry popcorn.* [SOURCE: Southern Illinois University.]

are inclined to be rather large and broad and the ears long and slender, with a comparatively small number of rows of kernels.

3. Sweetcorn *(Zea mays saccharata)* is characterized by a translucent, horny appearance when immature and more or less crinkled, wrinkled, or shriveled appearance of the endosperm when dry. It is unusually high in sugar content because of the presence of one recessive gene which prevents the conversion of some of the sugar into starch.

4. Flour corn *(Zea mays amylacea)* is characterized by having practically no horny or vitreous endosperm and little or no dent. This also is known as soft corn. The grains can easily be ground into meal. It is grown by the Indian tribes in the far Southwest, and in South America. It is one of the oldest types of corn, and is frequently found in graves of the ancient Incas and Aztecs.

5. Popcorn *(Zea mays everta)* is characterized by an excessive proportion of horny endosperm and the small size of the kernels and ears. Rice popcorn has pointed kernels and the pearl type has rounded kernels.

6. Waxy corn is a type of dent corn. It contains almost all branched, amylopectin starch, while common corn starch is a mixture of amylose and amylopectin. The endosperm is unusual in that it is possible to substitute it for tapioca starch. This was done on a commercial scale

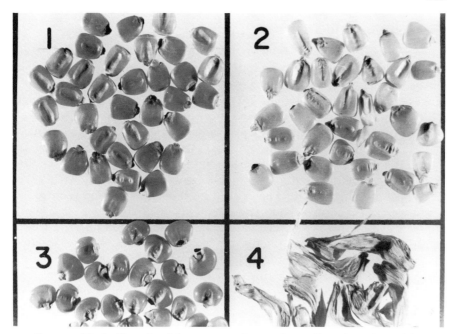

FIGURE 19-3. *Kernels of four types of corn: 1 = yellow dent, 2 = white dent, 3 = flint, 4 = pod corn.* [SOURCE: Southern Illinois University.]

for making adhesives during World War II. Since then additional industrial uses have been found for waxy corn. It is used for instant puddings, as a thickening agent in soups, and for other types of dessert foods.

7. Pod corn *(Zea mays tunicata)* has each kernel enclosed in a pod or husk, with the ear itself also enclosed in husks. Pod corn is not grown commercially, but is of interest in studies of the origin of corn.

Corn is classified also according to color. The first such classification was made in 1860 by the Chicago Board of Trade; corn was classified as white, yellow, and mixed. Prior to 1919, 23–40 percent of the corn received in Chicago was classed as mixed corn. By 1945 mixed corn constituted only 4–5 percent of the receipts, with yellow corn about 80 percent. White corn, which in 1917 comprised 25 percent of the corn delivered in Chicago, dropped to 7.3 percent and in 1978 was less than 1 percent of the total production. Some pure white corn is desired by corn millers for making cereal and hominy; yellow corn has come to be preferred for feeding farm animals because of its higher vitamin content. Yellow corn has greater vitamin A activity, but still exceedingly large amounts of yellow corn would have to be fed in order to supply the daily vitamin A requirements of livestock.

Adaptation and Distribution

Because of breeding to widen its adaptation, hybrids or cultivars exist such that corn can be grown in all 50 states and Canada, as well as in most countries of the world. Length of maturity among the diverse types ranges from 50 to 330 days. It will grow from 58°N in Canada through the tropics to 40°S, from below sea level to 13,000 feet above sea level, and with annual precipitation levels ranging from 15 inches (38 cm) to more than 200 inches (508 cm).

Corn is a warm-season crop. It requires a temperature of 50°–55°F (10°–13°C) for germination and mean summer temperatures of 70°–90°F (21°–32°C) for optimum growth. The critical level of precipitation required to produce a crop of corn without irrigation is about 15–20 inches (38–51 cm). Moisture availability is most critical during a period including a few weeks before to a few weeks after tasseling and pollination. Some have calculated that corn needs fully one-half of its total water for the growing season during a 6-week period bracketing the pollination stage. The water requirement of corn is relatively low; a relatively small number of pounds of water is required for each pound of dry matter produced. The acre requirement for water is high, however, because of the potential high acre production.

Best soils for corn production are fertile, well-drained loams with a high water-holding capacity to supply water during the critical period. Corn will grow adequately within the pH range 5.5–8.0.

Following the discovery of the Western Hemisphere and of corn, this new grain was grown throughout the civilized world, wherever the climate

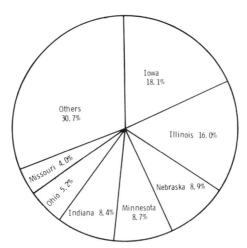

FIGURE 19–4. *Leading states in United States corn area (acreage). Seven states account for over two thirds of the total acreage, and two states for about one third of the total acreage.* [SOURCE: Southern Illinois University.]

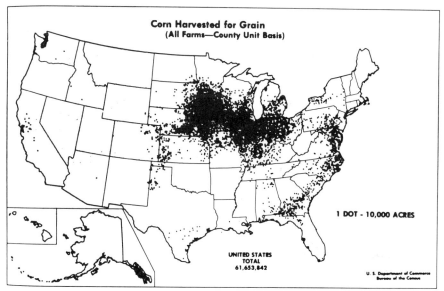

FIGURE 19–5. *Distribution of corn harvested for grain, shown geographically. (All farms—county unit basis.)* [SOURCE: U.S. Department of Commerce.]

and soil were suitable. In the United States the greatest acreage is in Iowa, Illinois, Nebraska, Minnesota, Indiana, Ohio, and Missouri (Figure 19–4). These seven states account for about 70 percent of the acreage; Iowa and Illinois alone plant 34 percent of the United States total acreage. Geographic distribution is shown in Figure 19–5. Early-maturing hybrids have been

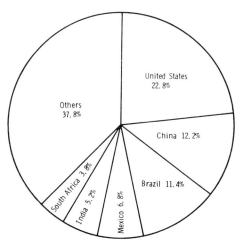

FIGURE 19–6. *Leading countries in world corn area (acreage).* [SOURCE: Southern Illinois University.]

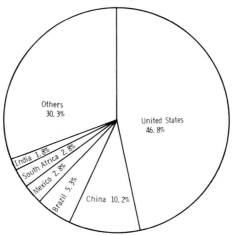

FIGURE 19-7. *Leading countries in world corn production. The United States continues to dominate, producing nearly half of the world total.* [SOURCE: Southern Illinois University.]

developed that are suitable for growing in Montana, North Dakota, and southern Canada; long-season hybrids are suitable for the extreme South.

World corn acreage and production is dominated by the United States, with nearly 23 percent of the acreage and nearly 47 percent of the production (Figures 19–6 and 19–7). Other corn-producing countries are China, Brazil, Mexico, India, and South Africa.

Uses and Farm Disposal

The primary worldwide utilization of corn is in food products. In the United States, however, this is mostly in the form of animal products resulting from the conversion of corn fed to farm animals. Much of the corn crop never leaves the farms on which it is grown. Especially in dairy areas significant amounts are cut for silage or harvested as fodder.

The corn crop contributes most to the production of beef, pork, dairy products, and poultry products, with much of these Corn Belt products going to other areas. In recent years the amount of corn used for feed has been about 65 percent of the total crop, about 25 is exported, and about 8 or 9 percent is used for food, alcohol, and seed.

Less than 9 percent of the corn crop is industrially processed. Processed corn contributes to the manufacture of many products, from breakfast foods, cornmeal, flour, and grits to cornstarch, corn syrup, corn sugar, corn oil, and alcohol and distilled spirits. Around 1–2 percent of the total production of corn is ground into dry cornmeal and grits. Such basic materials from corn as lactic acid, acetone, and butyl and ethyl alcohol are used in the production of hundreds of different products. One of the most recently found uses of corn is in production of a highly absorbent product called "super slurper." This is a chemically combined form of cornstarch and a type of plastic resin. It will absorb about 5,000 times its weight in water and may find use in such things as bandages, disposable diapers, and possibly as a soil conditioner or seed-coating material.

The United States produces well over 500 million gallons (1893 million liters) of corn syrup annually, with 99 percent used domestically. Corn oil is used in margarine manufacture to the extent of 218 million pounds (99 million kg); shortening manufacture uses 4 million pounds (1.8 million kg) each year. Exports of corn and cornmeal each year account for over 3.5 million hundredweights (159 million kg).

In the United States per-capita annual consumption of corn food products is approximately 8 pounds (3.6 kg) of meal, 2 pounds (0.9 kg) of cereal, 30 pounds (13.6 kg) of syrup, 5 pounds (2.3 kg) of sugar, and 2 pounds (0.9 kg) of starch.

Production Practices

Despite the fact that as the years have passed, corn land has had much of its natural fertility exhausted, has been subjected to wind and water erosion, and has become increasingly subjected to insects, diseases, and weed pests, the average acre yield of corn has increased enormously. This striking increase is credited to more effective production practices as well as to the use of hybrid seed. Seven improved practices are worthy of special mention; there are others almost equally important.

Seven Important Production Developments

HYBRID CORN. First in importance, and one of the most unusual developments in all agricultural history, is the general planting of hybrid corn seed. So outstandingly superior were the first hybrid corns that farmers turned to them almost as rapidly as adapted hybrids could be developed and the seed increased. Not only did the acre yield double within a few years, but these hybrids also possessed a stalk strength and resistance to insects and diseases formerly unknown.

INCREASE IN FERTILIZER USE. Corn growers have learned the advantage of having their soil tested to determine nutrient needs and of using fertilizers accordingly. Soil tests have become more refined. Few factors have had as great an impact on corn production as the extensive use of nitrogen.

INCREASED PLANTS PER ACRE. With the increased use of fertilizers, better control of weeds with herbicides, and improved cultural methods, acre yields can be increased markedly by planting higher populations than has been the practice. More than 20,000 plants/acre (49,419/ha) have been found optimum in many situations where yield potential is high.

MECHANIZED EQUIPMENT. In the heart of the Corn Belt virtually 100 percent of the corn crop is harvested with combines, picker-shellers, or pickers. The great majority is field-shelled, rather than being picked and stored as ear corn. These methods of harvest would be impossible except for the greater stalk strength of hybrid corn. The availability of mechanical power makes possible not only rapid mechanical harvesting, but also improved practices in seedbed preparation, planting, and cultivation.

SOIL-CONSERVING PRACTICES. On rolling land subject to erosion, not only has the power required to prepare the seedbed, plant, cultivate, and harvest been greatly reduced, but by contour plowing and cropping, and by minimum tillage practices, the productivity has been increased as a result of the soil and water resources being conserved.

INSECTS AND WEEDS CONTROLLED WITH CHEMICALS. Within the past 30 years insect and weed pressures have increased throughout the entire corn-growing area. Great advancements have been made in the development and release of effective insecticides and herbicides.

TIMELINESS OF PRODUCTION OPERATIONS. Better timing of all production operations, including seedbed preparation, planting, cultivation, and harvesting has been made possible by the availability of mechanical power and of larger field units.

Rotations vs. Continuous Corn

In much of the Corn Belt, corn is grown continuously for several years on the same field. In some situations, it is advisable to grow corn in rotation with other crops. The proper choice depends principally on the soil conditions and the management level.

Crop rotation offers the advantages of better labor distribution, less erosion on sloping fields, better weed control, and possibly fewer insect and disease problems. Some of the common rotations in the Midwest involving corn are corn–small grain; corn–small grain–legume or grass-legume; corn–corn–small grain–legume or grass-legume; and corn–soybeans–small grain–legume or grass-legume.

On some soils and in some regions, corn can be successfully produced continuously by heavy fertilization, return of crop residues to the field, pesticides for insect and weed control and attention to soil and water conservation. Corn can actually return more organic matter from cornstalks and roots than an average sod crop. If continuous corn is desirable and is the most profitable system, special attention should be given to maintaining soil structure and to the soil slope. The large number of field operations in row cropping gradually may destroy the soil structure. If the slope is such that erosion cannot be kept within reasonable limits, continuous cropping to corn is not advisable. Minimum tillage practices, which have been widely accepted in some areas, help to maintain soil structure and reduce erosion potential in these situations. Special cropping procedures, such as strip-cropping, growing alternate strips of corn and sod on sloping land, have been used with success.

There is some evidence to suggest that insects and diseases are not noticeably worse under continuous corn, as is commonly believed. An exception is the Western corn rootworm, which is more severe under continuous cropping. Corn grown continuously on one of the Morrow Plots at the University of Illinois for 100 years has shown no indication that insects and diseases are worse than on other plots where corn was rotated with oats and clover. It has been shown under these conditions that the most profit from field crops could be from corn grown continuously, rather than in a rotation. Other research studies have shown that yields from corn grown continuously under a high level of management have been nearly equal to those grown in rotation on deep, medium-textured soils of the Corn Belt. On other soils and in other regions, yields from corn grown continuously often have been inferior to those in rotation.

Tillage and Seedbed Preparation

Tillage practices used by farmers have changed considerably during the last 20 years. Most have not gone all the way to once-over practices or no-tillage, but there has been a general reduction in the amount of secondary tillage before planting and in postplanting tillage. Overall, a strong trend toward fewer trips across the field has occurred.

CONVENTIONAL TILLAGE. In conventional tillage, the field is disked or the residues chopped, plowed with a moldboard plow in the fall or spring, disked once or twice, and then harrowed or rolled. These operations result in a firm, finely pulverized seedbed. However, conventional seedbed preparation is expensive and because of many trips over the field excessive soil compaction may result. Water infiltration may be reduced, runoff and erosion increased, and drainage problems magnified.

DEEP TILLAGE. Benefits from deep plowing, greater than 12 inches (30 cm), may occur on some soils from incorporating residues, breaking up a more shallow plow-pan, or improving chemical conditions in part of the soil. Deep tillage may increase the soil volume in which plant roots can proliferate and thus obtain more subsoil moisture during periods of stress. Experiments on subsoiling before corn have rarely shown a yield response, and when there was an increase, it was not sufficient to justify the subsoiling expense.

MINIMUM TILLAGE SYSTEMS. It is now recognized that in many situations it is not good to till the soil any more than is necessary to obtain a satisfactory seedbed and control weeds. Pesticides can be used to control problem insects and weeds. Minimum tillage systems, which range from one less field operation to no-tillage, have come into widespread use. Some are considered as follows:

1. Plowing can be done in the conventional way, but most or all of the secondary operations, such as disking and harrowing, are omitted. The rough seedbed enhances water infiltration and reduces runoff and erosion; compaction of the soil is reduced because of fewer trips. The method involving planting directly, without secondary tillage, in

a plowed seedbed is called "plow-plant." It works best on silt loams and sandy loams.

2. With larger disks and tractors having more power, fields may be disked only without plowing. Much of the crop residues are left on or near the soil surface. This furnishes protection from wind erosion, as well as water erosion and runoff.

3. A chisel plow may be used to increase the depth of primary tillage but leave the surface rough. In addition to the obvious erosion and runoff control advantage, this offers the additional advantages of breaking up the "plow-pan" by deeper tillage, and speeding seedbed preparation operations prior to planting. The use of chiseling has increased greatly in the last 10 years.

4. Secondary tillage may be done in the row only by the wheels of the planter or tractor or by small rotary tillers. The seedbed between rows is left in rough condition. "Wheeltrack" planting may be a good system on light-textured soils with good tilth.

5. Strip tillage may be used, where a band near the row is prepared with a till-planter, sidewinder, or a rotary tiller.

6. Stubble-mulch tillage reduces runoff and erosion by protecting the soil surface from puddling and from exposure to wind. Stubble-mulch tillage may be done with a one-way disk plow, such as is common in the Great Plains, or with sweeps operated a few inches under the soil surface, lifting and loosening the soil, but not turning it. Special furrow openers have been devised for planting through residues.

7. No-tillage involves no seedbed preparation, except at planting when enough tillage is used to place the seed into the soil. All of the plant residue, which may be crop stubble or a grass sod killed near planting time with a herbicide, is left on the surface. This system allows corn to be grown on sloping land previously considered unsuited to row-cropping. Advantages from no-tillage planting are seen most often on sloping, well-drained soils of the lower Midwest and South. In the North, lowered soil temperatures at planting may limit its usefulness. Conventional tillage has given better results on poorly drained soils.

The Kentucky station and the Dixon Springs (Ill.) station generally have shown slight to moderate yield advantages for no-tillage planting over conventional. Other stations have noted slightly lower yields with no-tillage. These variable results apparently are due to variations in climate and soil characteristics and conditions.

Fertilizers

Virtually 100 percent of corn harvested for grain receives some type of commercial fertilizer, except in drier regions. Much of the acreage in the Northeast, Great Lakes, and West also receives animal manures. More total fertilizers are used on corn than on any other crop grown in the United States.

Nitrogen, phosphorus, and potassium, the macronutrients, are more likely to limit yield than other nutrients. The secondary elements calcium and magnesium may be limiting at a pH of 5.5 or lower. Zinc, iron, and other micronutrients may be deficient on some soils. The amount of nutrients removed by grain and stover at the 150 bu/acre (93.9 Q/ha) yield level is shown in Table 19–1. Fertilization rates should not be based on nutrient removal alone, but should be increased to compensate for fertilizer losses by leaching, erosion, or conversion to unavailable forms.

NITROGEN. More than 1 pound (0.45 kg) of nitrogen is needed to produce 1 bushel (0.25 Q) of grain, including the associated stover (see Table 19–1). Some of the required nitrogen may come from release of nitrogen from soil organic reserves, from the residue of a legume crop used in rotation, or from animal manures. However, in most instances application of a considerable quantity of nitrogen from commercial fertilizer sources is needed for excellent yields. The most common practice is to apply part of the nitrogen before planting or at the time of planting, and to apply the remainder as a sidedress application after the crop is up. It is not unusual in the Corn Belt to apply 150–200 pounds of nitrogen per acre (168–224 kg/ha), most of which is sidedressed. Anhydrous ammonia has been the most widely-used nitrogen source since the mid-1950s. Nitrate forms of nitrogen are leachable, and ammonium sources should be used when leaching losses may occur. Fall nitrogen applications should not be made until the soil temperature will remain below 50°F (10°C), the level at which conversion of ammonium nitrogen to the leachable nitrate form will be negligible.

PHOSPHORUS. A 150-bushel (93.9 Q/ha) corn crop will remove about 35 pounds (15.9 kg) of phosphorus, about 80 pounds (36.3 kg) of P_2O_5. Phosphorus fertilizer may be broadcast and plowed under, banded below and

TABLE 19–1. Nutrients in Grain and Stover of 150-Bu/ Acre (93.9 Q/ha) Corn Crop

| | Amount Contained in: | |
| | Grain | Stover |
Element	lb (kg)	lb (kg)
Nitrogen	115 (52.2)	55 (24.9)
Phosphorus	28 (12.7)	7 (3.2)
Potassium	35 (15.9)	140 (63.5)
Calcium	1.3 (0.6)	35 (15.9)
Magnesium	10 (4.5)	29 (13.2)
Sulfur	11 (5.0)	8 (3.6)
Iron	0.1 (0.05)	1.8 (0.8)
Zinc	0.17 (0.08)	0.17 (0.08)

Source: Barber, S. A., and R. A. Olson. "Fertilizer Use on Corn," Changing Patterns in Fertilizer Use, edited by L. B. Nelson (Madison, Wisc.: Soil Science Society of America, 1968).

to the side of the seed, placed with the seed, or a combination of these. When the rate of phosphorus exceeds 25 pounds/acre (28 kg/ha) broadcasting and plowing under is advisable. Up to 25 pounds/acre (28 kg/ha) can be banded 2 inches (5.1 cm) below and 2 inches (5.1 cm) to the side. This procedure decreases phosphorus "fixation" in the soil, a process that ties up this element in an unavailable form. No more than 5 pounds/acre (5.6 kg/ha) should be placed in direct contact with the seed, often called "pop-up."

POTASSIUM. Most of the potassium is best broadcast and plowed under. Small quantities may be placed in direct contact with the seed, but too much will cause a "salting" or osmotic effect, which will result in reduced germination and emergence.

Application rates of all fertilizers are best based upon soil test results.

Hybrid Seed Selection

Corn breeders have developed hybrids especially adapted to most every area where corn can be grown satisfactorily. The range of adaptability for each hybrid often is rather narrow; a hybrid may be specific for particular environmental conditions.

Of prime importance in hybrid selection are length of maturity, tolerance to stress conditions, and resistance to pests. More specifically, the grower should be interested in additional characteristics such as yield potential, adaptation to his conditions, stalk strength, and height of ears on the stalk. Some of this information may be available from the local seed dealers; other may have to be obtained by consulting results of testing by state experiment stations and extension service personnel.

Large seed sizes have shown greater seedling vigor early, but most often the effect of seed size on yield is minor. Flat seed shapes have not shown an advantage over round seed; rounds, at a generally lower cost, are as good if suitable planting equipment is available. In recent years the plate-less, air or vacuum-type planters will plant any seed size or shape with precision.

Seed treatment with an appropriate fungicide is used almost universally by seed companies. This fungicide gives tolerance to seed rots and early seedling diseases. It is best to buy seed that have been subjected to a good cold germination test, a procedure which is now common for many seed companies. Seed with good cold test results will withstand 2 weeks or more of cold, wet soil conditions often encountered in early spring planting.

Planting Methods

In general, planting may be by either one of two methods; surface-planted or listed. Except for rather limited areas of the West, corn is surface-planted in rows 40 inches (102 cm) or closer. In the past most of the acreage was

FIGURE 19–8. *A corn-planting operation in a conventionally prepared seedbed. An increasing number of larger planters, such as this 12-row planter, are coming into use. The one shown also is a cyclo (air or plateless) planter which can plant kernels of any size or shape.* [SOURCE: International Harvester Company.]

check-rowed with great accuracy, so that it could be cultivated both ways of the field. With an increased acreage planted on the contour and in view of the greater speed with which a drilled or hill-dropped crop can be planted, fewer acres now are check-rowed. Hill-dropping refers to dropping 2, 3, or 4 kernels at regular but not precisely spaced intervals in the row. This method gives better lodging resistance and better insurance of an adequate stand. With the development of hybrids with greater stalk strength, hill-dropping has been replaced largely by drilling, in which corn is singly spaced within the row.

Listing usually is preferable to surface planting on the rolling, deep, porous soils on the eastern edge of the Great Plains. Further east, throughout most of the corn-producing area, germination and seedling growth are likely to be slow, poor stands are obtained, and lower yields result from listing than when the crop is surface-planted.

In the Corn Belt, corn is commonly planted with 4-row planters with an increasing number of 6-, 8-, and 12-row planters coming into use.

Depth of planting will vary between soils, geographic conditions, varying temperatures, soil moisture at planting, and planting method (conventional vs. no-tillage). For most Corn Belt situations, a depth of 2 inches (5.1 cm) is ideal. Early-planted corn should be planted 0.5 to 1 inch (1.25 to 2.5 cm) more shallow. In dry soil, planting should be about 3 inches (7.6 cm) in clays to as much as 5 inches (12.7 cm) in sands.

The permanent (coronal) corn roots develop at nodes below the soil surface where the moisture and air conditions are most favorable. During germination the first internode elongates, the amount of elongation being relative to the depth of planting. Thus, depth of planting, within limits, does not influence the location in the soil of permanent roots.

Planting Date

Earlier planting may be the single production factor by which the largest number of farmers could increase corn yields. While there are several advantages of early planting, the major one is that the critical moisture-requirement period during tasseling and silking generally will be reached before drought and moisture stress occur in mid-summer.

The best date to plant may be as early as late February or early March in parts of Alabama, Mississippi, and Georgia to as late as early June in the extreme North. Planting should be delayed one day for each 13 miles to the north. In the central Corn Belt, most corn is planted from mid-April to mid-May. May 1 appears to be the optimum for this region.

In addition to calendar date, consideration should be given to soil temperature at planting. Risk is involved when planting before the soil temperature has reached 50°F (10°C) at a 2-inch (5.1-cm) depth at 7 A.M. or 55°F (13°C) at a 4-inch (10.2 cm) depth at 1 P.M.

Planting Rate, Population, and Row Width

There is no universal planting rate; the range is 10,000–30,000 kernels/acre (24,710–74,129/ha). The general recommendations in the Corn Belt for a full-season hybrid are planting rates of 20,000–26,000 kernels/acre (49,419–64,245/ha) to produce final stands of 18,000–24,000 plants/acre (44,478–59,303/ha). To achieve a population of 20,000 plants/acre (49,419/ha), a kernel should be dropped about every 10 inches (25 cm) in 30-inch wide rows (76-cm rows) and about every 8 inches (20 cm) in 38- or 40-inch wide rows (97- or 102-cm rows). The kernels dropped should be 10–20 percent greater, 2,000–4,000 more kernels/acre (4,942–9,884/ha), than the desired final population. Lack of germination and mortality from pests generally will average 10–15 percent.

Before the widespread use of hybirds, the population commonly used in the Corn Belt was 10,000–12,000 plants/acre (24,710–29,652/ha). When hybrids became dominant, stands up to 16,000/acre (39,535/ha) became common. During the 1950s and 1960s, with heavier fertilization and the development of high-stress hybrids (tolerant of high populations without lodging and producing a high percentage of barren stalks) average populations of 20,000/acre (49,419/ha) became common. Populations of 24,000 plants/acre (59,303/ha) are not uncommon in the Corn Belt.

The optimum population in drier regions and in the South, without irrigation, may be as low as 12,000 plants/acre (29,652/ha). The population should be increased by at least 10 percent when corn is cut for silage.

It is said that the original row spacings of 38–42 inches (97–107 cm) were determined by the width necessary for a mule or horse to pass between the rows for cultivation. Very narrow row spacings of 18–24 inches (46–61 cm) give more nearly equal space per plant in all directions. Efficiency of water use and light (radiation) intercepted by leaves indicate that corn spaced more nearly equidistant in narrow rows could yield more than corn in wider rows.

Published reports show anywhere from 0 to over 90 percent yield increase with narrow row spacings as compared to the conventional wide spacings. Most increases range from 3 to 20 percent. More specifically, corn yields usually increase 5–10 percent as row widths are narrowed from 40 inches (102 cm) to 30 inches (76 cm).

Dugan *et al.,* in a review of 115 published papers on corn populations, reported that yield increases resulted in most instances up to 20,000 or more plants per acre (49,419/ha). More recently, Yao and Shaw in Iowa compared 21-inch (53-cm) and 42-inch (107-cm) rows, both irrigated and nonirrigated, at 14,000 and 28,000 plants/acre (34,594 and 69,187/ha) populations. Nonirrigated corn yields were 4.8 percent and 6.0 percent higher for 21-inch (53-cm) rows at the two populations, respectively, for a three-year period. Irrigated yields were 10.5 percent and 4.5 percent higher for 21-inch (53-cm) rows for the two populations over a two-year period.

Based on the published evidence of yield response to narrow row spacings, row widths gradually decreased in the 1960s and early 1970s, but there has been little further decrease in recent years. Kansas growers continue to use the narrowest rows, averaging just under 32 inches (81 cm) with more than half the growers using rows about 30 inches (76 cm) or narrower. Of the seven major corn-producing states, Iowa growers use the widest, averaging over 36 inches (91 cm). Adaptations in present machinery or investment in new machinery is necessary for narrow-row culture. Also, a producer must be more dependent on effective, season-long herbicides for weed control.

Corn for Forage

High populations of corn, 100,000–400,000 per acre (247,097–988,386/ha) planted with a grain drill have produced green weight yields of 40 tons (90 MT/ha) or more per acre. The high tonnages have stimulated interest in the utilization of the high yields of green material. Solid plantings have been found useful for green chop feed during midsummer when other green feeds are less plentiful and less productive. Two crops of green chop can be obtained by seeding early, late April or as early in May as possible, and harvesting in mid-July, planting again as soon as possible, and harvesting again in September or early October. Severe lodging may occur during storms, making harvest difficult.

Weed Control

Mechanized Cultivation

The primary objective of cultivation is to control weeds. Cultivating more frequently or deeper than is necessary to kill weed seedlings is likely to cut roots and reduce corn yields. Also, deep cultivation brings weed seeds into the surface soil area where they germinate, making additional cultivation necessary. Timeliness of cultivation is recognized as important; weeds can be destroyed best when the seedlings are just emerging.

Klingman studied the interaction of cultivation and soil type and concluded that soil tillage, except that needed for weed control, was of no benefit to corn if the soil was porous and well aerated. Yields of "weed-free" corn were slightly improved on a heavy, clay-loam soil by one early cultivation in a dry season but not in seasons with favorable rainfall.

Machinery specifically designed for weed control in row crops first appeared about 1850 with the wheel cultivator. This implement gradually evolved into the horse-drawn riding cultivator. Today the most effective implements for early cultivation are the rotary hoe and the rotary cultivator. The rotary cultivator combines tillage wheels near the row with a shovel in the center between rows. Cultivation should be early, as the weeds are emerging; shallow cultivation with these implements can kill weeds without disturbing the crop if proper soil conditions exist. Shovel- or sweep-type cultivators can be used later in the season if necessary. There usually is no advantage of cultivating corn after it is about 24 inches (61 cm) tall when weeds have been controlled effectively up to that point.

Chemical Weed Control

The use of herbicides for selective weed control in corn was initiated in 1944 when it was discovered that 2,4-D would control many broadleaf weeds without damage to the corn. Herbicides are now available that will selectively control most species of weeds in corn. The herbicide or herbicide combination used should be selected on the basis of weeds present, stage of corn and weed growth, soil type and characteristics, geographic area, other control measures available, and succeeding crop in rotation. Specific recommendations should be obtained from the nearest state experiment station or county extension office.

Corn herbicides may be applied either preplant, preemergence, or postemergence. A preplant herbicide application offers the opportunity to apply and incorporate the herbicide, insecticide, and fertilizer at the same time. Some of the leading preplant herbicides are atrazine, butylate, simazine, cyanazine, EPTC, and alachlor, used alone or in combinations.

Preemergence herbicides may be applied at the time of planting in the same operation, or separately after planting is completed up until emergence. This is probably the most widely used application method. The most widely used pre-emergence herbicides are atrazine, simazine, cyanazine, propachlor, alachlor, and metolachlor.

Postemergence herbicides can control weeds not controlled earlier by cultivation or by preplant or preemergence chemicals. An overall spray can be used up until corn is 6–10 inches (15–25 cm) tall; on larger corn a directed spray should be used, minimizing risk of corn injury. The addition of a surfactant or nontoxic crop oil will help enhance postemergence activity. Widely used postemergence herbicides include 2,4-D, dicamba, atrazine, and cyanazine.

Weed control in corn produced by no-till methods requires a contact as well as a residual herbicide. Paraquat is used widely for providing the contact kill of vegetation (sod crop, cover crop, or weeds) at planting time. Atrazine, used in combination with paraquat, is the most commonly used residual herbicide for long-season control of germinating weeds. Others registered for use in no-tillage systems are simazine, cyanazine, and alachlor.

Insects

Several hundred insects sometimes affect corn, feeding on seed, roots, leaves, or stalks. About 25 insects are particularly troublesome, reducing annual corn yields by about 12 percent.

Insects may affect the seed or seedling, below-ground parts, or above-ground parts. Principal insect pests attacking the seed or seedling are seed-corn maggot, seed-corn beetle, wireworms, cutworms, webworms, flea beetles, and white grubs. Corn rootworms (northern, southern, western), corn root aphids, and the grape colaspis feed on the roots. Affecting the leaves, stalks, or ears are European corn borers, Southwestern corn borers, corn earworms, grasshoppers, cutworms, armyworms, chinch bugs, corn leaf aphids, thrips, and cereal leaf beetles.

Larvae of the rootworms feed on the roots, causing plants to lodge and thus reducing crop yield. Larvae of the corn earworm feed on the buds or central shoots of young corn plants, stunting them and reducing the yield. Later the worms go down through the silks of the ears and destroy many of the kernels. Sometimes they chew off the silks and prevent pollination. Their feeding in ears also allows fungus diseases to get started, which increases the total damage. The European corn borer reduces the yield and increases the cost of harvesting by causing broken stalks, poor ear development, and dropped ears. The southwestern corn borer larvae feed on leaves of young corn plants and retard their growth. They also damage the ears and girdle and tunnel the stalks, causing lodging. Preventive or control measures are as follows:

1. Hybrids are available which have tolerance or resistance to European corn borer, chinch bugs, aphids, and corn earworms (long, heavy, husk covering).

2. Good management practices, such as adequate fertilization, can result in vigorous plants that are less susceptible to insect injury.
3. Crop rotations tend to suppress or lessen the severity of insects such as western corn rootworms, northern corn rootworms, and root aphids. Crop rotation is not effective in controlling wireworms, cutworms, webworms, and grubs. The rootworm has caused severe damage in recent years and has developed resistance to previously effective chlorinated hydrocarbon insecticides.
4. Destroying all crop residues, such as by deep plowing, is effective in lessening the severity of corn borers and root aphids.
5. Several effective insecticides are available. For rootworms, cutworms, grape colaspis, and other soil insects, a granular insecticide banded in the row at planting time can be effective. Seed treatment can effectively control the seed-corn maggot and seed-corn beetle. Other insecticides may be applied as poison baits, or sprayed or dusted on the foliage as needed. An appropriate insecticide should be applied to the foliage to control grasshoppers, armyworms, chinch bugs, thrips, corn leaf aphids, and corn borers when sufficient damage is noted to warrant treatment.

Diseases

More than 25 diseases affect corn to some extent. Many corn diseases appear only sporadically, but sometimes losses in individual fields may exceed 30 percent. Generally, losses are due to lowered grain quality, decreased value of fodder, and decreased yield. The average annual loss to the United States corn crop caused by diseases has been approximately 12 percent. Annual losses in the Corn Belt range from 7 to 17 percent, and world corn production losses from disease average nearly 10 percent.

Nonparastic or noninfectious diseases may result from chemical or mechanical injury, genetic abnormalities, unfavorable climatic and soil conditions, nutritional deficiencies or inbalance, too much water, or too high or too low temperatures. Parasitic or infectious diseases are caused by fungi, bacteria, viruses, or nematodes. The more important infectious diseases may be grouped as follows:

1. *Seed rots and seedling blights*—The seed may rot and germination and stand may be poor, or seedlings wilt and die.
2. *Leaf spots and blights*—Round or oval to elongate dead areas develop on the leaves. Leaf damage reduces the production of carbohydrates to be translocated to the grain and results in immature, chaffy ears. The two major ones are northern leaf blight and southern leaf blight, both caused by the fungus *Helminthosporium*. Southern leaf blight caused devastating losses in 1970 throughout corn-producing regions of the United States to hybrids produced by using the "Texas type" of cytoplasmic male sterility.

3. *Stalk, root, and ear rots*—Stalks break readily and are discolored or hollow inside. Among the diseases, stalk rots are the most destructive and cause the greatest losses. Ear rots result in kernels, ears, and cobs being moldy and rotten. Ear rots are the second-most destructive and costly corn diseases in the Corn Belt. The two major stalk rots are diploidia stalk rot, which is most destructive in the Corn Belt, and gibberella stalk rot, which is widely distributed but causes most damage in the northern and eastern parts of the United States. Ear rots often are caused by the same fungi that attack stalks.

4. *Smuts*—Silvery galls, filled with black spores, may occur on any above-ground part. Injury, as a result of either mechanical or herbicidal damage, provides favorable sites for infection. Smut causes little damage in corn, although it is widespread.

5. *Rusts*—Small, reddish-brown pustules appear on leaves. The rusts ordinarily are not serious in the Corn Belt states.

6. *Virus diseases*—The principal ones are corn stunt and maize dwarf mosaic. Corn stunt is transmitted by leafhoppers and maize dwarf mosaic by aphids. Plants are stunted or dwarfed and bushy, with leaves often mottled or discolored.

Preventive or control measures include:

1. *Resistant or tolerant hybrids*—Hybrids with strong stalks are less susceptible to stalk rots. Corns with a good husk cover are less subject to ear and kernel rots. Certain hybrids possess some resistance to smut and also to leaf blights.

2. *Crop rotation*—While not very effective for controlling most diseases, this practice may be useful when the pathogen is specific for a crop and is strictly soil borne.

3. *Field sanitation*—Destroying crop residues by plowing, for example, may result in less overwintering of disease organisms.

4. *Soil management*—Balanced fertility lessens the effects of some diseases. Stalk rots and northern leaf blight often are more severe when potassium is deficient or nitrogen too abundant.

5. *Fungicides*—Treatment of the seed with fungicides such as captan and thiram can control some of the seed rots and seedling blights. This is now done by the seedsman. Large-scale spraying ordinarily is not done except in breeding nurseries and high-value seed crops.

The United States has successfully used a disease-monitoring program through cooperation with area extension specialists, who make regular observations. A recent development in large-scale disease detection is the use of remote sensing. This provides a tool with which pathologists or agronomists can rapidly observe and inventory crop conditions.

Harvesting and Storage

There are several different ways to harvest, dry, store, and feed corn. Corn can be harvested with varying moisture content and the crop delivered

FIGURE 19–9. *Corn harvesting procedure. A modern combine equipped with a corn head. Over three fourths of the corn in the Corn Belt is field-shelled.* [SOURCE: Allis Chalmers Corporation.]

to a grain wagon or truck as ear or shelled corn. The corn can be stored in the ear, as high-moisture shelled or ground ear corn, or as dry shelled corn. Some methods of handling a corn crop may save labor, allow for earlier harvest, and reduce field losses. But some methods without good management may increase handling costs per bushel. The costs of harvesting, drying, and storing a corn crop vary with the method of harvesting, size and type of equipment used, acreage operated, and several other factors.

HARVESTING. In the 1930s most corn was harvested by hand, with a total labor requirement for production and harvesting of over 100 man-hours per 100 bushels (25 quintals). Harvesting is now completely mechanized in all major corn-producing regions of the United States. The labor requirement has dropped to 1–6 hours per 100 bushels (25 Q).

Up until the early 1960s the corn picker was the most common harvesting implement in the central Corn Belt, but now it has largely disappeared. Corn pickers still are used some in the northern Corn Belt. Most pickers have been replaced by picker-shellers and combines with corn heads. More than three-fourths of the corn in the Corn Belt is field-shelled with either picker-shellers or combines. Field losses when harvesting with a combine with a corn head often are as much as 20 percent less than losses with a picker and 30 percent less than those with a picker-sheller. Combine-harvesting is the most economical method if more than 8000 bushels

(2000 Q) of corn are produced and if the combine is used to harvest other crops, such as soybeans and small grains.

Corn is physiologically mature at a moisture level of 30–32 percent. Early harvesting, at a relatively high moisture content, offers the advantages of less lodging losses, less chance of weather delays, less ear drop, and fewer shelled grain losses from the harvesting operation. The disadvantage is cost of drying grain to a safe level for storage.

The ideal moisture level, when harvesting should begin, for either a picker, picker-sheller, or combine is 25 percent, with a preferred range of 21–28 percent. Corn to be stored as high-moisture corn should be harvested at 28 percent moisture.

STORAGE AND DRYING. During the 1950s and early 1960s the U.S. government's agricultural policy included large grain reserves or large volumes of stored grain, and there was little emphasis on on-farm storage. Since that time on-farm storage has expanded greatly because of the inability of the grain trade to accept the volume of corn as offered and the heavy losses imposed by high-moisture discounts. In addition, on livestock farms, with the crop to be stored for later feeding, drying equipment for shelled corn is a must.

The moisture content of shelled corn ordinarily must be reduced to 13 percent or less for safe storage without spoilage by molding. Ear corn can be stored at 20–25 percent, depending on the size of the crib and temperature. Shelled corn at a moisture level higher than 13 percent has an allowable storage time of from less than 3 days—30 percent moisture and stored at 75°F (24°C)—to more than 1,100 days—15 percent moisture and stored at 35°F (2°C)—depending on the moisture level and storage temperature. Corn at 17 to 22 percent moisture, for example, can be stored in a conventional grain bin for a considerable period if the temperature is kept at 50°F (10°C) or less. Stirring and aerating grain in the bin also can help prevent spoilage.

Corn can be brought to a safe storage moisture level by custom-drying, but the trend is toward on-the-farm drying facilities. The principal drying systems are:

1. *In-storage drying*—Grain is dried and stored in the same structure with either heated or unheated air. This is used for relatively small volumes of grain. Layers of shelled corn may be added as the previous layer becomes dry.
2. *Batch-drying*—This method is suitable for corn at 18–30 percent moisture. Each day's harvest is dried by forcing air through a perforated floor; then the batch is moved to a separate storage facility. This method involves more labor and equipment than does in-storage drying because of the extra loading, unloading, and transport steps. Drying time is 6–10 hours. The bin used for batch drying can be used as an in-storage drying bin at the end of the season when it is no longer needed for batch drying.

3. *Continuous flow*—The grain travels in a continuous stream through the dryer at a rate necessary to dry to the desired moisture level. This system is suitable for handling large volumes of grain.

4. *Dryeration*—This involves a combination of drying and aerating. The corn is dried to 16–18 percent moisture, at which point it is transferred to a temporary storage bin. A fan is used to cool the grain and dry it further to a level safe for the final storage bins.

In addition to the above methods, low-temperature drying may be used when the moisture content is no greater than 24 percent. Air for drying is about 5°F (3°C) above outside air; this is suitable for small quantities of corn. Low-temperature drying can hold large quantities of wet corn until a regular dryer is available.

Solar energy for grain drying is being researched thoroughly. Several experimental solar drying units have been developed and used successfully. This method of grain drying is likely to gain importance in the 1980s.

HIGH-MOISTURE CORN. When harvested as high-moisture corn the crop can be taken out of the field earlier with fewer harvesting losses and no drying expense. This method of harvest fits into the fall harvesting schedule, usually coming between corn silage harvest and regular corn grain harvest. Feeding trials have indicated an advantage of high-moisture corn in some livestock rations. Grain at 28–32 percent moisture levels can be stored suitably in an airtight silo with no problems. If high-moisture grain or ears are to be stored in a conventional silo, it should be ground so that the material packs well.

Use of grain preservatives is a recent development, allowing high-moisture corn to be stored without drying or placing in an airtight silo. Most commercial preservatives are either propionic acid or a mixture of propionic and acetic acids. These mild, organic acids inhibit the growth of the fungus organisms that cause spoilage.

Corn Improvement

The one improvement with the single greatest impact on corn was the development and widespread use of hybrid corn, discussed previously. More recently, a number of other developments in corn improvement have occurred. Among the newer developments which promise present or future increases in corn production, quality, or corn uses are:

1. *High-lysine corn*—The opaque-2 gene gives the kernel a higher content of lysine, an essential amino acid present in only small amounts in regular corn. Unfortunately, current high-lysine hybrids have a slightly to moderately lower yielding potential, have softer kernels and are less resistant to diseases, and dry down more slowly in the field. It may not be practical at present to grow high-lysine corn unless the corn is to be fed to swine on the farm where produced. But this corn type has great potential for future improvement of animal nutri-

tion, and especially for human nutrition in countries where corn is a staple item in the diet.

2. *Prolific corn*—Prolific hybrids, which produce more than one ear per plant, are in the developmental stages at present. The main stalk often produces two ears, and the plant has a tendency to tiller, with each tiller producing a small ear. Prolifics have the flexibility to adjust to widely variable growing conditions from year to year.

3. *Liguleless, upright-leaved corn*—Corn without a ligule in the junction of the blade and sheath has an upright leaf orientation. Pepper et al. at Iowa State noted a yield advantage of upright leaf types over horizontal leaf types at high populations (high leaf area indices).

4. *Superstrong-stalked corn*—Although scientists have doubled stalk strength in the last 12 years, the future offers even more stalk improvement. A super strong stalk would allow a return to field drying before harvesting to a moisture level suitable for storage. Stalk strength is measured by pounds necessary to crush a stalk section placed under hydraulic pressure, and by the pressure needed to penetrate the skin or rind with a needle. Zuber from Missouri has selected types requiring 25–30 pounds of pressure to penetrate, as compared to 5–10 pounds for ordinary commercial hybrids.

5. *Tassel corn*—Ordinarily, flowers in the tassel have only male parts, but some experimental types have perfect flowers in the top and produce scattered kernels in the tassel. Some experimental types have produced 2,000 kernels per plant, as compared to 800 or 900 on a commercial ear. Tassel types would dry to a harvestable moisture level quicker, but would be more susceptible to attacks by insects, birds, and other pests.

6. *Brown midrib mutants*—Brown midrib mutant corns have a 30–40 percent lower lignin content than normal types and higher digestible dry matter. When used for silage, they furnish higher nutritional quality to ruminant animals. Lodging often is a problem; they should be chopped for silage early.

Popcorn

The bulk of the commercial popcorn of the world is produced in restricted areas within the United States. Popcorn enters into commerce so extensively, however, that it is used almost universally throughout the United States. People in other parts of the world are not generally familiar with popcorn and consume but little.

What Makes Popcorn Pop?

The popcorn kernel is horny or corneous throughout, except for the germ and a small core of soft starch behind the germ. The moisture content of the kernel must be exactly right and heat must be applied at a definite degree of rapidity if the kernel is to pop perfectly. If the moisture content is only a little too high or too low, the corn pops poorly and many kernels may not pop at all. When heat is applied and the pressure of the steam formed within the heated kernel becomes greater than the kernel structure can resist, the hard, horny kernel suddenly explodes with great force, and inverts. Of first importance to commercial corn poppers is the popping volume, or popping expansion. Corn which when popped under optimum conditions fails to increase at least 24 volumes is likely to be discounted in price in commercial markets.

No open-pollinated cultivars have a popping volume that approaches that of the better hybrids. Corn 8 to 10 years old may pop as well as the first year if not allowed to become too dry. Moisture content is the most important factor in obtaining satisfactory results; it should be 13.5 percent or a little more for most hybrids.

Production Areas and Practices

Highest acreages are in Nebraska, Iowa, Indiana, Kentucky, and Ohio (Fig. 19–10). Approximately 200,000 acres (80,940 ha) are harvested annually in the United States with the yield per acre ranging from 2,500 to 3,000 pounds (2,800–3,360 kg/ha). Within each state, the bulk of the acreage is concentrated in relatively small areas. Popcorn is often overproduced, so that a high price in a given year is likely to be followed by an acreage increase and low prices in years immediately following. Because special storage and curing bins and other special equipment are necessary for efficient popcorn production and satisfactory marketing, "in-and-out growers" are likely to find popcorn less profitable than field corn.

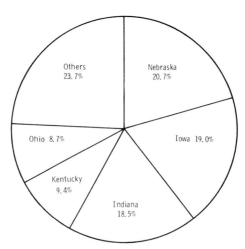

FIGURE 19–10. *Leading states of the United States in popcorn area (acreage). Five states have over three fourths of the acreage.* [SOURCE: Southern Illinois University.]

FIGURE 19-11. *Two kinds of popcorn, 1 = yellow, 2 = red.* [SOURCE: Southern Illinois University.]

TABLE 19-2. Moisture Content and Popping Volume (Iowa)

Moisture (%)	10.00	12.2	12.8	13.5	16.5	18.4
Popping volume	27.5	32.0	34.5	36.0	31.1	18.0

Much of the crop is produced under contract, with the seed to be planted and the price to be paid agreed on in advance of planting. Much research on popcorn has been under way for a considerable period at both the Indiana and Iowa stations.

Hybrid popcorn seed became commercially available in the early 1940s. Hybrids give better yields, higher popping volumes, and improved texture and flavor. The results from popcorn breeding work at the Iowa and Indiana stations have been outstanding, with practically the entire commercial popcorn acreage planted with hybrids from inbreds developed at these two stations. The popcorn hybrids now most grown are incompatible with field corn pollen, so losses to growers through crossing with field corn are eliminated.

Sweetcorn

Sweetcorn usually is thought of as a horticultural rather than a field crop. Most of the sweetcorn acreage is grown as a field crop, however, with the same methods of production applied as for the dent corn. The United States harvests more than 600,000 acres (240,000 ha) annually for

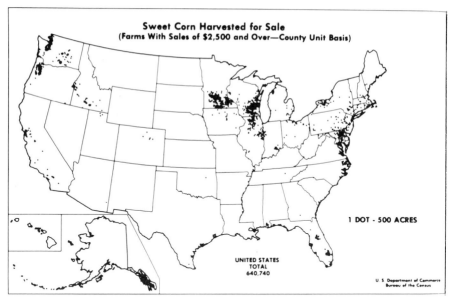

FIGURE 19–12. *Sweet corn harvested for sale in the United States, shown geographically. (Farms with sales of $2500 and over—county unit basis.)* [SOURCE: U.S. Department of Commerce.]

fresh market and for processing. Wisconsin, Minnesota, and Illinois lead in production, with most grown for processing. Florida leads in acreage grown for fresh use. Geographic distribution is shown in Figure 19–12.

Fresh sweetcorn is available at a price throughout much of the year. Florida and south Texas ship in the winter and early spring months, with California, Georgia, Alabama, and South Carolina supplying the market a little later. The early summer market is supplied by the states a little further north, but still below the Corn Belt and the high consumption areas.

Home gardens supply the very highest quality sweetcorn for fresh consumption during the midsummer months. Many superior sweetcorn hybrids for the home gardener have been developed, tender and with excellent flavor. Ranging in maturity from 75 to 120 days, sweetcorn is available at its best through three months or longer.

Sweetcorn to be processed usually is snapped in the morning and delivered to the cannery as promptly as possible. Sweetcorn deteriorates very rapidly after picking. It is particularly important that the harvested corn be kept cool. Under high-temperature conditions the corn may lose as much as half its sugar and much of its flavor within a few hours after picking. After the ears are mechanically husked, trimmed, and cleaned, the whole kernels may be cut from the cobs for "whole-kernel" canned corn. To produce "cream-style" canned corn, the kernels are cut across the top and the kernel substance scraped out, leaving most of the pericarp, or hull, attached to the cob.

A ton (0.9 MT) of ears will yield 800 pounds (363 kg) or more of cut corn. The U.S. average acre yields are about 8,000 pounds (3,629 kg). Most of the commercially canned sweetcorn is grown under contract, the cannery providing the seed and determining when the crop is to be planted and harvested.

The sweetcorn cultivars considered standard a few years ago have almost entirely disappeared. The sweetcorns now generally grown are hybrids, whether grown as a commercial crop or in the home garden. The first hybrid sweetcorn was released from the Connecticut station in 1924. 'Golden Cross Bantam,' released by the Purdue station in 1933, continued for many years to be the standard against which all other hybrids were compared. Great progress has been made in recent years in development and release of superior hybrids with uniformity, tenderness, good flavor, resistance to insects and diseases, and high-yielding ability. In the early 1960s geneticists at Illinois developed a "supersweet" corn, 'Illini X-Tra Sweet,' said to be one-fourth sweeter when picked than other sweet corns, and four times as sweet 24 hours after picking. This is because of much slower conversion of sugar to starch in the hours following picking. Since the release of 'Illini X-Tra Sweet,' a number of other supersweet hybrids have been developed.

REFERENCES AND SUGGESTED READINGS

1. ALDRICH, S. R., W. O. SCOTT, and E. R. LENG. *Modern Corn Production,* 2nd ed. (Champaign, Ill.: A&L Pub., 1975).
2. BARBER, S. A., and R. A. OLSON. "Fertilizer Use on Corn," *Changing Patterns in Fertilizer Use,* edited by L. B. Nelson, (Madison, Wisc: Soil Science Society of America, 1968).
3. BARGHOORN, E. S., *et al.* "Fossil Maize from the Valley of Mexico," *Harvard Univ. Bot. Museum Leaflets,* 16:229–240 (1954).
4. BATEMAN, H. P., and W. BOWERS. *Planning a Minimum Tillage System for Corn,* Univ. of Ill. Ext. Circ. 846 (1962).
5. BLEVINS, R. L., G. W. THOMAS, and P. L. CORNELIUS. "Influence of No-Tillage and Nitrogen Fertilization on Certain Soil Properties after 5 Years of Continuous Corn," *Agron. Jour.,* 69:383 (1977).
6. DUNGAN, G. H., *et al.* "Corn Plant Population in Relation to Soil Productivity," *Advances in Agronomy,* Vol. 10 (New York: Academic Press, 1958, pp. 435–473). ed. A.G. Norman.
7. ELKINS, D. M. *et al.* "No-Tillage Maize Production in Chemically Suppressed Grass Sod," *Agron. Jour.,* 71:101–105 (1979).
8. EL-TEKRITI, R. A., *et al.* "Structural Composition and *In Vitro* Dry Matter Disappearance of Brown Midrib Corn Residue," *Crop Sci.,* 16:387 (1976).
9. FAY, B. "New Enthusiasm for Prolifics," *Crops and Soils,* 26(3):12 (1973).
10. HIRNING, H. J., E. F. OLIVER, and G. C. SHOVE. *Drying Grain in Illinois,* Ill. Coop. Ext. Serv. Circ. 1100 (1974).
11. HOFF, D. J., and H. J. MEDERSKI. "Effect of Equidistant Corn Plant Spacing on Yield," *Agron. Jour.,* 53:295–297 (1960).

12. INGLETT, G. E. *Corn: Culture, Processing, Products,* (Westport, Conn.: AVI, 1970).

13. JENKINS, M. T. "Influence of Climate and Weather on the Growth of Corn," *Climate and Man* (USDA Yearbook of Agriculture, 1941).

14. JOSEPHSON, L. M., and F. BERGGREN. *Producing White Corn for the Milling Industry,* Tenn. Agr. Exp. Sta. Special Rep. 1 (1978).

15. JUGENHEIMER, R. W. *Corn Improvement, Seed Production, and Uses* (New York: Wiley, 1976).

16. KLINGMAN, G. C. *Weed Control: As a Science* (New York: Wiley, 1961).

17. LARSON, W. E. "Tillage Requirements for Corn," *J. Soil Water Cons.,* 17:3–7 (1962).

18. LARSON, W. E., *et al. Subsoiling in the North Central States,* Mimeographed Rep. Iowa State Univ. (1963).

19. LENG, E. "The Early Growth of Corn," *The Farm Quarterly,* (1968). (Spring). Vol. 23, No. 2, pp. 80–87.

20. MANGELSDORF, P. C. *et al.* "Domestication of Corn," *Science,* 538–548 (Feb. 7, 1964).

21. MARTIN, J. H., W. H. LEONARD, and D. L. STAMP. *Principles of Field Crop Production* (New York: Macmillan, 1976).

22. NELSON, L. R., *et al.* "Corn Forage Production in No-Till and Conventional Tillage Double-Cropping Systems," *Agron. Jour.,* 69:635 (1977).

23. PARKS, W. L., *et al. Plant Population and Row Spacing for Corn,* Tenn. Agr. Exp. Sta. Bul. 392 (1965).

24. PEPPER, G. E., R. B. PEARCE, and J. J. MOCK. "Leaf Orientation and Yield of Maize," *Crop Sci.* 17:883 (1977).

25. PHILLIPS, S. H., and H. M. YOUNG, JR. *No-Tillage Farming,* (Milwaukee: Reiman Associates, 1973).

26. ROSSMAN, E. C., and R. L. COOK. "Soil Preparation and Date, Rate, and Pattern of Planting," In *Advances in Corn Production,* edited by W. H. Pierre, S. A. Aldrich, and W. P. Martin (Ames: Iowa State Univ. Press, 1966).

27. SCHWART, R. B., and L. D. HILL. *Costs of Drying and Storing Shelled Corn on Illinois Farms,* Ill. Coop. Ext. Serv. Circ. 1141 (1977).

28. SHRADER, W. D., and W. H. PIERRE. "Soil Suitability and Cropping Systems," In *Advances in Corn Production,* edited by W. H. Pierre *et al.* (Ames: Iowa State Univ. Press, 1966).

29. SPRAGUE, G. F., and W.E. LARSON. *Corn Production,* USDA-ARS and Minn. Agr. Handbook No. 322 (1975).

30. STICKLER, F. C. "Row Width and Plant Population Studies with Corn," *Agron. Jour.,* 56:438–441 (1964).

31. Univ. of Illinois. *Diseases of Corn in the Midwest,* North Central Regional Ext. Pub. No. 21; Circ. 967 (1967).

32. Univ. of Illinois and USDA. *A Compendium of Corn Diseases* (St. Paul, Minn.: American Phytopathological Society, 1973).

33. USDA. Agr. Stat., 1977.

34. USDA-ARS, *Losses in Agriculture,* Agr. Handbook 291 (1965).

35. VAN FOSSEN, L., and E. S. STONEBERG. "Harvesting, Drying, and Storing Corn," *Iowa Farm Science,* 17:2 (1962).

36. VAN DOREN, D. M., JR., and G. J. RYDER. "Factors Affecting Use of Minimum Tillage for Corn," *Agron. Jour.,* 54:447–450 (1962).

37. WATSON, S. A. "Corn, How Many Uses Can You Name?" *Crops and Soils,* 30(3):10 (1977).

38. WEATHERWAX, P. "History and Origin of Corn," In *Corn and Corn Improvement* (New York: Academic Press, 1955).

39. YAO, A. Y., and R. H. SHAW. "Effect of Plant Populations and Planting Pattern of Corn on Water Use and Yield," *Agron. Jour.,* 56:147–152 (1964).

40. ZUBER, M. S., and M. S. KANG. "Corn Lodging Slowed by Sturdier Stalks," *Crops and Soils,* 30(5):13 (1978).

CHAPTER 20
THE SORGHUMS

SORGHUM provides a valuable and important grain crop throughout a large area in the Southwest and an important forage in a much larger area. As a world crop, sorghum is grown principally as a food grain; in the United States it is grown primarily as feed for livestock and poultry.

Changes in recent years in United States sorghum production have been fully as striking as those of any other major crop. These changes have included:

1. Development of dwarf, combine types of grain sorghum.
2. Development of hybrids, which became available about 1955.
3. Expansion of grain sorghum into short-season environments.
4. Development of disease-resistant, insect-tolerant, and bird-resistant cultivars.
5. Development of forage sorghum having palatable seed and low prussic acid potential.
6. Increased use for industrial purposes.
7. Improved cultural practices including growing grain sorghum in narrow rows and artificial drying of grain sorghum.

Origin and History

Sorghum, *Sorghum bicolor,* is believed to have originated in Africa, but in India the culture of this plant goes back beyond recorded history. It was grown in Assyria as early as 700 B.C., but apparently it did not reach China until sometime in the thirteenth century, and the Western Hemisphere much later.

Sorghum probably first came into the United States in the early part of the seventeenth century from Africa, but the crop was not extensively grown until much later. A cultivar of sweet sorghum, 'Black Amber,' once grown extensively, was introduced by way of France in 1853. Several cultivars important in this country as late as 1936 were introduced from South Africa in 1857.

There are a number of distinct types of sorghum. Durra reached California from Egypt in 1874; kafir, from South Africa in 1876; and milo, from Africa about 1885. Shallu came from India in 1890; feterita, hegari, and sudangrass, from the Sudan region of Africa in 1906, 1908, and 1909, respectively.

Adaptation and Distribution

Because of their drought-resistant qualities the grain sorghums are of outstanding importance in the southern Great Plains area, with Texas, Kansas, and Nebraska leading in production (Figure 20–1). In the United States approximately 15 million acres (6.1 million ha) of sorghum are grown annually for grain, 2 million (0.8 million ha) for forage, and less than 1 million (0.4 million ha) for silage. Sorgo or sweet sorghum is grown for fodder and silage and to a much lesser extent for syrup making. Geographic distribution of sorghum harvested for all purposes except syrup is shown in Figure 20–2.

Sorghum does best in the southern half of the United States, where the temperatures are uniformly high during the growing season. They make little growth at temperatures below 60°F (16°C). If planted too early, a poor stand results and seedlings make a slow growth, so that it is difficult to keep the crop clean. At the other end of the season, the minimum temperature for growth is about 20°F higher than that of corn. Extremely high temperatures which injure corn pollen and result in poorly filled ears are not equally injurious to sorghum.

The main concentration of grain sorghum is in areas where the rainfall is insufficient and the temperatures are too high for satisfactory corn production. Sorghums are well adapted to areas where the annual precipitation is only 17–25 inches (43–64 cm). During periods of extreme drought the plant becomes somewhat dormant but does not wither and die, growth being resumed when rain comes. It can retain water more effectively than is true of corn, perhaps because of the waxy cuticle. The water require-

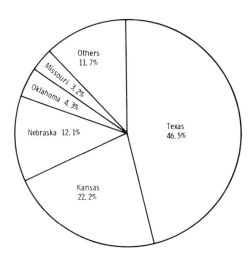

FIGURE 20–1. *Leading states in United States grain sorghum area (acreage). Texas alone accounts for nearly one half of the acreage.* [SOURCE: Southern Illinois University.]

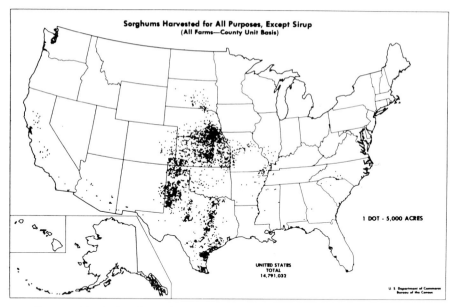

FIGURE 20–2. *Sorghums harvested for all purposes in the United States, shown geographically.* [SOURCE: U.S. Department of Commerce.]

ments per pound of dry matter produced by sorghum is considerably less than that of corn and of most other crops.

The Plant

The sorghum plant is best described as a coarse grass, different types and cultivars varying greatly. The growth height ranges from 2 to 15 feet (0.6–4.6 m) or more, but the height of commonly grown grain sorghum hybrids generally is 2–5 feet (0.6–1.5 m). In many respects the structure, growth, and general appearance are similar to that of corn: stalks are grooved on one side between the nodes; grooved internodes alternate from side to side; a leaf is borne at each node, on the grooved side, with the leaf sheath and blade arrangement also much like that of corn. Young sorghum plants can be distinguished from young corn plants by the serrated leaf margin of sorghum.

The buds that form at the nodes, at the base of the groove, often develop into grain-producing tillers. The tillers develop their own roots but remain attached to the old crown. The culms or stalks of some cultivars and types of sorghum are juicy and sweet, some are juicy but not sweet, and some neither juicy nor sweet. If the pith is not juicy, the midrib of the leaf is white in color, because of the air spaces in the tissues; when the air spaces are filled with juice the color is more neutral.

In contrast to corn, both the male and female flowers are borne in a panicle at the end of the culm. The panicle may be loose and open, or relatively dense. About 95 percent of the flowers are self-pollinated.

The color of the seed varies for different cultivars and types from white through buff and red to dark brown. The color pigments may be in either the pericarp or the testa, or both, but not the endosperm. The endosperm is white and the sorghums have the same vitamin A deficiency as white corn, except for carotene-containing yellow endosperm types which are now being produced in the United States. The size of the seed for different types and cultivars varies considerably, with a range of approximately 12,000–60,000 seeds per pound (26,455–132,275/kg).

Heat and Drought Resistance

Large groups of hygroscopic cells occur in single rows along each side of the midrib of sorghum leaves, causing the leaves to fold instead of roll, as do corn leaves. The folding of sorghum leaves takes place much more rapidly and completely than the rolling of corn leaves.

Perhaps the most significant difference between the leaves of sorghum and corn is the heavy bloom of white wax, which usually covers the blades and leaf sheaths of the sorghum and aids in water-use efficiency. There also is evidence that the transpiration rate of sorghum is lower than that of corn under conditions of high evaporation, and that the sorghum root system is more efficient in extracting water from a dry soil.

Prussic Acid Poisoning

Sorghums often have a relatively high content of the glucoside dhurrin which, when it breaks down, releases a poisonous substance known as prussic acid or hydrocyanic acid (HCN). The content is influenced by environmental factors, and the release of HCN is influenced by freezing. Young plants only a few inches in height, branches in the leaf axils on injured plants, and new shoots from the crown at the soil surface have been shown to contain more than twice as much as the leaves of older plants.

Numerous cases have been reported of cattle being poisoned when grazing sorghum. When the crop is cut and cured, or is silaged, the danger is eliminated. Less trouble is experienced in the South than farther north.

Sudangrass normally contains less than half as much of the glucoside as do most sorghums. Many years ago the Wisconsin station developed and released a low HCN-potential cultivar, 'Piper,' which still is recommended in many areas.

Botanical Classification and Cytology

Opinions on the botanical classification of the many cultivated forms of sorghum are diverse. The genetics and cytological evidence seem to indicate that the cultivars of sorghum and the grass sorghums with the somatic chromosome number of 20 constitute one large and varied species. The following classification is generally used:

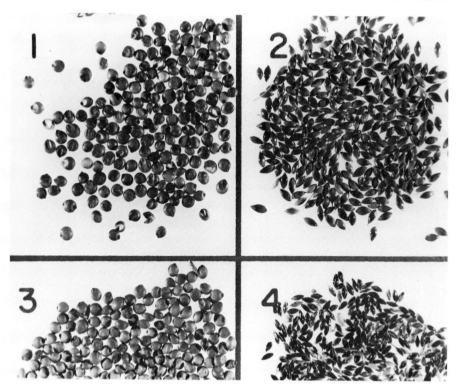

FIGURE 20-3. *Cultivated forms of sorghum. Grain of: 1 = milo, annual grain sorghum; 2 = sudangrass, annual grass sorghum; 3 = annual forage sorghum; 4 = johnsongrass, a perennial considered by most as a serious weed pest.*

I. Annual sorghums *(Sorghum bicolor)*
   A. Forage or sweet sorghum (sorgo; sweet or saccharine)
   B. Grain sorghum (nonsaccharine).
      1. Milo
      2. Kafir
      3. Feterita
      4. Hegari
      5. Hybrid derivatives
      6. Miscellaneous groups, such as durra, shallu, and kaoliang
   C. Broomcorn
   D. Grass sorghums: Sudangrass *(Sorghum bicolor,* formerly *S. sudanense)*
II. Perennial sorghum
   A. Johnsongrass *(Sorghum halepense)*

Sorghum flowers normally are self-fertilized, but they cross-pollinate readily when different cultivars or types are grown in close proximity. The sweet, juicy sorghums cross with the dry-pith grain types, or the grain type cultivars with sudangrass, and so on. A great deal of planned crossing has been done between cultivars in different type groups, with the result that the cultivars now most extensively grown do not necessarily fit into the preceding classification pattern.

Type Characteristics

THE SWEET SORGHUMS, OR SORGOS. The stalks or culms are packed full of sweet juice. With the ready availability of sugar and syrups of various kinds, the practice of growing sweet sorghum for syrup making has been largely discontinued, especially in the North.

The sweet, juicy stalks are very palatable to livestock when it is fed either cured or as silage. Cultivars range from very early (suitable for growing in the far north) to exceedingly late (suitable for growing only in the Gulf area). Seeds of some of the sweet sorghums may be bitter because of a high tannin content; in feeding value they are much below that of the grain cultivars. Among the most recently released cultivars for syrup production are 'Dale' and 'Theis,' both developed at the U.S. Sugar Crops Field Station, Meridian, Mississippi.

Grain Sorghums

The pith of grain sorghum stalks may be either dry or fairly juicy, and usually is not sweet. The fodder of many grain cultivars is not particularly desirable as feed. The stalks usually are rather short, with the heads (and individual seeds) larger and usually more compact than for the sweet sorghums. Within the grain group are several types. Kafir has relatively large, flat, dark-green leaves, with the stalks relatively thick and juicy. The kafirs are grown both for grain and forage. Hegari is similar to kafir. Its high proportion of leaf and sweeter juice make it preferable for forage.

Milo is less juicy than either kafir or hegari, and the leaves and stalks are somewhat smaller. Milo plants have a tendency to tiller considerably, average somewhat earlier, and are regarded as more drought resistant than kafir. Milo is a grain type; and stover has relatively low value. Feterita also is strictly a grain sorghum. Because of its earliness it often is used as a late-planted emergency crop. It sometimes is used as an early-planted crop with the idea of getting the crop made before severe drought develops.

Other types of grain sorghum important in other parts of the world, but not in the United States, are kaoliang, durra, and shallu. Kaoliang is grown almost to the exclusion of other types in Korea, Japan, China, and Manchuria, where it is used primarily for food. Durra is the type most grown in North Africa and the Near East as a food crop. Shallu is most extensively grown in India, with durra also grown there, although it is somewhat less important in most areas.

BROOMCORN. The essential characteristic of broomcorn is the long, fibrous seed branches, 12–36 inches (30–91 cm) long, used for making brooms and whiskbrooms. The plant has scant foliage, dry pith, and woody stalks,

FIGURE 20–4. 'Theis' sweet sorghum, a newer cultivar developed at the U.S. Sugar Crops Field Station, Meridian, Mississippi. Sweet sorghum can be used for syrup-making or for livestock feed. [SOURCE: Kelly C. Freeman, USDA-SEA, U.S. Sugar Crops Field Station.]

and is a light seeder. U.S. acreage has declined drastically in recent years. From 100,000 acres (40,470 ha) harvested in 1969, the acreage had declined to just over 7,000 acres (2833 ha) in 1974, when USDA estimates were discontinued. Much of the broomcorn needs are met by importing from Mexico.

Sorghum Uses

FEEDING LIVESTOCK. Sorghum has many uses in different parts of the world. In the United States, it is used principally as a feed grain for livestock, which accounts for 90 percent of the total grain sorghum produced.

Grain sorghum is approximately equal to corn in feeding value. It averages about 2 percent higher in protein and 1 percent lower in fat. All classes of livestock gain as rapidly on grain sorghum as on corn. Corn is a slightly more efficient feed for beef cattle and hogs; that is, slightly more sorghum is required per pound of gain. Sorghum has about the same feed efficiency as corn for laying hens, broilers, lambs, and dairy cows. Grain sorghum, in general, is highly palatable to livestock, although some cultivars with dark-colored grains are somewhat bitter. Grain should be rolled, ground, cracked, or popped for cattle. It is not economical to grind the grain if it is self-fed to hogs or sheep. Unthreshed heads, however, should be ground when fed to sheep. For poultry, sorghum should be fed whole, cracked in scratch feed mixtures, or ground in mash.

Grain sorghum is a carbohydrate feed and, like corn, is most efficiently utilized by livestock when supplemented with protein. Grain sorghum may be a more economical feed than corn when the two grains are purchased; an often lower price for sorghum compensates for its slightly lower feeding value.

The most efficient way to use sorghum forage is as silage. Sorghum silage may have a slightly lower feeding value than corn silage on a pound-for-pound basis when fed to fattening animals because of its lower percentage of grain. In addition, more of the grain may pass through the animals undigested.

HUMAN FOOD. Sorghum is not an important food in the United States, but in a considerable part of India, China, and Africa it is a most important food crop. In some areas as much as 75 percent of the cultivated acreage is reported as devoted to sorghum for human food, with the grain eaten in some form at each meal. In recent years, geneticists have identified the genetic factors resulting in high-lysine grain sorghum after screening 9,000 sorghums from all over the world. This breakthrough will enable plant breeders to develop high-lysine cultivars to replace some of the presently used cultivars, improving quality of grain sorghum as a food crop and the nutrition of millions of the world's people where sorghum is the food staple.

Another food product from sorghums is syrup, extracted from sorgos or sweet sorghums in limited amounts in some areas of the United States.

INDUSTRIAL UTILIZATION. Sorghums have a number of important industrial uses. Starch can be extracted and used in wallboard, paper and cloth sizing, and adhesives. Some of the starch is converted to dextrose and used in fruit canning and confection industries. Alcohol is produced from sorghum grain in quantities comparable with those obtained from wheat and corn.

Breeding and New Cultivars

It is probable that within recent years no field crop has been more influenced by the efforts of the plant breeder than the sorghums. Because sorghums are grown for grain, fodder, silage, pasture, syrup, and brooms, the sorghum breeder has varied objectives. They include breeding sorghums for greater productivity, adaptation to mechanical harvesting, earlier maturity, resistance to lodging and shattering, disease and insect resistance, resistance to bird damage, improved seed and forage quality, and low prussic acid potential.

The breeding methods of sorghum have been similar to those with other self-pollinated crops—introduction, selection, and hybridization. After about 1925 hybridization became the principal procedure by which new cultivars of sorghum originated. After the report of cytoplasmic male sterility in 1954 it was possible to produce hybrid sorghums on a commercial basis more economically.

Combine Sorghums

Of greatest importance has been the production of high-yielding, dwarf grain cultivars, in this way making it possible to combine-harvest the crop. Interest in combine sorghums had its beginning with the release of the 'Martin' cultivar in 1941. Other combine-type sorghums followed in rapid succession. By 1953, 98 percent of the sorghums harvested for grain were of the combine types. Harvesting grain sorghum with combines requires only one-eighth the labor required to harvest by hand heading, previously the common method.

Root Rot Resistance

Notable results have been obtained in developing resistance to destructive diseases. *Periconia* root rot, known also as milo disease and formerly attributed to *Pythium,* provides a good example. Through a long period of years milo was the basic grain sorghum of the Southwest. Most of the older cultivars have disappeared because of their susceptibility to the *Periconia* root rot. Milo would have been doomed except for the development of disease-resistant cultivars. The technique of breeding for resistance to milo disease is simple. Resistance is inherited by a single gene with resistance partially dominant.

Chinch Bug Resistance

All milos and combine types derived from milo parents released before 1945 were subject to chinch bug injury. 'Dwarf Kafir 44–14', 'Redlan', and 'Dwarf Kafir 60', released in Texas, were the first combine cultivars safe for growing where chinch bugs abound. Commonly used cultivars and hybrids vary in susceptibility to injury from this insect.

Decreased Poison Resistance

Improvements have been made in sorgo and sudangrass by the development of low prussic acid cultivars. 'Rancher' from the South Dakota station has a low HCN-potential and has replaced much of the 'Black Amber' from which it came. 'Piper' sudangrass previously mentioned has replaced previously-used sudans throughout much of the North. It is the only sudangrass currently recommended by the Illinois station, for example.

Lower Tannin

The grain of the forage sorghums contains enough tannin to make it bitter and somewhat distasteful to livestock. The 'Atlas' cultivar from the Kansas station was the first of the forage cultivars without bitter grain. This cultivar has been used as a parent in further breeding developments.

Hybrid Cultivar Choice

The first consideration in choosing a hybrid cultivar to plant is the use to be made of the crop—whether for grain alone, for grain and forage, or primarily for forage alone. The relative earliness of the cultivar often is an important factor in determining the best time to plant. Larger yields usually result when the crop reaches maturity shortly before the first killing frost, unless it is a matter of getting the crop made before extreme drought conditions develop, or of avoiding chinch bug damage. Experiment stations periodically release recommendation lists.

Production Practices

Many think of sorghum as a substitute for corn—as the grain and forage more important than any other feed crop throughout a large area where corn production is limited. Production practices considered here will apply more especially to the central and southern Great Plains area where production is concentrated.

In Rotations

Sorghums do not fit well into many rotations, and the usual benefits of a rotation may not particularly apply. In some areas where the soil is inclined to be light or sandy and the rainfall limited, sorghum often is planted year after year on the same land, or it is alternated with a year of fallow.

In areas where both winter wheat and sorghums are suitable, a rotation of sorghums–barley–wheat–wheat or wheat–fallow–sorghum is reported as giving good results. When sorghum follows winter wheat there is the opportunity for thorough seedbed preparation and excellent control of weeds.

Some researchers have suggested the undesirability of seeding winter wheat on land that had been in sorghum the same year. Winter wheat requires considerable nitrates relatively soon after planting. Available nitrates are particularly deficient following a sorghum crop and soil moisture often is depleted. Spring-planted small grains do much better than winter wheat following sorghum but crops which make most of their growth still later in the season, such as corn, do still better.

In Relation to Other Crops

It has been shown that cotton, corn, oats, wheat, and other crops following sorghum are likely to yield less than when following some other crop. This fact is supported by rotation experiments in Arkansas, Kansas, Nebraska, and Alabama. In these experiments the yields of crops following sorghum were reduced an average of 15 percent as compared to yields following corn. A factor contributing to the lower yields of crops following sorghum is the greater depletion of soil moisture because of the high acre production of sorghum and the unusual ability of the crop to extract moisture from the soil. The continued growth of sorghum until killed by frost also tends to deplete available moisture and plant nutrients. But it is believed that one of the most important factors is the high sugar content of sorghum roots and stubble. With an abundance of readily available carbohydrates in the soil, microorganisms are stimulated to multiply rapidly. These organisms compete with the crop plants for the available nitrogen.

The deleterious effects of sorghum are temporary, lasting for only a few months, or until after the sorghum residues have decomposed. In humid areas and under irrigation the use of nitrogen fertilizers usually corrects the injurious aftereffects of sorghum.

Fertilizer Use

Results from the use of fertilizer in dryland areas are rather uncertain because of the wide variation in available moisture from season to season. In most cases soil moisture in these areas rather than nutrients is the limiting factor. Where water is available for irrigation and high yields are obtained, fertilization usually is profitable. Fertilizer trials on the Texas High Plains for a nine-year period indicated yield responses to nitrogen were most frequent. Second greatest response was from phosphorus. There was little response to potash. Similar results were obtained in New Mexico. In the more humid sections sorghum responds to fertilizer use in much the same way as corn.

Seedbed Preparation

In much of the Corn Belt and the South, the operations used for corn and cotton also are applicable for sorghum.

In the central and northern Great Plains, sorghum is planted on fallow land or after wheat or sorghum. The best summer fallow tillage methods are those that help store soil moisture and prevent wind erosion.

Commonly used methods of land preparation are listing, chiseling, plowing with either disk or moldboard plows, and disking, or a combination of these methods. In the more humid and irrigated areas, plowing is common, while in drier areas, some other seedbed preparation method generally is used. In semiarid regions, stubble-mulch and one-way disk methods are common. On some of the medium-to-light-textured soils where the lister planter is used, grain sorghum may be grown without any land preparation with very little reduction in yield. Researchers at the High Plains Station at Plainview, Texas, obtained good results with minimum tillage. Only slightly lower yields were harvested from sorghum planted after chisel plowing just once, as compared to seedbed preparation involving plowing, disking, and listing.

Time of Planting

If sorghum is planted before the soil has warmed, stands are likely to be poor and weed control difficult. Sorghum should be planted only when the soil at planting depth has reached about 70°F (21°C). This date will be from late April to June, depending on the location. This usually is 10–20 days after the usual date of planting corn. The general rule is to plant just late enough to allow the cultivar to mature before frost. Under some conditions planting as early as possible is desirable, to mature the crop before severe drought conditions develop or to reduce the danger of chinch bug damage. Sorghum may be planted well into July in some areas when used in a double-cropping system following wheat or another small grain.

Planting Methods, Row Width, and Population

In the past most sorghum was planted in 40- or 42-inch (102- or 107-cm) rows on dry land and 24–36 inches (61–91 cm) apart under irrigation. Row-spacing studies in humid areas or with irrigation have shown consistently higher yields from 20- and 30-inch (51- and 76-cm) rows as compared to 40-inch (102-cm) rows. More recently much of the combine type of grain sorghum has been planted with the grain drill, plugging enough of the grain spouts to place the rows 10–18 inches (25–46 cm) apart. Planting in this way is regarded as somewhat risky unless the field conditions are favorable, the seedbed well prepared and free of weeds, and the seed planted at the proper rate and time. An important factor has been that planting with the grain drill and harvesting with the combine require no equipment other than that used in wheat production. Some increase in yield usually has resulted from the narrower row spacing, perhaps because of better spacing of plants; more shade; more efficient use of water, nutrients, and light; and reduced evaporation from the soil.

When planted in the wider rows, much of the crop has been listed. Loose-ground furrow-openers also have been used extensively and with equally good results.

Optimum plant population is most often 60,000–120,000 plants/acre (148,258–296,516/ha) with 4–6 plants/foot (13–20/m) of row for 30-inch (76-cm) rows.

Robinson reported on a study in southern Minnesota, the humid northern Corn Belt, where row spacings of 10, 20, 30, and 40 inches (25, 51, 76, 102 cm) each at populations of 78,408, 156,816, and 313,632 seeds/acre (193,744, 387,487, and 774,974/ha) were compared at three locations. A linear trend for increased yield occurred as rows narrowed from 40 to 10 inches (102 to 25 cm). Population or spacing in the row had little effect on yield. In a Kansas experiment, Stickler and Laude found that 78,000 plants/acre (192,735/ha) gave higher yields than 52,000 (128,490/ha), and a significant plant population times row spacing interaction occurred.

Planting Rates

Because of the wide variety of soil, fertility, and moisture conditions in areas where grain sorghum is grown, only general recommendations for rate of planting can be given. As a general rule, the per acre planting rates are 2–4 pounds/acre (2.2–4.5 kg/ha) in dryland conditions, and 6–10 pounds (6.7–11.2 kg/ha) in humid or irrigated regions. If a rate of 6–8 pounds/acre (6.7–9.0 kg/ha) with a 30-inch (76-cm) row spacing were used, 8–10 seeds would be dropped per foot (26–33 seeds/m) of row, with an expected final population of 4–6 plants/foot (13–20 plants/m) of row. This would be expected to result in optimum yields if moisture is adequate.

Weed Control

Sorghum may present special problems as to weed control. Germination in the field is lower and the seedlings are less vigorous than those of corn, thus affording less competition to weeds. A good start toward adequate weed control is preparing a fine, firm seedbed. This is particularly important when sorghum is drilled. Once a good stand is established, shading largely controls weeds.

Weeds may be controlled early, when weeds are small, with a rotary hoe or harrow. Later cultivations are with a conventional row cultivator.

Chemical weed control has become an important part of production, with more than 40 percent of the grain sorghum acreage treated with some type of herbicide. Some of the herbicides recommended for corn can be used on sorghum acreage, but other corn herbicides can be injurious to sorghum. In general, sorghum is more sensitive to herbicides than is corn.

Irrigation

Sorghum is a drought-tolerant crop, but it responds well to irrigation. Producing a crop requires 20–25 total inches (51–64 cm) of water; as much as 20 inches (51 cm) of irrigation water may be necessary in semiarid or arid regions. For best results, grain sorghum should be irrigated before periods of heaviest water usage—booting and heading. Sorghum responds best to irrigation if enough water is applied during these stages to prevent any sign of drought stress. A few heavy waterings are just as effective as more frequent light waterings. Generally, water is applied only two to four times after planting. It is not advisable to apply water after the grain reaches the dough stage.

Harvesting

GRAIN. Practically all the grain sorghums are harvested with combines; in the central and northern part of the Great Plains harvesting is done after frost has stopped further growth, or as soon as the crop is ripe where the growing season is longer. The sorghum plant remains alive until killed by frost, preventing natural dry-down rates comparable to corn. Weathering, which causes the grain to become soft following several rains, may occur in humid regions when harvest is delayed. Chemical dessicants have been used successfully to kill the plant and speed the drying rate before a frost. Flaming the upper portion of the stalk also kills the plant and

FIGURE 20-5. *Harvesting of grain sorghum. Practically all of this crop is harvested with combines in the United States. Harvesting may be delayed until after the first killing frost, or chemical dessicants may be used to kill the plants and speed drying of grain.* [SOURCE: Allis Chalmers Corporation.]

starts the drying process. Sorghum grain will thresh free from the head when the moisture content is 25–30 percent.

FORAGE. When the crop is harvested for forage, harvest is best delayed until the plants approach maturity. Acre production may increase nearly 40 percent between the time the plants first head and full maturity. Not only is a greater tonnage obtained from the more mature plants, but also the prussic acid content is lower, the plants are more palatable, and there is less danger of deterioration.

Grain Sorghum Storage

In the Great Plains states, a moisture content of under 13 percent is essential for safe storage. Eleven percent is safer in the more southern regions where the humidity is high. In recent years, some high-moisture grain sorghum has been stored without molding losses after treatment with preservatives such as propionic acid.

When artificial drying is necessary, forced-air drying is the best method for reducing the moisture content of grain sorghum. If the grain is not too damp or the humidity not too high, unheated air may be used. Under such conditions, drying operations may have to be limited to the drier parts of the day and be extended over a longer period than if heated air is used.

Hot-air drying allows harvesting of the crop before the onset of winter weather and before the stalks become lodged. Air temperature should be kept under 200°F (93°C) if ths grain is to be fed. If the grain is to be milled, air temperature should not exceed 140°F (60°C). If it is to be used for seed, maximum temperature should be 110°F (43°C).

In areas of high grain production in the Great Plains states, storage facilities usually are insufficient to hold the crop. Therefore, much of the grain is piled on the ground for a short period after harvest. Considerable drying takes place in these piles.

Stored grain that becomes infested with insects should be fumigated. Generally, grain sorghum needs higher rates of fumigant than wheat does because the seeds pack together more closely. The same insects that attack small grains attack stored grain sorghum.

Insects, Diseases, and Other Pests

Insects may cause crop losses averaging 9 percent annually. Chinch bugs, sorghum midge, corn earworm, and sorghum webworm can cause severe injury to grain sorghum, with the first two probably the most serious. Chinch bugs, small black sucking insects, cause damage by withdrawing fluid from cells and providing openings in plant tissue for invasion of fungi and bacteria. The midge feeds in florets and prevents grain formation.

Although not as widespread and serious, other insects including corn leaf aphid, cutworms, armyworms, and southwestern born borers, can be damaging on occasion.

Reduction in yield of grain sorghum as a result of diseases may average as much as 10 percent annually. With the use of high-quality seed treated with fungicides, seed rots and seedling diseases have been reduced. They are more severe in a cool, wet spring in the northern areas where seed is planted early.

Principal diseases causing grain sorghum losses are smuts, root and stalk rots, and foliar diseases. Smuts, some of the most serious diseases, are caused by fungi. The three prevalent smut diseases are covered kernel smut, loose kernel smut, and heat smut. Infestations are characterized by seeds or entire heads being replaced by a mass of dark-colored spores. The smuts may be controlled effectively by appropriate seed treatment and by use of resistant cultivars. There are several stalk and root rots, with charcoal rot and *Periconia* root rot the most damaging. Resistant cultivars generally are available. Foliar diseases caused by fungi include anthracnose, rough spot, leaf blight, zonate leaf spot, gray leaf spot, target

A   B

FIGURE 20-6. *Head types as related to bird resistance. (A) Compact head type, which may be susceptible to bird feeding. (B) Open panicle head, which is more bird resistant. Bird resistance may also result from the presence of tannin in the grain.* [SOURCE: Southern Illinois University.]

spot, sooty stripe, and rust. Foliar diseases caused by bacterial are bacterial stripe, bacterial streak, and bacterial spot.

Control measures for the fungus and bacterial foliar diseases include seed treatment, crop rotation, and use of resistant cultivars.

Various kinds of birds, including blackbird, can cause great yield losses by feeding on grain sorghum before it is harvested. Losses have been minimized somewhat by the development of bird-resistant cultivars, sometimes designated as "BR" in the cultivar name. Bird resistance may be attributed principally to the presence in the grain of tannin, which dissipates somewhat as the grain approaches maturity. Open panicle head types often are more bird resistant than are compact head types.

Sudangrass

Sudangrass is an annual grown widely in the United States for pasture, green chop, silage, and hay. The importance of sudangrass has increased as a result of the development of numerous sorghum–sudangrass hybrids, which have a higher yield potential than does sudangrass.

Adaptation

Sudangrass is a warm-season, heat-and drought-tolerant grass. Plants will stay alive during periods of drought even after close mowing or heavy pasturing and will resume growth quickly when rains come. Introduced from Africa in 1909, the sudangrass seed supply was increased rapidly in the Southwest. It soon spread to all parts of the United States. Sudangrass is now grown where it was first thought to be wholly unadapted. Its short growing period permits it to produce good crops of hay as far north as Michigan and New York.

Sudangrass does best on rich loam, but it has been grown successfully on almost every type of soil from heavy clay to light sand. Sudangrass does not do well on cold, wet soils. It is not especially sensitive to soil acidity and grows well on soils with a pH as low as 5.5.

The Plant

Sudangrass, when planted in rows, may reach a height of 6 feet (1.8 m) or more; when broadcast or planted in close drills, its growth usually is 4–5 feet (1.2–1.5 m). The leaves are rather numerous and long. The coarseness of the stem depends on the thickness of planting, soil fertility, and available moisture. When thin-planted on fertile, moist soil a single seed may result in 50–100 stems through tillering. Seeds usually are pale yellow

FIGURE 20–7. Sudangrass is grown widely in the United States for pasture, green chop, silage, and hay. Sudangrass has been crossed with forage sorghums, resulting in high-yielding hybrids.

to mahogany in color. The sorghum-sudangrass hybrids resemble sudangrass in growth, but the hybrids are taller and have larger stems and leaves. They have become one of the most popular summer annual grazing crops in the Gulf states.

Cultural Practices

Sudangrass is not seeded until the soil has become thoroughly warm, usually about two weeks after corn planting. This date may vary from April 1 to mid-July.

For pasture or green chop, seeding of half an area may be delayed to space production over a longer period. For grazing, green chop, and hay the rate of seeding broadcast is 20–30 pounds/acre (22–34 kg/ha) in humid areas; 12–15 pounds/acre (13–17 kg/ha) in dry areas; and 15–20 pounds/ acre (17–22 kg/ha) in irrigated areas. For silage, sudangrass may be planted at the rate of 5–25 pounds/acre (6–28 kg/ha); the amount of seed depends on rainfall as well as row spacing.

FIGURE 20–8. *When forage sorghum is to be harvested for silage, a greater production and a crop of better quality is obtained if near maturity at time of harvest.*

FIGURE 20-9. *Summer annual grasses, such as the sudan-sorghum hybrid shown here, provide high quality feed in the summer when managed properly. This grass is at the proper height to begin grazing.* [SOURCE: University of Tennessee Agricultural Experiment Station.]

Sudangrass for Pasture

By far the greatest use of sudangrass in the United States is for pasture or green chop. Its greatest value is to supplement permanent and rotation pastures. On fertile soil and with fair rainfall it will carry from two to three mature animals per acre (5–7 animals/ha) for a mid-summer period of 2.5–3 months. Sudangrass and sorghum-sudangrass hybrids usually are ready for grazing 5–6 weeks after planting. To avoid prussic acid poisoning, sudangrass should not be pastured until it is at least 18–24 inches (46–61 cm) high. For green chop, the first cut should be made just before the heading stage to ensure good regrowth.

Sudangrass for Hay and Silage

For highest hay yields, the crop should be harvested when the seed is in the soft-dough stage. Because curing is difficult at this stage, it is more practical to harvest at the boot stage when the plants are 30–40 inches (76–102 cm) high. Sudangrass and the sorghum–sudangrass hybrids make acceptable silage for beef and dairy cattle. They have about 90 percent as much feed value as corn silage. If harvested at the soft-dough stage for making silage, it is unnecessary to wilt the forage or to add a preservative, particularly if sweet-type cultivars or hybrids are used.

The Millets

Of the four generally recognized types of millet, three are grown for forage and one for grain. Although all are known as millets, no two belong

FIGURE 20-10. *Foxtail millet, a summer annual grown principally for forage in the northern Great Plains.* [SOURCE: Southern Illinois University.]

to the same genus. Foxtail, Japanese, and pearl types are grown mainly for forage and proso for grain. Some of the millets are utilized for bird seed. Pearl millet is the most widely grown.

All the millets grow best in areas where summer temperatures are high. They are most grown in areas approaching the semiarid. They usually are seeded relatively late but may escape serious drought injury because of the short growing period required.

The foxtail millets, *Setaria italica,* have been mostly grown in the northern Great Plains. Pearl millet, *Pennisetum glaucum,* has become the leading summer annual supplementary pasture crop in the lower Southeast. Japanese millet, *Eschinochloa frumentacea,* is occasionally grown in New York and Pennsylvania and in other Eastern states. The foxtail millets are characterized by slender, erect stems which usually make a growth of 2–4 feet (0.6–1.2 m). They usually are late-seeded, often as an emergency crop to replace a spring-sown crop that has failed. Pearl millet makes a very rank growth, often to a height of 10–15 feet (3.0–4.6 m). The plants tiller freely and the stems are rather coarse, harsh, and pithy. Japanese millet makes a rather coarse growth, with stems up to 4 feet (1.2 m). The stems of the proso *(Panicum miliaceum)* cultivars are woody and bear comparatively few leaves. The seeds are larger and more abundant than those of the other types. There are a considerable number of cultivars of the millets, especially of the foxtail and proso types.

FIGURE 20–11. *Pearl millet, the most widely-grown millet for summer forage, performs better than sudangrass or sudan-sorghum hybrids in some parts of the South.* [SOURCE: Southern Illinois University.]

REFERENCES AND SUGGESTED READINGS

1. AMATOR, J., *et al. Sorghum Diseases,* Texas A & M Univ. and Texas Agr. Ext. Serv. B-1085 (1969).

2. BROADHEAD, D. M., O. H. COLEMAN, and K. C. FREEMAN. *Dale, A New Variety of Sweet Sorghum for Sirup Production,* Miss. Agr. Exp. Sta. Info. Sheet 1009 (1970).

3. BROADHEAD, D. M., *et al.* "Sorgo Spacing Experiments in Mississippi," *Agron. Jour.,* 55:164–166 (1963).

4. BROADHEAD, D. M. *et al. 'Theis'—A New Variety of Sweet Sorghum for Sirup Production,* Miss. Agr. Exp. Sta. Info. Sheet 1236 (1974).

5. BURTON, G. W., and J. B. POWELL. "Pearl Millet Breeding and Cytogenetics," *Advances in Agronomy,* vol. 20 (New York: Academic Press, 1968). ed. A. G. Norman.

6. FRIBOURG, H. A. "Summer Annual Grasses and Cereals for Forage," In *Forages* (Ames: Iowa State Univ. Press, 1973). Edited by M. E. Heath, D. S. Metcalfe, and R. F. Barnes.

7. JENSEN, M. E., and J. T. MUSICK, *Irrigating Grain Sorghum,* USDA Leaflet 511 (1962).

8. LEONARD, W. H., and J. H. MARTIN. *Cereal Crops* (New York: Macmillan, Co., 1963).

9. MALM, N. R., and M. D. FINKNER. *Fertilizer Rates for Irrigated Grain Sorghum on the High Plains,* New Mex. Agr. Exp. Sta. Bul. 523 (1968).

10. MARTIN, J. H., W. H. LEONARD, and D. L. STAMP. *Principles of Field Crop Production,* 3rd ed. (New York: Macmillan, 1976).

11. Matz, S. A. *Cereal Science* (Westport, Connecticut: AVI, 1969).

12. Miller, F R., and R. W. Bovey. "Tolerance of *Sorghum bicolor* (L.) to Several Herbicides," *Agron. Jour.,* 61:282–285 (1969).

13. Millis, D. E., *et al. Grain Sorghum for Illinois,* III. Coop. Ext. Serv. Publ. AG-1974 (1974).

14. Moats, R. W., bibliographer. *Sorghum: Bibliography of World Literature, 1930–1963,* Biol. Sciences Communication Proj., George Washington Univ., Rockefeller Foundation (Metuchen, N.J.: Scarecrow, 1967).

15. Musick, J. T., and D. W. Grimes. *Water Management and Consumptive Use of Irrigated Grain Sorghum in Western Kansas,* Kans. Agr. Exp. Sta. Tech. Bul. 113 (1961).

16. Nelson, L. R., *et al.* "Storage of High Moisture Grain Sorghum (*Sorghum bicolor* (L.) Moench) Treated with Propionic Acid," *Agron. Jour.,* 65:423 (1973).

17. Plaut, Z., *et al.* "Effect of Soil Moisture Regime and Row Spacing on Grain Sorghum Production," *Agron. Jour.,* 61:344–347 (1969).

18. Quinby, J. R. "The Natural Genes of Sorghum" In *Advances in Agronomy,* vol. 19 (New York: Academic Press, 1967). ed. A. G. Norman.

19. Robinson, R. G. "Sunflower-Soybean and Grain Sorghum-Corn Rotations Versus Monoculture," *Agron. Jour.,* 58:475–477 (1966).

20. Robinson, R. G., *et al.* "Row Spacing and Plant Population for Grain Sorghum in the Humid North," *Agron. Jour.,* 56:189–191 (1964).

21. Robertson, W. *et al. Fertilizing Grain Sorghum on the Texas High Plains,* Texas Agr. Exp. Sta. MP-940 (1969).

22. Stickler, F. C., and H. H. Laude. "Effect of Row Spacing and Plant Population on Performance of Corn, Grain Sorghum, and Forage Sorghum," *Agron. Jour.,* 52:5 (1960).

23. Stickler, F. C., and A. W. Pauli. *Yield and Yield Components of Grain Sorghum as Influenced by Date of Planting,* Kans. Agr. Exp. Sta. Tech. Bul. 130 (1963).

24. Toler, R. W., *et al.* "Identification, Transmission, and Distribution of Maize Dwarf Mosaic in Texas," *Plant Disease Reporter,* 51:777 (1967).

25. USDA. Agr. Stat. (1977).

26. USDA-ARS. *Losses in Agriculture,* Agr. Handbook 291 (1965).

27. USDA. *Growing Grain Sorghum,* Leaflet 478 (1966).

28. USDA. *Sudangrass and Sorghum-Sudangrass Hybrids for Forage,* Farmers' Bul. 2241 (1969).

29. Wall, J. S., and W. M. Rose (Eds.). *Sorghum Production and Utilization* (Westport, Conn.: AVI, 1970).

30. Woodle, H. A. *Grain Sorghums,* Clemson Agr. Coll.-USDA Circ. 285 (1960).

CHAPTER 21
WHEAT AND RYE

No civilization worthy of the name has ever been founded on any agricultural basis other than the cereals. The ancient cultures of Babylon and Egypt, of Rome and Greece, and later those of northern and western Europe, were all based upon the growing of wheat, barley, rye, and oats. Those of India, China, Inca, Maya, and Aztec—looked to corn for their daily bread. . . .

Wheat is the world's most widely cultivated plant. The wheat plants growing on the earth may even outnumber those of any other seed-bearing plant species, wild or domesticated. . . . Apparently this grain was one of the earlier plants cultivated by man. Carbonized kernels of wheat were found recently . . . at the 6,700-year-old site of Jarmo in eastern Iraq, the oldest village yet discovered—a village which may have been one of the birthplaces of man's agriculture. . . . The resemblance between the ancient and modern grains is remarkable. There were two types of kernels in the Jarmo site; one almost identical with a wild wheat still growing in the East and the other almost exactly like present-day cultivated wheat of the type called Einkorn. Evidently there has been no appreciable change in these wheats in the 7,000 years since Jarmo.

Paul C. Mangelsdorf

Wheat

WHEAT is the leading food crop in many parts of the world. Production tends to be concentrated in certain areas, but wheat consumption is worldwide.

World Production and Distribution

On a world basis wheat is one of the four most important food crops; the other three are rice, corn, and sorghum (including millets). We often think of China as a rice-, millet-, and soybean-consuming nation, but China devotes more area to wheat than most other countries, and has nearly twice the acreage of Canada.

On the basis of both acreage and production, wheat is the world's leading crop. Wheat occupies about 65 percent greater acreage, and production is about 20 percent greater than rice, the second leading world crop. Most of the world crop is grown in regions of cold winters; India, Egypt, and North Africa are exceptions.

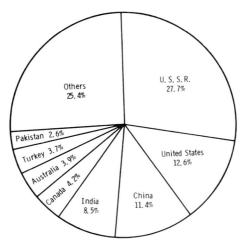

FIGURE 21-1. *Leading countries in world wheat area (acreage). Three countries account for more than one half of the total.* [SOURCE: Southern Illinois University.]

The leading countries in both acreage and total production are the USSR, the United States, China, India, Canada, Australia, Turkey, and Pakistan (Figures 21–1 and 21–2). World production by country is shown in Figure 21–3; the USSR produces nearly 100 million MT and the United States about 50 million MT, with total world production of more than 400 million MT.

Wheat is more easily stored and transported than most food crops. This has been an important factor in making wheat the leading grain in interna-

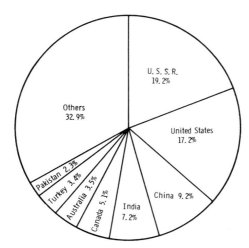

FIGURE 21-2. *Leading countries in world wheat production.* [SOURCE: Southern Illinois University.]

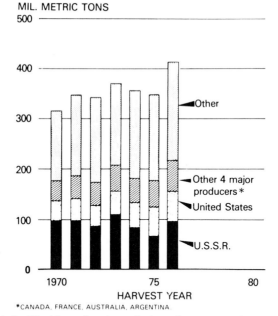

FIGURE 21–3. *World production of wheat by country.* [SOURCE: USDA.]

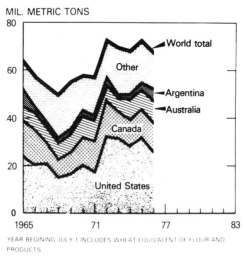

FIGURE 21–4. *World exports of wheat and flour by country.* [SOURCE: USDA.]

tional trade. World exports of wheat and flour by country are shown in Figure 21–4. The United States accounts for at least one-third of total world exports.

U.S. Production, Distribution, and Adaptation

In the United States wheat ranks second to corn on the basis of acreage. Geographic distribution is shown in Figure 21–5. The leading states in order of amount of acreage are Kansas, North Dakota, Oklahoma, Texas, Montana, Nebraska, and Washington (Figure 21–6). In the United States about 80 million acres (32.4 million ha) of wheat are grown annually, with about 70 million acres (28.3 million ha) harvested for grain. Harvested acreage and yield per acre are up 25 percent from the 1959–1961 period, and total production has increased more than 65 percent in the same time period (Figure 21–7). Wheat is a very important export grain. The United States has exported more than 1 billion bushels (27.3 million MT) during most every year since 1972 (Figure 21–8).

Wheat has a wide climatic and soil adaptation range; it is an important crop in both humid and subhumid areas. The crop is grown on a great variety of soils, ranging from the stiff clays in the New England states to the fine soils of volcanic ash origin on the Pacific Coast. In general, soils dark in color and relatively high in organic matter are less well suited for wheat growing than are the heavier soils.

In North America most of the crop is produced in regions having an

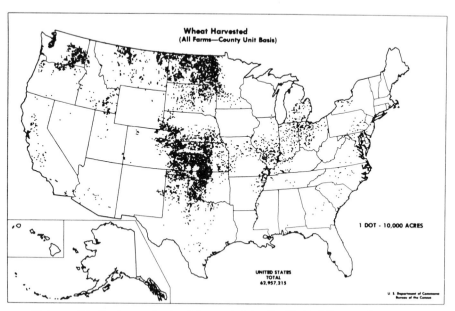

FIGURE 21–5. *Acres of wheat harvested in the United States, shown geographically.* [SOURCE: U.S. Department of Commerce.]

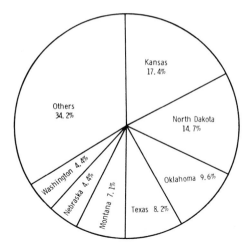

FIGURE 21-6. *Wheat area (acreage) in the United States. Four states have one half of the acreage.* [SOURCE: Southern Illinois University.]

annual rainfall of less than 30 inches (76 cm) and minimum winter temperatures of 0°F (−18°C) or below.

The Plant

Common wheat, *Triticum aestivum,* belongs to the tribe *Hordeae.* Wheat is a winter annual grass that grows to a height of 1–4 feet (0.3–1.2 m),

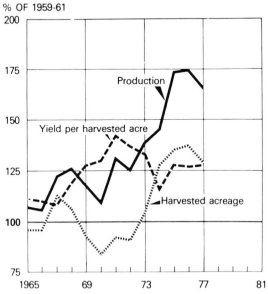

FIGURE 21-7. *Wheat acreage, yield, and production in the United States.* [SOURCE: USDA.]

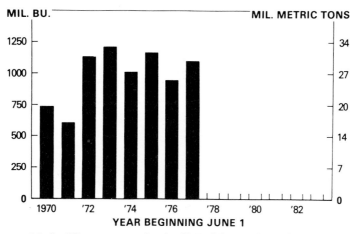

FIGURE 21-8. *Wheat exports by the United States. Note the large increase since 1970–71.* [SOURCE: USDA.]

depending on the cultivar and the environment. When wheat germinates, it produces a whorl of temporary roots and the plumule. The permanent roots are produced at the nodes of the new seedling. Each plant may produce 1–100 tillers of potential grain-bearing culms, depending on spacing and soil and climatic conditions. The number of tillers is usually 2–5 under most field conditions. Tillering enables wheat to make full use of favorable environmental conditions. One leaf is produced per node. Early when nodes are closely compacted, the wheat plant has a leafy appearance. As the plant grows, the internodes elongate and the leaf sheaths also elongate, with some overlapping. In the collar region of each leaf are clawlike, medium-size, hairy auricles.

The flowering head is a spike, ordinarily with each joint of the rachis bearing one sessile spikelet. In each inflorescence or head there are ordinarily 14–17 spikelets, each of which normally contains 2–3 florets, which upon fertilization and maturation become caryopses or grains. The florets are normally self-pollinated. One head generally bears 25–30 grains. The grain or caryopsis is naked after combining, threshing free from the lemma and palea coverings. The grain has a characteristic crease or groove, whose depth varies with the wheat type, and a hairy tip or brush. Color of the grain varies with the type. Each kernel consists of the germ or embryo, bran, and the starchy endosperm.

History and Origin

The growing of wheat, in one form or another, goes far back into prehistoric times. Some of the wheats were cultivated as early as 7000 B.C. in some parts of the world. Wheats found in archaeological ruins indicate that this was a valuable crop in Egypt, Greece, and Persia many thousands of years ago.

FIGURE 21-9. *The flowering spike of soft red winter wheat at anthesis.* [SOURCE: Funk Seeds International.]

Apparently some of the wild wheat types originated in southwestern Asia. Early explorers introduced this crop into Europe, and colonists brought it to North America. Records indicate that wheat was produced and sold in Massachusetts and Virginia in the early 1600s.

The origin of common wheat is still the subject of much speculation, but several theories have been proposed. Mangelsdorf suggested that wheat had its origin in the Caucasus–Turkey–Iraq area and that in the evolution of our common wheat, wild einkorn evolved into einkorn, which when crossed with a wild grass gave rise to Persian wheat. Persian wheat has 14 chromosomes, double the chromosome number of the einkorns. It is the theory that when this wheat crossed with the grass *Aegilops squarrosa,* our common wheat with 21 chromosomes resulted. One of the most recent theories is that the intercrossing of wild types resulted in emmer wheat, which crossed with another wild type to produce spelt wheat, which in turn mutated to form common wheat.

Classification of Wheat

Mangelsdorf accepts the species classification of Vavilov, the noted Russian geneticist and botanist, who with his colleagues brought together for study more than 31,000 samples of wheat from all parts of the world. Vavi-

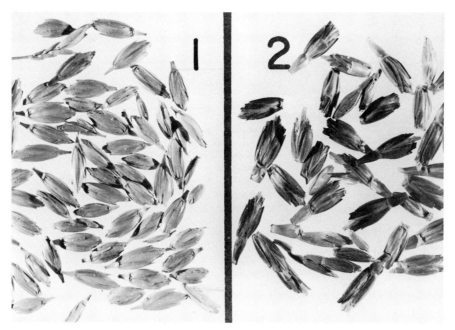

FIGURE 21–10. *Grains of two primitive wheats that may have contributed to the development of common wheat: 1 = emmer, 2 = spelt.*

lov recognized 14 species, divided into three distinct groups on the basis of chromosome number—7, 14, and 21.

The theory is that all the 14- and 21-chromosome wheats resulted from the 7-chromosome wheats and related grasses through hybridization followed by chromosome doubling.

The wheat classification by Vavilov is shown in Table 21–1. The geographic distribution, and earliest evidence on the existence of the different wheat species as set out by Mangelsdorf are shown in Table 21–2.

Wheat Classes and Production Areas

Wheat is classified by types into seven commercial market classes: (1) hard red spring, (2) amber durum, (3) red durum, (4) hard red winter, (5) soft red winter, (6) white (including club), and (7) mixed wheat. Classes are based on the color and texture or the geographic areas in which the wheat is produced. Most of the classes have two or three subclasses and each subclass contains five numerical grades and the sample grade.

Winter wheats normally have small germs, deep creases, and rounded cheeks. Spring types have larger germs, shallow creases, and angular cheeks. Durum has a pointed germ. The planted acreages of winter wheat, durum (spring-planted), and other spring wheats are shown in Figure 21–11. Geographic distribution of some of the wheat classes is shown in Figure 21–12.

TABLE 21–1. Vavilov's Classification of Wheats, with Some Characteristics

Group Number	Species of *Triticum*	Common Name	Chromosome Number	Domesti- cated	Grain
I	*T. aegilopoides*	Wild einkorn	7	no	in hull
II	*T. monococcum*	Einkorn	7	yes	in hull
III	*T. dicoccoides*	Wild emmer	14	no	in hull
IV	*T. dicoccum*	Emmer	14	yes	in hull
V	*T. durum*	Macaroni wheat	14	yes	naked
VI	*T. persicum*	Persian wheat	14	yes	naked
VII	*T. turgidum*	Rivet wheat	14	yes	naked
VIII	*T. polonicum*	Polish wheat	14	yes	naked
IX	*T. timopheevi*	—	14	yes	in hull
X	*T. aestivum*	Common wheat	21	yes	naked
XI	*T. sphaerococcum*	Shot wheat	21	yes	naked
XII	*T. compactum*	Club wheat	21	yes	naked
XIII	*T. spelta*	Spelt	21	yes	in hull
XIV	*T. macha*	Macha wheat	21	yes	in hull

HARD RED SPRING. Hard red spring wheat is grown principally in the northern Great Plains states where there exists a fairly deep black soil and hot dry summers. These environmental characteristics are believed important in the production of a high grade of spring wheat suitable for milling. There is a small acreage in the Western states.

Hard red spring wheat, mostly spring-seeded, occupies about 20 percent of the total U.S. wheat acreage (Figure 21–13). The grain is hard in texture and high in protein and is used for bakery bread.

The rapid change in recent years in the cultivars generally grown has been one of the marvels resulting from plant breeding. The cultivars previously grown were largely replaced following 1907 by 'Marquis,' which in 1929 occupied about 87 percent of the total hard red spring wheat acreage. Principally because of differences in resistance to currently predominating races of rust, 'Marquis' gave way to 'Ceres,' the 'Ceres' to 'Thatcher,' and most recently, the 'Thatcher' to cultivars such as 'Waldron,' 'Era,' 'Lark,' 'Fortuna,' and 'Chris.'

DURUM WHEAT. Durum cultivars, mostly spring-seeded, occupy about 5 percent of the total United States wheat acreage (Figure 21–13). Durum is grown mainly in the Northern Great Plains, particularly North Dakota, with small acreages in some of the Western states.

Spikes are compact and laterally compressed and bear grains whose lemmas generally have long, stiff awns. The grain is amber or red in color, kernels are long and pointed, hard in texture and vitreous, higher in protein than any other class. This wheat is used for semolina flour, from which macaroni, spaghetti, and similar products are made.

HARD RED WINTER. Hard red winter wheat, which is fall-seeded, is grown principally in the Central Plains and Northwest. It is the most impor-

TABLE 21-2. Geographic Distribution of Different Wheat Species, with Earliest Evidence of Existence

Species Number (see Table 21-1)	Geographic Distribution	Earliest Evidence
I	Western Iran, Asia Minor, Greece, southern Yugoslavia	Pre-agricultural
II	Eastern Caucasus, Asia Minor, Greece, central Europe	4750 B.C.
III	Western Iran, Syria, northern Palestine, northeastern Turkey, Armenia	Pre-agricultural
IV	India, Central Asia, Iran, Georgia, Armenia, Europe, Mediterranean area	4000 B.C.
V	Central Asia, Iran, Mesopotamia, Turkey, Abyssinia, southern Europe, U.S.	100 B.C.
VI	Dagestan, Georgia, Armenia, northeastern Turkey	No prehistoric remains
VII	Abyssinia, southern Europe	No prehistoric remains
VIII	Abyssinia, Mediterranean area	Seventeenth century
IX	Western Georgia (Soviet)	Twentieth century
X	Worldwide	Neolithic Period (300 to 2300 B.C.)
XI	Central and northwestern India	2500 B.C.
XII	Southwestern Asia, southeastern Europe, U.S.	Neolithic Period (300 to 2300 B.C.)
XIII	Central Europe	Bronze Age
XIV	Western Georgia (Soviet)	Twentieth century

Source: Mangelsdorf, P. C. "Wheat," *Scientific Amer.* (July 1953), pp. 50–59.

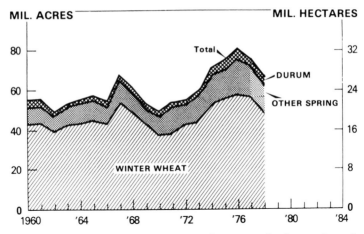

FIGURE 21–11. *Planted acres of winter, durum, and other spring wheats in the United States.* [SOURCE: USDA.]

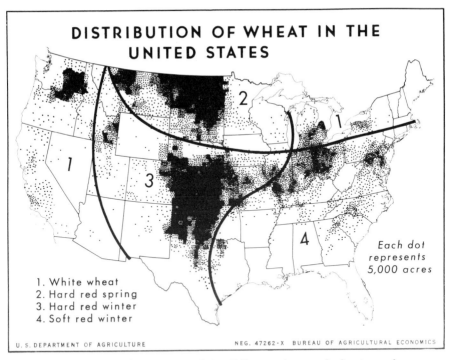

FIGURE 21–12. *The acreages of the different classes of wheat are shown to be sharply concentrated in widely different parts of the United States. The dividing lines are indications only, and are not absolute.* [SOURCE: USDA.]

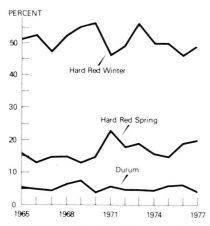

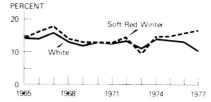

The U.S. produces more classes of wheat in volume than any other country. Of the two bread wheats, Hard Red Winter usually accounts for about half of the total wheat crops, and Hard Red Spring about a fifth. The cookie–pastry soft wheats, Soft Red Winter and White, provide about a fourth, and Durum, the pasta (macaroni and spaghetti) wheat, 5 percent.

FIGURE 21–13. *Wheat production by class as a percentage of total production in the United States.* [SOURCE: USDA-SEA.]

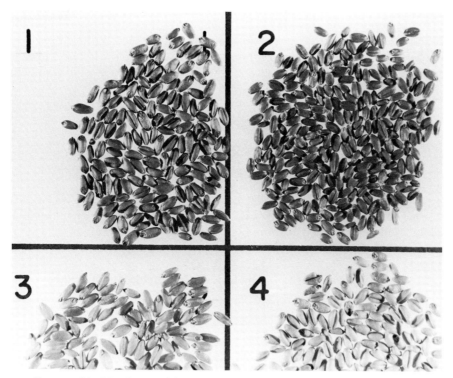

FIGURE 21–14. *Grain of four common commercial classes of wheat: 1 = soft red winter, 2 = hard red winter, 3 = durum, 4 = white.* [SOURCE: Southern Illinois University.]

tant class on the basis of acreage, occupying almost 50 percent of the total U.S. wheat acreage (Figure 21–13).

The grain, which is hard-textured and high in protein, is used for bread. Some of the leading cultivars are 'Lancer,' 'Centurk,' 'Scout,' 'Cheyenne,' 'Triumph,' and 'Eagle.'

SOFT RED WINTER. This wheat is grown almost exclusively east of a line running from Milwaukee to Dallas. Most of the area is subhumid to humid climate, not conducive to the production of wheat with a hard kernel texture. It is fall-seeded and accounts for about 15 percent of the total U.S. wheat acreage (Figure 21–13).

The grain is soft to semihard in texture, depending on the geographic area of production. The flour is suitable for cakes and cookies, pastries, and general purpose use. Some of the leading cultivars are 'Arthur,' 'Arthur 71,' 'Abe,' and 'Timwin.' Fully 75 percent of all soft red winter wheat cultivars were developed at Purdue University.

WHITE WHEAT. White wheat, mostly fall-seeded, is produced in two regions: (1) Western states, principally Washington, Oregon, and Idaho, and (2) Eastern states, mainly Michigan and New York. These wheats are grown on about 10 percent of the total wheat acreage, with a small percentage of that acreage in white club wheat (Figure 21–13).

White winter wheat grain ranges from soft to hard and usually has low protein. The flour is used for pastry and general purposes. White club wheat kernels are small, white or red, and laterally compressed. The grain is soft in texture, and yields flour suitable for cakes, cookies, and breakfast cereals.

Wheat Production Practices

ROTATIONS. Wheat regularly is grown with other crops in humid, subhumid, and irrigated areas, but in semiarid regions often it is grown continuously or alternated with summer fallow.

Definite crop rotations are not a part of the farming practice in the important wheat-producing areas; this is the result of climatic hazards, the lack of an easily established legume, and most important, the lack of a crop that can compete successfully with wheat for the use of land, labor, and capital.

Continuous wheat has not shown declining yields during many years in the Great Plains states, where rainfall ranges from 18 to 25 inches (46–64 cm). However, sometimes it is wise to use a rotation for weed, disease, and insect control.

In the central Wheat–Corn Belt and eastward, a common rotation is corn, oats, wheat, and legume-grass. A variation from this is to follow corn with soybeans instead of the oat crop, so that the wheat cash crop has the benefit of the legume. Sometimes a shorter rotation of wheat, legume, wheat is used. In the spring wheat area of the northern Great Plains it was for many years a common practice to fallow before wheat. A present practice in a considerable part of this area is for wheat to follow wheat or another small grain, or to follow grain sorghum, corn, sugarbeets, or some other cultivated crop. In semiarid regions where the precipitation averages less than 15 inches (38 cm), fallowing before wheat continues to be a common practice.

The benefits of fallowing were demonstrated many years ago in Kansas, Oklahoma, Colorado, and Nebraska. At Fort Hays, Kansas, wheat cultivar comparisons reported for each of 25 years averaged 25.9 bushels/acre (1,740 kg/ha) following fallow, as compared with 22.7 (1,525 kg/ha) on cropped land. At the Oklahoma station, for a nine-year period, wheat yields after fallow averaged 35.8 bushels/acre (2,406 kg/ha), as compared with 33.2 (2,231 kg/ha) on cropped land following spring-seeded oats. At the U.S. Dryland Field Station, Akron, Colorado, the nine-year average after fallow was 23.6 (1,586 kg/ha) as compared with 15.0 (1,008 kg/ha) after corn. At the Nebraska North Platte Station, 16 years of comparisons gave an average of 30 bushels (2,016 kg/ha) after fallow and 17.8 (1,196 kg/ha) after corn. Wheat failed entirely following corn in six years during the period.

Fertilizers

Wheat grown in most regions of the country responds to fertilization; the probable exception is wheat grown on fertile soils without irrigation in semiarid regions.

The rate of nitrogen applied may range from 30 to 120 pounds/acre (34–134 kg/ha), but rates of 30–80 pounds/acre (34–90 kg/ha) are more common. Newer semidwarf, strong-strawed wheat cultivars can utilize higher rates of nitrogen without lodging. For winter wheat, it is a common practice to apply a mixed fertilizer, containing some nitrogen, soon before or during the time of planting. Wheat is top-dressed with the remaining nitrogen in the early spring. Nitrogen applied in the fall is essential for good vegetative growth, and that in the spring for reproductive growth. The importance of available soil nitrogen to the wheat crop has long been recognized. Not only may the acre yield be greatly influenced but also the protein content of the grain. The wheats best for flour milling are high in protein. High availability of nitrogen during the reproductive stages of growth are necessary for a high-protein grain.

Wheat may be fertilized with phosphate, P_2O_5, at 30–50 pounds/acre (34–56 kg/ha) and with potash, K_2O, at 40–50 pounds/acre (45–56 kg/ha) in regions of high rainfall. These normally are applied during seedbed preparation or at the time of planting.

The Seedbed and Seeding

In the Great Plains the method of seedbed preparation is tied closely to moisture conservation. The land is often prepared by chiseling or plowing with a one-way disk plow in semiarid regions. In the humid areas

FIGURE 21-15. *Wheat lodging, as shown in the foreground, can result when the crop is fertilized with high rates of nitrogen fertilizer, which results in a taller or ranker growth. Newer semidwarf, strong-strawed cultivars can utilize higher rates of nitrogen without lodging.* [SOURCE: Southern Illinois University.]

seedbed preparation depends on the place of wheat in rotations. Disks or spring-tooth harrows may be the main implements used when wheat is to follow a row crop, such as corn or soybeans. When a grass-legume cover crop precedes wheat, it must be plowed under three to four weeks prior to seeding wheat.

The time of plowing has been particularly important. Plowing as early in the summer as conditions permit almost invariably has given improved yields. In southwest Kansas early plowing has given yields from 25 to 50 percent greater than when the land was plowed late. Late plowing just prior to seeding wheat may result in a cloddy, loose, weedy seedbed with lower moisture. When early plowing is practiced, clean tillage can result in a well-prepared weed-free seedbed.

Stubble mulching came into prominence as a practice to prevent wind and water erosion and to conserve moisture. In the seriarid regions stubble mulching often increases yield over clean tillage. In humid areas yield reductions often are noted.

In the Corn Belt and eastward much winter wheat is seeded following corn or soybeans without plowing. In some areas, such as western Kentucky, considerable wheat acreage is seeded with ground equipment or from airplane in standing row crops, most commonly corn or soybeans, before the row crops are harvested. Such a method requires no seedbed preparation and allows an earlier start for wheat if moisture conditions are favorable for establishment.

Date and Rate of Seeding

The optimum date for seeding wheat will vary with differences in climatic and soil conditions. With Hessian fly damage a probability with some cultivars, the "fly-free" date may be the determining factor.

Midseason seeding of winter wheat generally is most favorable. Early seeding produces poorly tillered, erect, weak plants with poor root systems subject to Hessian fly and certain diseases. Late seeding results in poor tillering and more chance of winter injury. In the East the winter wheat average date of seeding ranges from mid-September in the Northern states to early November in the South. In the Western states, from Montana to Oklahoma, where Hessian fly is seldom a factor, the date ranges from mid-September to mid-October. Fall moisture often is a determining factor. In the High Plains of Texas and Oklahoma, most acreage is seeded by September, but may be seeded between late August and mid-October. In the Pacific Northwest, winter wheat is seeded in mid-September. Early seeding of spring wheat generally gives best results. The usual practice is to seed as soon as the soil can be worked into a satisfactory seedbed in the spring, which may be as early as March in Colorado and Nebraska to May in the more northern regions.

The optimum seeding rate varies from more than 90 to as little as 20 pounds/acre (101–22 kg/ha) according to the moisture conditions, geographic regions, and wheat type. In the East, where winter types are grown, 90 pounds/acre (101 kg/ha) is a common seeding rate. Durum and other spring wheats in the Northern Great Plains most often are planted at 60 pounds/acre (67 kg/ha). The 60-pound rate also applies to much of the winter and spring wheat in the Central Plains, but as little as 20 pounds/acre (22 kg/ha) may be seeded in the western, semiarid parts of this region. Winter wheats in the Northwest are seeded at about 45 pounds/acre (50 kg/ha), while good results have been obtained under favorable conditions with as little as 30 pounds (34 kg/ha). Spring wheats are seeded at a 60 to 75-pound (67 to 84 kg/ha) rate in that area.

Method of Seeding

In early history wheat was sown broadcast for want of satisfactory seeding machinery. Now almost all wheat is drilled. Drilled wheat usually gives higher germination, more uniform stands, less winter injury, and higher yields. Wheat generally is seeded with a common surface drill or a furrow drill; the latter is more common in the drier regions. The types

FIGURE 21–16. *Most wheat is seeded by drilling in narrow (6-, 7-, or 8-inch or 15-, 18-, or 20-cm) rows. In the drilling operation shown here, winter wheat is being seeded following corn harvest.* [SOURCE: International Harvester Company.]

of surface drills are hoe, shoe, and single or double disk, with the single disk probably most common. The hoe drill should be used only on clean land. Most drill rows are 6, 7, or 8 inches (15, 18, or 20 cm) apart, but wider spacings may be used in dryland conditions.

Pasturing Winter Wheat

The value of wheat as pasture for livestock is often a major consideration. In some areas wheat is sown only for grazing. In other areas wheat is sown for both grazing and as a grain crop with the grain crop being of secondary importance. Young wheat plants have a high protein content and are a highly nutritive pasture, hay, or silage crop. Fall-sown wheat is more frequently grazed than spring-sown wheat. In some areas up to 80 percent of the winter wheat is grazed. In Kansas, for example, about two-thirds of the winter wheat acreage is pastured.

Winter wheat can be pastured in the late fall, winter, and early spring (before jointing stage), apparently with little detrimental effect on grain yield. Researchers at Nebraska concluded that removing excess fall growth through grazing could result in greater tillering and thus even higher yields than wheat not grazed. Pumphrey, however, posed some new questions about grazing the newer semidwarf wheat cultivars with exceptional tillering and yielding abilities.

Harvesting

Practically 100 percent of the United States wheat crop is now combine harvested. The binder or header of past days are no longer used. For direct combining the moisture content of the wheat should be 14 percent or less for safe storage, unless artificial drying facilities are used, in which the harvesting date can be hastened. Kernels are fully developed when the moisture content has dropped to 35 percent, but combining must be delayed until more moisture is lost to avoid kernel damage. If there are considerable green weeds, which could affect the combining operation, the field may be sprayed with a chemical herbicide or desiccant a few days before harvest. An alternative procedure is to cut and windrow and then combine with a pickup attachment on the combine.

Diseases, Insects, and Other Pests

As has been stated, the rapid changeover in wheat cultivars grown in recent years has been partially a result of disease losses. One of the most important objectives in producing new cultivars of wheat has been to obtain cultivars resistant to the important wheat diseases, especially the rusts. Annual losses from disease in the United States is about 14 percent of the planted wheat crop. The most common and destructive diseases of wheat in the United States are rusts, smuts, root rots, powdery mildew,

FIGURE 21–17. *Practically all of the United States wheat crop is combine harvested. The direct combining operation shown here is contrasted to the alternative procedure in some areas of cutting, windrowing, and then combining with a pickup attachment.* [SOURCE: Allis Chalmers Corporation.]

mosiac and other virus disorders, scab, and *Septoria*. Rusts can reduce the number and size of kernels and result in badly shriveled kernels. Smuts cause replacement of the floral parts and grain with a mass of black spores. *Fusarium* and *Helminthosporium* organisms may cause root and crown rots which result in decreased vigor and stand losses. The scab fungus attacks the seedlings and also the heads of plants. Premature drying of the spikelets can result in shrunken, scabby kernels. Powdery mildew may kill portions of leaves and result in stunted or lodged plants. *Septoria* appears as lesions on the leaves or glumes, and heads may be dwarfed and kernels shriveled. Mosaic symptoms appear as mottling and striping of leaves, stunted plants, and possible failure to head. Preventive or control measures include resistant cultivars, fungicidal and hot water seed treatments, use of fungicide sprays on the foliage, crop rotation, and control of weed alternate hosts and volunteer wheat.

Of more than 100 insects that attack wheat, the ones that infest and damage this crop most often are Hessian fly, cereal leaf beetle, grasshopper, wireworm, wheat stem sawfly, armyworm, cutworm, greenbug and other aphids, chinch bug, jointworm, and wheat stem maggot. In addition, the Angoumois grain moth is a problem in stored grain. Insects cause an esti-

FIGURE 21–18. *Wild garlic has become a serious weed pest in wheat fields. Aerial bulblets of wild garlic heads are shown in the foreground. Garlicky wheat is undesirable for milling and can result in sizable discounts.* [SOURCE: Southern Illinois University.]

mated 6 percent damage of the U.S. wheat crop annually. Resistant cultivars are available for Hessian fly and wheat stem sawfly, while insecticides
will control most others mentioned.

Wild garlic as a weed has become a serious pest in recent years. The
problem arose when wheat buyers became more conscious of garlic infestation and began enforcement of discount schedules for garlicky wheat,
which is undesirable for milling. Discounts are based on the number of
garlic bulblets in a 1,000-gram sample. Bulblets are about the same size
as wheat and not easy to separate by cleaning. The best way to avoid the
problem is to obtain a vigorous stand of wheat, and to use herbicides such
as 2,4-D when necessary. Herbicides should be applied when the wheat
is about 4–8 inches (10–20 cm) high, after it is well-tillered but before
jointing.

Utilization

REGIONAL QUALITY LABORATORY. The USDA maintains four regional
quality laboratories cooperating with the several states. These laboratories
determine the suitability of cultivars for specific milling and baking properties. The quality of any kind of wheat cannot be expressed in terms of
a single property, but depends on several milling, baking, processing, and
physical dough characteristics, each important in the production of bread
or pastries. Similarly, the quality of durum wheat depends on certain semolina or macaroni properties. Almost all cultivars grown in the United States
now have satisfactory milling and baking qualities. This situation is the
result of close cooperation between wheat breeders and quality laboratories
which evaluate agronomically promising lines for their utilization potential before release to the farmers.

WHEAT FLOUR. Nearly all the wheat processed in the United States is
made into flour and accompanying by-products. When made into white
flour, 25–30 percent of the total weight of the grain goes into by-products,
such as bran, middlings, and shorts. All these are much higher in protein
content than the grain itself. These wheat by-products now are used mostly
in mixed feeds. Millers and bakers use great care in their selection of
the wheat for milling. *Hard red winter wheat* and *hard red spring wheat*
generally are regarded as making the best flour for high-quality bread.
This flour has unusually high gluten qualities, which means it is very
elastic. Flour from *soft red winter wheat,* produced in the more humid
eastern area, has a somewhat lower percentage of protein than that of
some other wheats and is regarded as better suited for cake and biscuit
flours than for bread making. Flour from *durum wheat* is not suitable
for bread making. It is used especially in the manufacture of macaroni,
spaghetti, and other similar products.

Milling. When flour is to be made from wheat, the grain is first washed,
dried, and scoured to remove fuzz from the surface and to remove foreign
material. The washed wheat then is conditioned or tempered by the addi-

tion of water to toughen the bran. Tempering usually takes four to six hours, with 15–16 percent moisture. The grain then passes through several sets of corrugated rollers, flattening and crushing rather than grinding, in order that the seed coat, the bran, and the germ with its high oil and protein content can be separated from the high-starch endosperm portion. The next step is to separate out as much of the bran as possible. The starch portion again goes through additional sets of rollers for further refining; after each of these treatments more of the bran is separated out. After the bran separation process is completed the starch material is bolted, with the very finest of the starch particles passing through fine cloth sieves. The coarser material again goes through the rollers for further reduction, with further sifting and regrinding to separate out all the flour. The average straight flour yield is 70 to 74 percent. Freshly milled flour is improved in color and baking quality by using chemical bleaching agents. Flour often is enriched to restore or supplement certain vitamins and minerals that are largely lost when by-products such as bran are removed during milling.

Improvement

WHEAT BREEDING OBJECTIVES. The objectives of plant breeders include increased yields, improved disease and insect resistance, greater drought resistance and winterhardiness, stronger straw, better shatter resistance, photoperiod insensitivity, improved quality, and improved seed size, weight, and appearance. Long-term objectives include transfer of characters from other genera, accumulating favorable genes that affect yield components and other characters not simply inherited, and utilization of heterosis, altering the species by chromosome addition, substitution, or amphiploidy. Heyne and Smith have outlined the basic methods used in breeding wheat. Through wheat breeding, average yields have been increased by 70 percent since 1950.

Recommended Wheat Cultivars. The cultivars of wheat seeded on the larger acreages usually are those best adapted to the prevailing environmental conditions. However, new cultivars constantly are being developed by federal, state, and private breeders. The state stations and the USDA test new cultivars and compare them with the old and thus are able to recommend the best cultivars for a given locality.

Semidwarf Wheats. Semidwarf wheats generally show higher yield potential than standard cultivars. This is especially true under conditions of high fertility and moisture levels conducive to high yields. The Maize and Wheat Improvement Center in Mexico, CIMMYT (Centro Internacional de Mejoramiento de Maiz y Trigo), has released several semidwarf wheat cultivars with the primary purpose of increasing yields. More recently, acceptable quality characteristics have been bred into these cultivars.

There have been numerous others developed and released in recent years by the USDA, state experiment stations, and commercial companies. The rapid adoption of semidwarf cultivars by Mexican farmers and the export to other countries of the world represents one of the most sweeping changes in cultivar types in recent times. This new form has straw 5–10 inches (13–25 cm) shorter, resists lodging, and tolerates and responds to high rates of nitrogen fertilizers and rapid irrigation.

The trend toward use of shorter, stiffer-strawed cultivars in the United States is shown by noting a list of available semidwarfs. Some of the semidwarf cultivars used in the United States are soft red winter—'Timwin,' 'Beau,' 'Blueboy,' 'Blueboy II'; hard red spring—'Era,' 'Lark,' 'World Seeds 1809,' 'Bounty 208,' 'Olaf,' 'Bonanza,' and 'Shortana'; hard red winter— 'Sturdy,' 'Santanta,' 'Chanute,' 'Pronto,' and 'Caprock'; and white—'Nugaines,' 'Gaines,' 'Hyslop,' 'Coulee,' 'Luke,' 'Twin,' 'Springfield,' and 'Paha.'

Hybrid Wheat. Ever since hybrid corn was so spectacularly successful, plant breeders have attempted to duplicate these results with other crops. Hybrid corn can be easily produced because it is monoecious, having male and female flowers separate but on the same plant, and is naturally cross-pollinated. Hybridization of small grains has been much more difficult because such flowers are normally self-pollinated and perfect, with both male and female organs within the same florescence. To achieve effective hybridization of small grains, plant breeders had to have available a male sterility characteristic, to make possible mass cross-pollination. Presently, both genetic and cytoplasmic male sterile lines are available. In addition to the male-sterile characteristic, a restorer gene must be carried so that the subsequent plants grown from hybrid seed will be fertile and thus produce a crop. Farmers using hybrid wheats need to buy new seed every year because, like hybrid corn, the second generation seed will not breed true, resulting in plant variability and lower yields.

Hybrid cultivars did not become available to growers until after 1974, and the number of hybrids available commercially is still limited. Hybrids have the potential to increase yields by 20–30 percent. Hybrid "seed wheat" is produced by growers who plant alternate swaths of male and female parent wheats. The hybrid "seed" is produced in female swaths, which were male sterile, after having been cross-pollinated with male plants.

Rye

Until early in the 1900s, world production of rye amounted to nearly one-half that of wheat. The USSR at that time produced more than half of the world's crop and Germany produced almost a fourth. About 96 percent of the world's rye was grown and consumed in Europe. In the more recent years, rye is less favored as a food crop than formerly because of a general preference for breads from wheat.

Origin and Classification

Compared with wheat, rye is a relatively new crop. The earliest cultivation of rye appears to have been in western Asia and southern Russia possibly about 4000 B.C. Rye belongs to the tribe *Hordeae* and to the genus *Secale*. The genus and species of cultivated rye is *Secale cereale*.

Adaptation-Distribution

Rye can stand all kinds of adverse environmental conditions except heat. It is more winter-hardy than any of the small grains and is better suited to soils low in fertility, and especially to sandy soils.

The leading countries in world acreage, as well as production, are the USSR, with more than 50 percent, Poland, West Germany, East Germany, and Turkey (Figure 21–19). Rye occupies about 16 million hectares in the world. Approximately 3 million acres (1.2 million ha) of rye are seeded annually in the United States, but less than 1 million (0.4 million ha) are harvested for grain. In the North Central and Western states rye may be grown for grain but also is used for hay, for pasture, as a companion crop, or sometimes as a smother crop. In the Eastern states, rye is grown for pasture, or green manure. The states leading in rye acreage are North Dakota, Georgia, South Dakota, Minnesota, and Nebraska (Figure 21–20). Geographic distribution is shown in Figure 21–21. The average yield of the rye harvested for grain is about 20 bushels/acre (1,254 kg/ha).

The Plant

Rye is a tall-growing winter annual grass; the height often exceeds 5 feet (1.5 m), taller than other small grains. The root system is very exten-

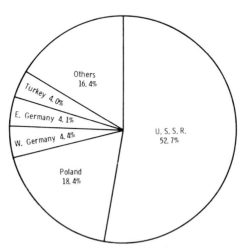

FIGURE 21–19. *Leading countries in world rye area (acreage). The USSR plants over one half of the total.* [SOURCE: Southern Illinois University.]

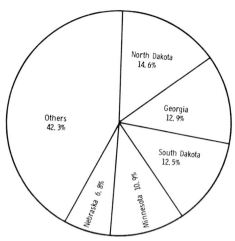

FIGURE 21-20. *Leading states in United States rye area (acreage).* [SOURCE: Southern Illinois University.]

sive, more so than other grains. The inflorescence is a spike, and spikes are nodding or drooping. Beards or awns are short and irregular. In contrast to the other small grains, rye normally is cross-pollinated. The caryopsis is naked, being separated from the lemma and palea when combined.

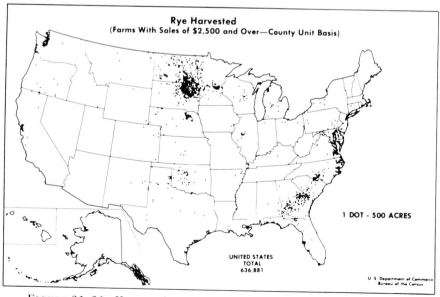

FIGURE 21-21. *Harvested acreage of rye in the United States, shown geographically.* [SOURCE: U.S. Department of Commerce.]

FIGURE 21–22. *An area of winter rye in the field. These tall-growing plants produce spike inflorescences which are nodding or drooping.* [SOURCE: Southern Illinois University.]

Uses

In the United States about one-half of the grain crop is used for livestock feed. About 20 percent is used for making alcoholic beverages and for food, and the remainder is used for seed.

Rye is the best suited of any of the cereals for general use as a pasture crop. It provides considerable fall as well as early spring grazing. It also is valuable as a winter cover and green manure, especially on sandy soils.

Cultivars

Common rye includes both spring and winter cultivars, though spring cultivars seldom are encountered. 'Rosen' was one of the first improved cultivars of rye to have general distribution; it is well adapted to the Corn Belt and eastward. 'Balbo' is another early cultivar which is still used. 'Dakold,' which originated at the North Dakota station through a natural selection, is a very winter-hardy cultivar in North Dakota and that area. Other cultivars include 'Emerald,' released in Minnesota; 'Pierre,' in South Dakota; and 'Raritan' in New Jersey. Spring rye is rarely grown. However, two cultivars are grown in the United States, 'Prolific' and 'Merced.' At one time there was some interest in 'Tetra-Petkus,' a tetraploid cultivar introduced from Germany. It produces vigorous growth and large kernels. The kernels are unsuited for cleaning and processing equipment.

Winter rye is used as a winter pasture crop throughout much of the South. The 'Abruzzi' cultivar is well adapted in the South.

The newest cultivars of winter rye, first described in 1970–1976, include 'Athens Abruzzi,' a forage type from Georgia; 'Coloma' from Wisconsin; 'Kodiak' and 'Puma' from Canada; 'Hiwassee,' a forage type from Tennesee; 'Okema,' from Oklahoma; and 'Rymin' from Minnesota.

Production Practices

The general procedures in growing winter rye are similar to those common in the production of winter wheat. Rye may not be fertilized, although it has responded to fertilizers on the less fertile soils.

The best time to seed rye depends on the location and use to be made of the crop. It may be seeded two or three weeks later than the time of seeding winter wheat. When grown for pasture or as a green manure cover it may be seeded as early as August. The period through which seedings can be made with good results is much longer than for other small grains.

Rye is seeded at the acre rate of 56–70 pounds/acre (63–78 kg/ha) in the Western states, 78–112 pounds/acre (110 to 125 kg/ha) in the Northeastern states, and 70–84 pounds/acre (78 to 94 kg/ha) in the South.

VOLUNTEER RYE. It generally is regarded as inadvisable to grow winter rye for grain on farms where winter wheat is produced; it is almost impossible to keep the rye from becoming mixed into the wheat. When winter rye has been grown on a given area it is almost certain that volunteer plants will appear in following crops.

Insects and Diseases

Insects attacking rye are the same as for other small grains; annual loss is small. Diseases of rye cause an annual loss to the crop of about 3 percent. Principal losses are caused by ergot and smuts. Ergot is a fungus that infects the flowers at blooming time, and replaces the normal grain with ergot bodies, masses of sclerotia. These bodies are poisonous to livestock and humans.

Triticale

Triticale is a hybrid grain produced by crossing wheat and rye, and is said to be the first man-made cereal. The development of a fertile triticale hybrid grain is said to be comparable to the theoretical development of a fertile mule. The name "triticale" is derived from the combination of scientific names of the two grains. The first field acreages were in southern Manitoba in the 1960s.

Triticale has the potential as a feed grain and in the milling industry. It combines some of the milling and baking qualities of wheat with some of the nutritional characteristics of rye (higher lysine, higher protein, and better amino acid balance than wheat). Under some conditions, it outyields either wheat or rye.

Although present triticale cultivars offer some promise, several characteristics need to be improved before this grain has a marked impact on American and world agriculture. Most cultivars are tall and lodging is a problem; breeders are presently working to develop semidwarf types. Other areas of concern are daylength sensitivity, ergot susceptibility, and deficiency of gluten quantity and quality, resulting in a weak dough structure.

REFERENCES AND SUGGESTED READINGS

1. BRIGGLE, L. W., and L. P. REITZ. *Wheat in the Eastern United States,* Agro. Info. Bul. 250 (1962).
2. DENNIS, R. E., *et al.* Growing Wheat in Arizona, Ariz. Coop. Ext. Serv. Bull. A32 (1976).
3. ELDER, W. C. *Grazing Characteristics and Clipping Responses of Small Grains,* Okla. Agr. Exp. Sta. Tech. Bul. 567 (1960).
4. HEHN, E. R., and M. A. BARMORE. "Breeding Wheat for Quality," In *Advances in Agronomy,* vol. 17 (New York: Academic Press, 1965). ed. A. G. Norman.
5. HEYNE, E. G., and G. S. SMITH. "Wheat Breeding," *Wheat and Wheat Improvement,* Monograph 13 (Madison, Wisc.: American Society of Agronomy, 1967).
6. JOHNSON, V. A., and J. W. SCHMIDT. "Hybrid Wheat," In *Advances in Agronomy,* vol. 20 (New York: Academic Press, 1968). ed. A. G. Norman.
7. KNAKE, E. L., and M. D. MCGLAMERY. *Wild Garlic,* Ill. Coop. Ext. Serv. Circ. 1109 (1975).
8. MCCALLA, T. M., and T. J. ARNY. "Stubble Mulch Farming," In *Advances in Agronomy,* vol. 13 (New York: Academic Press, 1961). ed. A. G. Norman.
9. MANGELSDORF, P. C. "Wheat," *Scientific Amer.* (July 1953). By permission. pp. 50–59.
10. MARTIN, J. H., W. H. LEONARD, and D. L. STAMP. *Principles of Field Crop Production,* 3rd ed. (New York: Macmillan, 1976).
11. MATZ, S. A. *Cereal Science* (Westport, Conn.: AVI, 1969).
12. MORRIS, R., and E. R. SEARS. "The Cytogenetics of Wheat and Its Relatives," *Wheat and Wheat Improvement,* Monograph 13 (Madison, Wisc.: American Society of Agronomy, 1967).
13. PETERSON, R. F. *Wheat: Botany, Cultivation, and Utilization* (New York: Wiley–Interscience, 1965).
14. PUMPHREY, F. V. "Semidwarf Winter Wheat Response to Early Spring Clipping and Grazing," *Agron. Jour.,* 62:641 (1970).
15. QUALSET, C. O. and W. W. STANLEY. *A Comparison of Small Grains for Winter Grazing,* Tenn. Agr. Exp. Sta. Bull. 438 (1968).
16. REITZ, L. P. "Wheat in the United States," USDA-ARS Agric. Info. Bull. 386 (1976).
17. SCHLEHUBER, A. M., and B. B. TUCKER. "Culture of Wheat," *Wheat and Wheat Improvement,* Monograph 13 (Madison, Wisc.: American Society of Agronomy, 1967).

18. SCOVILLE, J., and J. A. HODGES. *Practices and Costs on Wheat Farms in Western Kansas,* Kan. Agr. Exp. Sta. Circ. 268 (1950).
19. Sixth Hard Red Winter Conference, Stillwater, Oklahoma (1950).
20. SOMSEN, H. W., and K. L. OPPENLANDER. *Hessian Fly Biotype Distribution, Resistant Wheat Varieties and Control Practices in Hard Red Winter Wheat,* USDA-ARS-NC-34 (1975).
21. TSEN, C. C. (Ed.). *Triticale: First Man-Made Cereal* (St. Paul: American Association of Cereal Chemists, 1974).
22. USDA Agr. Stat. (1977).
23. USDA-ARS. *Losses in Agriculture,* Agr. Handbook 291 (1965).
24. USDA-Ext. Serv. *New Crop Cultivars, 1970–1976,* ESC-584, No. 13 (1977).

CHAPTER 22

OATS, BARLEY, AND RICE

Among the other discoveries of interest was a small, unpainted jar found 1 meter below the red-earth stratum. As I was taking it out my left palm became black. I spread the contents on a newspaper. Here were some seed preserved by charring. . . . Botanical investigation disclosed the fact that this barley is of the 6-rowed variety and this, so far as it is known here, is the first actually brought to light in Mesopotamia. . . . The 6-rowed type is the characteristic prehistoric barley. . . . that was taken along by the Anglo-Saxons on their migration from their original homes to the British Isles, . . . In view of the discovery of the 6-rowed barley at Kish the conclusion is now warranted that this cereal, so important in the development of agriculture, was first brought into cultivation at a prehistoric date in Mesopotamia, where the wild species also occurred, and that the cultivated species was diffused from that center to all other countries of the Near East, Egypt and Europe.

HENRY FIELD

The "Small Grains" Defined

WHEN we speak of the "small grain" crops we usually have in mind the cereals: wheat, oats, barley, and rye. These are all grasses grown primarily for their edible seed, classifying them as cereals. Corn and the sorghums also are of the grass family and are cereals, but they are not included in the small grain group. In some parts of the world the millets are grown primarily for their edible seed and therefore are cereals; in the United States they are grown as forage. Buckwheat and soybeans also are grown for the seed, but they are not small grains. Flax, although not a grass and not a cereal, is sometimes included in the small grain group because of its general production characteristics. Rice is a grass grown for its edible seed and therefore it is a cereal, but it is not usually thought of in the small grain group, perhaps because its adaptation and production requirements are so different.

The relative importance of the different small grains in the United States since 1949, when considered on an acre basis, is shown in Table 22–1, together with the acreage of corn and sorghum harvested for seed and of flax, rice, buckwheat, and soybeans.

New Small-Grain Cultivars

The most outstanding feature of small grain production in recent years is the extreme rapidity with which new cultivars have replaced those previ-

TABLE 22–1. Harvested Acreage of the Small Grains, Oats, Wheat, Barley, and Rye. Also (for comparison) Flax, Corn for Grain, Grain Sorghum, Rice, Buckwheat, and Soybeans (1,000 acres)

Crop	1949	1954	1960	1969	1976
Winter wheat	54,414	39,156	39,996	36,696	49,535
Spring wheat	21,496	15,123	11,900	10,859	21,289
All wheat	84,931	54,279	51,896	47,555	70,824
Oats	35,324	42,291	26,588	18,003	12,392
Barley	9,190	13,183	13,856	9,388	8,417
Rye	1,418	1,717	1,688	1,334	804
Flax	4,813	5,589	3,342	2,704	954
Corn (for grain)	83,366	80,269	80,678*	63,188*	71,085
Grain sorghum	6,325	11,218	15,601†	17,040†	14,877
Rice	1,857	2,405	1,595	2,128	2,501
Buckwheat	269	149	48	‡	‡
Soybeans	10,138	16,971	23,655	40,857	49,443

* All corn.
† All sorghum.
‡ Estimates discontinued.

ously grown. This applies to all the different small grains. In the past the acreage planted to a new cultivar increased only gradually through a considerable time.

But with the steadily increasing number of strains of disease organisms, such as the rusts, and with an increased knowledge of these different organisms and of techniques in breeding for disease resistance, newly released cultivars quickly become those most extensively grown—only to drop out with equal rapidity as new disease strains to which they are susceptible built up. The methods of obtaining new small grain cultivars has gone through several important changes: (1) the introduction of cultivars from other parts of the world, (2) mass selection, (3) the isolation of superior pure lines, and (4) the hybridization of different cultivars and strains.

Great progress has been made in recent years in the improvement of small grains by combining characteristics of different cultivars, hybridizing to get new combinations of desired characters such as disease resistance, resistance to lodging or to winter-killing, and increased productivity. Parent cultivars to be used are selected carefully on the basis of their known characters. By crossing such cultivars it is possible to combine in one strain the desirable characters. It was by this method of breeding that the many superior new small grain cultivars of recent years were produced.

Small-Grain Production Practices

The different small grains require somewhat similar seedbeds and, in general, the method of preparing the seedbed and seeding is similar. However, whereas land to be seeded to wheat and barley usually is plowed, a large part of the oat seedbed is not plowed. This is especially true when the oat crop follows corn. In areas of relatively low rainfall a great deal of wheat is seeded on land fallowed for a season as a means of increasing the available soil water.

Spring grains usually are seeded just as early in the spring as the soil and seasonal conditions permit; early seeding in almost all cases gives larger yield returns. In seeding winter grain, the ideal is to permit a vigorous growth and good fall establishment without making a rank growth that becomes "soft," when there is increased danger of winter-kill. On the other hand, it is desirable to delay seeding of susceptible cultivars of winter wheat sufficiently long to prevent the Hessian fly from depositing its eggs on the young wheat plants; the "fly-free date" is determined and reported by the different states throughout the winter wheat area.

Most small grains are seeded with a grain drill, except that throughout the Corn–Oats Belt some of the oat acreage is broadcast-seeded after disking or harrowing and without plowing. This method of seeding is favorable to getting the seed into the ground as early as possible. The small grain crop usually is permitted to stand until the moisture content is down to a favorable level for combining direct. Some of the small grain crop, however, is windrowed before the grain is dry enough to combine or store, the pickup combine being used to complete the harvest as soon as the crop is cured.

Weather, Insects, and Diseases

The successful production of small grain crops, like other agricultural production, depends on the success with which the grower meets and overcomes a number of hazards—weather, insects, and plant diseases.

WEATHER. The more or less limited rainfall in the Great Plains area, where such a large proportion of the land is given over to small grains, has become increasingly important as more and more land has gone into small grain production in this area. An additional weather hazard is freezing temperatures. Spring wheat in western Canada and in parts of our Northwest is susceptible to damage during much of the growing season. When spring wheat matures late in the northern areas, there is always the danger that the yield and quality will be injured by early fall frosts.

There are both winter and spring cultivars of the different small grains, with great differences in the relative winter hardiness of the different crops and of the different cultivars within each. In general, winter rye is the most hardy, followed in order by winter wheat, winter barley, and winter oats. There are four major causes of winter-killing: heaving, smothering, physiological drought, and the direct effect of low temperatures.

Heaving of the soil lifts plants so that the roots are broken and exposed to the air by the process of alternate freezing and thawing. The soil freezes down rather deeply, expanding as it does so. Thawing then starts at the top, and the soil releases its hold on the roots of the plant and settles as the ice melts out of it. In subsequent cold periods the soil starts freezing at the surface and expands as it freezes downward, thus pulling on the roots, which are held by the frozen layer of soil below the sod layer. Poorly drained, heavy soils are especially liable to heave, but heaving may occur also on fairly well-drained soils. On heavy, poorly drained soils most of the loss of winter grain is the result of heaving.

Smothering is believed to be the cause of winter-killing when an ice sheet forms on the surface of the soil. Respiration cannot proceed without oxygen, and the layer of ice prevents the entrance of oxygen.

Physiological drought is the condition resulting when a plant cannot get water from the soil. A frozen soil may be physiologically dry. The transpiration of moisture in excess of the plant's ability to get it from the soil results in the death of the plant. It has been suggested that the reason for the more hardy types of small grains having narrow leaves is that it lessens transpiration and the danger of physiological drought. Observations have shown that grain exposed to the wind is more likely to be killed than that in protected places.

The direct effect of low temperatures is a mechanical injury caused by the formation of ice crystals within plant tissues. Planting the more winter-hardy cultivars is the most certain method of lessening loss from winter-killing. The more hardy cultivars have a comparatively small leaf area in the fall and a recumbent habit of growth. An erect growth habit indicates lack of winter hardiness. Winter-hardy cultivars usually show a less rapid rate of growth in the field in the fall than nonhardy cultivars. Great differences in relative winter hardiness have been shown for cultivars in each of the small grains.

INSECTS AND DISEASES. The greenbug, or aphid, is a threat to wheat and oats, especially in the Southwest. Temperature and moisture conditions play an important part in the damage caused by this insect in the growth and increase of parasites that feed on it. A cold, wet spring is injurious to the bug and favorable to the development of parasites. The Hessian fly, which destroys or damages winter wheat, is commonly found from Maryland to western Kansas. The chinch bug always is a threat to small grains, especially in the central South from central Ohio west to central Colorado. The grasshopper, which destroys or damages all grains and grasses, is found practically everywhere in the temperate zone in some seasons.

Of the plant diseases affecting small grains, the rusts are particularly damaging to oats and wheat and are found in practically all parts of North America where these grains are grown.

The diseases and insects affecting the small grain crops are further considered with the different crops.

Oats

The oats crop ranks fifth among the cereals on a world basis in acreage and production, being exceeded by wheat, rice, corn, and barley. In the United States oats rank third in acreage and production, behind corn and wheat, but is second only to corn as a feed crop, with most of the crop consumed on farms where produced.

Origin and Classification

As for most cultivated plants, the time when and place where oats became domesticated are not known. Cultivated oats apparently were unknown to the ancient Egyptians, Chinese, and Hebrews; many other plants were domesticated and used extensively before oats were so used. Early writers indicate that oats were evident from 900 to 500 B.C., but they grew as a weed in other grains. At some time in history, they began to be grown in a somewhat pure stand. In the first century A.D., the Roman historian Pliny wrote that the Germanic peoples ate oats as a porridge. This crop appears to have spread from Western Europe to other parts of the world. Shortly after 1600 the grain was brought to North America. Oats were first grown off the Massachusetts coast on one of the Elizabeth Islands, and on the continent in 1611 by the Jamestown colonists.

Most of the oat species known today were first described by Linnaeus, the Swedish botanist, about 1750. The bulk of the world crop is the common white oat, *Avena sativa,* believed to have been developed from the wild oat, *A. fatua.* The red oat, *A. byzantina,* is grown mainly in regions too warm for the best growth of the common oat type. This was derived from the wild red oat, *A. sterilis.* Closely associated with the common white oat is the side oat, *A. orientalis.* In close relationship to the common red oat of the South, and of South America, Australia, and the Mediterranean countries, is the species *A. sterilis algeriensis.*

The Plant

The oat plant normally makes a growth of 2–5 feet (0.6–1.5 m) with three to five or more hollow culms. The fibrous roots usually penetrate to a depth of 3–5 feet (0.9–1.5 m). The leaves average about 10 inches (25 cm) in length, and have prominent, papery, toothed ligules but no auricles. The flowers are borne in much-branched panicles, either spreading or one-sided; most cultivars are spreading types. Oats are the only small grain with a panicle instead of a spike. The flowers and resulting seed are produced on short small branches in spikelets, with usually 20–120 spikelets per panicle. Each spikelet contains 2–3 florets, except in the hull-less cultivars, which contain 4–8 florets. The kernels of the common cultivars are tightly enclosed within the lemma and palea; when the grain is threshed the kernels remain enclosed. The mature kernel constitutes 65–75 percent of the grain by weight.

Adaptation and Distribution

This crop is best suited to cool, moist regions. For best development it requires more moisture than any of the other small grains. Hot, dry weather when the grain is developing often results in poor filling of the grain and low yield. Hot, humid weather, on the other hand, favors the development of disease organisms to which oats are particularly susceptible.

Winter oats are less cold-tolerant than winter cultivars of wheat, barley, or rye, but relatively winter-hardy cultivars are available.

For best production, soils should be well-drained and reasonably fertile, but oats tolerate a fairly wide range of soil conditions. This crop is less sensitive to soil conditions than wheat or barley but more sensitive than rye.

Oat production is general throughout much of the temperate zones. Production is concentrated, however, in certain areas within these zones. World oat acreage is dominated by the USSR, with nearly 40 percent of the total, then the United States, Canada, Poland, China, and West Germany, in that order as to both acreage and production (Figure 22–1).

The United States crop is dominated by the North Central states, with well over 60 percent of the acreage in five states: South Dakota, Minnesota, Iowa, North Dakota, and Wisconsin (Figure 22–2). Geographic distribution is shown in Figure 22–3. The U.S. average yield is about 45 bushels/acre (1,613 kg/ha).

Uses

A greater proportion of the oat crop is fed directly to livestock than any other cereal. About 60 percent is used on farms where produced, and only

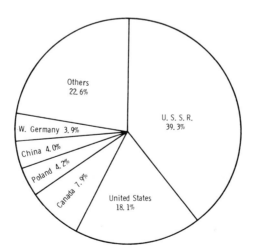

FIGURE 22–1. *Leading countries in world area (acreage) of oats. The USSR is by far the world leader.* [SOURCE: Southern Illinois University.]

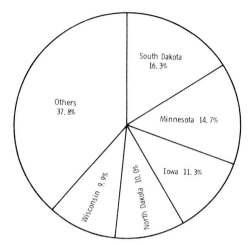

FIGURE 22-2. *Leading states in United States oat area (acreage). Well over 60 percent of the total acreage is located in five states.* [SOURCE: Southern Illinois University.]

about 40 percent sold off the farm. The oat grain is particularly valuable as feed for horses, dairy cows, poultry, and young breeding animals of all kinds. It is high in protein, fat, vitamin B_1, and in such minerals as phosphorus and iron.

Throughout the Corn–Oats Belt the oat crop is considered better suited

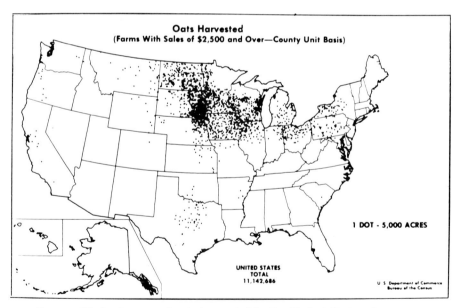

FIGURE 22-3. *Acreage of oats harvested in the United States, shown geographically.* [SOURCE: U.S. Department of Commerce.]

to the prevailing rotations and for farm use as a feed than any of the other small grains. In the South, winter oats are pastured to a considerable extent, and in the North the high acre production from animals grazed on this crop is well recognized. This practice also favors the successful establishment of legumes and grasses seeded in oats. Oats may be cut when in the milk or late dough stage for silage or hay. Oat hay is both nutritious and palatable, as well as high in acre production.

A small percentage of the oat crop in the United States is used in making breakfast foods. Medium quality oats will turn out about 42 pounds (19.1 kg) of oatmeal or rolled oats from 100 pounds (45.4 kg) of the grain in the hull. In Great Britain, and especially in Scotland, oats as a food is highly prized. Milled oats also are used in different forms of breakfast foods.

Rotations

Through a long period, probably the most common rotation in the Corn–Oats Belt was corn, one to two years, oats seeded to clover, alfalfa, and/ or grass one year, with the legume and grass left down for one to two years. This rotation frequently was varied to include a year of soybeans or of winter wheat. Farther South, oats may be rotated with lespedeza or sweetclover. In the South, winter oats may be used as a cover crop following soybeans, cotton, or corn. With the increased availability of disease-resistant cultivars, there has been a significant increase in the acreage of oats in the South, particularly for use as winter pasture.

Fertilizer Use

In the past oats in the Corn–Oats Belt were not heavily fertilized, if at all. More recently there has been an increased use of nitrogen, and in some cases complete fertilizers. It has become increasingly apparent that lack of available nitrogen to the young oat plant in the spring before the soil temperature is high enough to bring about soil nitrification has greatly retarded development.

In some areas it has been the rule to apply commercial fertilizer to other crops in the rotation rather than oats. In areas where legumes are established by seeding in the oats, an application of 200–250 pounds/acre (224–280 kg/ha) of superphosphate is recommended generally. In the Northeast and Mid-Atlantic states, the application of a complete or mixed fertilizer seems to be the general rule. Care should be exercised not to so stimulate the oats growth as to result in lodging, with a consequent low yield of grain and the loss of clover-grass seedings. More than 50 pounds of nitrogen per acre (56 kg/ha) may result in excessive lodging.

Seedbed Preparation and Seeding

In the Corn Belt, where oats most frequently follows corn, the seedbed usually is prepared by disking before seeding, and may be field cultivated.

When oats follow a sod crop or if there is heavy weed growth, plowing is necessary. Oats generally are seeded with a grain drill, but sometimes broadcast seeders are used for sowing oats on disked corn land in the spring. In such instances, the land may be disked again and harvested to cover the oat seed and to smooth the surface.

Spring oats should be seeded as early in the spring as the land can be prepared, generally by March 15 in most spring oats areas. In areas where winter oats are adapted, it is a general rule to seed three to four weeks before the first killing frost is expected.

The average seeding rate is about 70 pounds/acre (78 kg/ha), but varies from 100 or more pounds (112 kg/ha) under irrigated conditions to as little as 40–50 pounds/acre (45–56 kg/ha) under dryland conditions.

Harvesting

From 90 percent to less than 40 percent of the oats planted are harvested for grain, depending on the geographic region. Most of the remainder is grazed, or cut for silage or hay. Winter oats often are harvested for grain in May or June, and spring oats in July or August. A large percentage is harvested by combining. If there is a large amount of green material in the field, and to avoid lodging and shattering losses, oats may be first windrowed, cured for 2–3 days, and then combined with a pickup attachment. In livestock areas most of the oat straw is baled to provide bedding. It has been shown that when small grain is combined the chance for grass-legume seedings made in the oats is greatly improved if the oat straw and stubble are removed soon after harvest.

Oats utilized for silage should be chopped when the grain is in the milk stage, and for hay oats should be cut in the early dough stage.

Insects and Diseases

Insects cause an annual loss to this crop of less than 5 percent. Oats are less subject to insect damage than either wheat or barley. The Hessian fly does not bother oats, and the chinch bug generally prefers other grains. Insects most damaging include greenbugs, leafhoppers, billbugs, grasshoppers, crickets, and aphids, which transmit viruses that cause diseases.

The annual loss to oats caused by diseases is about 20 percent. The most serious diseases are loose smut, covered smut, stem rust, crown rust, Victoria blight, *Helminthosporium* leaf blotch, *Septoria* leaf blotch, halo blight, and mosaic diseases.

Diseases and Crop Improvement

The oat plant is particularly susceptible to disease organisms favored by the mild, humid environment where most of the oat acreage is found. The economic success of oat production is intimately tied with the availability of cultivars resistant to the more virulent of these diseases.

Through many years new oat cultivars were from European introductions. Then followed the pure-line method of breeding. Outstanding progress did not result, however, until an increased knowledge of genetics and plant pathology stimulated the crossing of oat cultivars as a means of transmitting resistance to rust, smut, and other diseases. The principal protection against losses from rusts is growing resistant cultivars. Although oat smut can be controlled by fungicidal seed treatment, new cultivars released in recent years are practically immune to loss from the smuts.

There have been two very rapid changeovers in the cultivars of oats generally grown in relatively recent years. The oat cultivars grown in the important North Central region during the period shortly preceding 1936 were mainly pure-line selections from 'Kherson' and 'Sixty Days,' such as 'Richland,' 'Iogold,' 'State Pride,' and 'Gopher.' Most of the Southern oats acreage was devoted to 'Red Rustproof,' 'Fulghum,' 'Lee,' and other similar cultivars. All these cultivars, in general, were weak-strawed and susceptible to the common races of crown rust, stem rust, loose smut, and covered smut. Yields, acreage, production, and quality were declining rapidly.

These new cultivars had 'Victoria' as one of the parents. A new oat disease, known as *Helminthosporium,* or Victoria blight, made its appearance, however, in 1944 and had become widespread and very destructive by 1946. All the Victoria-derivative cultivars were susceptible to this destructive disease. Fortunately, a new group of selections from crosses with 'Bond' as one of the parents was ready for naming and distribution at this critical time. 'Bond' was a cultivar introduced from Australia by the USDA. It was said to be the stiffest-strawed cultivar ever introduced, to have an even better type of crown rust resistance than that of 'Victoria,' and to be unsusceptible to *Helminthosporium. Helminthosporium*-susceptible cultivars practically disappeared in two years, being replaced by the 'Bond' derivatives.

NEW RUST RACES. With the large acreage planted to the 'Bond'-derived cultivars, races of rust that previously had been of no economic consequence multiplied rapidly, especially race 202 (formerly known as race 45) of crown rust and race 7 of stem rust. In certain years the 'Bond'-derived cultivars were greatly damaged by one or both of these races. New series of crosses have been made with parent cultivars resistant to these strains. By 1956 some 89 different races of brown rust had been recognized and classified.

It seems evident from the very nature of rusts attacking oats that breeding for crown and stem rust resistance will continue to be a major problem of the oat breeder. This is true to a lesser extent with the smuts. Also, much work remains to be done with resistance to other diseases.

Multilines of oats have been designed to break the chain of the crown rust disease. Each multiline consists of genetically different lines of oats mechanically mixed. Component lines have the same agronomic characteristics, but each line carries a different crown rust-resistance gene. Because epidemics of crown rust cannot develop as rapidly in the multiline as in pure-line cultivars, multilines should have a "longer" life.

Oat breeding programs which have resulted in development and release of the newer cultivars during the period 1970–1976 have been most concerned with these areas of improvement: (1) straw strength and lodging resistance; (2) test weight; (3) resistance to rusts, smuts, and other diseases; and (4) improved cold tolerance of winter cultivars. In the South, marked progress has been made in developing superior winter cultivars. With the injury causes by *Helminthosporium* disease and crown rust, selections from the old 'Red Rustproof' cultivars, which at one time had largely disappeared, are again being extensively grown. This type of oat has persisted for over 80 years in some areas of the South because of its great vigor and a degree of tolerance to rust and other diseases not found in other types.

Barley

Barley is an extensively grown cereal, with worldwide distribution and adaptation. It is the fourth most important grain in both the World and the United States as to acreage and production. It appears to have been cultivated long before some of the other cereals were known.

Origin and History

Barley is considered by some to be the most ancient cultivated grain. Excavations in Egypt and other countries have disclosed that barley was cultivated 5,000 to 10,000 years ago. The crop was grown by the ancient Babylonians, Chinese, and by the Swiss Lake Dwellers of the Stone Age. Apparently the 6-rowed types were cultivated earlier than 2-rowed types.

The early colonists first brought this grain to North America; it was grown off the coast of Massachusetts in 1602, and in Virginia in 1611. Early settlers carried it west. Other barley types were introduced into this country by the Dutch and Spanish.

Adaptation and Distribution

Barley is the most widely distributed of the cereals. It has been suggested that it is among the more widely distributed of the cultivated crops. It is grown near the Arctic Circle and also is important to within a few degrees of the Equator. Barley is one of the most dependable crops where drought, frost, and alkali are encountered. The crop is better suited to hot, dry conditions than to hot, humid areas. High temperatures and humidity favor diseases to which barley is susceptible, such as rust, mildew, and scab. Some of the spring types mature in as little as 60–70 days. The winter types are not particularly winter-hardy, with spring types being grown in the North.

The USSR dominates world barley acreage, with almost 40 percent of

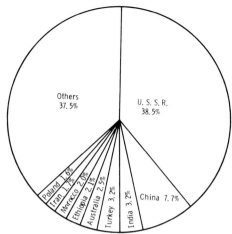

FIGURE 22–4. *Leading countries in world barley area (acreage). The USSR dominates world barley acreage.* [SOURCE: Southern Illinois University.]

the total (Figure 22–4). Other countries with appreciable acreage are China, India, Turkey, Australia, Ethiopia, Morocco, Iran, and Poland. In the United States, the leading states in order of greatest acreage, are North Dakota, Montana, California, Minnesota, Idaho, and South Dakota (Figure 22–5). These six states account for 75 percent of this country's total acreage. Geographic distribution is shown in Figure 22–6.

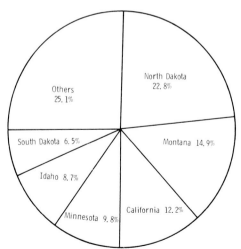

FIGURE 22–5. *Leading states of the United States in barley area (acreage). Three states plant one half of the total acreage.* [SOURCE: Southern Illinois University.]

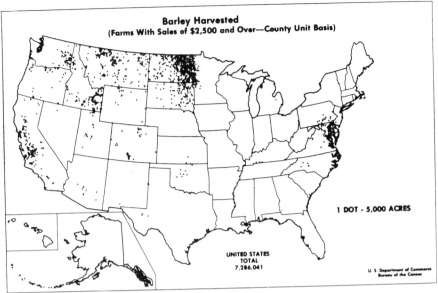

FIGURE 22-6. *Acreage of barley harvested in the United States, shown geographically.* [SOURCE: U.S. Department of Commerce.]

The Plant

Many of the vegetative features of barley are similar to those of wheat and rye. The ligule is short and inconspicuous, but auricles are very large and partly or entirely clasp the stem. The inflorescence is a spike, with a much-compressed zig-zag rachis. Three spikelets are attached at each node or joint of the rachis. In 6-rowed cultivars, all of these are fertile, while in 2-rowed types only the central spikelet is fertile. Each spikelet contains a single floret which produces one grain or caryopsis. The lemma and palea are firmly attached to the caryopsis in most barley types, and remain attached after combining. The lemma has a long, stiff awn or hood.

Types and Classification

TYPES. The barleys are divided roughly into spring barleys and winter barleys, and into hulled and hull-less types. Another division is on the basis of the number of rows of kernels: 6-rowed and 2-rowed. In addition, an important designation has been made according to whether the awns are smooth or barbed (rough).

The more important area of spring barley production in the United States includes the Dakotas, Minnesota, and Montana. In this area and in Canada the common 6-rowed barleys predominate and give the larger yield. There are limited areas where 2-rowed barleys outyield 6-rowed types and are extensively grown. Average yield of all types in the United States is approximately 45 bushels/acre (2,419 kg/ha).

In the southern part of Pennsylvania, Ohio, Indiana, and Illinois, and from central Missouri and Kansas to northern Texas, the barleys grown are almost entirely of the winter type. Farther south, typical spring barleys sometimes are seeded in the fall. In California, the cultivars used are typically spring in habit of growth but are sown in November or December. When the commonly grown "hulled" barley is to be used for human food, it may be "pearled" to remove the hulls.

Through the years, the heavily barbed awns were objectionable and no doubt deterred many from growing barley. Improved smooth-awned cultivars have been developed and distributed widely.

CLASSIFICATION. All the common 6-rowed and 2-rowed barleys commercially grown have a diploid chromosome number of 14. There are several species of wild barley that have 14 chromosomes and some have 28.

The cultivated barleys are listed as three species, or sometimes as subspecies or classes, based on the fertility of lateral spikelets: (1) *Hordeum vulgare*, the common 6-rowed types, (2) *H. distichum*, the two-rowed barleys, and (3) *H. irregulare*, irregular barley, in which lateral spikelets are much reduced in size.

CULTIVARS. Among the qualities sought in barley are high yields, resistance to shattering, suitability for combine harvest, smooth awns, good

FIGURE 22–7. *Six-row barley in a Illinois field. Note the kernel arrangement on the rachis and the long awns or hoods.* [SOURCE: Southern Illinois University.]

malting quality, stiff straw, disease resistance, and winterhardiness among winter cultivars. There also is considerable interest in developing adapted cultivars with high lysine or high protein.

Approximately 150 cultivars of barley are grown in the United States, and probably over 4,000 over the world. Cultivar recommendations vary greatly according to local environmental conditions and cultivar response. As for other crops, cultivar comparisons should be studied for a specific area.

The development of hybrid barley by a revolutionary genetic break-through promises increased yields per acre, but more significantly this method may be applied to other industrial crops such as rice, soybeans, and forage crops. Called the *balanced tertiary trisomic method,* it makes use of a genetic male-sterile gene and an extra chromosome. The extra chromosome regulates several critical genetic events but is eliminated automatically in one of the final steps of the method. 'Hembar,' the first successful commercial hybrid barley, was released by the USDA and the University of Arizona.

Uses

The largest domestic use of barley, accounting for more than 50 percent, is as a feed crop for livestock. The remainder is used for food, alcohol, or seed. Barley has about 95 percent of the feeding value of corn. To obtain the full feed value it is necessary to roll, crush, or grind the grain. It is fed extensively to hogs in areas where corn is not well adapted. In some areas of the United States, particularly the West, a considerable acreage of barley is utilized for pasture or hay. In irrigated regions of the West, it is often used as a companion crop for alfalfa.

MALTING. About 30 percent of the barley crop is used for malting. Malt is used in beer, distilled alcoholic products, malt syrup, breakfast foods, and coffee substitutes. Malt by-products include brewer's grain for dairy feeds, brewer's yeast for animal feed, human food, and fine chemicals, and distiller's grains and distiller's solubles for livestock and poultry feeds.

Only the best grades of barley are suitable for malting, and only certain barley cultivars can be used. Barley cultivars which produce a desirable combination of enzymatic activity and chemical composition upon malting are classed as malting varieties. Malting is a controlled limited germination process, designed primarily to produce or activate enzyme systems. In the respiration and germination processes and the removal of the rootlets, approximately 9 percent of the original weight of the barley is lost.

PEARLING. Pearling barley consists of grinding off the hulls and outer layers of the kernels. Barley pearled three times usually is sold as pot barley. Five or six pearlings remove all the kernel coating and practically all the embryo. This is the product sold on the market as pearled barley. One hundred pounds (45 kg) of barley will yield approximately 65 pounds

(29 kg) of pot barley, or 35 pounds (16 kg) of pearled barley. White, 2-rowed cultivars are most commonly used for this purpose.

Production Practices

It generally is recommended that the seedbed for barley be fall-plowed, especially for disease control. Some of the fungus diseases carry over in crop refuse in the field.

Barley does best when included in a rotation following a cultivated crop. The best yields of spring barley are obtained from early seeding—April in most areas. In California and Arizona, it is a common practice to sow spring cultivars in the fall or early winter; plants generally will not suffer winter damage. Most barley is seeded with a grain drill. The rate of seeding is most often 48–96 pounds/acre (54–108 kg/ha), depending on moisture conditions, type of barley, and geographic region.

Virtually all of the barley is combine harvested. A considerable portion of the acreage, particularly of malting types, is windrowed and combined from the windrow 3 or 4 days later. Such a method avoids some of the shattering losses. For malting barleys it also is important that the kernels not be broken or skinned in combining.

Insects and Diseases

Insects that attack barley are the same as those discussed for oats. Annual loss to barley caused by insects is estimated at 14 percent. Diseases are responsible for an estimated annual crop loss of 5 percent. The principal ones are covered smut, loose smut, stem rust, leaf rust, stripe rust, powdery mildew, scab, stripe, spot blotch, net blotch, bacterial leaf blight, scald, and various diseases. The main preventive or control measures are use of resistant cultivars, seed treatment, and crop rotation.

Rice

Rice, among the grains, is second to wheat in world importance as to acreage and production. However, some would argue that rice is the world's most important crop, a staple in the diet of a greater number of people. It is the principal food crop of about one-half the world's population. The leading countries in production are China, with more than one-third the total, India, Indonesia, Bangladesh, Thailand, and Brazil (Figure 22-8). In terms of acreage, the order of importance is India, China, Bangladesh, Indonesia, Thailand, and Brazil (Figure 22-9).

In the United States rice holds only sixth place among the grains as to acreage and production. Only six states produce rice, and the acreage is very small in Mississippi and Missouri. The states of Arkansas, Louisiana, Texas, and California account for 93 percent of the acreage (Figure 22-10). This amounts to about 2.5 million acres (1 million ha), yielding an

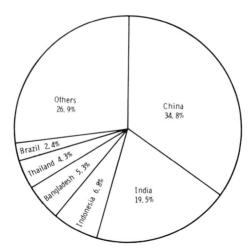

Figure 22–8. *Leading countries in world production of rice. Over one third of the total is produced in China.* [Source: Southern Illinois University.]

average of about 4,600 pounds/acre (5,152 kg/ha). Geographic distribution is shown in Figure 22–11.

The Plant

Rice, *Oryza sativa,* is an annual grass with culms usually growing to a height of 2–4 feet (0.6–1.2 m). Several tillers, usually four or five, may develop from each seed planted. The leaf sheaths are open, and the plant is characterized by the presence of very long ligules.

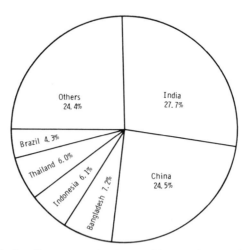

Figure 22–9. *Leading countries in world area (acreage) of rice. India and China have over one half of the world total.* [Source: Southern Illinois University.]

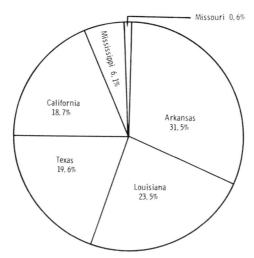

FIGURE 22–10. *Leading states in United States rice area (acreage). Four states have virtually all of the acreage.* [SOURCE: Southern Illinois University.]

The rice inflorescence is a loose panicle which may contain more than 100 one-flowered spikelets. The mature kernels are "hulled," or enclosed by the lemma and palea. They vary in color from white to amber and brown for most American cultivars.

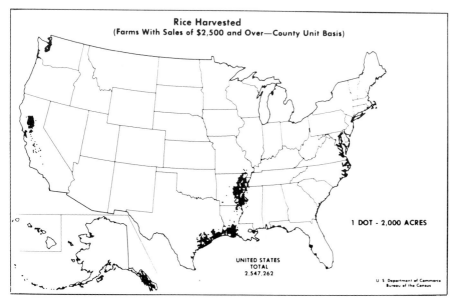

FIGURE 22–11. *Acreage of rice harvested in the United States, shown geographically.* [SOURCE: U.S. Department of Commerce.]

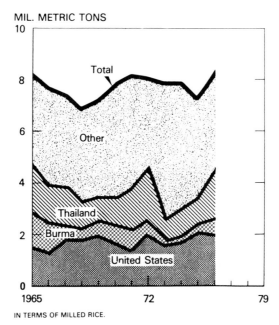

MIL. METRIC TONS

IN TERMS OF MILLED RICE.

FIGURE 22–12. *World exports of rice by country. Although only four states have appreciable rice acreage, the United States exports about 2 million metric tons each year.* [SOURCE: USDA.]

The most significant characteristic of rice is that it makes its best growth when the roots are submerged in water. Most other field crops are killed if submerged for only a few days. The rice plant transports oxygen to the submerged roots from the leaves, where it is used in respiration.

Cultivars and Types

The Orient produces hundreds of rice cultivars, but the United States has only a small number. Cultivars of rice grown in the United States are of the species *O. sativa.* In general, commercial cultivars are classified on the basis of length of the growing season, size and shape of the grain, and chemical character of the endosperm. On the basis of growing season cultivars in southern states are divided into four groups: (1) very early, 110–115 days; (2) early, 116–130 days; (3) midseason, 131–155 days; and (4) late, 156 days or more. The length of growing season is somewhat longer in California than in the South. These differences in length of growing season between the two areas are caused by daylength and temperature. Cultivars are divided into three grain-size and shape classes: short (pearl), medium, and long. The more slender long-grain cultivars are considered a fourth class. On the basis of chemical characters, the two types of rice are waxy (glutinous; endosperm contains no amylose), and common (ordinary starch; endosperm contains amylose as well as amylopectin). Very little waxy rice is grown in the United States.

A **Modern semidwarf before panicles emerge.**

Mature semidwarf.

B

C

FIGURE 22–13. *Flowering habit and inflorescence of rice. (A) Modern semidwarf rice cultivars.* [SOURCE: International Rice Research Institute.] *(B) Rice panicles at anthesis. (C) Mature panicles.* [Photographs of B and C with permission of the Rice Branch Research and Extension Center, Arkansas Agricultural Experiment Station.]

There are probably more than 8,000 different rice cultivars in the world, but fewer than 20 dominate the U.S. acreage. Hybridization programs conducted at rice experiment stations have resulted in cultivars that are most commonly grown in the United States. Among these are 'Bluebonnet,' 'Blue-

FIGURE 22-14. *Workers thresh grain from promising rice lines at the International Center of Tropical Agriculture (CIAT) in Colombia. The CIAT cooperates with the International Rice Research Institute to evaluate promising materials for Latin America and Caribbean conditions. These materials are then incorporated into a program of international rice trials for Latin America together with germplasm from national breeding programs to be tested under local conditions.* [SOURCE: International Center of Tropical Agriculture.]

FIGURE 22-15. *Individual plants in a rice breeding nursery at Arkansas' Rice Branch Research and Extension Center.* [Photograph with permission of Arkansas Agricultural Experiment Station.]

bonnet 50,' 'Starbonnet,' 'Texas Patna,' 'Calrose,' 'Dawn,' 'Belle Patna,' and 'Bluebelle.' In a consideration of the newer cultivars released and described during 1970–1976, the major attention was given to these characteristics: (1) high yielding ability, (2) short stature with lodging resistance, (3) good disease resistance, and (4) good milling and cooking qualities.

Production Practices

Crop rotation is practiced in most rice-producing areas of the United States. If rice is grown continuously in a field, the soil becomes depleted in fertility and organic matter and falls into poor physical condition. A weed problem may also develop. With better methods of weed control, increased fertilization, and better cultural practices, continuous rice is less of a problem. Several different rotations are followed. In Arkansas, for example, a common rotation is rice–soybeans, followed by fall-sown oats with lespedeza sown in the oats in early spring. In Arkansas and Mississippi, fish culture has, in some cases, been inserted into the rice rotation. In the Gulf Coast area the usual cropping pattern is to rotate rice with pasture for beef cattle. In California some farmers follow rice with spring- or summer-plowed fallow, on which wheat, oats, or vetch is fall-sown. After the grain crops, the field is returned to rice for several years. Spring-sown grain sorghum, field beans, and safflower may be grown one or two years and then the field returned to rice.

Most rice soils are either fall-, winter-, or spring-plowed, and the land is leveled with a land plane. Levees are constructed on the contour with levee disks or pushers or large diking machinery which divides the field into subfields. Final seedbed preparation, with a disk, harrow, or field cultivator, is done just prior to seeding.

The water requirement for rice is high, 1.5 to 8 acre-feet. The amount applied by irrigation will depend on the amount of rainfall during the growing season. Flood irrigation is used on rice grown in the United States. Upland, or rain-fed, rice is grown in some areas of the world where rain is a daily occurrence.

In the Southern states, most rice is sown from April 1 to May 30; in California from April 15 to May 15. Seeding rate in the South is about 90–110 pounds/acre (101–123 kg/ha) when the seed is drilled and 115–150 (129–168 kg/ha) when broadcast. Rice is sown with a grain drill, ground broadcast seeder, or with an airplane seeder. In the southern states, seed is often broadcast or drilled before flooding the field. When plants are 6–8 inches (15–20 cm) high, the land is flooded. As the plants grow, the depth of water is increased until the level reaches 4–6 inches (10–15 cm). Most of the rice acreage in California and much of it in the southern states is established by dropping presoaked seed with an airplane seeder onto a flooded field. Soaking seed increases the weight and allows them to sink to the soil rather than floating on top of the water. Fertilizer practices vary, but most rice soils will respond to complete fertilizer. Some of the shorter, strong-strawed rice cultivars will respond to nitrogen (N) applications of well over 100 pounds/acre (112 kg/ha) without lodging. An ammonium source of nitrogen is more efficient than a nitrate source.

Weeds reduce the yield and quality of rice in the United States by an estimated 15 percent each year. This loss is valued at $165 million. Aquatic and semiaquatic weeds such as barnyardgrass, coffeeweed, redweed, spike rush, and red rice are controlled by the use of appropriate herbicides, by controlling the water level, and by crop rotation. Herbicides most often used to control weeds in rice include propanil, molinate, bifenox, and phenoxy herbicides such as 2,4-D and silvex.

Most rice in the United States is harvested with self-propelled combines when the moisture content of the grain is between 18 and 22 percent. If harvested before this stage, yield will be reduced and delaying harvest may result in a considerable shattering loss. Fields are drained when heads have "turned over" and are beginning to turn brown. Harvest is about two weeks later. Rice harvested with a high moisture content is artificially dried until the moisture content is 12–13 percent, a safe level for storage.

Milling and Uses

Milling of rice begins with cleaning the field-run, or rough rice, removing chaff, weed seeds, and other foreign material. Next the hulling machine

FIGURE 22–16. *An aquatic weed control study with rice at Arkansas' Rice Branch Research and Extension Center.* [Photograph with permission of Arkansas Agricultural Experiment Station.]

removes the hulls from the kernel, leaving what we call brown rice. Then the brown bran coat and the germ are rubbed or milled off the kernel by hullers. The "brush" then removes the remainder of the innermost aleurone layer and adhering floury particles, a process known as "polishing." The milled rice is then graded. Milled rice also may pass through another step, being coated with glucose and talc powder and given a final polishing in a trumbal. Milling of rough rice gives about 64 percent whole and broken kernels, 13 percent bran, 3–4 percent polish, and 20 percent hulls.

For years the per-capita consumption of rice in the United States has varied from 5 to 7 pounds (2.3–3.2 kg), while that in some countries may be as much as 400 pounds (181 kg).

Rice is primarily an energy food. It is high in calories and contains a substantial supply of proteins. Milled white rice is deficient in calcium and somewhat lacking in iron. Various types of processed rice contain a large supply of calcium and iron. Rice is a source of vitamins; brown rice and converted rice are much higher in vitamin B_1 and somewhat higher in vitamin B_2.

Milled white rice makes up the bulk of the sales in the United States. Rice in the United States is consumed largely in the boiled state; some is sold in precooked or partly precooked condition. Rice also is used for break-

FIGURE 22–17. *The before and after of rice processing:* 1 = *rough rice after cleaning but before the hulls have been removed;* 2 = *hulled and polished long-grain rice ready for the consumer.* [SOURCE: Southern Illinois University.]

fast foods and as the starch in face powders. Rice bran and polish as well as the screenings and brewer's rice are used for cattle feed. The germ fraction of the bran provides rice oil often used in the manufacturing of soap. Rice hulls may be used for fuel, insulation material, cardboard, rayon, and linoleum.

Insects, Diseases, and Other Pests

The rice stink bug and the rice water weevil are the most damaging pests, causing annual losses of about 4 percent. Other insects include the rice stalk borer, rice leaf miner, fall armyworm, grasshopper, and chinch bug. Diseases cause an annual loss of 5 percent annually. Most damaging diseases are caused by fungi and include seedling blight, brown leaf spot, blast, stem rot, root rot, kernel smut, and brown-bordered leaf and sheath spot. In addition, the white tip disease is caused by a nematode, and

FIGURE 22–18. *Plants of buckwheat in the field. Plants begin to bloom 5 to 6 weeks after planting, and the crop matures quickly.* [SOURCE: Southern Illinois University.]

straighthead apparently results from some abnormal soil condition that develops around the roots after several weeks of flooding.

Other serious pests include muskrats, which can cause extensive damage to levees and canals, and blackbirds, which can cause extensive yield reductions through feeding.

Buckwheat

The acreage seeded to buckwheat is very small. Never an important food crop, the acreage seeded to buckwheat in recent years has declined, and the USDA has discontinued estimates.

Buckwheat, *Fagopyrum sagittatum* is an annual that makes a growth of 2–4 feet (0.6–1.2 m). The rather coarse, strongly grooved succulent stalks usually produce several branches. Buckwheat begins to bloom 5–6 weeks after seeding. When mature, the "seed" may be gray-brown, brown, or black, depending on the cultivar and type. The "seed" are actually one-seeded fruits called *achenes.*

More than one-half the acreage seeded to buckwheat has been in the states of New York and Pennsylvania, with significant acreages also in Ohio, West Virginia, Michigan, Wisconsin, and Minnesota. Buckwheat is best adapted to a cool, moist environment. When the plants are in bloom,

FIGURE 22–19. *These angular buckwheat "seed" are actually one-seeded fruits called achenes.* [SOURCE: Southern Illinois University.]

the crop is very sensitive to high temperatures and dry weather. Under these conditions flowers are likely to blast without forming seed.

Buckwheat usually is seeded relatively late in the season. A common recommendation is to seed sufficiently late so that general flowering will occur after the period of the highest summer temperatures.

More than 90 percent of the buckwheat acreage is harvested for the mature grain. The better-quality buckwheat is marketed for human food, with the poorer grades fed to livestock. Buckwheat is especially regarded as feed for poultry.

Buckwheat flour has a characteristic dark color, due to the small particles of hull that pass through the bolting cloths. Coarser bolting cloths generally are used for buckwheat than for wheat. One hundred pounds (45 kg) of high-quality buckwheat will yield 60–70 pounds (27–32 kg) of flour, 4–18 pounds (1.8–8.2 kg) of middlings, and 18–26 pounds (8.2–11.8 kg) of hulls. Buckwheat flour is used almost altogether in blends with wheat flour, and sometimes with cornmeal, to make buckwheat cakes or mixed pancake flours.

REFERENCES AND SUGGESTED READINGS

1. Adair, C. R. *Rice in the United States: Varieties and Production,* USDA Agr. Handbook 289 (1966).
2. Adair, C. R., *et al.* "Rice Improvement and Culture in the United States," In *Advances in Agronomy,* vol. 14 (New York: Academic Press, 1962). ed. A. G. Norman.
3. Atkins, J. G. *Rice Diseases,* USDA Farmers' Bul. No. 2120 (1972).
4. Barnes, G. *Control of Insects Attacking Rice.,* Ark. Coop. Ext. Serv. Pub. EL 330 (1977).
5. Bonnett, O. T. *The Oat Plant: Its Histology and Development,* Univ. Ill. Agr. Exp. Sta. Bul. 672 (1961).
6. Chandraratna, M. F. *Genetics and Breeding of Rice* (London: Longmans, 1964).
7. Coffman, F. A. "Origin and History," In *Oats and Oat Improvement,* Monograph 8 (Madison, Wisc., American Society of Agronomy, 1961), ed. F. A. Coffman.
8. Coffman, F. A., *et al.* "Oat Breeding," In *Oats and Oat Improvement,* Monograph 8 (Madison, Wisc.: American Society of Agronomy, 1961), ed. F. A. Coffman.
9. Compton, L. E., *et al.* *Norline Winter Oats,* Purdue Univ. Res. Bul. 799 (1965).
10. Cook, A. H., (Ed.). *Barley and Malt: Biology, Biochemistry, Technology* (New York: Academic Press, 1962).
11. Day, K. M., *et al.* *Performance and Adaptation of Small Grains in Indiana,* Purdue Res. Bul. 850 (1970).
12. Faw, W. F., and T. H. Johnston. *Effect of Seeding Date on Growth and Performance of Rice Varieties in Arkansas,* Ark. Agr. Exp. Sta. and USDA-ARS Rep. Ser. 224 (1975).
13. Field, H. *The Track of Man* (Garden City, N.Y.: Doubleday, 1953).
14. Frey, K. J., *et al.* "New Multiline Oats," *Iowa Farm Science,* 24:8 (1970).
15. Grist, D. H. *Rice* (London: Longmans, 1965).

16. HUEY, B. A. *Rice Production in Arkansas,* Ark. Coop. Ext. Serv. Circ. 476 Rev. (1977).

17. HUEY, B. A., and S. L. CHAPMAN. *Water Management for Rice Production,* Ark. Coop. Ext. Serv. Pub. EL 566 (1976).

18. International Rice Research Institute. *Symposium on the Major Insect Pests of the Rice Plants, 1964* (Baltimore: John Hopkins, 1967).

19. Iowa State University Coop. Ext. Serv. *Profitable Oat Production,* Pm-297 (1968).

20. LEONARD, W. H., and J. H. MARTIN. *Cereal Crops* (New York: Macmillan, 1963).

21. LUEKEL, R. W., and V. F. TAPKE. *Barley Diseases and Their Control,* USDA Farmers' Bul. 2089 (1965).

22. MARTIN, J. H., W. H. LEONARD, and D. L. STAMP. *Principles of Field Crop Production* (New York: Macmillan, 1976).

23. MATSUBAYASKI, M., *et al.* (Eds.). *Theory and Practice of Growing Rice* (Tokyo: Fuji, 1968).

24. MATZ, S. A. *Cereal Science* (Westport, Conn.: AVI, 1969).

25. MULLINS, T., *et al. World Rice Production, Disappearance, and Trade—1960 Decade with Projections to 1980,* Ark. Ag. Exp. Sta. Bull. 819 (1977).

26. PAINTER, C. G., *et al. Malting Barley, Idaho Fertilizer Guide,* Idaho Curr. Info. Ser. 270 (1977).

27. QUISENBERRY, K. S., and J. W. TAYLOR. *Growing Buckwheat,* USDA Farmers' Bul. 1835 (1939).

28. RAMAGE, T. R. "Balanced Tertiary Trisomics for Use in Hybrid Seed Production," *Crop Science,* 5:2 (1965).

29. SIMONS, M. D., and H. C. MURPHY. "Oat Diseases," *Oats and Oat Improvement,* Monograph 8 (Madison, Wisc.: American Society of Agronomy, 1961).

30. SMITH, R. J., JR. *Comparisons of Herbicide Treatments for Weed Control in Rice,* Ark. Exp. Sta. and USDA-ARS Rep. Ser. 233 (1977).

31. SMITH, R. J., JR, W. T. FLINCHUM, and D. E. SEAMAN. *Weed Control in U.S. Rice Production,* USDA-ARS Agr. Handbook No. 497 (1977).

32. USDA Agr. Stat. (1977).

33. USDA Agr. Stat. (1962, 1969).

34. USDA-ARS. *Barley: Origin, Botany, Culture, Winterhardiness, Genetics, Utilization, Pests,* Agr. Handbook 338 (1968).

35. USDA-ARS. *Losses in Agriculture,* Agr. Handbook 291 (1965).

36. USDA-Ext. Serv. *New Crop Cultivars* 1970–1976, 13, ESC-584 (1977).

37. WELLS, B. R., T. H. JOHNSTON, and P. A. SHOCKLEY. *Response of Seven Rice Varieties to Rate and Timing of Nitrogen Fertilizer in Arkansas,* Ark. Agr. Exp. Sta. and USDA-ARS Rep. Ser. 235 (1977).

38. WESENBERG, D. M., *et al. Malting Barley Production in Idaho,* Idaho Curr. Info. Ser. No. 276 (1975).

39. WIEBE, G. A. *Barley: Origin, Botany, Culture, Winterhardiness, Genetics, Utilization, Pests,* USDA Agr. Handbook 338 (1968).

40. WIEBE, G. A. *Classification of Barley Varieties Grown in the U.S. and Canada in 1958,* USDA Tech. Bul. 1224 (1961).

B. Oil and Fiber Crops

CHAPTER 23
SOYBEANS

THE soybean, *Glycine max,* is believed to be among the older plants cultivated by man. Most agree that the origin was in eastern Asia, possibly in
China. Wild ancestors of the cultivated soybean are found today in China
and Korea. Soybeans were mentioned in Chinese literature as early as 2838
B.C., but it is believed they were grown extensively in China and Manchuria
centuries earlier.

Production, Adaptation, and Distribution

Soybeans are grown satisfactorily under a wide variety of climatic and
soil conditions. Because of the ability to utilize atmospheric nitrogen when
the crop is properly inoculated, and perhaps because of other characteristics of the plant, soybeans give relatively better yields under unfavorable
soil conditions than corn and many other crops. In general, however, for
best production the soil and climatic conditions approximate those best
for corn. Specific cultivars are available that make the crop adaptable to
areas from Canada to the Gulf.

World Production

Prior to World War II China and Manchuria led the world in soybean
production. The war, however, stimulated a marked increase in the soybean
acreage in the United States because of the urgent need for more oil. By
1946 the United States was producing more than any other country, 38
percent of the world's crop; in 1968 it produced approximately 75 percent
and in 1976 about 66 percent. A history of production from the period
1930–34 to 1976 for selected countries is shown in Table 23–1.

The leading countries in soybean acreage are shown in Figure 23–1. The
United States presently has about 55 percent of the world's acreage and
66 percent of the production. The remarkable story in world soybean production in recent years is that of Brazil, which has moved up from virtually
nothing in the 1960s to their present level of third in acreage and second
in production (Table 23–1). China is now third in production and second
in acreage. Other countries with a much smaller share of the world total
are the USSR and Indonesia.

444

TABLE 23–1. Annual Average Production (000 Bu) of Soybeans by Periods of Years in the Principal Producing Countries of the World

	1930–1934	1945–1949	1960–1964	1968	1976
United States	16,603	208,885	660,582	1,079,662	1,264,890
China proper	231,327	190,248	278,000	240,000	348,333
Brazil	—	—	—	—	411,656
Japan	12,231	7,432	12,478	6,155	4,033
Korea	20,286	4,984*	5,660*	8,155*	10,816
Indonesia	9,731	6,393	13,507	16,718	21,083
World total (excluding USSR)	455,000	548,880	1,014,941	1,432,002†	2,534,326†

Source: USDA Agr. Stat. (1935, 1953, 1969, 1977).
* Republic of Korea only.
† Includes estimates from countries where data are not available; 1976 world total includes USSR.

Soybeans in the United States

First mentioned as a crop adapted to the Eastern United States in 1804, not more than eight cultivars of soybeans were grown prior to 1898. But in 1898 the USDA began to introduce many cultivars. In general, because the soybean is very sensitive to photoperiod differences, a given cultivar is limited in its climatic range.

At first soybeans were grown primarily as a forage crop, for hay and sometimes for silage and green manure. Soybeans were grown earliest in Pennsylvania and Massachusetts. As a forage crop, they occupied 50,000 acres (20,235 ha) in 1907 and this acreage reached a peak of 4.8 million acres (1.9 million ha) in 1940, before declining to 1 million (0.4 million ha) in 1953 and less than 0.5 million (0.2 million ha) in 1964. Previous to

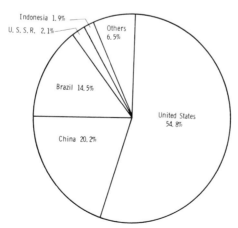

FIGURE 23–1. *Leading countries in world soybean area (acreage).* [SOURCE: Southern Illinois University.]

TABLE 23–2. United States Soybean Acreage, Yield, and Total Production by Five-Year Periods Av. 1962–1966; 1969; 1976.

	Acreage		Bushels	
Year	Planted Alone (000)	Harvested as Beans (000)	Yield per Acre	Total Production (000)
1925	1,539	415	11.7	4,875
1930	3,072	1,074	13.0	13,929
1935	6,966	2,915	16.8	48,901
1940	10,487	4,807	16.2	78,045
1945	13,007	10,740	18.0	193,167
1950	14,704	13,814	21.6	299,279
1955	19,669	18,559	20.0	371,276
1962–1966 (av.)	32,401	31,602	24.3	768,672
1969	42,100	40,857	27.3	1,116,876
1976	50,327	49,443	25.6	1,264,890

1930 the acreage harvested for beans generally was less than one-fourth the total acreage. In 1939, 40 percent of the total planted acreage was harvested for the bean; in 1945 it was 72 percent; by 1969 it was almost 100 percent. Table 23–2 shows the acreage and production history of soybeans in this country since 1925. The acreage planted, percentage harvested as beans, and total production has increased steadily during each period, with a marked increase in the early 1960s. However, the average yield per acre has not increased greatly since 1960; exact reasons for this "yield plateau" are the subject of much discussion (Figure 23–2).

States leading in area planted are in the order: Illinois, Iowa, Arkansas, Missouri, Indiana, Minnesota, Mississippi, and Ohio (Figures 23–3, 23–4). These eight states account for 70 percent of the total acreage. U.S. soybean acreage has continued to expand. From about 40 million acres (16 million ha) planted in 1968, the 1978 total exceeded 60 million (24 million ha) (Figure 23–5). The Eastern Cornbelt region accounted for the majority of

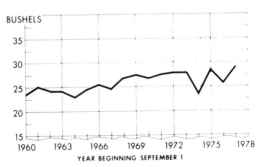

FIGURE 23–2. *United States yield per acre harvested. Average yield per acre has not increased appreciably since 1960. The reasons for this "yield plateau" need to be discussed.* [SOURCE: USDA-ESCS.]

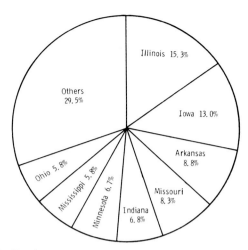

FIGURE 23–3. *Leading states in United States soybean area (acreage). Eight states account for 70 percent of the acreage.* [SOURCE: Southern Illinois University.]

the acreage, with the South Central and Southeast regions having smaller shares.

U.S. exports of soybeans and soybean products have a value of $5 billion (Figure 23–6). Prior to 1972 export value of soybeans ranged from $1 to $2 billion, but virtually exploded in 1973 and 1974.

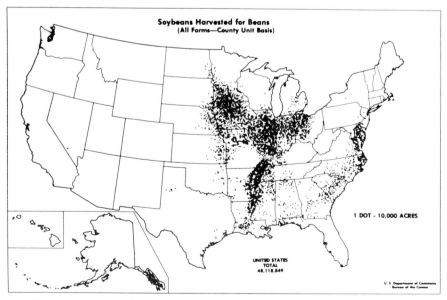

FIGURE 23–4. *Soybeans harvested for beans in the United States, shown geographically.* [SOURCE: U.S. Department of Commerce.]

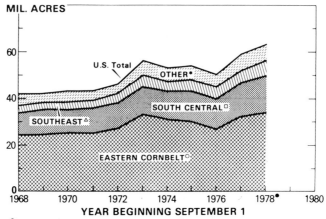

FIGURE 23–5. *Soybean acreage planted since 1968 in the United States by geographic region.* [SOURCE: USDA-ESCS.]

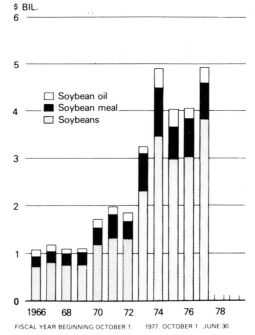

FIGURE 23–6. *United States exports of soybeans and products since 1966. Exports virtually "exploded" in 1973.* [SOURCE: USDA.]

The Plant

The soybean is an upright, branching, herbaceous, warm-season annual. The height ranges from 1 to 6 feet (0.3–1.8 m) depending on the cultivar and environment, with an average of 2–4 feet (0.6–1.2 m). Degree of branching from lower nodes depends largely upon spacing.

The seedling emerges in an epigeal manner, with the cotyledons being pulled above the soil and serving as leaves until true leaves are formed. Soybeans develop two unifoliolate leaves, with the remainder being palmately compound, trifoliolate.

The root system consists of a taproot, formed by the radicle, plus a large number of secondary and branch roots. The root system may penetrate the soil to a depth of up to 6 feet (1.8 m), but most of the roots are in the top foot (0.3 m) or less. As soon as root hairs are formed, nodules for fixing atmospheric nitrogen may begin to develop on the root system. Generally the most effective nodules are on the taproot and the larger secondary roots.

The plant may be determinate or indeterminate in growth habit. Determinate types stop most vegetative growth when flowering and pod formation begin. Most of the cultivars grown in the Midwest and North (shorter-season types in the first four maturity groups) are indeterminate, while those grown in the South are mostly determinate. Determinate cultivars for the North and Midwest presently are being investigated. These types would offer better lodging resistance and greater yield potential. However,

FIGURE 23–7. *Early in the history of soybeans in the United States, they were grown as a forage crop, for hay, silage, or green manure. The black-seeded soybean shown here is a hay type.* [SOURCE: Southern Illinois University.]

the shorter flowering period would make soybean production more risky. A droughty period during flowering could result in a crop failure.

Soybean plants bear typical legume flowers, white or purple, in terminal or axillary racemes, up to 35 flowers per cluster. Flowering in indeterminate cultivars begins near the base of the main stem and progresses upward. Flowers are mostly self-pollinated, but some crossing does occur. Up to 75 percent of the flowers abort and do not set pods, a fact that has challenged physiologists, plant breeders, and other interested in removing soybeans from the "yield plateau." Various soybean growth regulators have been tested as a means of reducing plant height and increasing pod set. Generally, none have consistently increased yield sufficiently to justify their cost.

Pods are 1- to 5-seeded, with most cultivars being 2- or 3-seeded. The number of pods varies from 2 to more than 20 in a single inflorescence cluster, and up to 400 on a plant. Stems, leaves, and pods most often are covered with dense, fine pubescence. The large seed generally are nearly spherical, and normally yellow to greenish-yellow, but some may be pale green, brown, or black. The hilum or seed scar varies from buff to yellow, black, or brown. The "edible" cultivars bear larger seeds, which are lighter in color and have a good flavor.

Production Practices

Soybeans in the Rotation

Soybeans have been raised successfully as an intertilled crop in many cropping systems. In the Corn Belt the soybean generally has followed corn. The ability of the soybean to obtain much of its nitrogen from soil air has caused most farmers to plant corn rather than soybeans after a clover or alfalfa crop.

In Iowa on livestock farms the usual rotation is lengthened when soybeans are grown. The four-year rotation of corn, corn, small grain, legume or legume-grass is frequently lengthened by a year, with the soybean crop most often going between the two corn crops. Cropping systems may vary from complex rotations of seven or more years to continuous soybeans. Although soybeans yield well when planted in successive years, this practice may increase problems with diseases, insects, weeds, and erosion. In areas where damage is caused by the soybean cyst nematode, especially race 4, continuous soybeans are not recommended. Crop rotation is the main method of avoiding severe damage from this latest nematode race.

In the Delta area of Arkansas, Louisiana, and Mississippi, as well as other parts of the South, the soybean fits in satisfactorily with cotton, rice, or corn in rotation. The crop provides additional income with little added investment. The soybean is well suited for clay soils, which are less suitable for cotton. Winter grains may be drilled behind the combine following soybeans. In Arkansas and some of the other southern states where soybeans started as a less important crop to use in rotations with more important cash crops such as cotton or rice, now soybeans have become the *main* cash crop.

Texas farmers in the irrigated High Plains section have successfully used a cotton–cotton–grain sorghum–soybean or a cotton–grain sorghum–soybean rotation. Cotton yields generally have been slightly higher when cotton followed soybeans.

Mid-Atlantic states farmers and Southern farmers often have used a double-cropping system in which soybeans follow winter grains, early potatoes, canning peas, and other vegetables. This practice produces two crops in one year on the same land and provides an opportunity for minimum tillage for soybean production, which reduces costs. From southern Illinois to points south, soybeans are often planted with a no-till planter in wheat or other small grain stubble soon after harvest without plowing.

The effect of soybeans on other crops in the rotation is important in determining what crop will follow beans. Yields of corn and winter wheat following soybeans average 5–10 percent higher than yields of the same grains following nonlegume crops. Schmid et al. in Minnesota found that corn yields following soybeans were higher than those following oats be-

FIGURE 23–8. *An experimental type of double cropping: Soybeans planted in tall fescue sod by no-tillage methods following removal of a cutting of hay. A more conventional double cropping system is soybeans following a winter small grain such as wheat.* [SOURCE: Southern Illinois University.]

cause of the residual nitrogen contributed by soybeans. Attention should be given as well to the crop preceding soybeans. When a small-seeded legume or grass-legume mixture exists in a rotation, corn can better utilize the residual nitrogen. Corn should follow the forage and soybeans should follow corn.

Soybeans generally are credited with leaving the soil in an unusually loose and friable condition as compared with the condition of the soil after other crops; such loose soils increase the danger of erosion loss. Moldenhauer and Wischmeier found soil erosion losses higher after soybeans than after corn in rotation studies on sloping land in Iowa. However, other studies in Iowa, Missouri, and Illinois have shown that land in soybeans is no more subject to erosion than land in corn if the beans occupy the same place in the rotation. The criticism that soybeans cause more erosion than corn may not be justified. By using the proper sequence of crops in rotation and employing sound cultural practices, no significant erosion differences should be observed.

Fertilizers and Lime

The soybean plant has the reputation of giving a poor response to direct fertilization except at low levels of fertility. It has been recognized fairly recently, however, that high soybean yields are associated with high fertility levels. The recent conclusions are that soybean response to high fertility levels are not greatly different from that of other crops considered to respond to direct fertilization. Thus, soybeans do best on soils of high fertility, whether the crop itself is fertilized directly or if the residual fertility level is high. Soybeans do have the ability to effectively utilize fertilizer residues that may not be available to other crops.

It is not unusual for soybean growers to apply 400 pounds/acre (448 kg/ha) or more of a mixed fertilizer such as 0–20–20. This crop is a particularly heavy user of potassium. Large amounts of nitrogen oridinarily are not applied because of the ability to fix atmospheric nitrogen, but in some areas a complete starter fertilizer, such as 6–24–24, containing a small quantity of nitrogen is used at planting. The nitrogen can stimulate early vegetative growth.

In some situations soybeans respond to micronutrients. On soils of pH 5.5 or below in the South, molybdenum applied as a seed treatment may give a yield response. Iron can be deficient on alkaline soils, and manganese deficiency can occur on sandy soils of the Atlantic Coastal Plain region.

Recently, Hanway and co-workers in Iowa observed yield increases from 5 to more than 12 bushels per acre (336 to 806 kg/ha) after foliar fertilization of soybeans. The ratio of fertilizer used was 10 parts nitrogen, 1 part phosphorus, 3 parts potassium, and 0.5 part sulfur. Best results were obtained with 2–4 sprayings of the foliar fertilizer with not more than 20–25 pounds of nitrogen per acre (22–28 kg/ha) in one spraying. Results of research elsewhere have been less consistent, and while some foliar fertilizers are available commercially, more conclusive work is needed before their use is recommended widely.

Soybeans grow best on soil with a pH of 5.8–7.0. It is likely that yield response will be obtained by liming a soil if the pH is below 5.8. If needed, lime should be applied broadcast and worked into the soil at least four months ahead of the soybean planting date.

Inoculation and Seed Treatment

It generally is recommended that soybeans be inoculated at planting time if a vigorous well-inoculated soybean crop has not been produced on the land in the last three to five years. The soybean does not cross-inoculate with any other legume. When adequate soil nitrogen is available, soybeans often make a satisfactory growth without nodulation. However, under these conditions a soybean crop will take from the soil more nitrogen than will almost any other crop that could be grown, in spite of the fact that it is a legume. When soil nitrogen is limited, increases in growth and yield from inoculation may be striking. In an early trial in Iowa the percentage increase of inoculated over uninoculated soybeans was 31 percent in acre yield, 11 percent in the protein content of the beans, and 47 percent in the protein production per acre. In several more recent research studies in Illinois, Minnesota, Arkansas, and Mississippi, little or no yield responses were obtained from inoculation on sites where soybeans had been grown previously, as long as 13 years ago. Responses often were obtained if the field had no previous soybean cropping history.

High-quality soybean seed do not ordinarily require fungicidal seed treatment, a means of controlling pathogens which carry over on the seed surface. But fungicidal treatment of low-quality seed, with germination less than 80 percent or infested with a disease such as pod and stem blight, often is beneficial. Such seed treatment should be applied several weeks ahead of planting. A relatively small proportion of soybean seed marketed is treated because this limits the seed use for planting only, and seed not marketed as seed beans cannot be marketed for grain. Chemical fungicides used for seed treatment may interfere with proper *Rhizobium* inoculation when used on soybeans being planted for the first time in new areas.

Planting Methods

The optimum width between rows varies from north to south, with the cultivar, with length of the growing season, and according to fertility of the soil. Row spacings as little as 18 cm (7 inches) (drilled) to as much as 102 cm (40 inches) are used. The row spacing selected is dependent largely upon the availability of equipment for soybeans and its adaptability to other crops, and the capacity for controlling the particular weed problems in a given field.

The potential advantage for growing soybeans in narrow rows has been recognized in much of the Corn Belt for as long as the crop has been of

FIGURE 23–9. *Soybean row spacing: A row-spacing experiment in Tennes-see. Note the wide-spaced rows in the foreground and the more narrow rows in the background.* [Photograph with permission of University of Tennessee Agricultural Experiment Station.]

economic importance. Recent trials in the Midwest show that soybean yields are consistently higher in narrow rows than the historial 102-cm (40-inch) rows. This increase is often in the 10–15 percent range. In the South, narrow rows generally have not given consistent yield increases.

If early, consistent weed control in narrow-row soybeans cannot be obtained with herbicides, it is advisable to plant in rows wide enough to allow cultivation. If weeds can be controlled early in the season, narrowing the rows will decrease weed growth by allowing a soybean canopy to develop quickly for shading. In recent years narrow-row soybean production has become more feasible because of better equipment, more consistent herbicides, and improvements in planting techniques. The stale seedbed technique, for example, couples a 3-week delay in planting after seedbed preparation with the use of effective herbicides at planting to result in more effective weed control.

Planting Rate

The soybean plant is capable of making maximum yields from a wide range of planting rates because of its ability to compensate for population variations. Planting too much probably is a more common mistake than planting too little seed. Rate recommendations given in pounds of seed per acre usually vary from 40 to 100 (45–112 kg/ha) or more. However, a recommendation should be based upon number of plants desired per foot of row rather than pounds of seed per acre because of great differences among cultivars in seed size. The final population desired ranges from 3 to 12 plants/foot (10–39/m) of row, depending on cultivar and row width. An almost universal recommendation is 8–12 seeds/foot (26–39/m) of row in 36- to 40-inch (91–102-cm) rows. As the row width is narrowed, planting rate should be adjusted to give these final populations in plants per foot of row: 30-inch (76-cm), 6–8 (20–26/m); 20-inch (51-cm), 4–6 (13–20/m); and 10-inch (25-cm) or less, 3 or 4 plants (10–13/m). Additional considerations in selecting the correct populations are susceptibility of the cultivar to lodging, and ability of that cultivar to branch.

Planting Date

May or early June is the best planting date in most soybean-growing states. The soil temperature should have warmed to 50–55°F (10–12°C). Soybeans have not shown the favorable response to extremely early planting as has corn. A delay in planting soybeans in the Midwest from early May to late May or early June will result in an average yield loss of 2–5 percent, but with corn the loss may be as much as 20 percent. Because many soybeans are daylength-sensitive (photoperiodic) and show a calendar maturity date response, a 3-day delay in planting may cause only a 1-day delay in maturity with selected cultivars in the North. If planted very late, late June or early July, an early cultivar would not be selected, but a medium- or full-season cultivar. This would enable adequate vegetative growth to occur before the daylength conditions (actually a critical length of the dark period) triggered the beginning of flowering.

In the Midwest, little yield variation has been observed from planting throughout May, but yields have decreased with June and July plantings. Early plantings have not produced higher yields than plantings in May or early June in the South and West. Planting before May 1 actually has resulted in height and yield reductions in the South. Plantings as late as mid-July are sometimes made in the South and lower Midwest where soybeans are double-cropped following harvest of a winter grain.

Effect of Environment on Seed Composition

The seed composition of the soybean is very important, especially the oil and protein content of the seed. Seed composition of the same cultivars has been shown to vary greatly when grown under different conditions. These differences have been attributed primarily to differences in climate rather than soil.

The seed of the same cultivar grown at widely varying locations has been shown by some researchers to vary in oil and protein content by as much as 1 to 12 percent in the same year. These differences were attributed to climatic variations. Cartter and Hartwig noted a decrease in oil and a slight increase in protein with a delay in planting; this was related to lower temperatures at the time of seed maturation. Howell and Cartter showed that plants at 84°F (29°C) produced seed with an oil percentage 2 to 3 points higher than plants exposed at 72°F (22°C). They produced soybeans with oil contents of 23.2, 20.8, and 19.5 percent when grown in a greenhouse at 85°F (29.4°C) 77°F (25.0°C), and 70°F (21.1°C), respectively, during the pod-filling stages.

Weed Control

Weeds cause average annual losses to the soybean crop of about 17 percent. Soybeans are not strong competitors with weeds early in the season. It is important to kill as many weeds as possible before the soybean crop is planted. The use of the rotary hoe about every 5 days following emergence and continuing until the beans have made a growth of about 6 inches (15 cm) has been particularly effective in controlling weeds.

Methods of minimizing or controlling weeds include (1) narrow rows, which will help control weeds later in the season by shading; (2) good seed, which will result in vigorous seedlings; (3) crop rotation, which will enable the use of herbicides with other crops not suitable for soybeans to control problem weeds; (4) mechanical cultivation; and (5) herbicides.

In addition to rotary-hoeing or using a spike-toothed harrow for control of weeds early, sweep cultivation can be used when soybeans are 8–10 inches (20–25 cm) tall. The cultivator should be set to run shallow because most soybean roots are in the top 12 inches (30 cm) of soil. About 70 percent of the U.S. soybean acreage is treated with some type of herbicide. Some states exceed this average, with the Illinois treated acreage being above

FIGURE 23-10. *Herbicide-treated plot of soybeans (left) as compared to a weedy control (right). About 70 percent of the United States soybeans acreage is treated with a herbicide.* [SOURCE: Southern Illinois University.]

80 percent. Herbicides may be applied preplant, pre-emergence, or post-emergence; sometimes combinations of these are used. One of the leading preplant soybean herbicides is trifluralin. Alachlor, chloramben, dinitramine, linuron, metribuzin, and naptalam are applied preemergence, while chloroxuron, 2,4-DB, and bentazon are leading postemergence chemicals. Sometimes glyphosate is used to control tall-growing weeds, such as johnsongrass, with the newer recirculating sprayers.

The principal grassy weeds in soybeans are crabgrass, foxtail, barnyardgrass, nutsedge, and johnsongrass. The broadleaf weeds posing the greatest problems include pigweed, lambsquarter, ragweed, cocklebur, morningglory, smartweed, jimsonweed, and velvetleaf.

Harvesting and Storing

Harvesting losses frequently amount to as much as 10 to 20 percent of the crop produced. Agricultural engineers have estimated that the average producer loses 8–10 percent. Four beans per square foot (43 beans/m²) of land surface means a loss of approximately 1 bushel per acre (67 kg/ha). A soybean loss calculating guide is shown in Table 23–3. Bushels per acre loss can be determined by counting the number of soybeans on the ground in a 10 square-foot (0.9 m²) area. The greater the ridging of rows with cultivating machinery, the greater the combine loss at harvest. It has been estimated that the expected acre loss at a 30-bu/acre (2,016-kg/ha) yield level may range from 1.5 bu (41 kg) with an excellent harvesting job to 3.8 bushels (124 kg) with an average harvesting job.

Virtually all soybeans are harvested with a combine. The moisture content at harvest should be 13 or 14 percent if beans are to be stored without artificial drying. Harvest can begin at a considerably higher moisture level without combine damage if drying facilities are available. Proper combine adjustment, such as proper cylinder speed, is one important factor in mini-

TABLE 23–3. Soybean Loss Calculating Guide*

No. of Soybeans	Bushels/Acre	No. of Soybeans	Bushels/Acre
10	0.25	110	2.75
20	0.5	120	3.0
30	0.75	130	3.25
40	1.0	140	3.5
50	1.25	150	3.75
60	1.5	160	4.0
70	1.75	170	4.25
80	2.0	180	4.5
90	2.25	190	4.75
100	2.5	200	5.0

* Loss in Bushels Per Acre from a Determination of Number of Beans in 10 Square Feet of the Field.

FIGURE 23–11. *Combine harvesting soybeans. Harvesting losses, including pre-harvest shattering and gathering losses, frequently amount to as much as 10 to 20 percent of the crop produced.* [SOURCE: Sperry New Holland.]

mizing seed damage and harvest losses. Cylinder speed should be reduced in comparison to that used for cereals. Also, the cutter bar should be run as close to the ground as possible. Losses from splits, broken beans, and unthreshed pods can be greatly reduced by harvesting as early as possible.

Most of the harvesting losses are attributed to preharvest shattering and gathering losses. Gathering losses refer to beans left on the stubble, or on uncut or lodged stalks. Both height of the first pod from the ground and lodging, which greatly influence gathering losses, vary greatly among the cultivars. Cultivars also vary as to resistance to shattering; some are available that are relatively shatter-resistant.

Beans can be stored up to three years if the moisture content is 12 percent or less. At 14 percent moisture, they can be stored for a few months, depending on the storage temperature.

Insects, Diseases, and Other Pests

Several insects may attack soybeans but relatively few are considered economically serious pests. The estimated annual loss from insects is about 3 percent. Most soybeans are grown without the application of an insecticide, and if insect buildup does occur, normally one application of an appropriate insecticide is sufficient.

Insects that may be damaging to the crop in certain situations include grasshoppers, leafhoppers, stinkbugs, Mexican bean beetles, velvet bean caterpillars, cabbage loopers, green clover worms, fall armyworms, wireworms, grubs, blister beetles, spider mites, and thrips. Some soil-borne insects may warrant application of a soil insecticide at planting time if past history indicates the need. Most insecticides are applied on an "as needed" basis, or when sufficient damage is observed.

About 50 diseases are known to attack this crop. Estimates of annual crop losses range from 12 to 14 percent of the crop. The fungus diseases are the most prevalent and include: phytophthora rot, brown stem rot, stem canker, pod and stem blight, frogeye leaf spot, downy mildew, brown spot, target spot, and purple stain. Bacterial diseases of importance are wildfire, bacterial blight, and bacterial pustule. Mosaic diseases are caused by viruses. Plant resistance to some diseases has been found. When available, use of resistant cultivars is the best preventive measure. Many diseases overwinter in plant residues or in the soil. The mode of overwintering permits effective control of the disease through deep plowing to bury residues, and through crop rotations. Certain diseases, such as brown stem rot, seldom cause damage except where fields have had soybeans for two or more consecutive years.

Another serious soybean pest is the nematode, a small wormlike organism. The most damaging and widespread nematode is the soybean cyst nematode, which attacks roots and results in stunted, yellowed plants. Damage may occur in more or less circular areas of the field. Resistant cultivars are available for race 3 of the organism, but none have been developed and released for race 4 resistance. Crop rotations of at least three or four years may be effective in lessening nematode severity.

Estimates of soybean crop losses from nematodes range from 2 to 10 percent of the crop, with the higher estimate probably more accurate since the severity of the soybean cyst nematode has been realized in recent years.

Utilization

Prior to 1920 the soybean was used primarily as a forage crop in the United States, for hay, silage, soilage, and pasture. With the war-stimulated demands for oil and the new industrial uses, a smaller and smaller percentage of the crop was harvested for forage. Now practically 100 percent of the acreage is harvested for beans.

Soybean Processing

The processing of soybeans on a large scale for the oil and high-protein meal depended upon the development of markets for these products. Such markets developed slowly at first. The earliest available record of soybeans processed for oil in the United States is that of a shipment from Manchuria in 1910 or 1911. In 1915 an expeller type of cottonseed oil mill in North Carolina processed locally grown soybeans.

Soybean processing may be said to have become established as an indus-

FIGURE 23-12. *Soybean processing results in a dual-purpose oil. This oil competes with the nonedible, drying oils such as linseed oil, and with the edible nondrying oils such as cottonseed oil. Most soybean oil is used in edible products.* [SOURCE: American Soybean Association.]

try in the early 1920s, when a number of plants were started. The large increase over 1942–1943 was in response to the government's appeal and program to obtain greatly increased production of soybean oil and other domestic oils because of the large wartime requirement and because foreign supplies had been cut off.

METHODS OF EXTRACTION. In the 1940s most of the soybean crop processed in this country was handled by mechanical oil extraction facilities, such as the hydraulic press and the continuous expeller or screw press. The 1950s brought a movement toward conversion to solvent extraction, and at present virtually all soybeans are handled in chemical solvent plants. Solvent extraction is more efficient and thus gives a higher yield of oil, about 11 pounds of oil per bushel of soybeans (5 kg per 27 kg). A bushel of soybeans also yields about 47–48 pounds (21 to 22 kg) of 44 percent protein meal, with an average loss of about 3 pounds (1.4 kg).

Oil and Meal Utilization

Soybean oil is a dual-purpose oil; it competes with the nonedible drying oils, such as linseed oil, and with the edible nondrying oils, such as cottonseed oil. More than 90 percent of the soybean oil produced is used in edible products, such as margarine, shortening, salad oils, cooking oils, and mayonnaise. The remainder is used industrially in paints, varnishes, linoleum, printing ink, soap, and rubber substitutes. Soybean oil, an edible oil, is intermediate as to degree of unsaturated fats. The common oils in order of most to least polyunsaturates are safflower, corn, soybean, cottonseed, sesame, peanut, olive, and coconut. The importance of soybean oil production in the United States in relation to other vegetable oils and animal fats can be seen in Figure 23–13. Soybean oil production is greater than production of all other oils and fats combined from domestic and imported materials.

Soybean meal is valued as a livestock and poultry feed for its high protein content; well over 90 percent of all soybean meal is used in this way. The relative importance of soybean meal for high-protein feed in comparison to other products is shown in Figure 23–14. A portion of the remaining processed meal is utilized as industrial protein in adhesives and in wall-

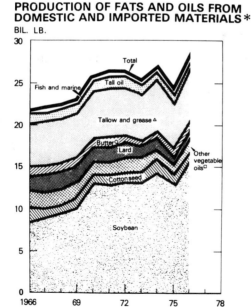

PRODUCTION OF FATS AND OILS FROM DOMESTIC AND IMPORTED MATERIALS *

BIL. LB.

* INCLUDES OIL EQUIVALENT OF EXPORTED DOMESTIC OILSEEDS. ᐃBOTH EDIBLE AND INEDIBLE KINDS. °FAT CONTENT. ◻INCLUDES CORN, OLIVE, PEANUT, SAFFLOWER, COCONUT, CASTOR, LINSEED, AND TUNG OILS.

FIGURE 23–13. *Soybean oil production in the United States in relation to other vegetable oils and animal fats.* [SOURCE: USDA.]

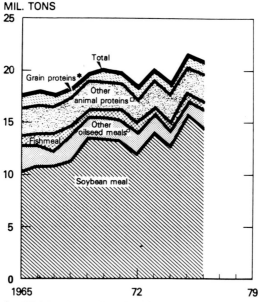

FIGURE 23-14. *The relative importance of soybean meal for high-protein feed in comparison to other products.* [SOURCE: USDA.]

board and paper coating. Soybean flour and grits are used in bread and bakery products, macaroni, spaghetti, baby foods, high-protein beverages, meat extenders, and dietary specialities.

As the world population increases, the soybean is destined to play an ever-increasing role in human nutrition. With the exception of the oil component, the soybean does not play a major role directly in the human diet of United States people. For centuries, however, soybeans and soybean foods have occupied a prominent place in the diets of people in the Far East.

Human soybean products gaining importance in this country are (1) soybean flour and grits, with 40–60 percent protein; (2) soybean protein concentrate, with more than 70 percent protein; and (3) isolated soybean protein containing 90–97 percent protein. It is projected that soybean demand and use will continue to expand and meat substitutes and meat extenders, made from soybeans, will gain wider acceptance in this country.

Cultivar Improvement

INTRODUCTION AND SELECTION. By 1920, more than 1,000 cultivars had been introduced from east Asia. Over half the soybean acreage of the United States prior to World War II was planted to these introduced cultivars and to selections from such introductions.

The USDA has collected soybeans from 120 places in Japan and Korea to serve as a new source of germplasm. Presently, the USDA has a collection of more than 3,500 soybean lines. Most of the early-maturing cultivars grown in the Midwest trace back to only six introductions from northern China. Because of such a narrow genetic base, there is a need to evaluate numerous plant introductions and include those most promising in a breeding program.

BREEDING FOR IMPROVEMENT. In an attempt to combine the more desirable characteristics of several cultivars into one, breeding programs were initiated at a number of experiment stations in cooperation with the USDA. A relatively high oil content of the seed became a prerequisite for the release of a new cultivar. More recently, with oil less in demand and with a relatively higher price for soybean meal, attention to the protein content of the beans is increasingly important. In addition to yield and protein and oil content, other characteristics noted in a description of cultivars released during 1970–1976 include resistance to certain insects, dis-

FIGURE 23–15. *One soybean improvement achieved through breeding has been the development of cultivars which resist shattering of the seed from the pod before harvest.* [SOURCE: American Soybean Association.]

eases, nematodes, lodging, and seed shattering. Also important in new cultivars are seed quality, and pods set high enough above the soil surface to
minimize harvesting losses. A consideration in lodging resistance is the
possible use of semidwarf cultivars. Recently, 'Elf,' a semidwarf cultivar,
was released by the USDA in cooperation with the Illinois and Ohio stations. This compact, determinate cultivar, which grows to a height of only
about 15 inches (38 cm), has the potential to increase yield by as much
as 10 or 15 percent when used in narrow or drilled rows at very high
populations. Pods set low, near the ground, may cause harvesting problems
with these compact types.

The first hybrid soybean may still be years away. The first major step
toward developing commercial hybrids, discovery of genetic male sterility,
has been accomplished. Currently the greatest problem toward large-scale
hybridization is attaining the amount of cross-pollination required for economic seed production. This may require complete re-engineering of this
normally self-pollinated plant.

The inherent sensitivity of the soybean to environment seriously limits
the latitudinal adaptation of a cultivar. Although a single cultivar may
be well-adapted to large belts in the heavy soybean-producing areas, in
the marginal areas, where environmental conditions are varied, cultivars

FIGURE 23-16. *Soybean cultivars have been classified into 10 maturity
groups. An early-maturing cultivar (left) as compared to a later-maturing
cultivar (right) can be seen in this cultivar evaluation trial.* [SOURCE: Southern Illinois University.]

TABLE 23–4. Maturity Groups, Adaptation, and Cultivars by Group

Group	Maturity Ave. Days From Planting	Representative or Newer Cultivars	States Where Adapted
00	120	Ada	Mich., N.D., S.D., Minn.
0	126	Evans, Grande, Swift, Vansoy, Wilkin	Mich., N.D., Wisc., S.D., Minn.
I	126	Hodgson, SRF 150, Steele	Mich., Wisc., S.D., Ohio, Neb., Minn., Iowa, Ind., Ill.
II	130	SRF 200, Wells, Amsoy, Beeson	Mich., Wisc., S.D., Ohio, Neb., Mo., Minn., Iowa, Ind., Ill.
III	131	SRF 307P, Williams, Woodworth, Calland	Ohio, Neb., Mo., Kan., Iowa, Ind., Ill., Del.
IV	136	Bonus, Columbus, Cutler 71, Mitchell, Pomona, SRF 425, SRF 450	Va., Tenn., Mo., Md., Ky., Kan., Ind., Ill., Del.
V	139	FFR 556, Forrest, Mack, McNair 500	Va., Tenn., Okla., Mo., Miss., Md., Ky., Kan., Del., Ark.
VI	148	Coker 136, FFR 666, Hood 75, Lee 74, McNair 600, Pickett 71, Tracy	Va., Tenn., S.C., Del., Okla., N.C., Miss., Md., La., Ky., Ga., Fla., Ala., Ark.
VII	156	FFR 777, Bragg	S.C., N.C., Miss., Va., Ga., Fla., Ala., Ark.
VIII	158	Cobb, Coker 338, Hutton	S.C., La., Ga., Fla., Ala.

frequently are found to respond differently in areas relatively near each other.

Soybean cultivars have been classified into 10 maturity groups from 00 to VIII. Table 23–4 shows the maturity groupings and average days to maturity, representative or newer cultivars in each group, and states where adapted. Cultivars in the 00 groups are earliest in maturity and are adapted to the northernmost parts of the United States and Southern Canada, while those in VIII are latest in maturity and are used mostly near the Gulf.

REFERENCES AND SUGGESTED READINGS

1. ABEL, G. H., and L. H. ERDMAN. "Response of Lee Soybeans to Different Strains of *Rhizobium Japonicum*," *Agron. Jour.*, 56:423 (1964).

2. ANTHOW, K. L., and R. M. CALDWELL. "The Influence of Seed Treatment and

Planting Rate on the Emergence and Yield of Soybeans," *Phytopathology,* 46:91 (1956).

3. Baldwin, F. L., R. P. Nester, and G. L. Morris. *Control Weeds Early in Soybeans.* Ark. Coop. Ext. Serv. Leaflet 444 (1976).

4. Barber, G., C. Hendrix, and C. R. Weber. "Our Experiences with Soybean Fungicides," *Proc. 2nd Soybean Conf.,* Chicago, Ill, Dec. 14–15, 1972. (Washington, D.C.: American Seed Trade Association, 1972).

5. Berg, G. L. (Ed.). *New Comprehensive Manual-Modern Soybean Production* (Ambler, Pa.: Anchem Products, 1967).

6. Brim, C. A. "Hybrid Soybeans," *Crops and Soils,* 25(2):12 (1972).

7. Browing, G. M. *USDA and Iowa Station Cooperating* (Chicago: National Soybean Processors Association, 1951).

8. Burnside, O. C., and R. S. Moomaw. "Control of Weeds in Narrow-Row Soybeans," *Agron. Jour.,* 69:793 (1977).

9. Burnside, O. C., and W. L. Colville. "Soybean and Weed Yields as Affected by Irrigation, Row Spacing, Tillage, and Amiben," *Weeds,* 12:109 (1964).

10. Caldwell, B. E., and G. Vest. "Effects of *Rhizobium Japonicum* Strains on Soybean Yields," *Crop Science,* 10:19 (1970).

11. Cartter, J. L., and E. E. Hartwig. "The Management of Soybeans," In *Advances in Agronomy,* Vol. 14 (New York: Academic Press, 1962). ed. A. G. Norman.

12. Caviness, C. E. "Nodulation of Soybeans Following Rice," Ark. Agr. Exp. Sta. *Ark. Farm. Res.* 15(6):12 (1966).

13. Chapman, S. R., and L.P. Carter. *Crop Production, Principles and Practices* (San Francisco: Freeman, 1976).

14. Elkins, D. M., *et al.* "Effect of Cropping History on Soybean Growth and Nodulation and Soil Rhizobia," *Agron. Jour.,* 68:513 (1976).

15. Ham, G. E., V. B. Caldwell, and H. W. Johnson. "Evaluation of *Rhizobium Japonicum* Inoculants in Soils Containing Naturalized Populations in Rhizobia," *Agron. Jour.,* 63:301 (1971).

16. Hanway, J. J. "Foliar Fertilizing of Soybeans," *Crops and Soils,* 29(7):9 (1977).

17. Hartwig, E. E. "Reinoculation of Soybeans," *Miss. Agr. Exp. Sta. Farm Res.,* 27:1 (1964).

18. Hinson, K. *Soybeans in Florida,* Fla. Agr. Exp. Sta. Bul. 716 (1967).

19. Howell, R. W. "Physiology of the Soybeans," In *Advances in Agronomy,* Vol. 12 (New York: Academic Press, 1960). ed. A. G. Norman.

20. Howell. R. W., and J. L. Cartter. "Physiological Factors Affecting Composition of Soybeans II. Response of Oil and Other Constituents of Soybeans to Temperature Under Controlled Conditions," *Agron. Jour.,* 50:664 (1958).

21. Jeffery, L. S. *et al. Control of Johnsongrass in Soybeans.,* Tenn. Agr. Exp. Sta. Bul. 574 (1978).

22. Jones, G. D., J. A. Lutz, Jr., and T. J. Smith. "Effects of Phosphorus and Potassium on Soybean Nodules and Seed Yield," *Agron. Jour.,* 69:1003 (1977).

23. Kust, C. A., and R. R. Smith. "Interaction of Linuron and Row Spacing for Control of Yellow Foxtail and Barnyardgrass in Soybeans," *Weed Science,* 17:489 (1969).

24. Lynch, D. J., and O. H. Sears. "The Effect of Inoculation Upon Yields of Soybeans on Treated and Untreated Soils," *Soil Sci. Soc. Amer. Proc.,* 16:214 (1952).

25. Martin, J. H., W. H. Leonard, and D. L. Stamp. *Principles of Field Crop Production.,* 3rd ed. (New York: Macmillan, 1976).

26. Moldenhauer, W. C., and W. H. Wischmeier. "Soybeans Permit Erosion in Corn-Soybean Rotation," *Crops and Soils,* 21(6):20 (1969).

27. Murphy, W. J., *et al. Predicting Flowering and Maturity of Soybeans,* Mo. Sci. and Tech. Agron. Guide 9 (1976).

28. Norman, A. G. "The Nitrogen Nutrition of Soybeans," *Soil Sci. Soc. Proc.,* 8 (1943). pp. 226–228.

29. Parker, M. B., and H. B. Harris. *Molybdenum Studies on Soybeans,* Ga. Coastal Plain Sta. Res. Bul. 215 (1978).

30. Pendleton, J. W., and E. E. Hartwig. "Management," In *Soybeans: Improvement, Production, and Uses,* Monograph 16, edited by B. E. Caldwell (Madison, Wisc.: American Society of Agronomy, 1973).

31. Robbins, P. R., and J. L. Strom. *The Economics of Narrow Row Corn and Soybeans,* Purdue Univ. Coop. Ext. Serv. EC-309 (1966).

32. Schmid, A. R., A. C. Caldwell, and R. A. Briggs. "Effect of Various Meadow Crops, Soybeans, and Grain on Crops Which Follow," *Agron. Jour.,* 51:160 (1959).

33. Scott, W. O., and S. R. Aldrich. *Modern Soybean Production* (Champaign, Ill.: S&A Publications, 1970).

34. Siemens, J. C., and H. J. Hirning. *Harvesting and Drying Soybeans.* Ill. Coop. Ext. Serv. Circ. 1094 (1974).

35. *Soybean Farming* (Chicago: National Soybean Processors Association, 1961).

36. Thompson, H. E., and J. C. Herman. *Guidelines for Profitable Soybean Production.* Iowa Agr. Exp. Sta. Pm-441 (1968).

37. TVA. *Soybean Production, Marketing and Use,* Bul. Y-69 (1974).

38. USDA Agr. Stat. (1977).

39. USDA-ARS. *Losses in Agriculture,* Agr. Handbook 291 (1965).

40. USDA-Ext. Serv. *New Crop Cultivars,* 13, 1970–76, Publ. ESC-584 (1977).

41. Wax, L. M. "Weed Control," In *Soybeans: Improvement, Production, and Uses,* Monograph 16, edited by B. E. Caldwell (Madison, Wisc.: American Society of Agronomy, 1973).

42. Wax, L. M. "Weed Control for Close-Drilled Soybeans," *Weed Science,* 20:16 (1972).

43. Wax, L. M., and J. W. Pendleton. "Effect of Row Spacing on Weed Control in Soybeans," *Weed Science,* 16:462 (1968).

Chapter 24
COTTON

COTTON is regarded as an important cash crop in the United States. Despite the development and widespread use of several synthetic fibers, domestic demand for cotton continues to be great. This is due in part to the extensive use of cotton for denim fabrics in the "age of jeans."

Cotton contributes not only fiber, but food and feed as well. In addition to lint, the seed is a by-product that furnishes high-grade oil and a protein concentrate.

History

Cotton was grown for its fiber in ancient times, long before there were any written records. Archaeological finds and the earliest written records indicate that cotton was known in Mexico as early as 5000 B.C., Pakistan in 3000 B.C., Peru in 2500 B.C., and in India at least by 2000 B.C. In the Digest of Ancient Laws, ascribed to Manu, 800 B.C., cotton is referred to so often and in such a way as to indicate that it must have been known in India for generations, both as a plant and as a textile. From 1500 B.C. until about 1500 A.D. India was the center of cotton production and manufacture. Early European travelers reported seeing "wool trees" in India.

Cotton fabrics discovered in Arizona ruins furnish evidence that this crop was utilized over 800 years ago in that state. When Columbus landed on the American continent he found Indians using cotton cloth. This contributed to his belief that he had reached India. Cotton was being grown in the British colonies of Virginia and the Carolinas shortly after they were founded. Soon cotton spread to other states and played a major role in development of the young country. Large-scale cotton production did not begin, however, until after Eli Whitney had invented the cotton gin in 1794.

Throughout the nineteenth century cotton production expanded steadily in the South. It moved west to Mississippi and then into Arkansas, Oklahoma, Louisiana, and Texas. Later it became an important crop in California, New Mexico, and Arizona.

Origin and Species

Cotton belongs to the genus *Gossypium* of the *Malvaceae*, or mallow, family. The genus *Gossypium* has been studied perhaps as intensely as any plant group, including all domestic and wild forms.

The number of species reported can vary, depending on the taxonomist and the system employed. A classification by Hutchison *et al.* in 1947 recognized 20 species, with the cultivated cottons grouped under four of these. The two cultivated American species, *G. hirsutum* and *G. barbadense,* cross readily and give fertile progeny, as do the two cultivated Asiatic species, *G. arboreum* and *G. herbaceum.* These species all bear a spinnable fiber called lint, which distinguishes them from the wild forms, which do not have this character.

A current listing of *Gossypium* types includes 34 species. Some species are diploid and some are tetraploid; some have a diploid chromosome number of 26, and others, including the cultivated American species, have 52 chromosomes.

There are many unknowns about the origin of different cotton species. Regarding the New World cottons, one hypothesis has been that the Asiatic and American cottons crossed and formed an amphidiploid containing a set of chromosomes from each parent. The theory has been advanced that the crossing occurred in Polynesia before the establishment of the Pacific Ocean in its present form. The alternative theory is that in a more recent geological epoch, man carried an Old World form to America, or vice versa, with the probability that *G. arboreum* of southwest Asia crossed with *G. raimondii,* the wild cotton of Peru. More recent cytogenetic work suggests that the Old World parent of the amphidiploid may have been an ancestral type nearer *G. herbaceum* than *G. arboreum.*

The cottons that concern growers in the United States are the short-staple American upland, the *G. hirsutum* species, and the long-staple cottons of the *G. barbadense* species.

The long-staple Sea Island cotton is not grown to a large extent in the United States; it is grown in the West Indies, in South America, and in the Nile Valley (Egypt and Sudan). The long-staple American Egyptian cotton is grown in the irrigated Southwest, in South America, and in the Nile Valley. These are both the *G. barbadense* species.

The presentation here concentrates mainly on short-staple American upland cottons of the species *G. hirsutum* native to southern Mexico and Central America and grown generally throughout the Cotton Belt and in all major cotton-producing areas of the world.

Adaptation and Distribution

The average summer temperature along the northern boundary of the Cotton Belt in the United States is 77°F (25°C). This temperature appears to be the limit, beyond which cotton production becomes unprofitable. The length of the growing season along the northern border is 180–200 days. The general rainfall throughout the South is 30–60 inches (76–152 cm). A rainfall of 6–16 inches (15–41 cm) in the spring, with a somewhat larger amount in the summer and a small amount in the fall is desirable. Cotton may be expected to produce crops worth harvesting, at least part of the time, in that part of the southern Great Plains where the normal rainfall is 17 inches (43 cm) or more, with approximately 75 percent occurring during the growing season.

The best soils for cotton production are the loams of the Mississippi Delta and the Southwest. Heavy clays tend to delay maturity and increase vegetative growth, resulting in more insect and disease problems. Soils should be well-drained, moderately fertile, and organic matter content high. The best soil pH is 5.8–6.5, but cotton is tolerant of a wide pH range. A summary of the most important requirements for good production would include (1) adequate moisture during the spring and summer, (2) adequate sunshine (high light intensity), and (3) a long, frost-free growing season.

In 1976 four Western states produced 29 percent of the total U.S. cotton on 13 percent of the acreage; three Southwestern states produced 43 percent of the total on 54 percent of the acreage; and 12 Southeastern and Delta states produced 28 percent of the cotton on 33 percent of the U.S. acreage. The total U.S. production was about 10.5 million bales on 10.9 million acres (4.4 million ha) harvested. Geographic distribution in the United States is shown in Figure 24–1.

Leading states in cotton acreage are Texas, with almost half of the total, Mississippi, California, Arkansas, Alabama, Tennessee, Louisiana, and Arizona, in that order (Figure 24–2). The states in order of total production are Texas, California, Mississippi, Arizona, Arkansas, and Louisiana. Aver-

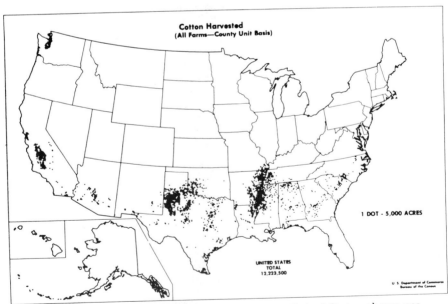

FIGURE 24–1. *Cotton acreage harvested in the United States, shown geographically.* [SOURCE: U.S. Department of Commerce.]

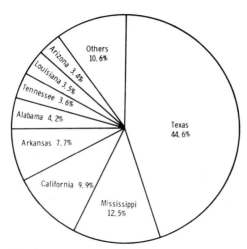

FIGURE 24–2. *Leading states in cotton area (acreage). Texas has almost one half of the United States total.* [SOURCE: Southern Illinois University.]

age acre yields in 1976 ranged from a high of nearly 1,200 pounds (544 kg) in Arizona to a low of about 250 pounds (113 kg) in Oklahoma, where considerable dryland cotton is produced. The U.S. average yield was about 465 pounds (211 kg), just under 1 bale/acre (0.4 bale/ha). Without question there is still a Cotton Belt in the South, but it is obvious from these statistics that much of the cotton production has shifted from the Old Cotton Belt to the Southwest and West.

Production Changes

World Production

The decade 1959–1969 saw a rapid and marked increase in world cotton production, particularly among the developing countries of Latin America, Africa, and Asia. The USSR, Eastern Europe, and mainland China increased their production by 4.7 percent during that decade. Although production had slowed somewhat by 1976, world production is nearly 60 million bales annually, 40 percent of which is by communist countries. In 1976 there were 31 countries which produced more than 100,000 bales annually. The world leaders in acreage are India, China, the United States, the USSR, Brazil, and Pakistan (Figure 24–3). The order of world production is the USSR, China, the United States, India, Pakistan, and Brazil (Figure 24–4).

Acreage expansion in foreign countries is in response to the profitability of cotton as a cash crop and an earner of foreign exchange. Government policies of the various countries also have been important in determining how much land and other resources would be devoted to cotton. In India, for example, the government provides producers with a guaranteed price,

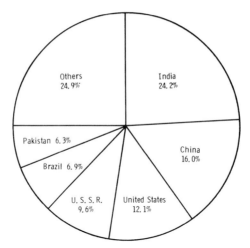

FIGURE 24–3. *Leading countries of the world in cotton area (acreage).*
India, China, and the United States have over one half of the world total.
[SOURCE: Southern Illinois University.]

subsidizes the cost of insect and disease control and of improved agricultural implements, and provides special credit facilities for purchasing seed
and fertilizer. When world prices go down, cotton acreage is reduced in
some foreign countries and the expansion slowed in others.

U.S. Production

The United States cotton crop once was about 60 percent of the world's
total production. For the 1960–1964 period, United States production was

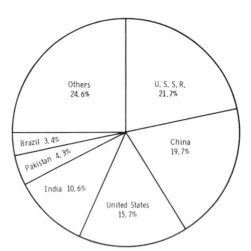

FIGURE 24–4. *Leading countries of the world in cotton production. The*
USSR, China, and the United States produce almost 60 percent of the world
total. [SOURCE: Southern Illinois University.]

only 19 percent of the world's production; in 1968 it was only 13 percent; in 1976 it was just over 15 percent. Because of decreased markets and accumulating surpluses an effort is being made to hold the American crop to 10–12 million bales/year. The total of United States cotton acreage harvested decreased from an average of more than 42 million acres (17 million ha) for the 1925–1929 period to about 23 million (9.3 million ha) for the 1950–1954 period, and 13 million (5.3 million ha) for the 1962–1966 period. Acreage held rather steady during the 1970–1978 period at 11 to 12 million (4.5 million ha) except for 1975–1976 (Figure 24–5).

Manmade Fibers Compete

Although cotton lint has been considered the most versatile fiber known, when all end uses are considered, economic and political conditions have resulted in increased competition by synthetic fibers. Manmade fiber production has increased considerably in the last 15 years. World output of manmade fibers is equivalent to somewhat over one-half the output of cotton. Most of the manmade fibers are produced in the United States, Western Europe, and Japan. The United States is the world's largest producer.

While present cotton consumption in the United States is still below the level of that in the early 1960s, there has been an upsurge in the 1970s. This success in competing with synthetic fibers and expanded use may be attributed to the development of new and improved all-cotton products to meet specific consumer requirements. This includes chemical finishing of cotton to modify existing properties or to impart entirely new properties. The newer chemical finishing processes produce cloths that are water and oil repellent, flame resistant, durable flame resistant, durable flame retar-

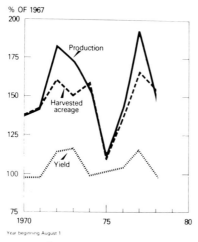

FIGURE 24–5. *Cotton production, acreage, and yield in the United States since 1970.* [SOURCE: USDA-ESCS.]

dant, mildew and rot resistant, sunlight resistant, able to be washed and worn and capable of retaining a permanent and durable press.

The Cotton Plant

The cotton plant has an indeterminate growth habit. Cotton may be said actually to be a tree (perennial), although in practice it is cultivated as an annual. It may grow to a height of only 6–8 inches (15–20 cm) and mature bolls, or to a height of 8–10 feet (2.4–3.0 m) depending on soil conditions and climatic factors. In general, the best crops are from plants that have made a growth of 4–5 feet (1.2–1.5 m).

The cotton plant is supported by a strong taproot, which may penetrate to a depth of 6–8 feet (1.8–2.4 m).

The cotton plant has two types of branching—vegetative and fruiting. The pattern of branch development can vary, depending on the species and cultivar, the climatic conditions, and cultural practices. A vegetative or fruiting branch can develop from a bud located at a node in a leaf axil. Occasionally a second axillary bud develops resulting in both a vegetative and a fruiting branch in the same leaf axil.

On the day of anthesis, flowers of upland cotton may be a rich, creamy white, or sometimes pure white, but they turn to a pinkish color during the afternoon and to a distinct red on the second day. Egyptian types have yellow flowers with a purple spot at the base. The corolla usually is released and drops to the ground on the second or third day after the flower has opened, leaving the small boll or "square" encased within three bractlets. As a rule, bolls reach full size within 20 days and require the remaining days for laying down the secondary wall of cellulose in the fibers and for maturing seed.

The usual number of segments, or carpels, of the bolls will vary from

FIGURE 24-6. *Cotton flowering and fruiting. (A) Upland cotton flowers are creamy white at anthesis, but turn pink and later red. Squares and bolls can also be seen here. (B) Open bolls of upland cotton shown here have five locks.* [SOURCE: Agricultural Information Office, Oklahoma State University.]

A   B

three to five. Upland cotton normally will produce from four- to five-lock bolls. The Egyptian cottons may show a higher percentage of three-lock bolls, with the crop generally running from three- to four-lock bolls. Each lock of a boll will contain from five to nine or more seed, the average being about seven.

Shedding of Bolls

The cotton plant under normal growing conditions will set all the bolls it can carry; any additional number are "shed." As many as 50 percent of the bolls of most cultivars are shed before reaching maturity. The shedding may take place before the corolla of the flower falls. Also, squares may shed before flowering. Conditions that can influence shedding are drought, waterlogged soils, deep cultivation late in the season, insufficient nutrients, and insect damage.

Pollination of the Flower

The cotton plant bears complete and perfect flowers, with the flowers usually self-fertilized. Fertilization with pollen from other plants may range from less than 5 percent to as much as 50 percent, depending on the activities of insects. There probably is little crossing, aside from that by insects.

The Seed

Within each boll of cotton there are from five to nine seeds per lock. In Upland cotton, the seeds have fuzzy, short fibers as well as longer fibers (lint). In Sea Island cotton the seeds are free from fuzz except at the tip. The seed coat is dark brown or black in color under the dense coat of fuzzy hairs which are not removed with the lint fibers in ginning. The embryo, including two large cotyledons, makes up almost the entire seed.

Fiber

Cotton fibers are single-cell specialized extensions of the epidermal cells of the ovules or seed. They usually begin growth about the day of anthesis and continue for approximately 25 days. The initial growth is known as the primary cell wall and consists of a very thin hyaline tube. After fiber elongation has been completed, the secondary wall, or inner part of the fiber, is deposited as thin concentric layers of cellulose. These give the fiber its body. Each cultivar of cotton is known for its particular length of fiber. Cotton planters are able to predict within one thirty-second of an inch (0.08 cm) the length of fibers they may expect to harvest from a given cultivar. Upland cotton cultivars generally have fiber lengths of ¾–1¼ inches (1.9–3.2 cm), while Egyptian cottons have fibers 1½ to 2 inches (3.8–5.1 cm) long or more.

FIBER QUALITY. The success with which cotton is used in spinning depends on such qualities as length, strength, fineness, and maturity.

The length of the fiber, closely associated with market value, is influenced not only by cultivar, but also by environmental factors during growth and by the end processing. The oldest measure for evaluating cotton fiber was "staple length"—samples combed out and measured. A photoelectric device known as the fibrograph came into use by cotton breeders; this device measures fiber length by scanning a fiber sample and giving a length–frequency distribution from which fiber lengths are readily obtained.

The methods of determining fiber strength in use through many years were extremely time-consuming. A rapid breaking-strength method, using what is known as the Pressley breaker, replaced methods previously used. The most recent device for testing fiber strength and elongation is the Stelometer. Much more rapid methods of getting a measure of fiber fineness have been devised than were formerly in use. A 50-grain (3.24-g) wad of cotton is placed in the Micronaire, and air is forced through it. The amount of resistance to the air is used as a measure of the fineness; fine fibers restrict the passage of air more than coarse fibers.

There are several ways to measure maturity. One method is called the Causticaire test. First, a 50-grain wad (3.24-g) of cotton from a sample is placed in the Micronaire and the measurement is recorded. Then another 50-grain (3.24-g) wad of cotton from the sample is treated with sodium hydroxide so that fibers swell and become as large as they would have if they had grown to maturity. This second sample is dried and also placed in the Micronaire and measured. Then the first measurement is divided by the second; the answer is the percentage of maturity.

Production

Cotton in the past was produced only with the expenditure of large amounts of labor, much of it a one-man and one-horse-plow proposition, followed by hand hoeing, thinning, and picking. A shortage of low-priced labor, caused by the migration of unskilled labor to the cities, led cotton producers to look for labor-saving machines and methods.

Large acreages of cotton now are produced under humid conditions in the South and Southeast and under subhumid or low-rainfall conditions in the Southwest and West. Particularly marked has been the increase in cotton production in the low-rainfall area of the West, which has been made possible by irrigation.

Important changes in cotton production practices have been occurring rapidly. Important details in efficient production vary greatly for different parts of the cotton area. Differences in climatic environment, especially as they relate to soil conditions and available moisture and daylength, vary the recommendations to be made. The student interested in cotton production will wish to obtain from his state university and similar sources the recommendations that apply especially to his local conditions.

A

B

FIGURE 24–7. *The geography of cotton production is represented by these photographs: (A) A cotton field in middle Tennessee, a portion of the "Old Cotton Belt."* [SOURCE: Southern Illinois University.] *(B) Much of the cotton production in recent years has moved to the low-rainfall Southwest and West, typified by this irrigated cotton field in New Mexico.* [Photograph with permission of New Mexico State University.]

Seedbed

The shape of the seedbeds prior to and after planting varies, depending upon soil types and climatic conditions. Generally, in the humid areas of the South and Southwest and irrigated areas of the West, optimum conditions can be obtained on raised or listed beds. To the other extreme, the final beds or soil in which the seed are planted are in the bottom of the furrows formed by listers. In the Southeast cotton is normally planted in flat rows. Cotton commonly is planted in furrows formed by listers in the western parts of Texas and Oklahoma.

Seedbed preparation in the irrigated areas of the Far West commonly includes stalk shredding followed by deep, flat, breaking and the land is bedded in February and early March. A heavy preplant irrigation is applied to the bedded land a month prior to planting. The cotton usually is planted in the beds in a manner similar to that in the Southwest and Mid-South areas. There is considerable interest in minimum tillage, by rebuilding old beds with a lister.

Fertilization

Throughout most of the Cotton Belt the crop most heavily fertilized is cotton. Much of the soil on which cotton is produced is relatively infertile; a profitable crop is dependent on adequate fertilization. In addition, proper fertilizer application hastens the crop to maturity and in this way may reduce boll weevil damage.

On most soils mixed fertilizer should be applied during or not more than 3 weeks before planting. On light soils part of the nitrogen should be applied as a side dressing. In the West to save labor nitrogen and other fertilizers may be applied successfully in irrigation water. Fertilizer banded at planting should be placed about 2–3 inches (5.1–7.6 cm) below and 2–3 inches (5.1–7.6 cm) to the side of the seed. Anhydrous ammonia should be placed deeper and farther to the side, especially when applied as a side dressing.

Cotton is tolerant to a wide soil pH range of 5.2–8.0. Lime usually is not applied. Plowing under animal manure, green-manure crops, and crop residues not only increases the organic matter content of the soil, but does much to improve its condition and water-holding capacity.

Planting

Cotton often is planted thicker than the desired final stand, and thinned mechanically with special cultivators or by flame when plants are 3–5 inches (7.6–12.7 cm) high. One reason for thick planting is that the cotton seedling is relatively tender and may not be able to push its way up through the soil, especially if the surface becomes crusted following rain. The swelling of a mass of seed, whether in a drill row or in a hill, often will break the soil crust, permitting the emergence of seedlings.

In the early days of cotton production, one of the most tedious and costly tasks was "chopping" cotton, thinning thickly planted cotton plants to a desirable stand by methodically chopping the extra seedlings by hand with a hoe.

The seeding rate varies between 8 and 25 pounds/acre of acid-delinted seed (9–28 kg/ha), depending on the row spacing and the moisture conditions of the region. In years past, undelinted, fuzzy seeds were planted, but procedures for delinting seed mechanically or with acid have facilitated the ease of handling and planting and have increased planting precision. Much of the cottonseed is treated with fungicides before planting. Although by no means a cure-all, seed treatment is an inexpensive means of (1) helping prevent seed rots and seedling diseases, such as "damping-off"; (2) saving expensive labor and replanting costs; and (3) improving emergence with more uniform stands of stronger plants.

A desirable final population is 20,000–50,000 plants/acre (49,419–123,548/ha) depending on spacings desired and the geographic area. A six-year planting rate study in Texas evaluated plant populations of 18,000; 33,500; 50,000; 64,000; and 77,400 plants/acre (44,477; 82,777; 123,548; 159,624; and 191,252 plants/ha) under irrigation. The highest stripper efficiency was obtained each year in the higher plant populations. There was no difference in yields for plant populations of 18,000; 33,500; and 50,000 and only slight decreases in yield for populations of 64,600 and 77,400 plants/acre. Buxton at Arizona found little difference in yield when plant populations ranged from 30,000 to 120,000 plants/acre (74,129–296,516/ha). Neither was there much effect on boll and fiber characteristics except boll size and fiber fineness.

Cottonseed is likely to rot if planting is followed by cold, damp weather. Most of the cotton is planted during March, April, and May. Early plantings are less susceptible to weevil damage. They fruit earlier and usually give higher yields. Cotton should not be planted until the soil has warmed up to 60°F (16°C) or more.

Factors to be considered in determining depth of planting include soil type, moisture, temperature, and seed vigor. For early stands shallow planting is preferred. The depth of planting in humid areas usually is 1–1.5 inches (2.5–3.8 cm); in the drier areas it may be as great as 3 inches (7.6 cm).

The distance between the rows has been standardized at 38–40 inches (97–102 cm). However, improved cultural practices, including increased use of fertilizers, irrigation, and herbicides, have resulted in renewed interest in row spacing. The Federal Cotton Acreage Control Programs also has caused many farmers to seek means of increasing yields to compensate for reduced acreage allotments. There has been considerable research with row spacings varying from 6 inches (15.2 cm) to various forms of skip-row planting. There has also been some interest in broadcast cotton.

Wanjura and Hudspeth studied irrigated cotton row spacings of 20, 21, and 24 inches (51, 53, and 61 cm) and two seed rows 14 inches (36 cm) apart on conventional 40-inch (102 cm) spaced beds on the Texas High Plain. Experimental data over a four-year period showed increases in yields ranging from 6 to 25 percent for close-row spacings over the conventional 40-inch rows (102 cm).

Research in Arizona indicates that narrower rows and higher plant populations have the possibility of producing higher yields with lower production costs. Such tests have shown boll size decreases, but little change in inherent cotton fiber properties, except for some reduction in fiber fineness. Switching from rows 40 inches (102 cm) apart to two rows on 40-inch (102-cm) beds, 10–14 inches (25–36 cm) between the two rows on a bed, consistently increased lint yield in Arizona trials.

Cultivation

Weeds in cotton may be controlled by cultural practices, flame, and herbicides.

CULTURAL PRACTICES. Weed and grass control in the dryland and irrigated areas of the Southwest has been a problem. The rotary hoe cultivator has been found an excellent tool for breaking soil crusts and destroying young weeds in the first month of top growth. The use of the rotary hoe may reduce the cotton stand by as much as 25 percent. To be effective, the rotary hoe must be used just as the weed seedlings are appearing at the soil surface. When the high-speed sweeps on each side of the rotary hoe are set flat and run shallow, cultivation can be 40 percent faster than a conventional cultivator with sweeps. The rotary hoe attachment can be used either in lister furrows or on beds.

FLAME. Flame cultivation uses propane or propane–butane gas burners. For midseason weed control, with more efficient burners than were formerly available, it is possible to begin flame cultivation when the plants are considerably smaller. It is important to have flame burners adjusted and positioned properly to avoid crop injury. With flame cultivation controlling the grass and weeds in the row, and the regular shovels of the standard cultivator used to control those in the middles, a greater degree of efficiency is now obtained in humid areas than has previously been possible.

Late-season control of grasses and weeds is very important when mechanical harvesters are to be used. The grass not only must be killed, but its aboveground growth must be destroyed as well; flaming does this effectively.

HERBICIDES. Large-scale use of herbicides for control of weeds in cotton began in the 1950s. Now herbicides are very widely used, with over 90 percent of the cotton acreage treated with a herbicide. The different times for application to the crop are referred to as preplanting, pre-emergence, and postemergence. Preplant herbicides are broadcast on the soil surface and incorporated into the top few inches of soil before planting cotton. Pre-emergence herbicides may be broadcast for full weed control, or else may be banded over the row at planting, and coupled with cultivation between rows for complete weed control. Postemergence herbicides generally are basally directed sprays after cotton plants are at least 3 inches (7.6 cm) tall. The principal herbicides used for cotton weed control include carbamates, toluidines, substituted ureas, and arsonates. Major weed pests include johnsongrass, crabgrass, pigweed, cocklebur, nutsedge, morning glory, and prickly sida.

Harvesting

The principal methods of harvesting cotton are machine stripping and machine picking. In addition, cotton may be picked or snapped by hand, but in most areas, hand picking and hand snapping have given way entirely to mechanical methods. Mechanical stripping first came into use about 1914 in the High Plains area of Texas, with many homemade sled strippers in use until about 1930. Two-row, tractor-mounted strippers were in use by 1943.

Harvest-Aid Chemicals

The need for preparation of the cotton plant for harvest became essential with the advent of the spindle picker and stripper harvester. Harvest-aid chemicals are now used on over 75 percent of the cotton acreage in the United States. The efficiency of mechanical harvesters and timing of harvest depend on reducing and removing foliage from the plant prior to harvest. For stripper harvesters the principal requirement of a harvest aid is to reduce the moisture content of the foliage; for the spindle harvester leaf moisture must be reduced or leaf fall must be induced prior to harvest. In addition to higher picking efficiency, there are other advantages to the use of defoliants: (1) less green stain and higher grade, (2) reduced boll rot, (3) quicker opening of bolls, and (4) reduction in number of insects that overwinter.

Defoliants induce leaf fall and are applied by ground machine or aircraft 7–14 days before harvest. Desiccants are applied for the purpose of rapidly drying the foliage; the dead foliage remains attached to the plant. Desiccants also are applied by ground machines or aircraft 1–3 days before harvest.

The growth habit of the cotton plant is such that harvest-aid chemical applications are only an adjunct to other cultural practices. Cotton may be grown and successfully machine-harvested without the use of harvest-aid chemicals. Cultural practices for highest yields per acre are associated with large amounts of foliage and production of bolls late in season. The advantage of applying harvest-aid chemicals to high-producing fields must be weighed against the disadvantages caused by loss of foliage. The development of the fiber and seed ceases when leaves are injured by the chemicals.

A

B

FIGURE 24–8. *Cotton harvesting—the old vs the new: (A) Cotton formerly was hand picked or hand snapped, (B) In most areas hand picking has given way entirely to mechanical harvesting.* [SOURCES: USDA-SEA, and Delta Branch, Mississippi State University Agriculture and Forestry Experiment Station.]

Mechanical Cotton Harvesters

The unavailability of labor in the World War II years resulted in an intensification of research and the more ready acceptance of new methods by cotton growers. Most important of all was the development and use of cotton pickers and strippers, making full mechanization of the crop a reality.

Several different machines for picking cotton have been developed, and now virtually all of the cotton is harvested mechanically. The pickers which have come into general use remove the fiber from the plants by means of mechanical fingers which are brought into contact with the cotton. In the rotating, spindle type of picker, mechanical fingers are brought in laterally to all parts of the cotton plant. The rotating motion of these fingers pulls the lint from the bolls. The forward motion of the picker is synchronized with the rearward movement of the spindles, so that the relative motion between the spindles and the cotton plant is zero. Mechanical pickers of the spindle type are used widely in the humid areas.

Strippers are most used in the extreme western part of the Cotton Belt. In this section cotton may be left on the plants until practically all the bolls are ripe and the leaves have fallen, making it possible simply to drive a stripper over the rows, stripping the bolls from the plants. The bolls and trash are separated at the gin. When a stripper is to be used, a storm proof type of cotton is strongly recommended.

Cotton cultivar characteristics have a very important relationship to a mechanized production program. Certain plant characteristics are particularly desirable for harvesting with spindle pickers and different characteristics for a stripper. Characters considered desirable for spindle picking include an upright growth, early fruiting, a fairly determinate growth

FIGURE 24–9. *Kinds of mechanical cotton harvesters: (A) A cotton stripper, being used to harvest broadcast cotton, can also harvest narrow row or regular cotton.* [SOURCE: Allis Chalmers Corporation.] *(B) A cotton picker, unloading cotton into a wagon for transport.* [SOURCE: International Harvester Company.]

A   B

A   B

FIGURE 24-10. *Newer cotton handling procedures: (A) Some cotton is stored as modules in the field until labor is available for transporting to the gin.* [SOURCE: University of Arizona Agricultural Experiment Station.] *(B) A modern module transporter.* [SOURCE: Lummus Industries, Inc.]

habit, bolls evenly spaced on the plant but beginning well off the ground, bolls which allow the cotton to fluff but at the same time cause it to stick in the bur for storm resistance, and leaves which shed readily when most of the bolls have matured. It is recognized that some of these characteristics are antagonistic to others.

For stripper harvesting, some of the characteristics considered desirable are a dwarf or semidwarf plant with short to medium fruiting branches that will fruit rapidly and mature fruit early and well off the ground. Also, the seed cotton should be closely held in the boll at maturity, because all or most of the bolls on the plant must be mature before the stripper enters the field.

Some work has been done with chemical growth retardants to suppress the late growth and fruiting in order to facilitate defoliation and expedite harvesting. At present, it appears that plant breeding, selecting lines with a more determinate fruiting habit, is a more promising procedure.

Cotton Ginning

The cotton grower is vitally interested in the ginning of his crop. It takes only minutes to gin a bale of cotton, but the expertise with which the ginning job is done determines the grade and staple on which the selling price of the cotton is based. Cotton ginning equipment is complicated. The

FIGURE 24-11 [OPPOSITE]. *Cotton ginning. (A) A modern saw gin outfit in operation. The saw gin is used for most upland cotton and is much faster than the roller gin. (B) A bale with securing bands or straps being ejected from the gin onto a conveyor for weighing and transfer.* [SOURCE: Lummus Industries, Inc.]

A

B

gin is primarily a mechanism for separating the lint from the cottonseed. There are two kinds of gins, saw and roller types.

Green or wet cotton, which often finds its way to the gin, cannot be satisfactorily handled until it is dried. Most gins are equipped with driers and lint cleaners. As the cotton comes from the initial drier, a cleaner removes some foreign material and fluffs up the fiber. Cleaners pick up a portion of the foreign materials and extractors remove additional sticks, stems, burs, hulls, and other trash.

The Saw Gin

This type of gin consists essentially of a cylinder of circular saws, each about 12 inches (30 cm) in diameter and spaced about 0.5 inch (1.3 cm) apart. These saws pass between specially designed metal strips, known as ribs, spaced to permit the saws to rotate but to prevent the cottonseed from passing through. These ribs are shaped like an elongated S and are nearly 0.375 inches (0.95 cm) wide on the face. They support a "roll" of seed cotton through which the saws are rotated at a speed of 400–800 rpm. The teeth on the saws act as hooks that catch and pull the lint from the seed.

The "doffer" removes the lint from the gin saws by directing through a nozzle a sharp blast of air in the same direction that the saws are rotating. This disengages all lint from the saw teeth.

A load of 1,200–1,500 pounds (544–680 kg) of seed cotton usually will yield about 500 pounds (227 kg) of lint (a bale) and 800–900 pounds (363–408 kg) of seed. The lint usually is baled in a rectangular bale about 27 inches (69 cm) thick, 54 inches (137 cm) long, and usually about 45 inches (114 cm) wide, although width varies. When cotton is to be shipped some distance the bales are put through a compress to reduce the size considerably.

The Roller Gin

American-Pima cottons yield a staple length of approximately 1.5 inches (3.8 cm). They are ginned on a *roller* gin, which has no saws. The fiber is removed by the use of special composition rollers which engage the lint and pull it from the seed. Roller ginning is a comparatively slow process. One of the principal merits of the saw gin is the speed with which large volumes of seed cotton can be handled. The cost of ginning Upland cotton has averaged about half the cost of ginning American-Pima.

Cottonseed

Cottonseed was at one time considered largely a waste product; at present it is a valuable part of the overall cotton crop. Production in 1977 exceeded 5 million tons (4.5 million MT). All parts of the seed are used—lint, fuzz or linters, hulls, oil, and meat. The percentage of oil in the seed can vary from 17 to 22. A ton of cottonseed will yield about 320 pounds (145 kg) of crude oil, 900 pounds (408 kg) of meal, 500 pounds (227 kg) of hulls, and 135 pounds (61 kg) of linters, with an average processing loss of 145 pounds (66 kg). Linters are used for stuffing, and for making cellulose-type products.

Cottonseed meal, a high-protein concentrate for animals, contains a toxic substance, gossypol, which limits use of this feed in livestock rations. Low-gossypol or "glandless" cotton cultivars have been developed that increase the value of the meal.

There has been a trend toward the solvent method of separating the oil from cottonseed, just as there has been in processing soybeans. Hydraulic processing leaves a significant amount of oil in the meal; with the solvent method practically all the oil is removed.

Cottonseed oil has many uses, among which are uses in margarine, salad and cooking oils, emulsions for medical purposes, soap, cosmetics, washing powder, linoleum, and oilcloth and as a paint base.

Insect Pests

The boll weevil came into southern Texas from Mexico as early as 1892 and by 1922 had spread throughout practically the entire Cotton Belt. About 40 percent of the losses attributed to cotton pests derive from the boll weevil. The adult punctures the bolls or squares and lays eggs; when the eggs hatch, the boll is destroyed. It is most destructive in humid areas which receive more than 25 inches (64 cm) of rainfall annually. Probably no other insect pest in the United States has been the subject of so intensive a research program.

The pink bollworm is known as the world's most destructive cotton pest; it is very destructive in the United States as well. It first appeared in the United States in 1917. Under heavy infestation the crop is inferior because of staining, shorter staple, lower yields, and less tensile strength.

The bollworm, the same insect as the corn earworm and the tobacco budworm, also may destroy the boll.

Other destructive cotton insects include aphids, cotton leaf perforators, spider mites, red spiders, thrips, cotton leafworms, garden webworms, and lygus bugs. Preventive or control measures include destroying old cotton stalks, often by deep plowing; destroying weeds and other alternate hosts for insects; heat treatment of seed (for pink bollworm); and use of insecticides. Among the insecticides in common use are endrin, parathion, methyl parathion, demeton, disulfoton, phorate, methomyl, and methidathion. Biological control of cotton insects is gaining importance. Predators, parasites, and other natural enemies play an important role in control. In addition, the application of the bacterium *Bacillus thuringiensis*, a biological control agent for looper-type insects, is recommended in some areas.

Diseases

Several different fungi and bacteria attack cotton in the seedling stage. Perhaps the best known and most widespread is damping off. Other diseases include anthracnose, a fungus disease, and bacterial blight (angular leaf spot). These diseases can be controlled by chemical seed treatment. Wet-weather blight and bacterial blight are the two main causes of leaf spot. Wet-weather blight usually is halted by drought; rotation helps control bacterial blight. In addition, some cultivars are resistant to certain races of the bacterial blight organism.

There are two kinds of wilt, *Fusarium* and *Verticillium.* Fusarium wilt can be controlled by planting wilt-resistant cultivars. Clean summer fallowing, rotation, and other cultural practices help reduce losses from both fusarium wilt and verticillium wilt. Some resistant cultivars are available for verticillium wilt in some regions, but not to the same extent as fusariums. The root-knot nematode is controlled by planting tolerant cultivars, practicing rotation, and fumigating. Root-rot damage is reduced by crop rotation, use of early-maturing cultivars, and heavy applications of nitrogen and barnyard manure. There are several kinds of boll rot. To prevent boll rot, sunlight and air must be available. In rank cotton, defoliation may be the answer. Seed treatment and rotation also may help prevent boll rot. Crazy top generally appears when plants do not get enough moisture. Crinkle leaf is caused by a lack of calcium. There are two kinds of cotton rust. The true rust is caused by a fungus; the other rust is physiological and is caused by a potash deficiency.

Cotton Improvement

The principal objectives in breeding cotton are high production of lint fiber, early maturity, resistance to disease and insect injury, adaptation to mechanical harvesting, and improvement of fiber quality.

The chief means used in the past to improve cotton was continued selection. In making selections the seeds of plants most closely approaching the ideal of the breeder have been saved separately and planted in different rows, from which the best plants from the best rows were carried over for futher testing and increase. Until recently, most of the leading cotton cultivars came about as a result of such selection. Selection must be on a broad genetic base, with continued selection for relatively long periods. The importance of genetic variability in the primary breeding material is self-evident; the breeder cannot bring out anything which is not present in the material with which he is working. Since 1945 several expeditions have gone into southern Mexico and Central America in search of new cotton germplasm.

Many cultivars, however, are believed to have originated from natural hybridization. Various types of controlled hybridization have been used, including intervarietal, interspecific, and backcrossing. Utilization of hybrid vigor in cotton by growing first-generation hybrids has been suggested. Several methods of producing F_1 hybrid seeds have been proposed, including hand pollination, natural cross-pollination, and use of genetic or cytoplasmic male sterility. The cytoplasmic-genetic system (male sterility) for hybrid cotton already has been developed. Ray at the Texas Station predicts exciting prospects ahead for cotton—hybrid cotton will be one of them.

Commercial open-canopy (okra-leaf) types of cotton, such as 'Pronto' and 'Louisiana Super Okra-5' from the Louisiana Station, are generating interest in some localities. These types allow greater light penetration to all parts of the plant and subsequently a lower incidence of boll rot. Because of a lesser shading effect, late-season weed problems might be more severe with open canopy types.

In a 1977 *Crops and Soils* article, Ray projected some additional changes and possibilities ahead for cotton: (1) cultivars which are more effective in maturing fiber at low temperatures, (2) shorter-season cultivars, (3) cultivars with greater herbicide tolerance for wider use in crop rotations (for example, cotton following corn where triazine herbicides had been used), (4) altered plant sizes and confirmations to reduce the trash content, due to air pollution and health problems in the cotton industry (smoothleaf types, closed-capsule concept, or altered bract shape), (5) cultivars with increased fiber strength because of changes in spinning systems in textile manufacturing, and (6) greater attention focused on improving the seed for feed and food, such as increasing oil content and protein percentage.

FIGURE 24-12. *The newer open-canopy (okra-leaved) cotton shown here allows greater light penetration to all parts of the plant and a subsequent lower incidence of boll rot.* [SOURCE: Southern Illinois University.]

REFERENCES AND SELECTED READINGS

1. ALLISON, J. R. *Cotton Irrigation Potential in Georgia,* Ga. Agr. Exp. Sta. Res. Bul. 111 (1972).

2. ANDERSON, R. F. *Utilization of Cotton and Other Fibers by Georgia Mills, Season of 1969–70.* Ga. Agr. Exp. Sta. Res. Bul. 112 (1972).

3. ANONYMOUS. *1978 Insect Pest Management for Cotton,* Ariz. Coop. Ext. Serv. Publ. Q11 (1978).

4. ANONYMOUS. "Researcher's Projection: Changes Ahead for Cotton," *Crops and Soils,* 30(3):18 (1977).

5. BIRD, K., *et al.* "The Technological Front in Food and Fiber Economy," *The U.S. Food and Fiber System in a Changing World Environment* (Washington, D.C.: National Advisory Commission on Food and Fiber, 1968).

6. BRIGGS, R. E. "Why Narrow Row Cotton?" *American Cotton Grower,* 6:4 (1970).

7. BRIGGS, R. E., and L. L. PATTERSON. *Narrow Row Spacing of Cotton,* Beltwide Cotton Production Research Conference, 1969.

8. BUXTON, D. R. "What's the Best Population for Narrow-Row Cotton?," *Crops and Soils,* 30(5):18 (1978).

9. DENNIS, R. E., and R. E. BRIGGS. *Growth and Development of the Cotton Plant in Arizona,* Ariz. Coop. Ext. Serv. and Agr. Exp. Sta. Bul. A-64 (1969).

10. FRIESEN, J. A., I. W. KIRK, and A. D. BRASHEARS. *A Cotton Stripper with Air-Belt Conveyors for Harvesting Variable Row-Width Research Plots,* USDA-ARS Publ. ARS-S-34 (1974).

11. *Genetics and Cytology of Cotton, 1956–67,* Southern Cooperative Serv. Bul. 139, (1968).

12. GRAVES, C. R., J. R. OVERTON, and T. McCUTCHEN. *The Performance of Cotton Varieties Stripped and Spindle Picked,* Tenn. Agr. Exp. Sta. Bul. 455 (1969).

13. HOLSTEEN, J. T., JR., and O. B. WOOTEN. "Weeds and Their Control," In *Advances in Production and Utilization of Quality Cotton* (Ames: Iowa State Univ. Press, 1968).

14. LAMBERT, W. R. III. *Cotton Insect Control,* Ga. Coop. Ext. Serv. Circ. 501 (1976).

15. LEWIS, C. F., and T. R. RICHMOND. "Cotton as a Crop," In *Advances in Production and Utilization of Quality Cotton* (Ames: Iowa State Univ. Press, 1968). edited by F. C. Elliot, M. Hoover, and W. K. Porter, Jr. 532 pp. Papers presented at Symposium.

16. LUCK, H. W. *Cotton Production in Tennessee,* Tenn. Agr. Ext. Serv. Publ. 432 (1973).

17. MARTIN, J. H., W. H. LEONARD, and D. L. STAMP. *Principles of Field Crop Production,* 3rd ed. (New York: Macmillan, 1976).

18. MOORE, V. P., and R. F. COLWICK. "Ginning Today's Cotton," *After a Hundred Years: Yearbook of Agriculture* (Washington, D.C.: USDA, 1962).

19. RAY, L. L., *et al. Cotton Planting Rate Studies on the High Plains,* Tex. Agr. Exp. Sta. MP-358 (1959).

20. SAUNDERS, J. H. *The Wild Species of Gossypium and Their Evolutionary History* (London: Oxford Univ. Press, 1961).

21. THOMAS, R. O. "Cotton Flowering and Fruiting Responses to Application Timing of Chemical Growth Retardants," *Crop Science,* 15:87 (1975).

22. USDA Agr. Stat. (1977).

23. USDA-AMS. *Cotton Quality,* 49:8 (1976).

24. USDA-ARS. *27th Annual Conference Report on Cotton Insect Research and Control,* Dallas, Tex. (1974).

25. USDA-ERS. *Synthetics and Substitutes for Agricultural Products—A Compendium,* Misc. Publ. 1141 (1969).

26. WALHOOD, V. T., and F. T. ADDICOTT. "Harvest-Aid Programs: Principles and Practices," In *Advances in Production and Utilization of Growing Cotton,* (Ames: Iowa State Univ. Press, 1968).

27. WANJURA, D. F., and E. B. HUDSPETH. *Effects of Close Row Spacing on Cotton Yields on the Texas High Plains,* Tex. Agr. Exp. Sta. PR-2266 (1963).

28. WEAVER, J. B., JR., J. MILLER, and D. WEAVER. *Tolerance and Weed Control Response of Three Cotton Genotypes to Three Levels of Two Dinitroaniline Herbicides,* Ga. Agr. Exp. Sta. Res. Bul. 213 (1978).

CHAPTER 25
MINOR OIL AND FIBER CROPS

MINOR OIL CROPS

Flax

THERE are two distinct types of flax, *Linum usitatissimum.* One is grown for the seed and the other for the fiber. Since production of flax for fiber was discontinued in the United States in the mid-1950s, this presentation focuses on seed flax, produced and processed for its oil.

Flax is an annual that grows to a height of 12–40 inches (30–102 cm). The flax flower has five petals and a five-celled boll or capsule, which when filled contains 10 seeds. The flax flower normally is self-pollinated. The flowers open at sunrise, and the petals usually fall before noon. The petals are pale blue, white, or pink, depending on the cultivar. Seeds are small, flat, shiny, and ordinarily dark brown in color.

Flaxseed contains 32–44 percent oil. In commercial crushing about 19 pounds (8.6 kg) of oil are obtained from a 56-pound (25.4-kg) bushel. Linseed oil is especially valued in paints for its excellent drying quality—its capacity to combine with oxygen to form a protective film resistant to weathering and wear. In recent years its importance has been declining because of latex paints and plasticlike substitutes. Linseed cake is prepared for livestock feed, either by grinding to meal or by making pellets suitable for outdoor feeding. The protein content of oil meal is about 35 percent. Although not suitable for spinning, straw from seed flax is used to some extent in upholstery, insulation, wallboard, twine, and paper production.

Through the more than 100 years for which flax was a pioneer crop, its center of production was always near the frontier. This westward drive was the result of the flax-wilt disease; flax production was not profitable on wilt-infected soils.

World flaxseed acreage is concentrated in India, with about 35 percent, and the USSR, with more than 25 percent of the total (Figure 25-3). Sizable acreages also are located in the United States, Canada, and Argentina. In the United States, the concentrated acreage is in North Dakota, with 50 percent, and South Dakota, with more than 30 percent of the total (Figure 25-4). Minnesota has the only other appreciable acreage. Geographic distribution is shown in Figure 25-5. The total U.S. acreage has declined gradually, from more than 3 million acres (1.2 million ha) planted in 1959 to about 1 million (0.4 million ha) at present. The yield averages less than

FIGURE 25–1. *Flax is an annual plant which bears five-celled bolls or capsules containing 10 seed each.* [SOURCE: Southern Illinois University.]

8 bushels/acre (502 kg/ha); this also has declined since 1968–1969, when the per acre average was about 13 bushels/acre (815 kg/ha). The declining production of flax in this country has resulted in increased imports. In recent years, the United States has imported 9–22 million pounds (4.1–10.0 million kg) of flax seed and 100,000–350,000 pounds (45,360–158,760 kg) of linseed oil.

The flax plant makes a short, open growth and has comparatively small leaves. It is a poor competitor with weeds. When flax follows small grain, as it often does, plowing the small grain stubble soon after harvest keeps many weeds from producing seed. Satisfactory results are obtained with flax when it follows corn or soybeans which have been kept fairly free of weeds and when the flax seedbed has been prepared without plowing. The use of selective chemicals on flax for weed control is becoming common.

The development of cultivars resistant to wilt and other flax diseases provides an outstanding example of the practical value of scientific re-

FIGURE 25-2. *Flax seeds are small, flat, shiny, and ordinarily dark brown in color. They yield linseed oil, valuable in paints and varnishes for its drying quality.* [SOURCE: Southern Illinois University.]

search. At the North Dakota station an experimental plot has been seeded to flax every year since 1894. It was on this plot that the first wilt-resistant cultivars were developed by Dr. H. L. Bolley, who in 1902 found a few wilt-resistant plants in the mixed flax cultivars grown.

Sunflower

Sunflower, *Helianthus annuus*, a member of the *Compositae* family, has been grown on a considerable acreage as an oil crop in Canada and more

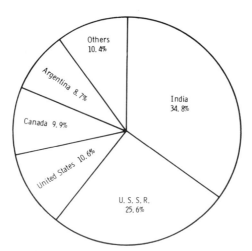

FIGURE 25-3. *Leading countries in world flaxseed area (acreage). India and the USSR have about 60 percent of the world total.* [SOURCE: Southern Illinois University.]

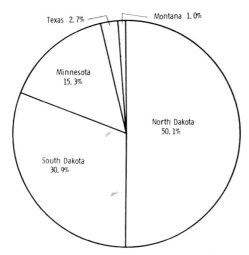

FIGURE 25–4. *Leading states in United States flaxseed area (acreage). Over 80 percent of the total acreage is located in the Dakotas.* [SOURCE: Southern Illinois University.]

recently in North Dakota, Minnesota, Texas, and California. Oilseed production started in 1943 in Canada and in 1947 in Minnesota. The world production of sunflowers for oil is about 3.5 million MT, down from 1974; the USSR is the world's leading producer. In the United States, North Da-

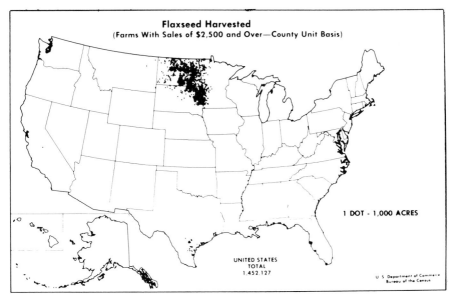

FIGURE 25–5. *Flaxseed harvested in the United States, shown geographically. Flax is grown for oil and not for fiber in the United States.* [SOURCE: U.S. Department of Commerce.]

FIGURE 25-6. *The sunflower is a member of the composite family, with dozens of fertile flowers and subsequently seeds borne in each flowering head. The ray flowers borne on the outside of the head are sterile flowers.* [SOURCE: Southern Illinois University.]

kota harvests more than 400,000 acres (161,880 ha) each year and Minnesota about 250,000 acres (101,175 ha). The yield of seed averages about 1,100 pounds/acre (1,232 kg/ha).

The standard classes of U.S. sunflowers are the following: Class I, edible and birdseed; Class II, oilseed; and Class III, mixed. The cultivated sunflower is a native of America, but high-oil sunflowers imported from the USSR in the 1960s increased the interest in sunflowers as an oil crop.

The cultivated sunflower varies in height from 3.5 to 20 feet (1.1–6.1 m) high. The oil types, such as 'Peredovik' and 'Peredovik 66,' are dwarf types for easy harvesting with a combine, and the seed contain about 45 percent oil. The oil, resistant to rancidity, is excellent for use in margarine, salad oils, and other foods. Sunflower oil is also used in making soaps and paints and for other industrial purposes. Types used for human food and birdseed, such as 'Mingren' and 'Arrowhead,' contain only 26–30 percent oil. Sunflowers are characterized by nutation, bending of the stem and tilting the heads toward the sun, accounting for the name "sunflower."

This oil crop can be grown where corn will do well. The seedbed should

FIGURE 25-7. *Seed of two major classes of sunflowers: 1 = oilseed; most high-oil types were imported from the USSR in the 1960s; 2 = edible and birdseed.* [SOURCE: Southern Illinois University.]

be prepared similar to that for corn. It can be planted with a corn planter equipped with sunflower plates, or with an air planter. The high-oil types are planted 4–8 inches (10–20 cm) apart in 20–30 inch-wide (51–76 cm) rows. In Canada some sunflowers are drilled in very narrow rows. Planting rate is ordinarily 4–5 pounds/acre (4.5–5.6 kg/ha).

Sunflowers are harvested with a combine equipped with a special sunflower adapter header needed to minimize loss from shattering and dropping heads. While statewide yields have averaged only 1,100 pounds/acre (1,232 kg/ha), experimental plots at several stations have shown that yields well over 2,000 pounds/acre (2,240 kg/ha) are possible. Illinois researchers recommend one hive of honey bees per 2 acres (0.8 ha) to satisfy the cross-pollination requirement and increase seed yields.

The major hazards to sunflower seed production are insects, birds, diseases, and weeds. The major insect pest in Minnesota is the sunflower moth, which results in severe risk during commercial production. Minnesota agronomists have concluded that sunflowers are not as profitable as soybeans, on land where soybeans can be grown. Illinois agronomists determined that sunflowers cannot compete with corn and soybeans as cash crops in that state.

FIGURE 25-8. *Sunflowers are characterized by nutation, bending of the stem and tilting the head toward the sun. This response to the sun is called heliotropism.* [SOURCE: Southern Illinois University.]

Sesame

Production of sesame, *Sesamum indicum*, is relatively important in the world, with more than 650,000 MT of oil produced annually. Acreage in the United States is small, with a few hundred acres in Texas, New Mexico, and California. Production is not enough to meet needs, and the United States imports more than 50 million pounds (22.7 million kg) of sesame seed annually.

Plants are herbaceous annuals which reach a height of 3–5 feet (0.9–1.5 m). They bear bell-shaped flowers that generally are white or very pale pink, but may be darker to almost purple.

Sesame seeds, whose oil content may exceed 50 percent, are borne in upright pods which may split open or dehisce at maturity. Some cultivars are nonshattering types. Seed are normally red-brown in color, but may be white, yellow, or black, depending on the cultivar. The oil does not turn rancid easily, and is used for making salad and cooking oils and in the manufacture of margarine. It also is used in the manufacture of perfumes, pharmaceuticals, and insecticides. Hulled seed is used in baking and other industries. Decorticated seeds are often sprinkled on various types of bread or rolls before baking.

FIGURE 25-9. *Sesame is an annual plant which bears bell-shaped, normally white flowers and upright pods which dehisce at maturity. Note the extensive podding along the main stem.* [SOURCE: Southern Illinois University.]

Safflower

In the United States, the oilseed crop safflower, *Carthamus tinctorius,* is best adapted to the semiarid and irrigated sections of the West. It is subject to severe disease injury when wet weather prevails during the blooming and filling period. The world production of safflower oil exceeds 300,000 MT, but U.S. production is limited; California is the leading state.

Safflower is a herbaceous annual and a member of the *Compositae* family. Plants grow 2–5 feet (0.6–1.5 m) high and bear sessile leaves with sharp spines. Spineless types are being developed. Flower heads have white, yellow, orange, or red petals. The fruit is a shiny, white or cream-colored, angular achene. The "seed" or fruit contains up to 40 percent oil. The light-colored oil is highly unsaturated and of excellent quality for edible purposes. It contains a large percentage of linoleic acid, an unsaturated fatty acid. It is used primarily in the production of margarine, salad oils, mayonnaise, and other food products. Safflower is an excellent

FIGURE 25–10. *Sesame seed are normally red-brown in color. Oil is used for edible products as well as in manufacture of perfumes, pharmaceuticals, and insecticides. The hulled or decorticated seeds are used in baking.* [SOURCE: Southern Illinois University.]

drying oil for use in paints and varnishes. White paints and varnishes do not yellow with age.

This crop may replace a small grain in rotations. It is usually spring-sown with a grain drill, and may be irrigated in arid areas. Safflower plants do not lodge easily nor does the seed shatter easily; thus, it is well-adapted to combine harvesting.

Castorbean

The castorbean, *Ricinus communis,* is not a true bean, being a member of the *Euphorbiaceae* or spurge family. Also called mole bean because of its use in controlling these small subterranean creatures, the crop is grown for its oil, used mainly for industrial purposes. The oil is not poisonous and was previously used as a purgative, especially for ailing children, but this use has largely ceased.

World production of castor oil exceeds 300,000 MT. The world leaders in production are Brazil, India, and the USSR. The United States is the leading importer of castor oil, with about 1 billion pounds (0.45 billion kg) of castor oil and 10,000 pounds (4,536 kg) of castorbeans brought in annually to meet two-thirds of its castor oil needs. Less than 20,000 acres (8,094 ha) of castorbeans are planted in the United States. The present acreage is concentrated in Texas, with limited acreages also in New Mexico, Kansas, and Nebraska. Virtually all of the U.S. castorbean acreage grown today is under contract with processors.

The castorbean plant is a short-lived perennial grown mostly as an an-

FIGURE 25-11. *Safflower, a composite, is an annual plant bearing sessile leaves which are often spiny. Flowering heads bear several one-seeded fruits containing a valuable oil.* [SOURCE: University of Arizona Agricultural Experiment Station.]

nual in the United States. Some cultivars grow to a height exceeding 8 feet (2.4 m), but dwarf cultivars 3–5 feet (0.9–1.5 m) tall are available. The leaves are large and palmately divided into five or more lobes. The flowers are monoecious, similar to corn, in that the pistillate and staminate flowers are borne separately on the same plant. The fruit is a three-celled capsule, often spiny. The seed is commonly mottled brown to reddish-brown and contains a prominent hilum or caruncle at one end. The hulled seed contains 35–55 percent oil. The oil is of high density and has a low congealing point. It is used in the manufacture of soaps, linoleum, oil cloth, printer's ink, plastics, and lubricants and solvents. It also it used widely in paints and varnishes. The castorbean and the press cake after the oil has been removed contain ricin, an albumin, and ricinine, an alkaloid, which are poisonous to humans, livestock, and poultry. The oil is not poisonous but the leaves and stems are toxic because of ricinine.

This crop is harvested with a self-propelled harvester or a combine with a special attachment.

FIGURE 25-12. *The "seeds" of safflower are actually one-seeded fruits called achenes, which contain up to 40 percent oil. The oil is highly unsaturated and an excellent oil for edible purposes.* [SOURCE: Southern Illinois University.]

Rapeseed

Several types of rape are grown in the United States, some of which is used for forage, but the most important one is summer rape, *Brassica napus* var. *annus,* the oilseed type. The world annual production of rapeseed oil is nearly 3 million MT. Production is centered in India, Poland, Canada, and France. The United States plants only about 2,000 acres (809 ha) each year, mostly in Montana and Minnesota. This is not enough to meet the needs, and in recent years the United States has imported 12 million pounds (5.4 million kg) of rapeseed oil and 7 million pounds (3.2 million kg) of rapeseed each year.

Rapeseed is about 45 percent oil. The oil is used in the production of rubber products, steel, brake fluid, and lubricants. It is used for edible purposes in some countries.

Mustard

Three sources of the commercial mustard seed, *Brassica* spp., are used in the United States: (1) imported seed, (2) seed domestically produced under cultivation, and (3) seed obtained from the small-grain and flax screenings (mustard grows as a weed in these crops). Mustard is an annual

FIGURE 25-13. *The castorbean plant may grow to a height exceeding 8 feet (2.4 m). Leaves are large and lobed and flowers are monoecious, with pistillate and staminate flowers borne separately but on the same plant.* [SOURCE: Southern Illinois University.]

FIGURE 25-14. *Castorbean fruit and seed: (A) The castorbean fruit is a three-celled, spiny capsule; (B) Seeds are mottled brown to reddish brown and contain a prominent hilum or caruncle at one end. Oil from the seed is used principally in various manufacturing processes.* [SOURCE: Southern Illinois University.]

A

B

FIGURE 25–15. *Summer rape yields an oil from the seed used in the production of rubber products, steel, brake fluid, and lubricants.* [SOURCE: Southern Illinois University.]

with wide geographic adaptation and distribution. It is grown in very much the same manner as the small grain and flax crops.

Most of the United States's domestic mustard is produced in Montana, with some production in Minnesota and California. Three types of mustard are grown—black, brown, and yellow. Besides yielding a thick glyceride oil, the yellow type is useful in condiments, where the ground seed or paste is used as a spread. The nondrying mustard oil has counter-irritant properties and is used for medicinal purposes; oil from the brown type is the best.

Crambe

Crambe abyssinica, like mustard and rape, is in the family *Cruciferae.* The seed contains 35–53 percent oil, a product which has potential to be used in plasticizers, rubber additives, waxes, and as a lubricant in the continuous casting of steel. Crambe oil has some of the same uses as imported rapeseed oil; thus, it is not a potential competitor to domestically produced oilseed crops.

Crambe has been grown in Canada and in a few states of the Midwest, Great Plains, and West. The plant is an erect annual which grows to a

FIGURE 25–16. *Crambe, a member of the mustard family, grows to a height of 2 to 3 feet (0.6 to 0.9 m) and bears one-seeded, round fruits.* [SOURCE: Southern Illinois University.]

height of 2–3 feet (0.6–0.9 m). It bears white flowers and produces numerous one-seeded round fruits; seed yield generally is 500–1000 pounds/acre (560–1,120 kg/ha).

This cool-season crop should be planted in the early spring, April in the Midwest, and can be harvested in early July. Seed may be planted with a wheat drill. In the Midwest, the crop can be harvested by direct combining. Availability of a market should be determined before planting.

Wormseed, Wormwood

Wormseed, *Chenopodium ambrosioides* var. *anthelminticum*, contains a volatile oil of disagreeable odor used in worm remedies for man and beast. It also is used in vermifuges. Oil is distilled from the leaves and tops of wormwood, *Artemisia absinthium*, and was formerly used for tonics and external proprietary medicines. It is no longer used as a worm remedy because of toxic properties.

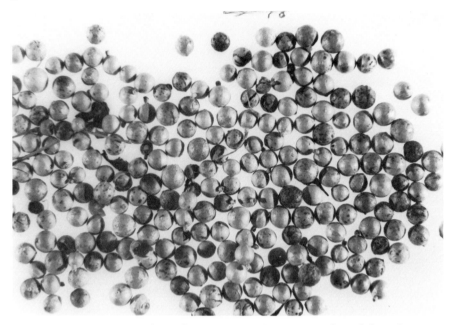

FIGURE 25-17. *Crambe seed contain up to 50 percent oil, useful in plasticizers, rubber additives, waxes, and as a lubricant in the continuous casting of steel.* [SOURCE: Southern Illinois University.]

Guar

Guar, *Cyamopsis psoralides,* is a summer annual legume that has been grown primarily in India and Pakistan for forage and green manure. It was introduced into this country around 1900. In the United States it is grown for the production of a gum called manogalactan, which is used in manufacturing paper, textiles, printing ink, and as a stabilizing and gel-forming material in some foods. The greatest acreage is in northwestern Texas, where normally it is planted in May and harvested in September.

Peppermint

Mentha piperita, a member of *Labiatae,* or the mint family, is grown for its volatile oil, which is used extensively for flavoring chewing gum, toothpaste, confections, and medicines.

The United States plants about 70,000 acres (28,329 ha) to peppermint and produces about 3.7 million pounds (1.7 million kg) of oil annually. More than half the acreage and production is in Oregon, with limited acreages also in Washington, Wisconsin, Indiana, and Idaho. The average acre yield of oil is about 50 pounds (56 kg/ha) with a value as much as $15 per pound ($33 per kg). The crop is produced with the greatest success on soils high in organic matter and does especially well on muck soils. It also thrives on sandy loam soils of the West.

Peppermint is propagated by rhizomes laid end to end in furrows about 3 feet (0.9 m) apart with a special machine or by hand. Newly established plantings can be cultivated effectively to keep the crop free from weeds, but as the rhizomes spread in all directions a good deal of hand weeding is necessary. It is essential that the ground be kept free of weeds, because the presence of weeds in the harvested crop tends to reduce the quality of the oil.

Harvesting is begun in July or August, when the plants are in full bloom. It is mowed, windrowed, and allowed to dry partially before chopping with a forage harvester. The oil is distilled from the chopped mint with steam.

Spearmint

Mentha spicata or *M. viridis* frequently is found growing wild throughout the eastern half of the United States. Like peppermint, it is easily grown on any fertile soil that is fairly moist, but the crop is especially well adapted to reclaimed swampland. It is propagated, cultivated, harvested, and distilled for its volatile oil in much the same way as peppermint. In addition, the leaves and flowering tops are widely used as a seasoning for meats and beverages.

Up to 25,000 acres (10,118 ha) are grown annually for oil in Washington, Indiana, Idaho, Michigan, and Wisconsin, with a production of more than 1.5 million pounds (0.7 million kg) of oil.

Perilla

Another member of the *Labiatae* or mint family is perilla, *Perilla frutescens.* The seed of perilla contain a drying oil of excellent quality used in paints, varnishes, and linoleum. The oil content of the seed averages about 38 percent. It is similar to linseed oil in odor and taste but higher in drying quality; thus, it is excellent for quick-drying paints and varnishes. Perilla is grown mainly in the Orient; there is no appreciable acreage in the United States.

MINOR FIBER CROPS

Several of the minor fiber crops are not grown in the United States because of climatic reasons. Moreover, because of labor costs in harvesting and processing, it is cheaper to import them. In addition, processing requires special equipment.

Fiber Flax

Among the textiles, linen cloth always has been highly regarded. Production of fiber flax has ceased in North America. All linen goods are imported. Special skill is required for spinning fine flax yarn and also for weaving the fine yarns into linen fabrics. Practically all of this kind of work is done in limited areas in Ireland, Scotland, northern France, and Belgium. The straw of seed flax is sometimes utilized in the United States for making cigarette paper, Bible pages, and other high-quality paper.

Hemp

Hemp, *Cannabis sativa,* was at one time grown rather extensively in the United States. Because of the large amount of hand labor required to harvest the crop and to separate the fiber, the acreage steadily decreased soon after the Civil War. As an emergency during World War II, when it was impossible to obtain Manila hemp from the Philippines, hemp production in the United States was greatly stimulated.

Today it is illegal to grow hemp in the United States without a license issued by the Commissioner of Narcotics because of its content of the strong narcotic resin cannabin and its use as marijuana.

Henequen, Manila Hemp, and Sisal

The fibers most familiar to American farmers have been those used to make binder and baler twine. The fibers used in binder and baler twine—henequen, sisal, and Manila—cannot be grown in the United States. These fibers, commonly called the hard fibers, are obtained from leaves of tropical plants. A great deal of hand labor is required in handling and processing these crops.

HENEQUEN. Henequen, *Agave fourcroydes,* is produced primarily in Mexico and Cuba. The plants are started in nurseries and later transferred to wide-spaced rows. The plants require four to six years to mature and attain a height of 5–10 feet (1.5–3.0 m). The lower, more mature leaves are harvested for the fiber; cuttings are made two to four times a year. A single plant may yield from 20–40 leaves in a season. The harvested leaves yield only about 3 percent by weight of marketable fiber.

MANILA HEMP. The fiber of Manila hemp, *Musa textilis,* is derived from a plant closely related to the banana that thrives principally throughout the Philippines, where it is called *abaca* by the natives. New plantings are started from root portions and are spaced in rows 8–10 feet (2.4–3.0 m) apart. A planting may continue to produce for 20 years or more, although replantings often are made after 12–15 years. The stalks vary in length from 8–14 feet (2.4–4.3 m). Manila hemp is excellent for the manufacture of marine ropes, as it does not swell when wet.

SISAL. Sisal, *Agave sisalana,* is very similar to henequen. Most of the sisal fiber is produced in East Africa and the West Indies. The plant consists of a large rosette of rigid, straight, fleshy leaves arising from a short trunk. The life period of the plant is seven to ten years. In harvesting, the lower, mature leaves are cut off at the base. The fiber is white to yellowish in color; its value for cordage is second only to abaca. Sisal is used principally in binder and baler twine but also in other types of cordage.

Ramie

Ramie, *Boehmeria nivea,* is a semitropical plant grown throughout the Orient. It is sometimes called Chinagrass. Some success has been obtained from small plantings in the Southeast and in irrigated valleys of southern

FIGURE 25–18. *Kenaf, a tall-growing annual, has shown promise as a jute substitute and has potential as a source of pulp for paper-making. Most recently, kenaf has been used as bean poles.* [SOURCE: Southern Illinois University.]

California. Ramie is a perennial that grows from root stocks, often to a height of 6–8 feet (1.8–2.4 m). From two to four crops may be cut per year. The fiber is similar in appearance and use to flax fiber, making strong cloth with a lustrous appearance. Ramie textiles are suitable for tropical wear. Chinese linen and Canton linen are made from ramie.

Kenaf

Kenaf, *Hibiscus cannabinus,* has shown some promise as a jute substitute. Considerable research was conducted on this plant during World War II, when jute and other fiber imports were cut off from the United States. Most recently, it has been evaluated for its potential as a source of pulp for paper-making, and for use as bean poles. Most of the small acreage remaining is in Florida.

Jute

Jute, *Corchorus capsularis,* is produced where labor is cheap, mostly India and Pakistan. The plant is an annual produced from seed. The primary use of jute is in the manufacture of burlap for bags and sacks. Other uses are in twine and backing for linoleum. The fiber as a textile is of poor quality. Much of this fiber is used, perhaps because of its low cost.

REFERENCES AND SUGGESTED READINGS

1. CHRISTMAS, E. P., *et al. Crambe—A Potential New Crop for Indiana,* Purdue Univ. Coop. Ext. Serv. Publ. AY-168 (1968).
2. CULP, T. W. *Sesame Production in the Mississippi Delta,* Miss. Agr. Exp. Sta. Bul. 674 (1963).
3. DENNIS, R. E., and D. D. RUBIS. *Safflower Production in Arizona,* UA Bul. A-47 (1966).
4. GRAVES, C. R., *et al. Production of Sunflowers in Tennessee,* Tenn. Agr. Exp. Sta. Bul. 494 (1972).
5. MARTIN, J. H., W. H. LEONARD, and D. L. STAMP. *Principles of Field Crop Production* (New York: Macmillan, 1976).
6. ROBINSON, R. G. *The Sunflower Crop in Minnesota,* Minn. Agr. Ext. Serv. Bul. 299 (1973).
7. SEALE, C. C., E. O. GANGSTAD, and J. F. JOYNER. *Agronomic Studies of Ramie in the Florida Everglades,* Fla. Agr. Exp. Sta. Bul. 525 (1953).
8. USDA. Agr. Stat. (1977).
9. USDA. *Growing Safflower—An Oilseed Crop,* Farmers' Bul. 2133 (1966).
10. USDA. *Safflower 1900–1960, A List of Selected References,* Library List 73 (1962).
11. USDA. *Sesame Production,* Farmers' Bul. 2119 (1958).
12. WEISS, E. A. *Castor, Sesame, and Safflower* (New York: Barnes and Noble, 1971).
13. WILSON, F. D., *et al. 'Everglades 41' and 'Everglades 71', Two New Varieties of Kenaf Exp. (Hibiscus cannabinus L.)* for Fiber and Seed, Fla. Agr. Exp. Sta. Circ. S-168 (1965).

C. Sugar Crops

CHAPTER 26
THE SUGAR CROPS

THE source of most of the sugar produced in the United States, about 57 percent, is sugarbeets and most of the remainder is sugarcane (Figure 26–1). A considerable share of the sugar used is imported. The bulk of the sugarbeets is produced in 16 Western and North Central states, with California, Minnesota, Idaho, Colorado, North Dakota, and Nebraska growing the larger acreages (See Figure 26–2). Geographic distribution of sugarbeets is shown in Figure 26–3. The world leaders in acreage and production are the USSR, with nearly one-half the world's acreage and one-third of the production; the United States; France; West Germany; and Poland (Figure 26–4).

Sugar from Puerto Rico and Hawaii derives entirely from sugarcane. The continental production of cane sugar is limited to Louisiana, Florida, and Texas. Louisiana leads in acreage, with more than 42 percent, and Florida is second, with 38 percent (Figure 26–5). Geographic distribution of sugarcane is shown in Figure 26–6. The world leaders in sugarcane production are Brazil, Cuba, and India, with smaller areas of production in Australia, the United States, Mexico, and the Philippines (Figure 26–7).

Sugar Legislation

For many years it has been the policy of the government to preserve within the United States the ability to produce a substantial portion of our sugar requirements. This has been done because sugar is an essential and vital food product, the supply of which on a worldwide basis has been marked by periods of alternating scarcity and surplus. In 1934 Congress initiated a sugar quota system, which is still the basis of our current Sugar Act and under which all domestic sugar-producing areas and friendly foreign sugar-producing countries supply the sugar needs of U.S. consumers. Each participating area and country receives an annual sugar marketing quota representing the amount of sugar the area or country can sell in the United States during the year. Allotments are set for individual farms or producers after a national sugar acreage is established.

The Sugar Act Amendments of 1965 changed the original Act, but the principal purposes have not been changed. In summary, they are (1) to assure U.S. consumers of an adequate sugar supply at reasonable prices,

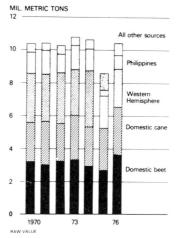

FIGURE 26–1. *Sources of sugar used in the United States. Sugar from do-mestic cane and beets are almost evenly split. A considerable share of the sugar used is imported.* [SOURCE: USDA.]

(2) to encourage foreign trade, and (3) to provide a healthy economic cli-mate for a competitive sugar industry in this country.

Sugarbeets

History

Many believe sugarbeets were eaten by the laborers who piled up the pyramid of Cheops, Egyptian Pharaoh, who lived 3,000 years before Christ.

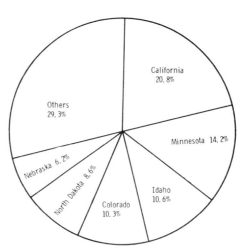

FIGURE 26–2. *Leading states in United States sugarbeet area (acreage). Six states have over 70 percent of the acreage.* [SOURCE: Southern Illinois University.]

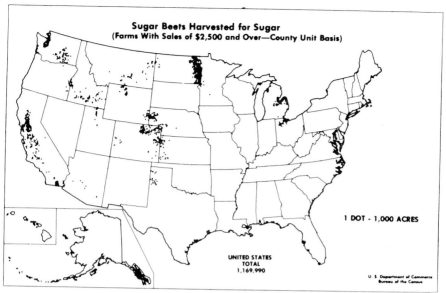

FIGURE 26–3. *Sugarbeets harvested for sugar in the United States, shown geographically.* [SOURCE: U.S. Department of Commerce.]

Apparently, the beet grew in wild state in parts of Asia, and at an early time it was cultivated in southern Europe as well as in Egypt. Early in the 1600s it was demonstrated in France that the beet root yielded a juice similar to the syrup of sugar. Another century passed before extraction of the sugar from the beet was considered to be practical on a commercial

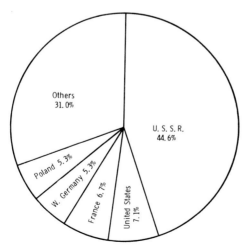

FIGURE 26–4. *Leading countries in world sugarbeet area (acreage). The USSR has nearly one half of the world total.* [SOURCE: Southern Illinois University.]

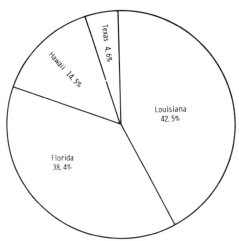

FIGURE 26-5. *Leading states in United States sugarcane area (acreage).*
Louisiana and Florida are greatly dominant, with over 80 percent of the
total. [SOURCE: Southern Illinois University.]

scale. The establishment of the beet sugar industry in Europe involved
the laboratory of a German chemist, Andreas Marggraf, in 1747; the backing of a Prussian king, Frederick William III; and the dynamic action of
the French Emperor, Napoleon. Marggraf found that beet sugar crystals
had the same physical and chemical characteristics as sugar from sugar-

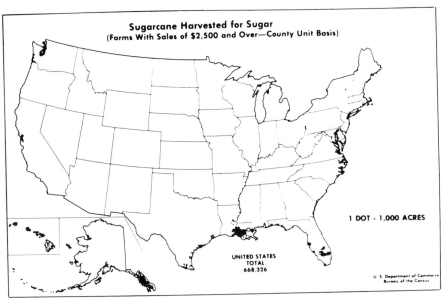

FIGURE 26-6. *Sugarcane harvested for sugar in the United States, shown*
geographically. [SOURCE: U.S. Department of Commerce.]

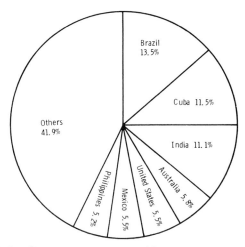

FIGURE 26-7. *Leading countries in world sugarcane production. Three countries produce over one third of the world total.* [SOURCE: Southern Illinois University.]

cane. The French found themselves cut off from the sugar of the West Indies. By decree Napoleon had several beet sugar factories erected. Although there were several setbacks, by 1854 the industry was operating on a large scale over most of the Continent and spread to the Near East, the Far East, and the Americas.

Although wild sugarbeets were found growing in central California, the sugarbeet industry began on the East Coast in 1830, first in Philadelphia. The first ventures failed because of lack of technical knowledge and skill. The first successful beet sugar factory was established in California in 1879. By 1900 there were 30 completely equipped sugarbeet factories in eleven states from New York to the West Coast.

The Sugarbeet Plant

The sugarbeet, *Beta vulgaris,* belongs to the goosefoot family, *Chenopodiaceae,* which also includes the major weed pest pigweed.

This plant is a biennial grown as an annual when used for sugar production. The first year it develops a rosette of leaves and a large root in which is stored food reserves. If it is allowed to grow the second year, it develops a flowering stalk and produces seed.

The elongated root which develops is divided into three regions: crown, neck, and root. Flowers are borne in clusters on the flowering stalk, which arises from a small rosette or spiral of leaves in the crown area. As the ovaries of the flowers start to ripen the perianths surrounding each flower fuse, with the result that a "seed ball" is formed containing several ovaries. What is normally looked upon as a beet seed is not one seed, but a fruit consisting of 2 to 5 seeds. This multigerm seed has been responsible for

FIGURE 26–8. *Monogerm seed of sugarbeet. Some cultivars have multi-germ seed, "seed balls" containing several ovaries. Cultivars with mono-germ seed present fewer planting and production problems.* [SOURCE: Southern Illinois University.]

one of the principal difficulties in sugarbeet production, the large amount of hand labor required to thin the stands in the field. Fewer problems have resulted since the development of cultivars that produce "monogerm" seed.

Large amounts of sugar are formed in the leaves. The greater part of this sugar may be used by the plant to maintain its period of rapid growth, but when vegetative growth is slower late in the season, a large part of the sugar is stored in the roots. The sugar content in the roots frequently exceeds 20 percent of the root weight. The crop is harvested near the end of the first year's growth, when the roots contain the maximum amount of sugar.

Climate and Soil

Beets thrive throughout the northern latitudes of the United States, in Canada, and as far south as the Mexican border, at elevations varying from below sea level to an altitude of 7,000 feet (2,121 m) and in a wide variety of soils. While they are produced on heavy soils sometimes, sandy loams or loams are best because of fewer difficulties in harvesting and cleaning the roots. The crop exhibits a unique tolerance to alkali, which is present in large areas of land in arid regions of the Western states.

The sugarbeet requires a plentiful and well-distributed supply of moisture during the growing season. As the season advances, more and more water is required. A large part of the Intermountain and West Coast sugarbeet acreage is grown under irrigation. The crop is favored by a long and moderately cool growing season. Warm days and fairly cool nights favor rapid growth. In the latter part of the season, progressively cooler nights, exhaustion of available nitrogen, and a decreased moisture supply slow up vegetative growth and accelerate sugar storage.

Production Under Contract

The production of sugarbeets is limited to areas not far removed from a sugarbeet processing plant. The cost of transporting beets is an important factor. Sugarbeets are grown under contract with a beet sugar factory. The contract, entered into yearly by the growers, bears stipulations as to the acreage to be grown, cultivar, planting, culture, harvest, quality, crop delivery, and price. The company ordinarily supplies the seed and furnishes technical assistance.

Rotations

Sugarbeets should be grown in rotation with other field or vegetable crops. However, they should not be grown in rotation with cole crops, because these crops are host for the sugarbeet nematode. Most sugarbeets are grown in rotations involving such legumes as alfalfa, red clover, or sweetclover; small grains; and such intertilled crops as corn, soybeans, or potatoes. The use of an intertilled crop between the legume and the sugarbeet gives time for the breakdown of the organic matter and aids in disease and weed control. Certain legumes, as well as weeds, promote the occurrence of soil-borne organisms causing damping off and black root.

Irrigation

Most of the sugarbeet acreage in the Western states is under irrigation. Preirrigation may be necessary to provide moisture for seed germination, to prevent soil crusting, and in many areas to control salinity in establishing a sugarbeet stand. When the beets are small, moisture demands are small. Overirrigation at early stages may tend to leach nitrates, waste water, and enhance seedling diseases. During periods of growth several light irrigations may be more important than a few heavy ones. As the root system develops, so do moisture demands. At midseason sugarbeets obtain most of their moisture from the upper 3 feet (0.9 m) of soil and heavier irrigations are required to supply this moisture. If the plants wilt early in the morning or show slow recovery in the afternoon as temperature and light intensity decline, irrigation is necessary. For maximum yields irrigations should be applied when 50 percent of the available moisture, as checked with a soil probe, has been removed by the crop. Moderate moisture stress just before harvest tends to increase sugar percentage without limiting sugar

FIGURE 26-9. *An irrigated field of sugarbeets in Colorado. Most of the sugarbeet acreage in the Western States is under irrigation.*

yield per acre. The last irrigation may be two to five weeks before harvest. Most of the irrigation is by the furrow method. Sprinkler irrigation, however, offers several advantages. Light, frequent irrigations required for germination and emergence may be applied efficiently. Nitrates are not accumulated during the growing season so there is no hazard of leaching into the root zone during the critical period just prior to harvest. Young plants, however, may be damaged by sprinkling with high saline water.

The Seedbed

To encourage prompt germination and vigorous growth the seedbed should be firm and fine but not overly compacted. The seedlings are small. Fall plowing to a depth of 8–12 inches (20–30 cm) usually is advisable except for friable soils. Chiseling or subsoiling will increase yield when there is a compacted layer of subsurface soil which interferes with water and root penetration. When the crop follows beans satisfactory results have been obtained by preparing the seedbed without plowing. Muck soils sometimes are so loose that thorough firming of the soil is of first importance.

Fertilization

Sugarbeets require high rates of fertilizer for maximum yield. Plants require considerable nitrogen for vegetative growth. When sugarbeets follow vegetables or alfalfa, less nitrogen fertilizer is necessary than when beets follow other crops. The profitable rate will vary from 0 to 150 pounds of nitrogen per acre (0–168 kg/ha) depending on soils, previous cropping history, nitrogen in the irrigation water, and other factors. Nitrogen should be provided in split applications, a small amount applied at planting time and the rest after. Too much nitrogen or nitrogen applied too late reduces the percentage of sugar in the beets because the sugar is used for new vegetative growth and is not stored. Petiole analysis is useful in determining the amount of nitrogen necessary for the crop.

Applications of phosphate at a rate up to 80 pounds of P_2O_5 (90 kg/ha) at planting time may be used where phosphate responses have been obtained on other crops. Phosphate is very important for early plant growth, and responses have been obtained from banding part of the phosphate fertilizer below and to the side at planting time. Most of the sugarbeets are grown on soils well supplied with potash. Muck soils, however, often require potash.

Micronutrients or secondary elements may be required in some localities. Boron and manganese deficiencies have occurred on mineral soils in the humid areas. These two elements are required under most Michigan conditions. Copper sulfate, when applied to some muck soils, and magnesium, when applied to some muck and some mineral soils, have had beneficial effects on the crop. Some sugarbeet fields require applications of zinc and iron.

Manure applied at the rate of up to 12 tons/acre (4.4 MT/ha) will meet some of the fertilizer needs, but generally supplemental commercial fertilizers, particularly phosphorus, are required.

Planting

Planting as early as the season permits is favored. Planting in California is often during the fall or winter months; in most other areas, beets are planted in March to May. Depth of planting should be comparatively shallow, about 0.75–1.25 inches (1.9–3.2 cm) deep in a firm, fine-structured seedbed. Sugarbeets may be planted in single or double rows on elevated beds 4–6 inches (10.2–15.2 cm) high, or they may be planted flat, as is common in humid areas. The width between single rows may vary from 18 to 32 inches (46–81 cm). When two rows are placed on a bed there is usually at least 36 inches (91 cm) between bed centers. Usually a four-, six-, or eight-row planter equipped with plates for sugarbeets is used. The planting rate should be adjusted to the conditions to give a final population of 100–120 beets per 100 feet (30 m) of row. About 5–8 pounds of monogerm seed are required per acre (5.6–9.0 kg/ha).

Thinning

Sugarbeets may be planted to stand, but this may not be desirable in view of the many hazards in getting an adequate final population. More commonly beets are planted thicker to assure a good stand or later thinned to the desired row spacing.

When unprocessed, multigerm seed were used, the blocking and thinning operations were almost entirely by hand. With precision planting and the use of monogerm seed, several types of implements for doing this thinning have come into use. These remove surplus plants and have largely replaced the across-the-row method of cultivation to thin the beets. The thinners are of two types: (1) rotary down-the-row, random stand reduction blockers, which chop a 2–4-inch (5.1–10.2-cm) gap every 5–10 inches (12.7–25.4 cm) of row; and (2) electronic or electric eye thinners. The latter have gained acceptance after the widespread use of precision planting of monogerm seed and chemical weed control.

Thinning is done when beets have reached the 8- to 10-leaf stage. Final spacing in the row should be 8–10 inches (20.3–25.4 cm), with a final population of 21,000–36,000 plants/acre (51,890–88,955/ha).

Weed Control

Weeds compete with sugarbeets for light, nutrients, and moisture. They also harbor insects and serve as a reservoir for viruses. Weed problems are not as severe after leaf growth of sugarbeets is sufficient to shade the soil surface, but winter annual weeds, including volunteer barley, require control.

Barnyardgrass, cocklebur, crabgrass, foxtail, kochia, lambsquarter, pigweed, sowthistle, sunflower, wild oats, and such perennial weeds as Canada thistle, field bindweed, milkweed, and quackgrass cause moderate to heavy reductions in yield and quality of sugarbeets.

Mechanical weeders, cultivation, and herbicides are used for weed control. The availability of effective herbicides has made possible elimination of most hand labor. The mid-1960s brought a transition from physical methods of weed control to chemical weeding. Preplant, pre-emergence, and early postemergence herbicides may be used.

Knife-type or sweep shovel-type cultivators are commonly used to supplement weed control obtained from herbicides. It is important to control seedling weeds early. Mechanical cultivation is effective in controlling weeds between rows but may prune crop roots if practiced too close to the row.

Harvesting

Harvest usually is delayed as long as possible to permit the crop to make the greatest accumulation of sugar. The browning of the lower leaves and yellowing of the remaining leaves generally indicates maturity.

FIGURE 26–10. *Mechanized harvesting of sugarbeets. Mechanical harvesters are available which will top beets, lift them from the soil, shake, and elevate to trucks. Sometimes the topping is a separate operation prior to harvest.* [SOURCE: Deere and Company.]

Early harvest permits the planting of another crop sooner and will result in a lower water requirement for growing beets. Later harvest usually results in a higher tonnage per acre with a higher percentage of sugar.

In the early days before World War II, hand-harvesting methods were used. This required 60–70 man-hours/acre (148–173 /ha) to lift roots from the soil, cut off the tops with a knife, windrow or pile the beets, and subsequently load by hand on wagons or trucks. By the late 1950s, virtually all sugarbeet harvesting operations were fully mechanized.

Sugarbeet tops must be removed before they are processed. The topping operation may be separate from digging and lifting, or it may be a part of a one-trip fully mechanized operation. Harvesters range from 1-row to 6-row. Mechanical harvesters are available which will top beets, lift them from the ground, shake to loosen the soil, and elevate them to loading bins or trucks. Modern beet harvesters are equipped with screens to remove clods, rocks, and trash. Each day they can harvest as much as 24 acres (9.7 ha) or 500 tons (454 MT) of beets.

The harvested beets are hauled in trucks to a beet-receiving station or directly to the processing plant. They may be processed immediately or may be stockpiled for later processing.

Cultivars and Improvements

The European sugarbeet cultivars were rapidly replaced in the decade 1930–1940 by American-bred cultivars. In 1925 the USDA and state agricultural experiment stations initiated sugarbeet breeding programs to develop disease-resistant cultivars. After disease-resistant cultivars were developed the plant breeders turned to other objectives, such as nonbolting, monogerm seed, storage ability of roots, sugar percentage, cold resistance, and earliness. Cultivars with sucrose content as high as 20–22 percent have been developed but gross yields are inferior to cultivars averaging about 18 percent. Adapted cultivars seldom bolt (produce a seed stalk) in the area where they are grown for sugar. The large beet sugar companies maintain breeding programs so that they can supply their contract growers with the best-adapted cultivars. Commercial American sugarbeet cultivars are identified by numbers. These numbers assigned apply to definite genetic entities. Methods in breeding used are mass selection, family breeding, development of synthetic cultivars, hybridization, and polyploidy. A major part of the sugarbeet acreage is planted to hybrid cultivars.

Processing

When sugarbeets arrive at the processing plant, they are conveyed to storage. Processing begins by conveying from the piles, washing thoroughly, and slicing beets into thin strips called "cosettes." The sliced beets are then cooked. The next step is separation of the solid matter and juice, the solid matter to be dried and shredded as "beet pulp." Hydrated lime is added to the raw juice to combine with impurities, after which the purified juice is removed by filtration. The juice is further clarified and decolorized and then filtered again. The purified juice is evaporated in vacuum distillation vats to bring about crystallization. When a batch has crystallized it is brown in color because of the presence of molasses. This batch then goes to a centrifuge for the separation of the heavier molasses, leaving only the white sugar crystals. The sugar is next dried, ground and packaged.

Insects and Diseases

Leading insect pests include leafhoppers, grasshoppers, webworms, and flea beetles. Other insects that may cause losses some years in selected geographic areas are armyworms, leafminers, grubs, cutworms, wireworms, tarnished plant bugs, aphids, and blister beetles. Armyworms, grasshoppers, and webworms cause their damage by feeding on the foliage; root aphids and wireworms attack the roots; and aphids and leafhoppers transmit virus diseases. The sugarbeet nematode is destructive by attacking the roots and stunting the plants.

FIGURE 26-11. *A storage pile of sugarbeets at the processing plant. Processing begins by washing, prior to slicing and cooking to extract the juice.*

Most of the insect pests can be controlled with organic phosphate or carbamate insecticides.

The most serious diseases are caused by fungi and viruses; they may result in losses exceeding 20 percent of the crop. The most serious fungus diseases are black root, which affects young seedlings and cercospora leaf spot. Others are crown and root rot and downy mildew. Virus diseases can be devastating in certain areas. Curly top, a virus disease whose vector is the beet leafhopper, does not cause the serious losses of years past because of the development of resistant cultivars. Virus yellows and mosaics sometimes appear.

Cultivars or hybrids resistant to most fungus and virus diseases mentioned are available; their use is the best preventive measure. In addition, the use of crop rotation, good cultural practices, seed treatment with fungicides, and the use of pesticide sprays are effective preventive or control measures.

Sugarcane

In Europe, sugarbeets are the principal source of refined sugar. Similarly, in the United States, the domestic production of refined sugar is greater from beets than cane. However, in this country cane sugar constitutes most of the total sugar consumed, including import supplies. On a worldwide basis, sugarcane supplies the majority of the world's refined sugar. Of the many plants that contain sucrose, from none other can it be obtained at so low a cost and in so highly purified a form as from sugarcane. Sugarcane has been said to excel over all other plants as a converter of the sun's energy and the carbon dioxide of the air into energy food and fiber.

Revolutionary changes have been made in the methods of producing and handling sugarcane, in the cultivars grown, and in processing methods. As produced in the past a tremendous amount of hand labor was required. World War II and the resulting unavailability of hand-labor forced mechanization. In addition, the recent development and use of sugar-enhancing chemical sprays applied to growing plants offer the promise of even higher sugar yields from an already high-producing crop.

Sugarcane production in the United States was developed around a cultivar that proved to be so extremely susceptible to an invading mosaic disease that it appeared for a time that the whole sugarcane industry might be doomed. The situation was saved by the introduction of disease-resistant cultivars from Java and India, followed by extensive breeding programs.

History

Sugarcane was grown as a cultivated crop in India long before the beginning of the Christian era. Botanical evidence suggests that cultivated types were developed there from the wild canes relatively low in sugar. The Chinese got cane from India and were growing it as early as 1766 B.C. It is probable that early native voyagers carried cane stalks as food and that by this means it went to Java and many of the islands of the Pacific, where it was found growing apparently wild by European explorers.

The growing of cane also spread westward from India into Persia and Arabia. It is believed that sugar was first produced in the Western Hemisphere before 1510 in Santo Domingo, after having been taken there by Columbus. The first recorded continental planting was in 1751, near the city of New Orleans. A marketable sugar is known to have been produced 40 years later. The first successful sugar crop was produced in 1794 on land which later became the location of the Louisiana Sugar Experiment Station.

The production of cane sugar prospered. Yields ranged from 16 to 20 tons/acre (35.8–44.8 MT/ha). One general type was grown, the noble cane, characterized by large stalk diameter, low fiber content, and fairly high sucrose content. But this noble cane became less and less productive because of diseases to which it was susceptible. By 1926 the average production of about 300,000 tons (272,100 MT) per year was down to 47,000 (42,629 MT). The rapidity with which sugar production came back with the introduction of disease-resistant canes from Java and India is indicated by the 199,000 tons (180,493 MT) produced in 1929.

The development of new cultivars has been necessary to meet the exacting requirements of the sugar industry. A cultivar must be early maturing, have good stubbling qualities, be resistant to insects and diseases, and stand relatively erect at maturity. A cultivar also should have a certain degree of cold tolerance and have a resistance to inversion of sucrose to simple sugars or maintain quality after cutting.

The Sugarcane Plant

Sugarcane is a tall-growing, tropical perennial grass requiring usually 8–24 months to reach maturity. The plant may grow to a height of 20 feet (6.1 m) or more. The stems are solid, and have prominent nodes, with one leaf formed at each node. Leaves are long but most often erect. The inflorescence, which may not develop until 12–24 months after planting, is a panicle called an "arrow." Flowering and seed formation are important only to sugarcane breeders because of vegetative propagation.

Planting

Sugarcane is field planted with vegetative stem pieces about 2–3 feet (0.6–0.9 m) long or with whole stalks. Each node or joint of the stalk has one bud or eye which can give rise to a new plant. The seed cane is first stripped of leaves and cut by hand or machine. Hand planting gradually is being replaced by mechanical planters. Some planters are equipped with stalk cutters, but the stalk sections or whole stalks normally are dropped end to end into furrows from tractor-drawn carts or by workers riding the planter. Sometimes two or three stalks or pieces are dropped side by side. Fully automatic planters are commercially available that can plant, fertilize, and cover in one operation, but most planting procedures still involve a crew of at least four persons. The stalks are covered to a depth of 2–4 inches (5.1–10.2 cm) and sometimes rows are cultipacked after planting. Distance between rows varies from 3 to 7 feet (0.9–2.1 m), depending on the locality.

In Florida and Louisiana, planting normally is done in the fall months, while the planting date in Hawaii may be any month of the year; spring planting is practiced in the more northern regions of adaptation.

As many as three crops may be harvested from each planting. Regrowth after harvesting the first crop is a stubble or ratoon crop.

Harvesting

Hawaiian sugarcane is allowed to grow for about two years before it is harvested. Elsewhere in the United States, it is ordinarily harvested 7–8 months after planting.

In Florida and Hawaii, standing cane is fired to burn off leaves prior to cutting stalks. Then the crop is cut, topped, and loaded. These operations may be by hand in some areas, but machines are available that will cut, top, and load in one operation. In Louisiana, the crop is cut, topped, and piled into "heap" rows by machine. Cane is allowed to remain in these piles for 2–4 hours; then the leaves are burned off after dry.

Harvested cane is loaded generally with grab-type loaders, on tractor-drawn carts, or large trucks or trailers and hauled to the mill for processing. It may be transferred to railroad cars or large tractor-trailer trucks for longer hauls.

Processing

Cane is prepared for milling by washing thoroughly and then passing through a series of revolving knives, or sometimes shredding. Juice is extracted by passing the cane between heavy, grooved rollers for crushing. Between crushings, water or a water–juice mixture is added to dissolve more of the sugar. Lime is added to the juice to raise the pH to 6.0–6.5 and to precipitate colloids and other nonsugar impurities. The juice is concentrated into a semisirup in vacuum evaporators, and then is crystallized by further evaporation in other vacuum pans. The mixture of molasses and sugar crystals is separated by spinning in a centrifuge. The blackstrap molasses by-product is used mainly for livestock feed. The light brown, raw sugar is ready for refining. A "sugar slinger" loads the sugar into railroad cars for transport to refineries, where it is to be purified into consumer sugar products. In the refining process, it is melted, decolorized with a carbon filter, and crystallized into familiar white, granular sugar.

Bagasse is what remains of the fibrous cane stalks after the juice is extracted. It is used for fuel in some factories. Bagasse from other cane factories is used as an ingredient in the manufacture of paper and building boards, as mulching material for plants, as litter or bedding for poultry and livestock, and as an ingredient in the manufacture of plastics.

Insects, Diseases, and Weeds

The sugarcane borer is the most injurious insect attacking sugarcane. Injury is caused by larval feeding and tunneling within the stalks. Other serious insect pests are sugarcane beetles, which gnaw ragged holes in young plants below the soil surface; wireworms, which bore into young plants just below the soil surface; and soil anthropods, which prune root hairs and gnaw on roots. Effective preventive or control measures include use of resistant cultivars, selection of the proper planting date, selection of soils with good drainage, and use of insecticides when necessary.

Diseases may reduce yields of sugar 20 percent or more in Hawaii and in most areas of the continental United States. Red rot and root rot, caused by fungi, and mosiac and ratoon stunting, caused by viruses, are among the most serious diseases. Effective preventive or control measures are the use of resistant cultivars when available, heat treatment of seed cane pieces before planting, and selection of well-drained soil sites.

Troublesome and damaging weeds in sugarcane include the winter weeds chickweed, henbit, Virginia pepperweed, and Carolina geranium, and the spring or summer weeds johnsongrass, itchgrass, crabgrass, junglerice, bermudagrass, and morning glory. Johnsongrass, a perennial that spreads by rhizomes, is a particularly serious problem, and in many areas is the most limiting factor in sugarcane production. It is important to have an effective weed control program early—in the weeks after planting and until sugarcane is established for shading the rows. Herbicides are in common use on sugarcane acreage. Among those used are terbacil, fenac, trifluralin, TCA, dalapon, silvex, 2,4-D, and simazine.

REFERENCES AND SELECTED READINGS

1. AYKROYD, W. R. *The Story of Sugar* (Chicago: Quadrangle, 1967).
2. BARNES, A. C. *The Sugar Cane* (London: Leonard Hill, 1964).
3. BEATTY, K. D., and I. E. STOKES. *Sugarcane Variety Research in the Imperial Valley.* USDA-ARS 34–108 (1969).
4. BUGBEE, W. M. *Storage Rot of Sugarbeet,* USDA-ARS Publ. NC-56 (1977).
5. Calif. Agr. Exp. Sta. *Growing Sugar Beets Without Hand Labor,* Ext. Serv. Publ. AXT-188 (1965).
6. CHAPMAN, S. R., and L. P. CARTER. *Crop Production Principles and Practices,* (San Francisco: Freeman, 1976).
7. COWLEY, W. R., and B. A. SMITH. *Sugarcane Trials in the Lower Rio Grande Valley of Texas,* Tex. Agr. Exp. Sta. B-1086 (1969).
8. CREEK, C. R. *Changes in Sugar Beet Production in Colorado,* Colo. State Univ. Exp. Sta. Bul. 530-S (1968).
9. DENNIS, R. E. *Sugar Beets in Arizona,* Univ. Ariz. Coop. Ext. Serv. Folder 110 (1965).
10. FENWICK, H. S. *Root Rots of Sugar Beets,* Idaho Coll. of Agric. Current Info. Ser. 295 (1975).
11. HEBERT, L. P. *Culture of Sugarcane for Sugar Production in Louisiana.* USDA-ARS Agr. Handbook 262 (1964).
12. HUGHES, C. G., *et al. Sugarcane Diseases,* Vol. I (London: Elsevier, 1961).
13. HUMBERT, R. P. *The Growing of Sugarcane* (New York: American Elsevier, 1968).
14. JOHNSON, R. T., *et al.* (Ed.). *Advances in Sugarbeet Production: Principles and Practices* (Ames: Iowa State Univ. Press, 1971).
15. KING, N. J., *et al. Manual of Cane-Growing* (New York: American Elsevier, 1965).
16. MARTIN, J. H., W. H. LEONARD, and D. L. STAMP. *Principles of Field Crop Production,* 3rd ed. (New York: Macmillan, 1976).
17. MARTIN, L. B. "The Production and Use of Sugarcane," In *Crops in Peace and War: Yearbook of Agriculture* (Washington, D.C.: USDA, 1950–51).
18. MATHERNE, R. J., *et al. Culture of Sugarcane for Sugar Production in the Mississippi Delta,* USDA-ARS Agr. Handbook 417 (1977). The Yearbook Committee Chairman, G. W. Irving, Jr.
19. Michigan Sugar Co., 1977 *Sugar Beet Growers' Guide,* Caro, Mich. (1977).
20. NICHELL, L. G. "Crop Improvement in Sugarcane: Studies Using In Vitro Methods," *Crop Science,* 17:717 (1977).
21. PEAY, W. E. *Sugar Beet Insects, How to Control Them,* USDA Farmers' Bul. 2219 (1968).

22. PRINGLE, G. E., and J. ZAPATA. *Practices Used on Puerto Rican Farms with High Production of Sugar,* Puerto Rico Agr. Exp. Sta. Bul. 212 (1969).

23. SHERROD, L. B., *et al. Nutritive Value of Seed Cane Toppings and Mill Cane Strippings with and without Supplemental Protein,* Hawaii Agr. Exp. Sta. Tech. Prog. Rep. 168 (1968).

24. STEVENSON, G. C. *Genetics and Breeding of Sugarcane* (London: Longmans, 1965).

25. ULRICH, R. *Plant Analysis in Sugar Beet Nutrition.* Am. Institute of Biological Sciences, Publ. 8 (1961).

26. USDA. Agr. Stat. (1977).

D. Drug and Miscellaneous Crops

CHAPTER 27

TOBACCO

TOBACCO, *Nicotiana tabacum,* is native to the Western Hemisphere, where it was being cultivated by the Indians when the continent was first visited by European explorers. There are many records supporting the widespread ancient growth and use of tobacco in the Caribbean, Mexico, and South America. Organized production apparently began about 1612 when John Rolfe at Jamestown, Virginia, grew a crop for export. The venture was successful and within five years large areas in Jamestown were planted to tobacco. By 1620 a large volume of tobacco was exported to England. U.S. cultivation extended from Virginia to Maryland in about 1631, and to Kentucky about the end of the eighteenth century.

The United States is the largest producer of tobacco in the world, but China has the greater acreage (Figures 27–1 and 27–2). Other leading producers are India, USSR, Brazil, and Bulgaria. The United States and China produce nearly 37 percent of the world's total tobacco (Figure 27–2).

Approximately 1 million acres (0.4 million hectares) of tobacco are planted annually in the United States with an average yield of just over 2,000 pounds/acre (2,240 kg/ha). Leading tobacco-producing states are North Carolina, Kentucky, South Carolina, Virginia, Georgia, and Tennessee (Figure 27–3). These states account for more than 90 percent of the U.S. acreage, with 44 percent in North Carolina alone. Geographic distribution is shown in Figure 27–4.

Federal regulations have restricted advertising and required a warning label on cigarette packages. These regulations have not influenced total tobacco use, which remains at about 1.2 billion pounds (0.54 billion kg) annually, virtually the same level as in 1965 (Figure 27–5). Neither have the regulations caused a decline in cigarette production and use; production has increased rather steadily since 1965 to a present level approaching 700 billion cigarettes per year.

Because of medical evidence which has raised serious questions about tobacco use and human health, alternate uses for tobacco have been sought. California researchers have extracted a high-grade protein from tobacco leaves, with a nutritional value greater than standard animal protein. They predict a bright future for tobacco as a primary nutrition source, but development and adoption of this alternate use remains to be seen.

FIGURE 27–1. *Leading countries in world tobacco area (acreage). China is the leader in acreage but not production.* [SOURCE: Southern Illinois University.]

The Plant

Tobacco is a warm-season, herbaceous annual, with a simple cylindrical stem 4–8 feet (1.2–2.4 m) in height, and alternate leaves arising directly from the central stem. As many as 25 leaves are borne per plant. Each leaf is an extremely broad, elongated structure, often 1 foot wide and 3 feet long (30.5 × 91 cm). The leaves are covered with a soft, downy, glandular pubescence, which makes the surface somewhat sticky. Flowers, nor-

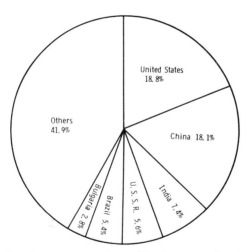

FIGURE 27–2. *Leading countries in world tobacco production. The United States and China produce well over one third of the total crop.* [SOURCE: Southern Illinois University.]

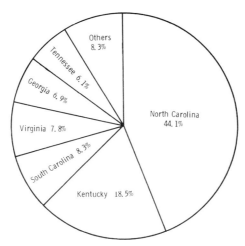

FIGURE 27–3. *Leading states in United States tobacco area (acreage). North Carolina has almost one half of the total acreage.* [SOURCE: Southern Illinois University.]

mally pink but varying from white to red among cultivars, are borne in racemes at the top of the plant. The flowers are normally self-fertilized, but cross-fertilization is easily effected by insects. Tobacco produces capsule fruit that split on reaching maturity to expose exceedingly small seed. Each capsule may bear several thousand seed; there are about 5 million seeds per pound (11 million per kg).

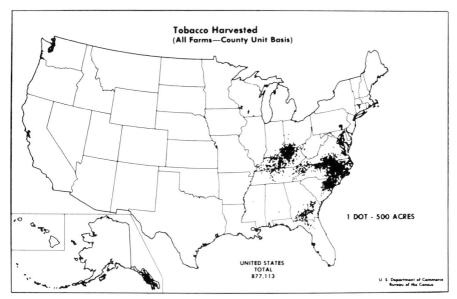

FIGURE 27–4. *Tobacco harvested on all farms in the United States, shown geographically.* [SOURCE: U.S. Department of Commerce.]

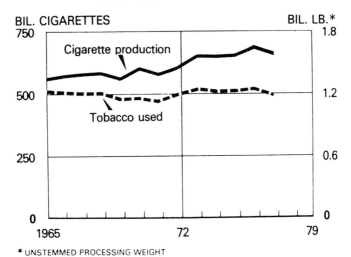

FIGURE 27–5. *Cigarette production and tobacco used in the United States. Tobacco usage has remained virtually the same since 1965 despite health warnings.* [SOURCE: USDA.]

Classification and Types

Tobacco belongs to the family *Solanaceae,* the night-shade family. There are classes, types, and grades. The differences between classes result not only from differences in curing methods, but also from variations in soil, cultural practices, and climate. Variations between types in any one class may consist of differences in color, body, quality in a general sense, and even in response to fermentation and aging in storage. Grade names indicate, for the most part, either directly or by implication, the uses made of the leaf. The seven general classes of tobacco based on curing methods and uses are (1) flue-cured, (2) fire-cured, (3) air-cured, (4) cigar filler, (5) cigar binder, (6) cigar wrapper, and (7) miscellaneous. The classes and type names of localities are shown in Table 27–1.

The flue-cured types, extensively grown in the Carolinas, Georgia, and Florida, are used mostly in the manufacture of cigarettes and as smoking and chewing tobaccos; they are also used for export. The bright yellow leaf color is due mainly to the soil on which it is grown and to the method of curing. Of the total tobacco used in the United States, the flue-cured class accounts for almost 50 percent (Figure 27–6).

Fire-cured types are grown almost exclusively in western Kentucky and Tennessee and in central Virginia. Their principal characteristics are their dark color, heavy body, and distinctive flavor imparted by the smoke of the open fires used in curing. The greater portion of these types is exported, but they also are used in the domestic production of snuff and as plug wrapper.

Table 27–1. Classes and Types of Tobacco Established by the U.S. Department of Agriculture

Type of Curing and Class	Type	Type Name or Locality
Flue-cured, Class 1	11A	Old Belt—Virginia and North Carolina
	11B	Middle Belt—North Carolina
	12	Eastern Belt—North Carolina
	13	Border Belt—South Carolina and Southeastern North Carolina
	14	Georgia and Florida
Fire-cured, Class 2	21	Virginia
	22	Eastern—Kentucky and Tennessee
	23	Western—Kentucky and Tennessee
Air-cured		
Class 3A (light air-cured)	31	Burley
	32	Maryland
Class 3B (dark air-cured)	35	One-Sucker
	36	Green River
	37	Virginia Sun-Cured
Class 4 (Cigar Filler)	41	Pennsylvania Seedleaf, or Broadleaf
	42	Gebhardt
	43	Zimmer Spanish
	44	Little Dutch
	46	Puerto Rico
Class 5 (Cigar Binder)	51	Connecticut Broadleaf
	52	Connecticut Havana Seed
	53	New York and Pennsylvania Havana Seed
	54	Southern Wisconsin Havana Seed
	55	Northern Wisconsin Havana Seed
Class 6 (Cigar Wrapper)	61	Connecticut Valley Shade-Grown
	62	Georgia and Florida Shade-Grown
Miscellaneous, Class 7	72	Louisiana Perique

Source: Chaplin, J. F., *et al. Tobacco Production.* USDA-ARS Agric. Info. Bul. 245 (1976).

Burley, a light air-cured type, is produced in the bluegrass section of Kentucky, in eastern Tennessee, and in southern Ohio. It is light in color and body and is used for plug tobacco and for blending in cigarettes. Burley accounts for about 35 percent of the total tobacco used in the United States (Figure 27–7).

In the portion of Kentucky and Tennessee lying between the Burley district and the fire-cured section, the types grown are mainly dark, air-cured tobaccos. They are raised for chewing and smoking tobaccos and for export.

For cigars, the filler-leaf type is produced mainly in Lancaster County, Pennsylvania; in the Miami Valley of Ohio; and in the Onanadaga district of New York. The binder-leaf type is grown in Wisconsin, New York, and Pennsylvania; the principal wrapper-leaf sections are in the Connecticut Valley area and in certain areas in Georgia and Florida.

Broadly speaking, local conditions of soil and climate and the cultivar,

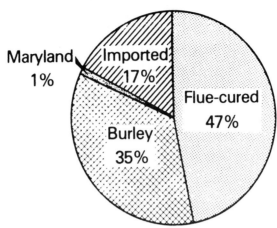

FIGURE 27–6. *Tobacco use by kind in the United States. Flue-cured tobacco, used extensively in the manufacture of cigarettes, accounts for almost one half of the total use.* [SOURCE: USDA.]

supplemented with appropriate methods of culture and handling, determine the predominant usage of the crop and therefore the type. Production of secondary grades in all types, on the other hand, is due to fundamental differences in properties of the leaves occupying different positions on the stalk. Quality within the grade is influenced by all these factors.

Cultivars

The number of tobacco cultivars is large. In general, cultivars fall into groups corresponding to the various commercial types of leaves. Growers realize the importance of authentic seed, true to type. Some tobacco-growing states have organizations to aid growers in obtaining dependable seed of standard cultivars and strains. Akehurst has given detailed information on the many cultivars and strains of tobacco and on tobacco breeding.

A number of stations have regularly conducted tobacco cultivar tests. The Coastal Plain station in Georgia, among others, has found a widely variable nicotine content among cultivars. Kentucky researchers have searched for ways to reduce the nictoine content of a given cultivar or strain. They found a virus which can lower nicotine levels, but the virus also stunted plant growth and reduced leaf yield.

Climate and Soil

Tobacco can be grown successfully on a great variety of soils in all latitudes from southern Canada to the tropics. But soil type and condition and the climate probably affect the quality of the product to a greater extent than of any other crop.

Light sandy and sandy loam soils, because of low water-holding capacity and low soluble mineral matter, tend to produce a leaf of relatively large

FIGURE 27-7. *Two commonly grown classes of tobacco. (A) Burley, class 3A, light air-cured, is produced in Kentucky, eastern Tennessee, and southern Ohio.* [SOURCE: University of Tennessee Agricultural Experiment Station.] *(B) Flue-cured, class 1, is grown extensively in the Carolinas and Georgia.* [SOURCE: University of Georgia, College of Agriculture.]

size, light in color and body, fine in texture, and weak in aroma. Heavier soils that contain more silt and clay tend to produce a leaf of smaller size, dark color, heavy body, and strong aroma. A soil reaction of pH 5.0–6.0 is considered desirable for tobacco.

Rotations

There is a tendency in some regions to plant continuous tobacco to make maximum use of relatively small areas of the very best soil. Although systematic rotations have not come into general use, there are several recommendations. The rotation selected depends on the tobacco type and the area in which it is grown. Nematode-susceptible crops, such as cowpeas, soybeans, and sweetpotatoes, should not be included in the rotation. Legume winter cover crops immediately preceding tobacco have not always given good results. Crops with which tobacco is more successfully rotated include corn, cotton, small grains, and grass sods of two or more years. The green manure crops rye, crimson clover, and hairy vetch are considered suitable in some areas for use immediately before tobacco. Variations in the crops used in rotation with tobacco generally give better disease and nematode control than if only one other crop is used in rotation.

Workers in Kentucky, Pennsylvania, and Maryland have shown that disease-resistant cultivars of tobacco can be grown satisfactorily on the same land year after year. This continuous cropping has gained favor among some growers in these states. With the use of small grain and winter legume cover crops and liberal fertilization, the fertility level can be maintained without severe erosion.

Fertilizers and Manure

Best results are obtained with an uninterrupted, relatively rapid rate of growth all the way from transplanting to the approach of maturity. The ideal is an adequate and well-balanced supply of essential nutrients throughout the earlier growing period, with a rapid decline as the crop approaches maturity. The tobacco crop is heavily fertilized, with Burley and flue-cured often receiving more than 1,000 pounds/acre (1,120 kg/ha), and cigar wrapper and binder often receiving more than 2,000 pounds/acre (2,240 kg/ha) of a complete fertilizer.

Lack of nitrogen can result in slow growth and a poor yield, but excess nitrogen results in delayed maturity and a low quality leaf. Phosphorus-deficient plants are stunted, abnormally dark green, and late maturing. Excessive phosphorus shortens leaf burn duration. Lack of potassium is a common cause of low quality in tobacco. Liberal potassium is needed for good burning quality of leaves. Source of potassium also is important. Large amounts of muriate of potash, KCl, should not be used; excessive chlorine is detrimental to the quality of cured tobacco.

The amount of fertilizer recommended depends upon the types of tobacco grown and the area. Fertilization most often should be based upon soil test recommendations. Flue-cured tobacco may require 35 pounds of nitrogen, 100 pounds of phosphate, 120 pounds of potash per acre (39, 112, and 134 kg/ha). Burley may use up to 100 pounds of nitrogen and 300 pounds of potash per acre (112 and 336 kg/ha). Some of the soils where Burley is produced in Kentucky and Tennessee are high in native phosphate.

Manure may be used for most types of tobacco at rates of 10 tons/acre (22 MT/ha) or more. It should be applied 3 or 4 weeks prior to transplanting and incorporated thoroughly. Fertilizer rates should be reduced accordingly when manure is used.

Seedbeds

Attempts at direct sowing of seed in the fields, including pelleting of individual seeds to facilitate drilling, have been largely unsuccessful. Tobacco usually is seeded in beds six to twelve weeks before time to set the plants in the fields. Soil selection for the seedbed is important. The best soil is well-drained, mellow, sandy loam of high fertility. Manure may be plowed under, or fertilizer applied liberally well in advance of seeding.

For the seedbed a cold frame usually is used. The bed should be sterilized for the control of soil-borne diseases, weed seed, and certain insects. In years past, the common method of sterilization involved burning large quantities of brush or wood in the seedbed. Steam sterilization also has been used. For best results a minimum of 30 minutes of steaming is required. The temperature of the surface soil rapidly rises to 212°F (100°C) and a temperature of 140°F (60°C) is considered effective to a depth of 6 inches (15 cm).

A number of chemicals are now being largely substituted for wood fires or steaming. Among these are calcium cyanamide, sodium methyldithiocarbamate and methyl bromide, probably the most common. Methyl bromide comes in a pressurized can and should be used at a rate of 1 pound per 100 square feet (4.9 kg/100 m²) of seedbed. The bed should be covered with a polyethylene cover when methyl bromide is used, since it is released from the can in a gaseous form.

The recommendation for seeding is 1 or 2 level teaspoons of seed for an entire bed 9 × 100 feet (2.7 × 30.5 m) in size. Since the seed is so small, it should be mixed with sand, sifted wood ashes, cornmeal, or another inert material in order to promote even distribution over the bed area. Seed should be scattered three or more times for good distribution of seed in all parts of the bed. The soil is often rolled after seeding to firm the seed in contact with the soil. Then the bed is covered with glass or a polyester or cotton cover fastened to the frame to protect seedlings from cold winds and insects and to keep the soil surface from drying.

Plants should be watched carefully to see that plantbed diseases, such as wildfire, blue mold, and damping off, and insects, such as flea beetles, aphids, and cutworms, do not result in high seedling mortality.

The production of transplants is the only major aspect of many flue-cured tobacco systems not mechanized. Equipment is available for mechanized planting, harvesting, and curing but a large amount of hand labor is needed for producing and handling transplants. Georgia workers investigated a more mechanized transplant-producing system, which involved precision seeding of coated seed, trickle irrigation, and an elevated solid cover put in place with a machine. Initial performance of this system was encouraging, but there are several problems to be solved.

Field Planting

When plants have developed 4 to 6 leaves and are 5–6 inches (12.7–15.2 cm) in height, they are ready for the field. Tobacco plants are set in rows that vary from less than 3 feet (91 cm) to more than 4 feet (122 cm) between rows, depending on the region and type of tobacco. Spacing between plants in the row varies from 10 to 18 inches (25–46 cm). It requires from less than 6,000 to more than 10,000 plants/acre (14,800–24,700/ha). Some high-density experimental plantings done mechanically have used up to 32,000 plants/acre (79,000/ha). Planting should be done as soon as possible after the last killing frost, when there is an abundance of moisture. There are three common methods of transplanting: (1) hand setting, the oldest method; (2) hand transplanter, which gives a more uniform stand of plants; and (3) a tractor-drawn transplanter, which is the most desirable.

Experimental work at the Kentucky Station has shown that no-till planting of Burley into a chemically controlled grass sod can be successful but requires closer supervision. Such a method is yet to be proven for use by tobacco farmers.

Topping and Suckering

Topping consists of breaking off the crown of the plant below the flowering head. This prevents the plant from producing seed and allow carbohydrates and nutrients to go toward vegatative (leaf) growth rather than reproductive growth. A topped plant yields more high-quality leaves and ripens more uniformly. Topping causes thickening of the leaves and increases the body of the leaf, but it also causes an increase in nicotine content, as much as double that of an unaltered plant. With a normal growing season, tobacco is ready to top about 8 or 9 weeks after transplanting. When the seedhead begins to form, it is time to top for maximum results. Plants may be topped by hand, or with mechanical cutters.

When the tops are broken out, the buds in the axis of the leaves start development, producing suckers. These are removed as soon as they are large enough to be pulled. It may be necessary to remove the suckers several times; some stations report best results from suckering at 7–10-day intervals. Sucker growth can be suppressed satisfactorily with chemicals such as maleic hydrazide and with various oils. Kentucky researchers found that plots treated with MH-30, which gave the best sucker control of those treatments tested, produced significantly higher yields with greater acre

Table 27–2. Influence of Topping and Suckering Practices on Yield and Value of Burley 21 Leaf*

Topping or Suckering Practice	Avg. Yield (Lb/Acre)	Value ($/cwt)	Value ($/Acre)
No topping	2375	64.06	1521
No suckering	2567	65.79	1689
Hand-suckering	2774	63.61	1764
Chemical suckering with MH-30	3038	64.30	1953

* Average of two harvest dates.

Source: Sims, J. L., and W. O. Atkinson. "Yield and Value of Burley 21 Tobacco as Influenced by Nitrogen Nutrition, Suckering Practice, and Harvest Date," Ky. Agron. Notes, 3:4 (1970).

value than plants not topped or suckered (Table 27–2). However, date of harvest, early or late, influenced the yield and value of tobacco with different suckering practices.

Harvesting and Curing

Tobacco is ready to harvest when leaves turn a lighter shade of green to slightly yellow. Some classes of tobacco, such as flue-cured, cigar-wrapper, and cigar-binder, are harvested by picking off each leaf as it matures. This is called the priming method. Priming starts with the lower 2 to 4 leaves, and proceeds up the plant, with 2 to 4 additional leaves being picked every 5–10 days. Historically, most priming has been a hand or manual operation. Presently, mechanical harvesting of flue-cured tobacco is widespread. Once-over tobacco harvesters are available which strip leaves from plants and accumulate them in a trailer. Primed leaves may be strung to a lath or stick about 4 feet (1.2 m) long and taken to the barn for curing. More recently, more efficient handling and curing procedures have been adopted. Where hand stringers once worked, now tobacco is often loaded by armfuls of loose leaves into bulk curing racks which are hoisted mechanically into curing barns. Mobile bulk curing barns are now available; they can be towed to the field, loaded, and then towed back to the heating units. Solar-heated curing barns are being used on a limited scale.

The usual method of harvesting classes such as burley is to cut the entire plant near the soil surface when middle leaves show the first tinge of yellow. It is recommended that plants be placed immediately on tobacco sticks or laths stuck in the ground by piercing the stalk near its base with a removable metal "spearhead" placed on the end of the stick. Each stick will hold about 6 plants. Plants are allowed to wilt in the field for a period of a few hours up to 4 or 5 days; wilted plants are easier to handle without damage. The laths carrying the plants are placed on a rack and hauled to the curing barn.

FIGURE 27–8. *Harvesting tobacco. Flue-cured tobacco is harvested by picking off each leaf as it matures. This priming method starts with lower leaves and proceeds up the plant. Primed leaves are placed in a trailer to be taken to the barn for curing.* [SOURCE: University of Georgia College of Agriculture.]

There are three principal methods of curing tobacco: (1) air curing, (2) flue curing, and (3) open-fire curing. Most of the Burley and much of the cigar tobaccos are air-cured without the use of artificial heat. The tobacco sticks, containing 6 or more plants, are placed 10–14 inches (25–36 cm) on tier rails in a well-ventilated curing barn. A barn may be high enough to have three or more levels of tiers. Curing is controlled by regulating ventilation. During cool weather, a nonsmoking fire, such as coke or gas burners, may be necessary. Air curing of tobacco is slow, and when the weather is unfavorable the process may take several weeks.

The distinctive feature of flue curing is that the barn is provided with a system of large pipes, or flues, that carry off fuel gases so that the smoke does not come in contact with the tobacco. Artificial heat is used throughout the cure, which is completed within a few days. The conventional flue-curing barn is being replaced with loose leaf handling and the bulk curing systems as discussed previously.

The use of open fires is confined to "dark tobacco" districts. Open, hardwood fires burning in the barn for a few days are sufficient to dry the tobacco, and the smoke imparts a characteristic taste desired in the resulting chewing or snuff products.

FIGURE 27–9. *Harvesting tobacco. (A) The entire burley tobacco plant is cut near the soil surface and placed on tobacco sticks stuck in the ground by piercing the stalk near its base with a removable metal "spearhead" placed on the end of the stick. (B) Each stick will hold about 6 plants. Plants are allowed to wilt in the field before hauling to the barn for air-drying.* [SOURCE: Southern Illinois University.]

FIGURE 27-10. *Barn for air-curing Burley tobacco. The barn should be located on an open, well-drained area for good ventilation. Tobacco stalks should be spread on the sticks, and leaves should be hanging down, not doubled up, for best results. Tobacco should be spaced throughout the barn so that air can circulate evenly.* [SOURCE: Ira E. Massie, University of Kentucky.]

Stripping and Packing

Tobacco types such as Burley, where the entire stalk has been cut, must be prepared for marketing after it is thoroughly cured. When the midribs are completely dry (the stalks may still be green), tobacco should be taken down from the tiers and piled in heaps. It is important that leaves are allowed to become moist by the absorption of moisture from the air, or they are placed in rooms where moisture is supplied by artificial means, such as dampening the plants in cellars, steaming, or sprinkling the butts of the plants. This makes the tobacco sufficiently pliable to be handled without breaking. This is known as being "in case" or "in order." In packing tobacco down into heaps, it should be taken off the sticks and packed with the tips of the leaves together and the butts out in bulk at least 6 feet (1.8 m) high. The bulk may be covered to keep the crop from drying and breaking and keep it clean until leaves are stripped.

In the stripping process the leaves are removed from the stem, sorted into grades or groupings of like quality, and tied into hands or bundles. A hand is a bunch of 15–30 leaves tied together, usually with another leaf. A knowledge of the various leaf groupings will enable one to do a good job of stripping. Leaves produced on a stalk of Burley from bottom to top are flyings, lugs, leaf, and tips. Flyings, bottom leaves, often show a great deal of injury. Lugs, the broadest and largest leaves on the plant, are the most valuable. The leaf group has larger stems and tends to fold up when cured. The tips are leaf groups under 16 inches (40.6 cm) in length.

Burley and Maryland tobaccos usually are sold looseleaf, tied into hands. Cigar-filler and binder leaves are made into bundles or bales, wrapped with heavy paper, and tied. These bundles usually weigh 30–60 pounds (13.6–27.2 kg). Sometimes Maryland tobacco is sold in hogsheads weighing 700–800 pounds (318–363 kg). Most flue-cured tobacco is now sold looseleaf without tying.

Grading and Marketing

Tobacco leaves vary in value probably more than the leaf of any other plant. This must be kept in mind when grading for market. The size of the leaf, the body, the size of midribs and veins and chemical composition all materially affect value. Color is closely associated with certain conditions within the leaf and has a definite relation to value. Color, therefore, is one of the main factors used in grading. Body, thickness of the leaf, soundness, and uniformity of size also are important factors that must be kept in mind when grading.

FIGURE 27–11. *Burley tobacco in the warehouse ready for auction. After Burley is thoroughly cured, leaves are stripped from the stem, sorted into grades or groupings of like quality, and tied into hands or bundles. A hand is a bunch of 15 to 30 leaves tied together at the base, usually with another leaf. Properly sorted, neatly prepared, and correctly arranged lots of tobacco sell at the highest current market price.* [SOURCE: Ira E. Massie, University of Kentucky.]

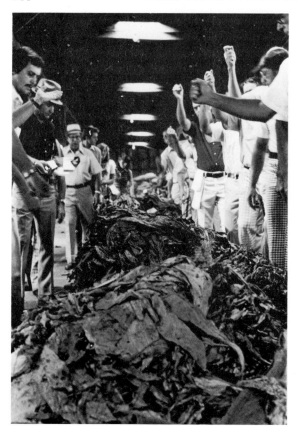

FIGURE 27–12. *The tobacco auction scene has changed little over the years. After grading and labeling, individual lots of tobacco are laid out on warehouse floors for buyers to inspect. During the sale, buyers pass along the rows and bid competitively on each pile of tobacco that meets their requirements. An auctioneer may sell hundreds of individual lots per hour.* [SOURCE: University of Georgia, College of Agriculture.]

Most of the tobacco crop is sold by the auction system on warehouse floors. The marketing season may be several months after harvest or even the year following its production. After grading and labeling, individual lots of tobacco are laid out for buyers to inspect. During the actual sale, buyers pass along the rows and bid competitively on each pile of tobacco that meets their individual requirements. An auctioneer may sell several hundred individual lots per hour.

Insects and Diseases

Growing tobacco plants are commonly injured by hornworms, tobacco budworms, and flea beetles. Hornworms devour entire leaves and are the most widely distributed and destructive. Tobacco budworms feed at the top of the plants, cutting holes in the young leaves. The flea beetle fills the leaves with small holes. Other insects damaging growing tobacco plants include cutworms, grasshoppers, and aphids. The tobacco or cigarette beetle attacks stored or manufactured tobacco.

Tobacco is affected by diseases caused by each of the three major organisms—fungi, bacteria, and viruses. The most serious fungus diseases are black shank, black root rot, anthracnose, blue mold, and southern stem rot. Bacterial diseases include wildfire, blackfire, and bacterial or Granville wilt. The most serious of the virus diseases is tobacco mosiac, which is present wherever tobacco is grown. Seed treatment, treatment of seedbeds, the use of resistant cultivars, suitable rotations, and good sanitation practices all must have attention when controlling a wide range of diseases.

REFERENCES AND SUGGESTED READINGS

1. AKEHURST, B. C. *Tobacco* (London: Longmans, Green, 1968).
2. CHAPLIN, J. F., *et al. Tobacco Production,* USDA-ARS Agric. Info. Bul. 245 (1976).
3. CUNDIFF, J. S. *A New Production System for Tobacco Transplants in Field Beds,* Ga. Coastal Plain Res. Rep. 241 (1977).
4. LUCAS, G. B. *Diseases of Tobacco,* 2nd ed. (New York: Scarecrow, 1965).
5. MCMURTREY, J. E., JR. *Tobacco Production,* USDA Agr. Info. Bul. 245 (1961).
6. MARTIN, J. H., W. H. LEONARD, and D. L. STAMP. *Principles of Field Crop Production,* 3rd ed. (New York: Macmillan, 1976).
7. MASSIE, I. E., G. A. EVERETTE, and J. H. SMILEY. *Burley Tobacco Production,* Ky. Coop. Ext. Serv. Circ. 616 (1968).
8. MASSIE, I. E., G. A. EVERETTE, and J. H. SMILEY. *Tobacco Plant Bed Management.* Ky. Coop. Ext. Serv. Circ. 546-B (1966).
9. MASSIE, I. E., and J. H. SMILEY. *Harvesting and Curing Burley Tobacco,* Ky. Coop. Ext. Serv. Publ. AGR-14 (1974).
10. MILES, R. L., and C. ROLAND. *Growing Flue-Cured Tobacco in Georgia,* Univ. Ga. Coop. Ext. Bul. 599 (1969).
11. NICHOLS, B. C., and J. E. MCMURTREY, JR. *Priming Tests with Burley Tobacco,* USDA and Tenn. Agr. Exp. Sta. Bul. 285 (1958).
12. NICHOLS, B. C., R. L. DAVIS, and J. E. MCMURTREY, JR. *A Comparison of Sulfate and Muriate of Potash for Production of Burley Tobacco Seedlings.* USDA and Tenn. Agr. Exp. Sta. Bul. 342 (1962).
13. PARKS, W. L., and L. M. SAFLEY. *The Effect of Irrigation and Nitrogen Upon the Yield and Quality of Dark Tobacco,* Tenn. Agr. Exp. Sta. Bul. 394 (1965).
14. RHODES, G. N., and J. N. MATTHEWS. *Burley Tobacco Production in Tennessee,* Tenn. Agr. Ext. Serv. Publ. 358 (1976).
15. SAFLEY, L. M., and H. C. SMITH. *Cultural Practices for Fire-Cured Tobacco,* Tenn. Agr. Exp. Sta. Bul. 462 (1969).
16. SIMS, J. L., and W. O. ATKINSON. "Yield and Value of Burley 21 Tobacco as Influenced by Nitrogen Nutrition, Suckering Practice, and Harvest Date," *Ky. Agron. Notes,* 21:4 (1970).
17. The Tobacco Institute. *Kentucky and Tobacco,* Tobacco History Series, 5th ed. (1976).
18. USDA. Agr. Stat. (1977).
19. USDA-ARS. *Stored Tobacco Insects, Biology and Control,* Agr. Handbook 233 (1971).
20. WHITTY, E. B., *et al. Flue-Cured Tobacco Guide,* Agr. Ext. Serv. Circ. 269A (1969).

Chapter 28
POTATOES

The potato has had much to do with shaping human society. Two cultures in particular, each in its own way, offer striking examples of its influence. One is Ireland, whose history has been bound up with the potato for more than 300 years. The other is Peru, where for some 2000 years the potato was not only the staple of life but a spiritual symbol as well. . . .

It is in Ireland, . . . that one finds the clearest evidence of the influence which a cheap foodstuff can exercise on a society. The potato reached Ireland around 1588. . . . It provided a maximum of sustenance with a minimum of labor put out. The output of an Irish acre was sufficient to supply a married couple and four children with all they needed in the way of potatoes and still leave enough for the pigs and hens, notwithstanding that the average man consumed 12 pounds, the wife 10 pounds and each child 5 pounds per day. . . .

Then in 1845 and 1846 came the total destruction of the potato crop by the previously unknown fungus *(Phytophthora infestans)*. The tale of this fatal mold and of the Great Famine that followed has been told too often to need repetition here. . . . Death and immigration reduced the population of Ireland from 9 million to 6½ million within 6 years and in a couple of decades it had fallen to 4 million.

R. N. SALAMAN

White Potatoes

THE potato, *Solanum tuberosum,* known both as the Irish potato and as the white potato, is related to such crop plants as the tomato, eggplant, and tobacco. It also is related to some well-known weeds: horse nettle, buffalo bur, and jimsonweed. The potato is believed to have been cultivated first in Peru. It was carried to Spain, Italy, and Germany soon after the conquest of Peru about the middle of the sixteenth century. It was introduced into Ireland about 1586 and was cultivated as a field crop before 1663. The potato became the staple food in Ireland. The Irish Famine was the result of the late blight disease, which destroyed the entire potato crop in 1845 and 1846. It is said that potatoes were first brought to the United States in 1719 by Scotch–Irish immigrants who established a settlement at Londonderry, New Hampshire. During the early years, commercial potato production was centered in New England and the Mid-Atlantic states before it gradually spread westward.

Production Importance

The importance of the potato as a food is generally recognized. It now is grown in most countries of the world. In addition to its extensive use as food, the potato is used both in agriculture as a livestock feed and in industry. The leading country in both acreage and production is the USSR, with nearly half the world's acreage and almost 40 percent of the total production (Figure 28–1). Other countries with sizable acreages and production are Poland, India, and the United States.

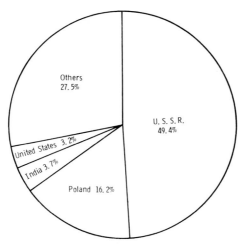

Others
27. 5%

U. S. S. R.
49. 4%

United States 3. 2%

India 3. 7%

Poland 16. 2%

FIGURE 28–1. *Leading countries in world white potato area (acreage). The USSR has nearly one-half of the total.* [SOURCE: Southern Illinois University.]

Potatoes are grown commercially in nearly every state. Geographic distribution is shown in Figure 28–2. Throughout the United States there is not a single month in the year when potatoes are not being planted or harvested somewhere. Total U.S. production is more than 350 million bushels (9.5 million MT) annually for a value of more than $1 billion. Leading states in potato production include Idaho, Maine, North Dakota, Washington, Minnesota, Wisconsin, Oregon, Colorado, and California (Figure 28–3). Average production per acre in the United States increased from a national average of 80 hundredweight (3.6 MT) in 1940 to 257 cwt (9.9 MT) in 1976. In the United States there has been a decrease in per-capita consumption from the average 182 pounds (82 kg) for the period 1909–1913, to less than 125 pounds (56 kg) for the period 1965–1976 (Figure 28–4). Consumption of fresh potatoes has declined markedly in the last 10 years, but the rather steady per-capita consumption during that period is the result of a large increase in the use of processed potatoes.

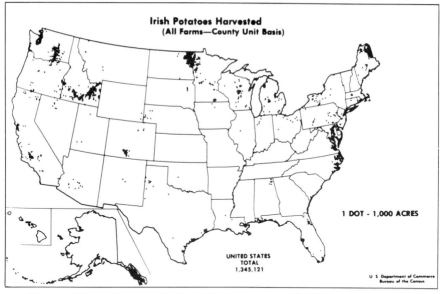

FIGURE 28-2. *Acreage of white (Irish) potatoes harvested in the United States, shown geographically.* [SOURCE: U.S. Department of Commerce.]

The Plant

The potato normally is a herbaceous perennial dicotyledon but is grown as an annual. The aboveground portion is an erect bush with hollow or pithy stems, except for the nodes, which are always solid. Plants grow to a height of 2–5 feet (0.6–1.5 m) or more. The leaves are compound, pinnate,

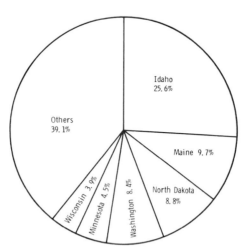

FIGURE 28-3. *Leading states in United States white potato area (acreage). Idaho has over one fourth of the United States total.* [SOURCE: Southern Illinois University.]

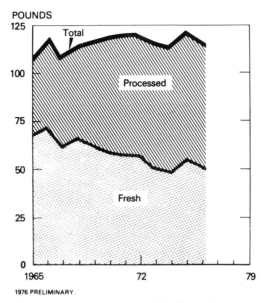

FIGURE 28–4. *Per capita consumption of fresh and processed white potatoes in the United States since 1965.* [SOURCE: USDA.]

FIGURE 28–5. *The potato plant, a perennial which is grown as an annual, bears pinnately compound leaves and clusters of flowers. Tubers originate as swellings on the ends of rhizomes, underground modified stems.* [SOURCE: Southern Illinois University.]

with opposite paired large and small leaves alternating. The root system is largely fibrous. The flowers are borne in clusters, usually appearing as though they were terminal and producing, when fertilized, a many-seeded boll. The tuber is a fleshy stem bearing buds or "eyes." Tubers originate as swellings on the ends of rhizomes; when rhizome tips come in contact with a growing medium, such as soil, tips start to enlarge.

The potato is propagated vegetatively—by whole seed or seed pieces. True seeds are planted only for purposes of potato breeding.

Growth Requirements and Culture

Climatic factors and soil properties have a great influence on yield and quality of potatoes.

Climate and Soil

The potato is a cool-weather crop. Profitable yields are possible where the mean growing-season temperature is 60°–70°F (16°–21°C) and the rainfall totals 12–18 inches (30–46 cm), well distributed over a period of three and a half to four months. Commercial production of spring and summer potatoes is best suited in the north temperate zone, where the climate is cool and moist; however, fall and winter potatoes are produced in California, Florida, and many other geographic areas.

Soils that generally produce the highest yields and best-quality tubers are friable, porous, well aerated, and well drained to a depth of at least 12 inches (31 cm). They cannot be grown where the subsoil is poorly drained. Preferred are well-drained sandy, silt, gravelly, and shale loams well supplied with organic matter and available nutrients. If well drained, peat or muck soils are particularly good for potato production. Potatoes should not be planted in soils that tend to stick to the tubers.

The potato is tolerant of soil acidity. It makes a satisfactory growth and yield in a soil with a pH range of 4.8–5.4. Potato scab is less prevalent in a relatively acid soil. It can be grown in a slightly acid soil if scab-resistant cultivars are selected.

Rotations

Potatoes should be grown in a planned crop rotation to keep the soil fertile, maintain a loose friable structure, check weeds, build up organic matter, and reduce loss from insect damage and diseases. Long rotations, potatoes planted three or more years apart on the same land, are used in some areas for reducing potato losses from soil-borne organisms causing diseases. In some areas, however, now it is considered highly desirable to abandon long rotations in favor of shorter ones. Legumes or grass–legume mixtures are the basis for short rotations in some localities. Rye, millet, or small grain-clover serve as cover or green manure crops in two-year or three-year rotations. The shorter rotations have been successful in western New York; in some areas of Long Island potatoes have been grown on the same area each year with a cover crop of rye seeded after harvest.

Fertilizers and Lime

Fertilization of the potato crop varies greatly from state to state and on different soil types within a relatively small area. In the Northeast, the Middle Atlantic states, and the South, where rainfall is high and there is much leaching, large fertilizer applications are necessary. Moreover, potatoes are heavy fertilizer users, and require an ample supply of nutrients to insure rapid, steady growth and proper tuber development.

The potato plant responds markedly to available nitrogen, except on muck or peat soils where nitrogen applications may not be necessary. The rate depends on the preceding crop, plant population, crop residues, and a number of other factors. If additional nitrogen is needed, it can be applied by sidedressing when tubers are just beginning to develop. Applying nitrogen in irrigation water is an effective method in regions where irrigation is used. Nitrogen should not be applied late in the growing season because of the resulting delay in maturity and a possible decline in dry matter.

The response to phosphorus depends entirely on the native supply of available phosphorus and the carryover from previous years. It is reported that many growers in Maine apply 180 pounds or more of P_2O_5 per acre (202 kg/ha) and growers in Arizona apply 450 pounds or more of P_2O_5 (504 kg/ha) under irrigated conditions.

The recommended use of potash ranges from none in some Western producing areas to as much as 450 pounds of K_2O per acre (504 kg/ha) in Eastern and Northeastern potato-growing sections.

Complete fertilizer generally is used in most commercial potato-producing sections. The rate of fertilizer varies from one section to another, depending on rainfall, soil properties, potato cultivar, spacing, and other growth factors. Growers in some areas may apply 2,000 pounds/acre (2,240 kg/ha) or more of a complete fertilizer on heavier soils, but only 500 pounds/acre (560 kg/ha) or less on sandy soils, where lack of moisture is a problem. As a general rule, Arizona recommends about 800–1,000 pounds (896–1,120 kg/ha) of a 1–3–0 ratio fertilizer, all applied at planting. On the dryland areas of Idaho, 200–300 pounds (224–336 kg/ha) of 16–20–0 or a comparable ratio is sufficient. Furthermore, the Idaho recommendation is that all P and K be incorporated thoroughly by plowing or disking before planting.

In most areas, all fertilizer is applied at planting time in bands along both sides of the row and at a slightly greater depth than the seed piece. The band should not be applied so that fertilizer comes in contact with the seed. When fertilizer rates of greater than 800 to 1,000 pounds/acre (896–1,120 kg/ha) are used, it is advisable to broadcast one-third to one-half the fertilizer before tillage operations and incorporate before planting. The remainder is applied at planting. No one ratio of $N–P_2O_5–K_2O$ will suit all soils or cultivars, but 1–1–1, 1–2–1, 1–2–2, and 2–3–3 are most widely recommended. If manure is used, it should not be applied just before planting since it favors scab disease incidence. Where needed, magnesium and calcium compounds usually are included. Some soils may need the micronutrients manganese, zinc, iron, and copper.

From work at the Rhode Island station, Odland and Allbritten concluded that under the conditions there, potato yields were little influenced by soil reactions between pH 4.8 and 6.1. Below 4.8 or above 6.1 there was a definite tendency toward decreased yields. If the reaction was kept at or below 5.0, scab was controlled. A soil reaction of 5.5 or above tended to increase scab.

In dryland and irrigated areas of the West, the soil reaction usually is above pH 7.0. It is seldom economical to increase acidity by using such soil amendments as sulfur. Where scab is a problem, the most important method of control is to plant resistant cultivars.

Seed

Growers should select satisfactory seed for planting. Among the characteristics of satisfactory seed potatoes are adapted cultivar, true to name, sound, not oversize, and free from disease. Commercial growers should always use certified seed, the best guarantee of freedom from seedborne diseases. Certified seed is grown chiefly in northern areas; such northern-produced seed is preferable, giving the largest yields of highest quality tubers with less chance of disease. Certified seed is produced on both irrigated and nonirrigated land. For years growers were prejudiced against seed stock grown under irrigation, but experiments have shown that irrigation has little or no effect on either the vitality of the seed or the vigor of plants grown from it. It seldom is profitable to use seed potatoes grown in certain areas of the South even if they are disease free.

SEED TREATMENT. The purpose of seed treatment is to kill the surface-borne disease-producing organisms. It is not effective against diseases borne internally or those present in the soil. Most seed treatments involve the use of an appropriate fungicide on cut seed pieces to prevent disease such as *Fusarium* wilt, or the use of disinfectants to prevent common scab.

Seed treatment may be unnecessary if certain other recommendations are followed. The use of certified seed, clean and disease-free, planted into a clean soil will avoid most diseases. Selecting potato cultivars that are resistant to one or more diseases is most effective.

Preparation of Seed Potatoes

Seed potatoes should be brought out of storage for a pre-plant warming period. They should be exposed to a temperature of 55°–65°F (13°–18°C) for 10 days to 2 weeks so that sprouts will begin to form. If tubers have been stored at temperatures too high, long sprouts may have developed.

FIGURE 28-6. *Seed potatoes. In preparation for planting, most tubers are cut into blocky pieces containing at least one "eye." Some growers prefer to plant small, whole tubers to reduce the risk of seed piece rots.* [SOURCE: Southern Illinois University.]

They can be de-sprouted without problems if done 7–10 days before planting.

Most tubers are cut into pieces. Several types of commercial automatic seed cutters are available. Pieces should be blocky, have at least one "eye," and weigh 1½–2 ounces (43–57 g) each; seed smaller than 1½ ounces (43 g) should not be planted. Tubers can be cut several days ahead of planting if stored at a temperature of 60°–70°F (16°–21°C) and at conditions of high relative humidity. Such conditions allow the cut pieces to suberize, sealing over or drying and corking of the cut surface. While some researchers claim that suberization of cut pieces is essential, planting freshly cut tubers is a common practice in Arizona, with no apparent ill effects. In some areas, however, suberized seed pieces are reported to reduce seed piece rot after planting. To minimize cut seed pieces from sticking together, some growers dust with phosphate or gypsum materials to dry the cut surfaces.

Cut seed is not better than whole pieces from the standpoint of yield, and cutting adds to the labor cost. Some growers have had excellent results from planting whole, small tubers, which may reduce seed piece rots.

Birecki and Roztropowicz studied the influence of seed size on the subsequent potato crop. They found that plants resulting from large seed tubers developed at least twice as many stems as plants from small seed tubers; this would result in a greater assimilation area and a higher yield (Table 28–1). Chucka and Steinmetz compared the yield from different sizes of tubers, both whole and cut, of three cultivars. The yield increased as size

TABLE 28–1. Effect of Seed Tuber Size on Subsequent Crop of Potatoes

Tuber Wt. (grams)	Eyes per Tuber	Stems per Plant	Eyes Sprouting (%)	Tuber Weight	
				Per Eye (grams)	Per Stem (grams)
Less than 40	6.51	2.47	37.6	4.6	12.1
40–75	8.56	4.09	47.8	6.4	13.4
75–100	9.37	5.09	54.3	8.5	15.7
More than 100	10.35	7.26	70.1	11.6	16.5

Source: Birecki, M., and S. Roztropowicz. "Studies on Seed Potato Size and Productivity," *Eur. Potato Jour.*, 6:1 (1963).

of piece increased for both whole and cut seed; whole seed gave somewhat higher yields than cut seed for pieces of similar weight; the number of stalks per plant and number of tubers per hill increased as size of piece increased; and there were more oversized tubers on plants from the smallest seed pieces.

Planting

The actual planting date depends on the length of the growing season for the region, time of the year when the crop can be marketed to best advantage, and seasonal weather throughout the growth period. Generally, early planting should be practiced to grow potatoes to better maturity. Planting should begin as soon as the soil warms to 45°–50°F (7°–10°C) and the seedbed can be prepared without danger of compaction.

The best planting date should provide cool, moist conditions when the plants are blossoming and setting tubers, and relatively short days in the season for maximum tuber development. These conditions are met in most of the important areas in the Northern and Northeastern states by planting between May 1 and June 15, in Central states in April, and in Southern states from November to February. Fall planting, which takes advantage of the cool growing season, is practiced in much of the South.

The fertility of the soil and its moisture-holding capacity, the climate, the cultivar, and the amount of fertilizer applied govern spacing within and between rows. In the North and Northeast, rows are spaced 32–36 inches (81–91 cm) apart, depending on equipment. In the early days, distances between seed pieces in the row was 15–20 inches (38–51 cm); in modern planting, the distance is most often 5–11 inches (13–28 cm). Most growers probably could best use a 6–8-inch (15–20-cm) spacing, but the specific recommendation depends on the cultivar and other conditions. In dryland areas of the West, the distance between rows may be as much as 42 inches (107 cm) with seed pieces spaced 14–30 inches (36–76 cm) apart in the row. Rows under irrigation are 34–36 inches (86–91 cm) apart and seed pieces 9–14 (23–36 cm). Planting one acre requires 15,000 to more than 35,000 seed pieces, or 12 to 32 cwt (545–1,453 kg) of seed potatoes, the exact amount depending on seed piece size and spacing.

Depth of planting varies with the soil type and the equipment used. Recommendations vary from 2 to 3 inches (5.1–7.6 cm) to 3–4 inches (7.6–10.2 cm). The greater depth is most often used in level culture. In some areas pieces are planted 3–4 inches (7.6–10.2 cm) below the soil and covered with only 2–3 inches (5.1–7.6 cm) of soil. Then rows are ridged after early cultivations so that tubers develop at 2½–5 inches (6.4–12.7 cm) under satisfactory temperature conditions, which favor high quality.

Four general types of potato planters are used—the automatic picker type, the assisted feed type, the automatic cup type, and the tuber unit planter. The automatic picker type is most widely used. The cup-type planters are well adapted to planting whole seed, and are used extensively in areas where the whole seed is preferred.

Cultivation and Chemical Weed Control

Callihan and Higgins in 1975 reported that weeds and associated cultivations cost Idaho potato growers an average of $42 per acre ($104 per hectare). Uncontrolled weeds reduce potato yields because of competition—losses averaging $20 per acre ($49 per hectare); cultivation often injures potato plants by severing roots and causes soil compaction, which results in average losses of an additional $22 per acre ($54 per hectare). The presence of large weeds also can cause considerable difficulty during mechanical harvest.

The use of selected herbicides for weed control is one of the most important labor-saving practices since the mechanized production of potatoes first became general. This weed-control procedure is also useful in avoiding root injury, reducing soil compaction, and allowing closer spacing of rows. Herbicides may be applied preplant, preemergence, postplant preemergence, or post emergence. Some of the most important herbicides used in potato production are EPTC, trifluralin, alachlor, dinoseb, and metribuzin.

Cultivars

Cultivars originate by four methods: artificial or controlled hybridization, mutation, selection of seedlings from naturally produced seed balls, and clonal selection. Once a cultivar is developed by any of the preceding methods, it is increased and maintained by clonal or vegetative propagation. Because all tubers from a given cultivar have originated from a common source, every plant of that cultivar is identical genetically. The statement that potato cultivars "run out" has little merit.

A comprehensive breeding program continued through a long period has had as its objectives improved culinary and storage qualities; high yield; and resistance to heat, drought, insects, and diseases. Many of the older cultivars were deep-eyed, dark-skinned, and relatively unattractive on the market. New cultivars now available not only give large acre yield, but also have bright skins, shallow eyes, and regular shapes.

There are many cultivars. A European classification described more than 170 cultivars; in the United States more than 300 named cultivars were grown commercially before the start of this century. Proper cultivar selection is essential to the production of a top quality crop. The one selected must be adapted to the soil and cultural conditions under which it is grown, and must fulfill the market requirements. Some are chosen because of resistance to specific diseases. Others are chosen as early-market potatoes and for specialized uses as chip manufacture, frozen french fries, starch recovery, and baking.

In 1975 more than 60 cultivars were certified in the United States. The 'Russet Burbank' cultivar leads all others, accounting for just under 40 percent of all seed grown. Idaho produces a majority of this amount. 'Kennebec' ranks second with 12 percent of the total seed; North Dakota, Minnesota, and Maine produce most of the seed. Others of importance include 'Katahdin,' 'Superior,' and 'Norchip.' These five cultivars represent 75 percent of the seed acreage. The next six cultivars are 'Norgold Russet,' 'Norland,' 'Red Pontiac,' 'Red LaSoda,' 'Monoma,' and 'Irish Cobbler.'

Harvesting

Potatoes should not be harvested until the vines are dead, which lessens the chances of injury from skinning and bruising. In the past, digging was done with a tractor-drawn machine that lifted the potatoes, separated

FIGURE 28–7. *Potatoes may be harvested with a self-propelled machine or with a tractor-pulled potato combine as shown here. Harvesting should not be done until potato vines are dead, as a result of natural, mechanical, or chemical means.* [SOURCE: J. I. Case.]

and sifted the soil, and dropped them on top of the freshly dug bed ready to be picked up and loaded by hand. Today the harvesting and handling procedure is almost full mechanized except in areas where soil topography or field size or shape is not adapted to large machinery. A single self-propelled machine is used to lift, sift the soil, remove debris and rocks, and load the potatoes for transport to the packing shed. The trend throughout the country is toward larger acreages on fewer farms and the use of larger, more complicated machinery.

Potatoes are harvested almost every month of the year in one state or another. Marketing competition results in some potatoes being harvested before they are mature. Because the yield increases rapidly right up to the time the plant matures, harvesting before maturity means a loss of market quality as well as acre yield. Many potatoes are harvested early for a high market price or because the growers cannot risk freezing temperatures.

KILLING POTATO VINES. The tops or vines should be dead before harvesting for these reasons: (1) facilitates earlier harvest, (2) promotes a firmer skin to reduce injury during harvesting, (3) controls tuber size and, (4) controls spread of virus diseases and of late blight to tubers.

When it is desirable to begin harvesting before natural death of the vines occurs, some means of destroying the top growth is necessary. Beaters with rubber flails or chains have been used for this purpose, but chemical dessicants are now in common use. Some of the most frequently used chemicals used for killing the vines are ametryne, dinoseb, and paraquat.

Storage

When ready to store, the tubers should be clean, mature, and free from injury. The proper temperature for storage depends on the use for which the tubers are intended. Tubers for seed should be kept at 38°–40°F (3°–4°C), while tubers to be processed into french fries or chips must be kept at 45°–50°F (7°–10°C). Since potatoes stored at the latter temperatures will begin to sprout in a few weeks, they must be treated with a sprout inhibitor if long storage is necessary. The storage area should have an adequate humidification system; a relative humidity above 95 percent in the air surrounding the tubers is recommended. There should be proper ventilation of the storage area to provide oxygen and eliminate the carbon dioxide and heat of respiration.

If potatoes have been stored for a long period at a low temperature, the tubers must be conditioned at a high temperature before handling. Low-temperature tubers are damaged easily.

Certain growth-regulating chemicals sometimes are used to control or reduce sprouting of potatoes. One of these, maleic hydrazide, is applied to the foliage approximately 6 weeks before harvest, and is translocated to tubers for sprout inhibition. A number of other chemicals are available for application to tubers being put into storage, or after a period in storage to avoid interference with wound healing.

Insects and Diseases

The potato crop is infested with many insects which cause large losses in yield and quality. Millions of dollars are spent annually for insect control. The insects that have the most potential for serious damage and losses in areas where they occur are: grasshoppers, leafhoppers, potato psyllid, aphids, potato tuberworm, wireworms, Colorado potato beetle, and flea beetles.

The estimated loss from potato pathogens is 19 percent. The most serious fungus diseases that contribute to this loss are late blight, about 4 percent; *Verticillium* wilt, 2.7 percent; common scab, 1.5 percent; early blight, 1 percent. Blackleg and ring rot are the most serious bacterial diseases, and leafroll and mosaics are virus diseases. Additional diseases are caused by nematodes, from plant injuries produced by insect feeding, and by unfavorable environmental conditions causing physiological abnormalities.

The most effective means of disease prevention is use of cultivars that are resistant to one or more diseases. Good cultural practices, such as crop rotation and the use of clean seed, are effective in preventing some disease losses. Bordeaux mixture was used for many years to control potato foliage diseases, especially late blight. Many new organic fungicides have largely replaced Bordeaux mixture because of greater effectiveness and lower phytotoxicity.

Sweetpotatoes

States leading in the production of sweetpotatoes are North Carolina, Louisiana, Texas, Mississippi, Georgia, and California. (See Figure 28–8.) North Carolina and Louisiana have more than 50 percent of the U.S. acreage, and the 6 states listed above have 80 percent of the acreage. Average yield per acre is about 115 cwt (5.2 MT).

Reasons given for the decline in commercial production are (1) less consumption of all starchy foods; (2) production not well mechanized, with much hand labor required; (3) retail prices so high that in many areas sale volume has declined; (4) high losses from rot and breakdown after purchase, which add to cost and discourage buyers.

The sweetpotato, *Ipomoea batatas* is a member of the *Convolvulaceae,* or morning-glory, family. The funnel-shaped flowers usually are rose violet to bluish in color. The vines, which may grow 15 feet (4.6 m) in length, are green or red to purple and often are somewhat hairy, especially at the nodes. The leaves usually are pear shaped and deeply lobed. The roots, the edible product, have white, yellow, salmon, red or purplish-red skin; the flesh is white or various shades of yellow, orange, or salmon. New shoots come from adventitious buds.

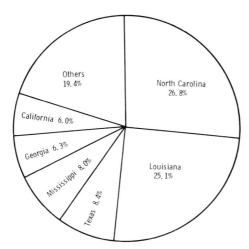

FIGURE 28–8. *Leading states in United States sweetpotato area (acreage).*
North Carolina and Louisiana have more than 50 percent of the total
acreage. [SOURCE: Southern Illinois University.]

Although a perennial, in the United States the plant functions as an
annual and is killed by cold weather each winter. Sweetpotatoes are grown
in most tropical and subtropical regions of the world.

Season and Soil

For high yield, sweetpotatoes require a frost-free growing season of four
to five months. Most of the crop is grown in areas that receive 40 inches
(102 cm) of rainfall or more annually; few are grown in areas with less
than 35 inches (89 cm) unless irrigated. They also thrive in warm weather,
doing best where the mean temperature is at least 70°–80°F (21°–27°C)
during the season.

Soils rated good to excellent for sweetpotatoes are moderately deep, friable, fine sandy loam, or loamy fine sand with a slightly acid pH. The soil
must be well drained. Excellent sweetpotato soils have surface layers more
than 1 foot (30 cm) in depth; those of 6- to 12-inch (15–30-cm) depths are
considered good, but not excellent. Heavy river-bottom soils and other clay
soils are unsatisfactory. In the South many sweetpotatoes are grown on
light sandy soil because the crop produced is of excellent quality, although
the acre yield may be relatively low.

Cultivars

Sweetpotatoes are used primarily for food, with the culls being used
for livestock feed. This food crop is divided into two groups, the so-called
dry-fleshed and moist-fleshed. The moist-fleshed cultivars are sometimes
called *yams,* although this is not strictly correct. The yam is not a sweetpotato; it produces tubers rather than an enlarged root of the sweetpotato

plant. A long list of cultivars has shrunk to a few. Cultivars now grown include Porto Rico, Acadian Goldrush, Centennial, Julian, and Allgold.

Plants and Cuttings

Sweetpotato plants for transplanting are produced by bedding mother roots (seed) in warm soil or sand and later removing the sprouts or plants that grow out from the buds on these roots. These also are called draws. The word *slip* generally is used for cuttings made from the early vine growth of transplants. Slips are transplanted in the same way as draws.

Seedbed and Transplanting

The soil should be prepared thoroughly before planting. The soil should be plowed 8–10 inches (20–25 cm) deep as far in advance of planting as possible. Sufficient time should be allowed for decomposition of the plowed-under organic matter before setting the slips in the field.

Mechanical transplanters save labor and do a more uniform and satisfactory job. Rows most often are spaced 42 inches (107 cm) apart and plants are spaced 6–15 inches (15–38 cm) apart in the row.

Time of planting affects the general shape of the root with some cultivars; if planted before the last of May a high percentage of the potatoes may be turnip-shaped, which is undesirable. The ideal shape is 5–7 inches (12.7–17.8 cm) long and tapering at both ends. July-planted potatoes usually are long for their diameter and their yield is likely to be rather low.

Rotation and Fertilizers

A legume should be included in the rotation system and turned under green. Other crops which may be included in a four- or five-year rotation are cotton, corn, grain sorghum, and temporary pastures. Sweetpotatoes should not follow such crops as watermelons, cantaloupes, peanuts, or cucumbers because of nematode buildup. Excessive amounts of nitrogen cause heavy vine growth and lower the yield. Apply 500–800 pounds/acre (560–896 kg/ha) of 4–12–8, 5–10–10, or 6–12–6 or equivalent, depending on fertility, 7–10 days before setting plants.

Harvesting

The sweetpotato root has a delicate skin that is easily bruised, broken, or cut. Wounds are followed by decay unless they are healed. If the root is allowed to become chilled the eating quality, storage property, and value of the roots for seed are all damaged. Sweetpotatoes do not mature and therefore should be harvested before killing frosts or as soon as most of the roots reach desired size. Usually four to four and a half months are required for maximum yields.

Until recently sweetpotato harvest was expensive and laborious. While harvesting procedures still are widely variable, mechanical harvesters are available that will dig, top, sort, and load roots into containers. Such equipment is labor-saving, but more bruising and skinning damage may occur.

Curing and Storage

The sweetpotato must be properly cured before it can be stored successfully. Curing hastens the healing of the wounds on the roots made during harvesting. While curing, the sugar content is greatly increased and the edible quality is improved. Curing involves placing the roots in a room where the temperature is held at about 85°F (29°C) and the relative humidity kept at 85–90 percent for 6–8 days. For curing to be most effective the roots must be given the proper temperature and humidity immediately after harvest. At the end of the curing period the temperature is lowered gradually to 55°–60°F (13°–16°C), where it is maintained for the rest of the storage period.

Diseases and Insects

Diseases include stem rot, black rot, frost rot (often called ring rot), root rot, scurf, and pox. In addition, soft rot may occur in storage. The sweetpotato is not seriously injured by many insects, but a root weevil is injurious, especially in the Gulf states. Other insects are sweetpotato leaf beetle, grasshoppers, flea beetles, and wireworms.

REFERENCES AND SUGGESTED READINGS

1. BIRECKI, M., and S. ROZTROPOWICZ. "Studies on Seed Potato Size and Productivity," *Eur. Potato Jour.,* 6:1 (1963).
2. BOYD, J. S., A. L. RIPPEN, and F. H. BUELOW. *Potato Storage Design and Operation,* Mich. State Ext. Bul. 585 (1967).
3. CALLIHAN, R. H., and R. E. HIGGINS. *Control Weeds in Potatoes With Herbicides,* Idaho Agr. Exp. Sta. Current Info. Ser. 290 (1975).
4. CHASE, R. W., and N. R. THOMPSON. *Potato Production in Michigan,* Mich. State Ext. Bul. 546 (1967).
5. CHUCKA, J. A., and F. A. STEINMETZ. *Potatoes,* Me. Agr. Exp. Sta. Bul. 438 (1945).
6. CLARK, C. K., and P. M. LOMBARD. *Description of and Key to American Potato Varieties,* USDA Circ. 741 (1951).
7. DUNTON, E. M. *Soil Fertilizers and Cropping Practices,* Va. Joint. Agr. Publ. 5 (1963).
8. EDMOND, J. B., *et al. Factors Affecting the Yield and Flesh Color of the Porto Rico Sweetpotato,* USDA Circ. 832 (1950).
9. HILDEBRAND, E. M., and H. T. COOK. *Sweetpotato Diseases,* USDA Farmers' Bul. 1059 (1959).
10. KEHR, A. E., *et al. Commercial Potato Production,* USDA Handbook 267 (1964).
11. KUSHMAN, L. J. *Preparing Sweetpotatoes for Market,* USDA Mktg. Bul. 38 (1967).
12. MONTELARO, J., *et al. Sweetpotatoes in Louisiana,* La. Coop. Ext. Ser. 1450 (1966).

13. O'BRIEN, M. J., and A. E. RICH. *Potato Diseases,* USDA Agr. Handbook 474 (1976).

14. ODLAND, T. E., and H. G. ALLBRITTEN. "Soil Reaction and Calcium Supply as Factors Influencing the Yields of Potatoes and the Occurrence of Scab," *Agron. Jour.,* 42:6 (1950).

15. PAINTER, C. G., *et al. Potatoes, Idaho Fertilizer Guide,* Idaho Agr. Exp. Sta. Current Info. Ser. 261 (1975).

16. PEW, W. D., *et al. Growing Potatoes in Arizona,* Ariz. Coop. Ext. Serv. Bul. A83 (1976).

17. Potato Association of America. *Potato Handbook* (New Brunswick, N.J.: Rutgers State Univ., 1967).

18. ROBERTS, C. R. *Kentucky Chipping Potato Research—1977,* Ky. Agr. Exp. Sta. Prog. Rep. 231 (1978).

19. SALAMAN, R. N. *The Influence of the Potato,* Scientific American (Dec. 1952) Vol. 187, No. 6, pp. 50–56.

20. SMITH, O. *Potatoes: Production, Storing, Processing,* 2nd ed. (Westport, Conn.: AVI, 1977).

21. SPARKS, W. C. *Potato Storage—Construction and Management,* Idaho Agr. Exp. Sta. Current Info. Ser. 297 (1975).

CHAPTER 29

PEANUTS AND OTHER EDIBLE LEGUMES

Peanuts

History

THE peanut, *Arachis hypogaea,* is not a nut; it is a legume closely related to the pea and bean. Its exact origin is not certain, but it was known as early as 950 B.C. and apparently was first found in Brazil or Peru. Later it was carried to Africa by early missionaries and explorers. The peanut is referred to in early literature as goober, pinder, and groundnut, the name still used in several parts of the world. Peanuts were brought to North America from Africa by slave traders in early colonial days. In this country, peanuts were not grown extensively until after the Civil War in 1865. For many years thereafter, the main production was confined to Virginia and North Carolina. Thomas Jefferson mentioned in his writings that peanuts were grown in Virginia, but apparently they were of little commercial importance.

The greatest factor that contributed to an expansion of peanut production in the United States in the early 1900s was the invention of machinery to plant, cultivate, harvest, and process peanuts. Its geographic distribution expanded further about 1920 when farmers in Georgia, Alabama, and Florida, due to the ravages of the cotton boll weevil, were forced to find a substitute for cotton and turned to peanuts.

George Washington Carver from Tuskegee Institute in Alabama contributed greatly to peanut utilization. He is said to have made more than 300 food and industrial products from the peanut before World War I. During World War II, a demand was created for a large volume of peanuts for oil, food, and feed, and thus production of this strategic crop expanded greatly. After the war, however, the demand dwindled, mainly because of a lower export volume.

This crop was grown first in the United States for fattening farm animals, especially hogs and poultry. Presently, peanuts are valuable sources of oil, plant proteins, and related food products. There is a renewed interest in the peanut as a valuable protein source for the world's malnourished people.

The Plant

The peanut is a warm-season annual. The plant has a central, upright stem with many lateral branches. When the branches are nearly erect

A   B

FIGURE 29–1. *The peanut is characterized by an unusual fruiting habit.* *(A) After pollination, a stalklike structure immediately behind the ovary elongates and forces the fertilized ovary into the soil. "Pegging" is essential, for pod formation will not occur aboveground.* [SOURCE: Southern Illinois University]. *(B) Pod formation on "pegs." Note also the taproot system.* [SOURCE: University of Georgia College of Agriculture.]

and most pods are produced in a cluster at the base, the peanut is a bunch-type; when horizontal with pods produced on these prostrate branches, the plant is a runner type.

This legume is characterized by a relatively deep tap-root system but also a well-developed lateral root system. Leaves are pinnately compound, each leaf of which commonly consists of four leaflets.

The most unusual characteristic of the plant relates to its flowering and

FIGURE 29–2. *Two types of peanuts: 1 = the large-seeded Virginia, 2 = the small-seeded Spanish.* [SOURCE: Southern Illinois University.]

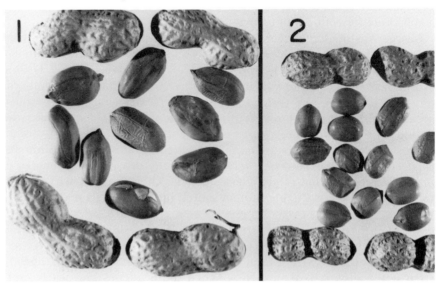

fruiting habit. Flowers are borne in leaf axils either singly or in groups up to 3 in number. It is not unusual for flowers to be borne as much as 3 inches (7.6 cm) below the ground. After pollination, each flower withers and the gynophore, a stalklike structure immediately behind the ovary, elongates and forces the fertilized ovary into the soil. This process is commonly called "pegging." The ovary matures underground into a pod, and the fertilized ovules into seeds or kernels. The mature pod contains 1–6 seeds, usually 1–3. Pod formation will not occur aboveground.

For commercial purposes, U.S. peanuts are grouped into four types:

1. Large-seeded Virginia with both bunch and runner-type plants.
2. True Runner types.
3. Small-seeded Spanish.
4. Early or short-season Valencia types.

Importance and Production

Peanuts constitute an important cash crop in parts of the South and Southwest. The increased need for oil for various uses in time of war has caused great expansion of peanut production during each period since 1860. Production was greatly stimulated as a cash crop during World War II, when the acreage practically doubled—up to 5 million acres (2 million ha), with 3.5 million acres (1.4 million ha) harvested for the nuts. As much as 1 million acres (0.4 million ha) has been harvested by hogs; considerably less has been used this way in recent years. Peanut hay, or more strictly speaking peanut straw, a by-product, has been used as a roughage for feeding farm animals.

In the United States during the last 20 years, the acreage of peanuts has held rather steady at approximately 1.5 million acres (0.6 million ha). In 1977 about 98 percent of all peanuts planted were harvested for nuts; thus, other uses, such as "hogging off," have declined considerably. The average yield in the United States is approximately 2,500 pounds/acre (2,800 kg/ha), with a total production of almost 4 billion pounds (1.8 billion kg) (See Figure 29–3.)

Approximately 42 million acres (17 million ha) of peanuts are harvested annually worldwide. Total world production is about 19 million tons (17 million MT) of nuts.

Distribution and Adaptation

The United States is fourth in world peanut area, with just over 3 percent of the acreage. Peanuts are most important in India, China, and Senegal. India and China alone have 50 percent of the world acreage (Figure 29–4). The United States is third in total world production with about 9 percent, exceeded by India with nearly 38 percent and China with almost 15 percent.

U.S. peanuts are grown in different parts of the South as far north as Virginia. The peanut is much more important in certain restricted areas

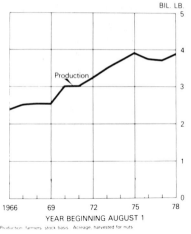

FIGURE 29–3. *Peanut production in the United States. The present annual production is almost 4 billion pounds (1.8 billion kg).* [SOURCE: USDA-ESCS.]

than it generally is throughout the South. Most U.S. peanuts are produced in three distinct districts. The Virginia–Carolina district grows the large Virginia and Runner types; the Georgia–Florida–Alabama district grows the Southeastern runner, Virginia, and Spanish types; the Southwestern district, including Texas and Oklahoma grows the Spanish type; and a small area in New Mexico produces a large share of all the Valencias. Leading states in peanut production are Georgia, Texas, Alabama, North Carolina, Oklahoma, and Virginia, listed in declining order of acreage. (See Figures 29–5 and 29–6.)

For high yields and superior quality peanuts require a moderately long

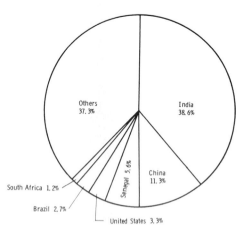

FIGURE 29–4. *Leading countries in world peanut area (acreage). India and China have one half of the total.* [SOURCE: Southern Illinois University.]

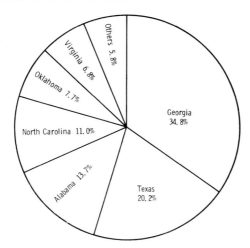

FIGURE 29–5. *Leading states in United States peanut area (acreage). Georgia and Texas have over one half of the total.* [SOURCE: Southern Illinois University.]

growing period of four to five months with a steady, rather high temperature and a moderate, uniformly distributed supply of moisture. Adequate moisture is needed especially during the period when the peanuts are forming, followed by dry conditions where the rainfall amounts to about 20 inches (51 cm) during the peanut growing season.

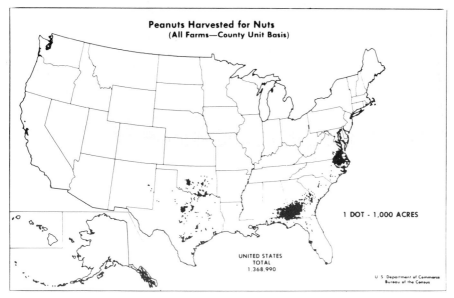

FIGURE 29–6. *Peanuts harvested for nuts in the United States, shown geographically.* [SOURCE: U.S. Department of Commerce.]

Well-drained, light sandy loam soils are especially well suited to the production of good market nuts. Light-textured soils offer less resistance to the penetration of the pegs that must enter the ground for pods to develop and allow the pods to form properly. In addition, at harvest it is more difficult to remove peanuts from the heavier-textured soils without excessive pod loss.

Peanuts thrive best on soils with a pH of 6.0–6.5. Soils containing a high proportion of organic matter are likely to produce discolored pods, making them unfit for some trade purposes. Moreover, the appearance and quality of the kernels are likely to be lowered. Large quantities of available nitrogen and potash in the soil usually are unfavorable to the production of high-quality peanuts.

Peanuts in Rotations

Although there is no single best rotation for peanuts, several factors should be considered. Peanuts should not be grown year after year on the same land. They should be treated as one of a number of crops grown, not as a specialty. Because the peanut can use fertilizer residues left in the soil it is desirable to follow some heavily fertilized crop such as corn. Since peanuts are considered a soil-depleting crop, large amounts of nutrients being removed from the soil, the rotation should include soil-building crops and cover crops. Crimson clover, oats, rye, and other crops should be planted after peanuts are harvested, as the land is left practically bare. When peanuts follow sweetpotatoes, soybeans, or other crops that leave large quantities of heavy-stemmed green vegetable matter at or near the surface of the soil, conditions are favorable for the winter carryover of the sclerotia of the southern root rot, a disease that causes heavy losses to peanuts. Crops such as soybeans grown immediately before peanuts may increase the nematode population. If peanuts follow corn, small grains, or sod there is less disease damage and yield and quality are superior.

Cultural Practices

Deep plowing is encouraged to bury completely all cover crops or other materials on the surface, which aids in disease prevention. Shallow soils seldom produce profitable yields. Light-textured soils subject to leaching and wind damage should be plowed just prior to planting. After plowing, land for peanuts should be disked and harrowed, and a fine seedbed prepared.

Peanuts are sensitive to fertilizer burn, so lime and most fertilizer should be broadcast and incorporated to avoid injury to germinating seedlings. Fertilization should be based on a soil test, but generally a complete fertilizer is used. Table 29–1 shows a potassium fertilizer response on a Troup fine sand at the Tifton, Georgia Station. Both Spanish and Runner peanuts responded to potassium, regardless of method of application. This soil had

TABLE 29–1. Effect of Rate and Method of Application of Potassium on Yield of Spanish and Runner Peanuts at Tifton, Georgia, Three-Year Average

Potassium Added (Pounds/Acre)	Spanish Peanuts*		Runner Peanuts*	
	Broadcast	Drilled	Broadcast	Drilled
0	2,198	2,068	1,884	1,836
30	2,541	2,440	2,220	2,476
60	2,591	2,379	2,413	2,554
90	2,775	2,270	2,315	2,662
120	2,611	2,491	2,522	2,669
150	2,743	2,213	2,742	2,510

* Yield of pods in pounds/acre.

Source: Walker, M. E., H. D. Morris, and R. L. Carter, *Ga. Res. Bul.* 152 (1974).

been neither cultivated nor fertilized for 15 years. The initial soil pH was 6.0.

Peanuts also respond well to residual fertilizer, resulting from high fertilization of a previous crop, such as corn. Research suggests that relatively deep placement of fertilizer may increase yields. Adequate calcium is essential to high peanut yields. Plants cannot absorb enough calcium through the roots to meet the needs; pegs can absorb calcium directly from the soil. It is a common practice to topdress plants at early bloom stage with gypsum ($CaSO_4$), or another calcium source, at a rate of 500–700 pounds/acre (560–784 kg/ha). Lime or gypsum broadcast in rows before planting is much less effective.

Planting

Federal, state, and industry researchers have produced high-quality peanut hybrids, selections, and introductions. Cultivars are chosen for yield, disease resistance, and improved market acceptability.

In early days peanuts were either hand-shelled or planted in the shell. Planting in the shell is wasteful because the seed in the tip of the pod usually germinates first and frequently pushes the shell with the basal seed above the surface. Moreover, unless moisture is abundant, germination in the shell is poor and results in a weak stand. Acreage increase and decrease in available labor has made hand shelling of seed stock impractical and growers depend on machine-shelled seed. It is important to select and use undamaged seed.

Treating seed with a fungicide is recommended. Peanuts cross-inoculate with both cowpeas and velvetbeans. Unless these crops or peanuts have been grown on the soil recently, inoculation is a recommended practice.

Planting dates vary with geographic location. This date should be selected for a warm, moist seedbed, as peanuts are very susceptible to unfavorable germination conditions such as a cold soil. In Texas planting may start as early as March 1 and may continue until July 12. In Oklahoma planting takes place from May 15 to July 1. In the Southeast most of the planting is done during April and May. Planting dates must allow sufficient time for the crop to develop before excessively cold weather in the autumn.

Commercial planting generally is done with a multiple-row peanut planter, with a corn or cotton planter equipped with peanut plates, or with an air planter. Various combinations of row and plant spacings may be used to secure the desired balance of plants. Variations in row spacing may greatly influence production and harvesting efficiency. Choice of row pattern may be influenced by row spacing of other crops grown in the area.

Increases in yields of as much as 25 percent often are obtained by planting rows closer together, by placing the seeds closer together in the rows, or by both. Table 29–2 shows average seeding rates in Georgia for various drill spacings and row patterns.

Peanuts should be covered to a depth of about 2.5–3 inches (6.4–7.6 cm) in light soil and 1–2 inches (2.5–5.1 cm) in heavier soil. Under dry conditions or in late planting the depth should be increased slightly.

Cultivation or other mechanical weed control procedures may be used. However, preemergence and postemergence herbicides have come into widespread use in recent years. The two most critical periods in weed control are when peanuts are very small and when they undergo pegging. Chemical weed control may be beneficial during the early and late periods of growth without incurring excessive damage to young plants or pegs as might result from cultivation.

Harvesting

Until about 1920, growing peanuts was a very laborious process, for the most part involving hand labor for shelling, planting, and harvesting. By

TABLE 29–2. Average Seeding Rates for Various Drill Spacings and Row Patterns

Peanut Type or Variety	Row Pattern	Row Spacing (Inches)	Drill Spacing (Inches)	Rate (Pounds/Acre)
Spanish	Std. or Mod. 2-row	24–36	2–3	80–100
Virginia bunch 67 and runners	Std. or Mod. 2-row	30–36	3–4	60–80
Large-seeded Virginia	Std. or Mod. 2-row	30–36	3–4	80–100
Spanish	4 Close-row	13–14–13	Outside 2.5	90–115
		13–16–13	Inside 3	
Virginia bunch 67 and runners	4 Close-row	13–14–13	Outside 3	95–125
		13–16–13	Inside 6	
Large-seeded Virginia	4 Close-row	13–16–13	Outside 6	110–120
			Inside 6	

Source: McGill, J. F., and L. E. Samples. *Peanuts in Georgia*, Ga. Coop. Ext. Serv. Bul. 640 (1969).

the mid-1960s more than 90 percent of the peanuts grown in the regions of greatest production were mechanized in terms of planting, digging, windrowing, and threshing.

Peanuts must be harvested at the right time. If harvested too early, the pods will be immature and will have a large proportion of shriveled kernels; if harvested too late, many of the pods may be lost through disease and unfavorable weather conditions. Vines may be clipped before digging, or mechanical equipment may cut the root system below the nut zone and lift the plants from the ground. Modern digger-shakers vary in size, design, and function. Those used commercially have combination digging, shaking, and windrowing capabilities. Inverting shakers invert the plant, leaving the pods exposed more uniformly to the drying air. Most peanuts produced are separated from the whole plants, cleaned, stemmed, and conveyed into a bulk bin on the combine. Green harvest is practical only when drying facilities are available. Cost of curing is higher for green combined peanuts and mechanical damage can be more severe. Pod moisture must be reduced to 8–10 percent for safe storage. Usually peanuts combined from the windrow need to be cured in mechanical dryers.

Diseases and Insects

Principal diseases are leaf spot, caused by fungi of the genus *Cercospora,* and southern blight, caused by the genus *Sclerolium.* Plant breeders have developed cultivars resistant to *Cercospora* leaf spot. Destructive insects include the Southern corn rootworm, potato leafhopper, tobacco thrip, white-fringed beetle, cutworm, velvetbean caterpillar, and cucumber beetle larvae. Several types of nematodes also cause yield reductions. The most widespread one is the northern root-knot nematode.

Utilization

The shell is 20–30 percent of the weight of the mature peanut. The kernel contains 43–50 percent oil and 25–30 percent protein. An average ton (0.9 MT) of clean nuts gives about 530 pounds (240 kg) of oil and 820 pounds (372 kg) of meal.

Utilization of peanuts in the United States differs greatly from that of other countries of the world. Two-thirds of the world crop is crushed for oil, while the U.S. crop is used primarily for food products. Peanut butter is by far the most important U.S. peanut product, with salted peanuts being second. Also important are peanuts in various confections, such as candies and bakery goods. The bulk of the Spanish and small runners is used in making peanut butter and confections. Virginias and Valencias are used largely for roasting in the shell and for specialty products. On the average, 200–400 million pounds (91–181 million kg) of peanuts are crushed for oil each year in this country.

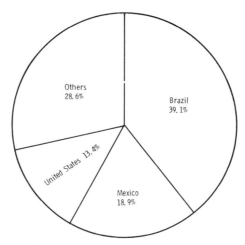

FIGURE 29–7. *Leading countries in world dry edible bean production. Brazil produces almost 40 percent of the total.* [SOURCE: Southern Illinois University.]

Dry Edible Beans

Dry edible beans are used for human consumption, either canned or in dry form. The seed contains more than 20 percent protein. In contrast to soybeans, dry beans are low in oil. The world leaders in dry edible bean production are Brazil, Mexico, and the United States. (See Figure 29–7.) In the United States, about 1.5 million acres (0.6 million ha) are grown annually with an average yield of about 1,200 pounds/acre (1,344 kg/ha).

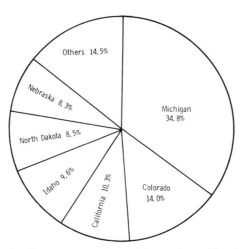

FIGURE 29–8. *Leading states in United States dry edible bean area (acreage). Michigan and Colorado have almost one half of the total.* [SOURCE: Southern Illinois University.]

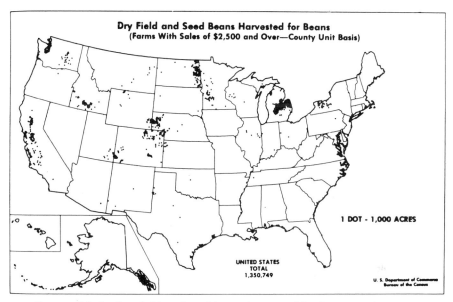

FIGURE 29–9. *Dry field and seed beans harvested for beans in the United States, shown geographically.* [SOURCE: U.S. Department of Commerce.]

Michigan grows more than one-third the total U.S. average. Other states with large dry edible bean acreages are Colorado, California, Idaho, North Dakota, and Nebraska. (See Figures 29–8 and 29–9.)

Plants of the common bean may be bushy or viny. The leaves are pinnately trifoliolate. Both leaves and stems are pubescent. The flowers may be white, yellow, or bluish purple. The pods are straight or distinctly curved and are 4–8 inches (10–20 cm) long. Seeds may be white, brown, pink, red, blue-black, or speckled in color.

Types

The five major classes of dry edible beans are (1) white (navy or pea), (2) pinto, (3) Great Northern, (4) kidney, and (5) lima. The first four are all *Phaseolus vulgaris.*

The two most widely grown beans are the pea or navy bean and the pinto bean. These two account for production of about 1 billion pounds (0.45 billion kg) in the United States. The navy bean is a small semitrailing plant with white flowers and small white seed. The great northern and small white are types similar to the navy bean. Navy beans are grown mainly in Michigan, while great northern beans are produced mostly on irrigated land in Nebraska, Idaho, and Wyoming.

The pinto bean is easily recognized by its buff color speckled with brown to tan spots. It is grown mostly in Colorado, but there are considerable acreages in Idaho, Nebraska, and Wyoming. The red kidney bean is a long-season bean, requiring about 140 days to mature. Pods splashed with purple

FIGURE 29–10. *The two most widely grown beans are the pinto bean (1) and the pea or navy bean (2). These two beans account for a United States production of about 1 billion pounds (0.45 billion kg).* [SOURCE: Southern Illinois University.]

are borne on bush-type plants. Seeds normally are pink when harvested but turn dark red when older.

The large lima, *P. limensis,* is a perennial grown as an annual, while the small or baby lima, *P. lunatus,* is a true annual. The beans are consumed to a larger extent as green limas than dry beans. Most of the acreage is in California.

Garbanzo beans, or chickpeas, *Cicer arietinum,* are produced to the extent of about 5 million pounds (2.3 million kg) almost entirely in California. There is some interest in the tepary bean, *P. acutifolius* var. *latifolius,* developed and grown by Indian tribes in the Southwest. The mung bean, *Phaseolus aureus,* provides the canned bean sprouts used in chop suey and similar foods. In the past, nearly all these beans used in the United States were imported. At the present time there is limited production in Oklahoma, California, and northern Texas.

Dry Edible Peas

The earliest colonists in the United States brought the field peas, *Pisum arvense,* with them from England. The pea is an annual herbaceous plant with slender succulent stems 2–4 feet (0.6–1.2 m) long. The leaf has one to three pairs of leaflets and terminal branched tendrils. Blossoms usually are white or reddish purple. The pods are about 3 inches (7.6 cm) long

FIGURE 29–11. *Dry beans growing in a compacted soil (left) as compared to a soil with few physical problems (right). The growth rate is slow, the population variable, and the rooting depth shallow on the compacted soil. Growth is normal on the soil at right, which was fall-plowed and minimum tilled.* [SOURCE: Michigan State University, Departmant of Information Services.]

and contain four to nine seeds. They differ from field beans in that emergence is hypogeal, the cotyledons not being pulled above the ground.

Field peas differ from garden peas, *P. sativum,* mainly in the sweeter and more delicate flavor of the latter. Although most of the garden and canning peas are utilized as green peas, some also are grown as dry field peas.

Dry edible peas are grown on 130,000 acres (52,611 ha) in the United States, with virtually all of the acreage in Washington and Idaho. The average yield of clean seed is about 1,500 pounds/acre (1,680 kg/ha). India is by far the world leader in dry pea acreage and production, with more than 2 million acres (0.8 million ha) devoted to this crop.

Most of the domestic pea production consists of the Alaska pea and other smooth, green-seeded types, which are used as edible dry peas or for seed. The smooth, yellow types are consumed primarily as dry, split peas, while the white types serve mainly as feed.

Field peas require a temperate to cool climate. They do best where high temperatures are seldom experienced during the growing season. Seed does not develop well in the more southern latitudes. They also do best under relatively high rainfall conditions.

FIGURE 29–12. *A disease-control study on dry beans at the International Center of Tropical Agriculture in Colombia.* [SOURCE: International Center of Tropical Agriculture.]

REFERENCES AND SUGGESTED READINGS

1. AMES, G. C. W. *Peanuts: Domestic, World Production and Trade.* Ga. Res. Rep. 215 (1975).

2. ANDERSON, A. L., *et al. Navy Bean Production: Methods for Improving Yield,* Mich. State. Ext. Bul. E-854 (1975).

3. BOND, M. D. *Producing Virginia-Type Peanuts,* Auburn Univ. Ext. Serv. Circ. P-5 (1964).

4. CHAPMAN, S. R., and L. P. CARTER. *Crop Production Principles and Practices* (San Francisco: Freeman, 1976).

5. COYNE, D. P., *et al. Field Bean Production in Nebraska,* Neb. Agr. Exp. Sta. SB 486 (1965).

6. ERDMANN, M. H., *et al. Field Bean Production in Michigan,* Mich. Agr. Ext. Bul. 513 (1965).

7. GREIG, J. K., and R. E. GWIN, JR. *Dry Bean Production in Kansas,* Kans. Agr. Exp. Sta. Bul. 486 (1966).

8. HARTZOG, D., and F. ADAMS. *Fertilizer, Gypsum, and Lime Experiments with Peanuts in Alabama, 1967–1972,* Auburn Univ. Agr. Exp. Sta. Bul. 448 (1973).

9. JACKSON, C. R. *A Field Study of Fungal Association on Peanut Fruit,* Ga. Res. Bul. 26 (1968).

10. MCGILL, J. F., and L. E. SAMPLES. *Peanuts in Georgia,* Ga. Coop. Ext. Serv. Bul. 640 (1969).

11. Martin, J. W., W. H. Leonard, and D. L. Stamp. *Principles of Field Crop Production,* 3rd ed. (New York: Macmillan, 1976).
12. Morrison, K. J., *et al. Growing Field Beans in Columbia Basin,* Wash. Agr. Ext. Bul. 497 (1962).
13. North Carolina Agricultural Extension Service. *Peanut Production Guide,* N.C. Ext. Circ. (1963).
14. Robertson, L. S., and R. D. Frazier. *Dry Bean Production—Principles and Practices.* Mich. State Univ. Coop. Ext. Bul. E-1251 (1978).
15. Robinson, R. G. *Dry Field Beans in Minnesota,* Minn. Agr. Ext. Serv. Bul. 310 (1964).
16. Tripp, L. D. *Peanut Production Guidelines,* Okla. State Univ. Publ. 2008 (1967).
17. Tucker, B. B., and L. Tripp. *Use of Fertilizer on Peanuts.* Okla. State Univ. Publ. No. 2219 (1968).
18. USDA. Agr. Stat. (1977).
19. Vaughan, C. E., and R. P. Moore. *Tetrazolium Evaluation of the Nature and Progress of Deterioration of Peanut* (Arachis Hypogaea L.) *Seed in Storage, Proc. Assoc. of Official Seed Analysts,* 60:104 (1970).
20. Walker, M. E., H. D. Morris, and R. L. Carter. *The Effect of Rate and Method of Application of N, P, and K on Yield, Quality and Chemical Composition of Spanish and Runner Peanuts,* Ga. Res. Bul. 152 (1974).
21. Woodroof, J. S. *Peanuts—Production, Processing, Products* (Westport, Conn.: AVI, 1973).
22. Wynne, J. C., W. R. Baker, Jr., and P. W. Rice. "Effects of Spacing and a Growth Regulator, Kylar, on Size and Yield of Fruit of Virginia-Type Peanut Cultivars," *Agron. Jour.,* 66:192 (1974).

Part VI
Forage Crops

CHAPTER 30
FORAGES

THE value of forage production in the United States is estimated to be $12 billion. Beef cattle rank among the top four income-producing commodities in 41 states, and dairy products are among the top 4 in 39 states. Thus, few feed or food resources are as important as forages to the nation's welfare. Despite this fact, few areas of crop production have been given such low priority for improvement. In addition to their value for livestock production, forages add beauty to the landscape, aid in preserving environmental quality, help control soil erosion and water runoff, and offer great recreational potential.

An early recognition of the high value of grass was indicated by the writer of the Psalms thousands of years ago. The theme of grazing runs throughout the books of *Genesis* and *Exodus*. In Biblical times, grass for cattle was a promise or reward for obeying the commandments of God, and the want of grass was recognized as a symbol of desolation.

Grasses are present on the earth in greater abundance than any comparable group of plants. There are grasses that are adapted from the humid tropics to the Arctic. Some are important components of marshy, swampland vegetation, while others inhabit desert regions where the annual precipitation is 5 inches (12.7 cm) or less.

Most civilizations have developed in grassland regions. It is probable that the human population of the world owes its present level of achievement to the abundance and widespread distribution of grasses. The discovery that green forage could be converted into cured hay, to be stored and used throughout a considerable time period, had a more important part in the development of civilization than most realize. It is associated with a stabilized type of agriculture.

A statement relative to the importance of grass, recognized through the years as a literary masterpiece that should be known to every student of agriculture, comes from the pen of John James Ingalls. It appeared in *Kansas Magazine* in 1872 and was reprinted in part in *Grass, the 1948 Yearbook of Agriculture.* The following brief excerpts include some of the most-quoted passages and indicate the enduring worth of Ingalls's statement.

Next in importance to the Divine profusion of water, light, and air—those three great physical facts which render existence possible—may be reckoned the universal beneficence of grass.

Lying in the sunshine among the buttercups and dandelions of May, scarcely higher in intelligence than the minute tenants of that mimic wilderness, our earliest recollections are of grass; and when the fitful fever is ended, and the foolish wrangle of the market and forum is closed, grass heals over the scar which our descent into the bosom of the earth has made, and the carpet of the infant becomes the blanket of the dead.

. . . Grass is The Forgiveness of Nature—her constant benediction. Fields trampled with battle, saturated with blood, torn with the ruts of cannon, grow green again with grass, and carnage is forgotten. Streets abandoned by traffic become grass-grown, like rural lanes, and are obliterated. Forests decay, harvests perish, flowers vanish, but grass is immortal. Sown by the winds, by wandering birds, propagated by the subtle horticulture of the elements, which are its ministers and servants, it softens the rude outline of the world. Its tenacious fibers hold the earth in its place, and prevent its soluble components from washing into the wasting sea. It invades the solitude of deserts, climbs the inaccessible slopes and forbidding pinnacles of mountains, modifies climates, and determines the history, character, and destiny of nations.

Unobtrusive and patient, it has immortal vigor and aggression. Banished from the thoroughfare and the field, it bides its time to return and when vigilance is relaxed, or the dynasty has perished, it silently resumes the throne from which it has been expelled but which it never abdicates. It bears no blazonry of bloom to charm the senses with fragrance of splendor, but its homely hue is more enchanting than the lily of the rose. It yields no fruit in earth or air, and yet, should its harvest fail for a single year, famine would depopulate the world. . . .

Forages in Relation to Other Feeds

Of the total U.S. acres in farms with sales of $2,500 and over (about 40 percent of the total land area), about 45 percent is in total cropland, but crops were harvested from only about 72 percent of total cropland. Total woodland is about 8 percent, and more than 50 percent is classified as pastureland and rangeland including cropland and woodland which is grazed. This amounts to 390 million acres (158 million ha) of permanent pastures and rangelands, 33 million acres (13 million ha) of woodland that is pastured, and more than 65 million acres (26 million ha) of cropland used only for pasture and grazing.

In general, wherever livestock production is an important enterprise the production of hay must be considerable. The acreage in hay usually correlates to the acres in pasture. Hay acreage is particularly important in the Dairy and Hay Belt in the North Central states and New England, where alfalfa mixtures, and red clover–timothy and mixtures lead in the acreages grown. The acreage of hay harvested annually as an average of the time periods 1959–1963, 1970–1973, and 1974–1976 is shown in Table 30–1, grouped as (1) alfalfa and alfalfa mixture and (2) all other hay. Production is shown by groups of states or geographic divisions.

The use of alfalfa and alfalfa mixtures has increased since 1970–1972 in the North Central region, with more than 15 million acres (6.1 million ha) total and more than 3 million acres (1.2 million ha) in Wisconsin and

TABLE 30–1. Acres of Alfalfa and Alfalfa Mixtures for Hay and All Other Hay*

Geographic Divisions†	Alfalfa and Alfalfa Mixtures (000)			All Other Hay (000)		
	1959–1963	1970–1972	1974–1976	1959–1963	1970–1972	1974–1976
New England (6)	215	201	191	1,563	815	814
Mid-Atlantic (3)	1,888	1,891	1,820	3,346	2,623	2,587
East North Central (5)	6,571	2,075	5,670	4,494	3,145	3,196
West North Central (7)	11,580	11,316	11,267	13,886	11,784	11,996
South Atlantic (8)	551	265	253	3,354	2,798	2,829
East South Central (4)	531	286	296	3,546	3,485	3,686
West South Central (4)	603	915	789	3,663	4,006	4,249
Mountain (8)	4,176	4,499	4,483	3,332	3,497	3,322
Pacific (5)	1,945	2,139	2,042	1,767	1,520	1,549
U.S. Totals	28,060	23,587	26,811	38,951	33,673	34,228

* Averages given for all periods.
† Number of states per region given in parentheses.

more than 2 million acres (0.8 million ha) in South Dakota and Minnesota in 1976. As shown in Table 30–1, much of this increase is in the eastern part of the North Central region, with an increase of more than 3 million acres (1.2 million ha) from the 1970–1972 period to the 1974–1976 period. Acreage of all other kinds of hay, excluding alfalfa and alfalfa mixtures, generally has not increased appreciably and has even declined in some regions since the 1959–1963 period. The importance of alfalfa and alfalfa mixtures is illustrated by the fact that there is more than three-fourths as much alfalfa acreage as of all other kinds of hay acreage in the United States, including wild hay. Table 30–2 shows a further analysis of hay acreage by geographic divisions and selected states in each division for 1974. While clover-timothy and mixtures remains the most important hay crop in New England, Mid-Atlantic, South Atlantic, and East South Central divisions, the acreage of alfalfa and alfalfa mixtures far exceeds that of other hays in the North Central, West South Central, Mountain, and Pacific divisions. The U.S. acreage of alfalfa and mixtures in 1974 more than doubled that of clover and mixtures.

Forage Use in Livestock Programs

Approximately 70 percent of the acreage from which crops are harvested produces feed for livestock. In a broad sense all feed consumed by livestock is classed as forage. In a more restricted sense however, we think of forages as the roughages—mostly hay, pasture, and silage.

More than 48 percent of the land in farms is used as range or pasture of various kinds. The yield of this land plus that used for silage and additional hay, provides the principal forage supply for livestock feeding. Cattle

Table 30-2. Acres of Hay Harvested for Divisions and for Highest-Acreage States Within Each Group, 1974*

Geographic Divisions and States	All Hay (000)	Alfalfa and Alfalfa Mixtures (000)	Clover-Timothy and Mixtures (000)	Lespedeza (000)	Wild Hay (000)
New England (6)	833	150	504	—	25
Vermont	405	81	227	—	13
Maine	181	14	134	—	11
Mid-Atlantic (3)	3,526	1,347	1,525	—	50
New York	2,012	777	839	—	35
Pennsylvania	1,412	523	648	—	13
East North Central (5)	8,106	4,231	2,220	13	46
Wisconsin	3,906	2,295	582	—	19
West North Central (7)	20,593	10,433	3,213	57	4,542
South Dakota	4,304	2,481	106	—	1,189
North Dakota	3,465	1,825	224	—	1,107
Nebraska	3,423	1,558	159	—	1,414
South Atlantic (8)	1,926	213	849	47	67
Virginia	632	78	425	13	20
East South Central (4)	2,842	240	1,393	205	220
Kentucky	1,170	154	782	91	35
West South Central (4)	4,075	647	271	64	689
Texas	2,036	231	40	7	202
Mountain (8)	7,070	4,332	760	64	1,296
Montana	2,119	1,231	254	—	401
Pacific (5)	3,097	1,698	450	—	375
California	1,501	946	97	—	118
U.S. totals	52,068	23,291	11,185	386	7,310

* Data shown for farms with sales of $2,500 and over.

and sheep producers are heavily dependent on forages as an economical source of feed.

Forages furnish approximately 55 percent of the total feed units consumed by all livestock (Table 30–3). The percentage of forage in the diets of some specific livestock is 92 percent for beef cattle other than those on feed, 90 percent for sheep and goats, 83 percent for dairy cattle other than milk cows, almost 80 percent for horses, and 63 percent for milk cows. All kinds of forage crops will continue to make a major contribution toward production or maintenance of most classes of livestock in the foreseeable future. Estimates indicate increased need for forage feed units. Yields of forages on the average have not increased as much as for most other crops. High losses occur in the conservation and utilization of forages.

FIGURE 30-1. *Beef cattle grazing a tall fescue-Ladino clover pasture on Class IV land. Forages are valuable for livestock production, add beauty to the landscape, aid in preserving environmental quality, and help to control soil erosion and runoff. Forages furnish about 55 percent of the total feed units consumed by livestock, and over 90 percent of the feed units for beef cattle.* [SOURCE: USDA, Soil Conservation Service.]

These losses are associated with poor quality and reduced acceptability of the harvested feed.

Complementary Benefits from Forages

There are a number of complementary benefits from forage production in addition to the low-cost, health-giving feed provided. Even the idle grass has value in soil protection and organic matter renewal. With forages, the animal is in a soil-conserving and soil-improving relationship to the land.

TABLE 30-3. Contribution of Different Kinds of Feed by Class of Livestock

Class of Livestock	Percent Consumed				
	Concentrates	Hay	Other Harvested Forage	Pasture	All Forage
Sheep and goats	9.7	4.7	3.7	81.9	90.3
Cattle on feed	73.5	13.4	7.7	5.4	26.5
Other beef cattle	7.9	14.6	4.4	73.1	92.1
Milk cows	37.1	23.5	19.5	19.9	62.9
Other dairy cattle	17.3	30.1	6.6	46.0	82.7
Horses	20.6	18.4	10.2	50.8	79.4
All livestock	45.4	12.2	6.1	36.3	54.6

AS A SOIL BUILDER. Much of the world's most fertile and productive soils was developed under grass vegetation cover. Roots, stolons, rhizomes, and litter are effective soil builders and soil stabilizers. A high-quality sod in the rotation results in high row-crop yields. A high-quality sod is one that results from a vigorous growth of grasses or legumes, with the forage so utilized through livestock that maximum row-crop yields are obtained when they follow the sod crops. The legumes have taproots which decay rather quickly. The "sod effect" results more particularly from the fibrous grass roots. Grass roots are capable of sponging up the nitrogen furnished by the legumes. They provide a much more stable type of organic matter than do the legumes. The rapidity and time of release of nitrogen by plowed-down sod somewhat parallels the growth curve and the need for nitrogen of summer-growing crops, such as corn. In a study at Lancaster, Wisconsin, corn grown in a rotation that included a legume sod outyielded continuous corn, regardless of how much nitrogen was applied. First-year corn following alfalfa yielded as much as corn fertilized with 150 pounds of nitrogen per acre (168 kg/ha). Researchers in the Midwest found that soybeans supply about as many pounds of nitrogen equivalent to a succeeding crop as the soybean crop yielded in bushels per acre.

AS A SOIL PROTECTOR. Not only do legume roots penetrate the subsoil and improve drainage, but the fibrous grass roots literally permeate the plow layer. Adequate use of forages is the key to improved soil tilth and internal drainage. In addition, adapted grass–legume combinations furnish the best protection on erodible, sloping soil surfaces. Erosion and runoff vary widely for different soils and with the degree of slope. The more outstanding differences, however, are between the runoff and erosion from land in row crops and in sod crops. It has been shown that in Ohio the soil loss from a Muskingum silt loam was 99.3 tons/acre (223 MT/ha) from corn, as compared with 0.02 (0.04 MT/ha) from a bluegrass sod; 40.3 percent of the total rainfall was lost as runoff when the land was in row crops, as compared with only 4.8 percent when in bluegrass sod. Approximately 7,500 years would be required to erode 1 inch (2.5 cm) of soil under bluegrass. In contrast, the lost of 99.3 tons/acre (223 MT/ha) under corn would require 1.5 years and would remove an inch (2.5 cm) of soil by erosion. In another study, in Iowa, on land sloping from 2–4 percent on ridges and valleys and from 12–18 percent on the sides, sheet erosion from continuous corn was 100 times that of erosion on a well-managed smooth bromegrass pasture. These studies demonstrate the effectiveness of grass sod crops in protecting the soil.

When sod crops are plowed the soil surface is protected as a result of the grass-root action in increasing soil particle aggregation, holding and binding the soil particles together and increasing percolation. Forages protect the soil in a number of ways, but probably their greatest contribution is in shielding the soil surface from the impact of raindrops.

CANOPY INTERCEPTION. Forages, with the dense covering of leaves and stems which they afford, provide a maximum canopy interception of rainfall. Indications are that rainfall interception increases directly with an increase in vegetative cover. It is recognized that the beating action of raindrops on unprotected soil surfaces results in a soil-sealing action, sharply reducing water percolation; the result is the loss of a large percentage of rainfall, with markedly increased soil losses from erosion.

Forages grasses and legumes have the future potential use for extracting food-grade proteins from the leaves. This is a relatively unexplored method of providing a primary food source to feed hungry people in developing countries.

Economic Aspects of Forage Production

Climatic and soil conditions determine where forage can be grown and affect the yield obtained. In the final analysis, however, economic considerations determine where, when, and how much forage should be grown and whether a given area should be devoted to grain, fiber, forage, or other crops. Forage crops must be viewed from the standpoint of how they fit into the farming pattern of a given farm unit. Usually the amount of labor, capital, and land is limited. To maximize profits it is necessary that each unit of resources be used in the manner in which it can be expected to bring the greatest return.

The amount of forage produced per farm varies greatly from one locality to another. In some areas nearly all the land area is devoted to grasses and legumes; in other areas very little of these crops is grown; and in still other localities there may be a nearly even balance between forages and grains. The principle of comparative advantage is the economic law which helps explain this regional specialization. A given crop should not necessarily be grown when and because absolute yields and income per acre are greatest, but rather where relative or comparative yields and returns are greatest.

Forages are complementary to other crops only when considered through a period of years. Any increase in the production of grain resulting from the nitrogen, organic matter, or other contributions of grasses and legumes to yield must come from the grain crops that follow the forages in rotation.

Legume Inoculation

Legumes store in their leaves and seed an abundance of protein, the basis of the protoplasm of all plants and animals. Legumes, therefore, are in demand as human food, as livestock feed, and as green manure.

The successful growing of legumes is dependent on the right kinds of rhizobia being present in the soil. Inoculation is the practice of adding the proper bacteria to the legume seed, or to the soil where the legume is to be seeded. The most common type of culture is a peat-based inoculant applied to the seed before planting. Other available forms are liquid cultures and granular inoculants. The latter is soil-applied, banded in the row at planting time. Pre-inoculation of seed, application of inoculant well in advance of planting, appeared in the 1960s. Rhizobia may not be viable by the time seed is purchased, depending on how the seed was stored after pre-inoculation.

During the initial stages of nodulation, the rhizobia are invaders and aggressors. At a later stage, however, they become true symbionts in a naturally beneficial relationship with the host plant. In exchange for food and a home for the rhizobia in the tumor-like nodules formed on plant roots, the rhizobia aid plant growth by making nitrogen available. The process by which the nitrogen of the air is converted into a usable form available to the plant is called *symbiotic nitrogen fixation.* Nitrogenase enzymes catalyze the conversion of free nitrogen to ammonia, which is then incorporated into amino acids and proteins.

One cannot be sure that a particular soil contains the right kind of rhizobia. Only about 25 percent of strains occurring naturally in a soil are considered desirable. Legume inoculation definitely should be practiced when

FIGURE 30-2. *An alfalfa root showing nodules. Well-nodulated forage legumes are involved in "symbiotic nitrogen fixation," in which nitrogen from the soil air is converted into a usable form available to the plant. Legume seed should be inoculated with the proper* Rhizobium *culture before planting when the legume to be established, or a legume in its cross-inoculation group, has not been grown on the land in recent years.*

TABLE 30-4. Legume Inoculation Groups

Common Name	Latin Name (Genus–Species)	Common Name	Latin Name (Genus–Species)
		ALFALFA	
Alfalfa	*Medicago sativa*	Tifton burclover	*M. rigidula*
Buttonclover	*M. orbicularis*	Yellow alfalfa	*M. falcata*
California bur-clover	*M. denticulata*	White sweetclover	*Melilotus alba*
		Hubam sweet-clover	*M. alba annua*
Spotted burclover	*M. arabica*		
Black medic	*M. lupulina*	Yellow sweetclover	*M. officinalis*
Snail burclover	*M. scutellata*	Bitterclover (sourclover)	*M. indica*
Tubercule bur-clover	*M. tuberculata*		
		Fenugreek	*Trigonella foe-numgraceum*
Little burclover	*M. minima*		
		CLOVER	
Alsike clover	*Trifolium hybrid-um*	Berseem clover	*T. alexandrinum*
		Cluster clover	*T. glomeratum*
Crimson clover	*T. incarnatum*	Zigzig clover	*T. medium*
Hop clover	*T. agrarium*	Ball clover	*T. nigrescens*
Small hop clover	*T. dubium*	Persian clover	*T. resupinatum*
Large hop clover	*T. procumbens*	Carolina clover	*T. carolinianum*
Rabbitfoot clover	*T. arvense*	Rose clover	*T. hirtum*
Red clover	*T. pratense*	Buffalo clover	*T. reflexum*
White clover	*T. repens*	Hungarian clover	*T. pannonicum*
Ladino clover	*T. repens (gigan-teum)*	Seaside clover	*T. wormskjoldii*
		Lappa clover	*T. lappaceum*
Sub clover	*T. subterraneum*	Bigflower clover	*T. michelianum*
Strawberry clover	*T. fragiferum*	Puff clover	*T. fucatum*
		PEA AND VETCH	
Field pea	*Pisum arvense*	Purple vetch	*V. atropurpurea*
Garden pea	*P. sativum*	Monantha vetch	*V. articulata*
Austrian Winter pea	*P. sativum* (var. *arvense*)	Sweet pea	*Lathyrus odoratus*
		Rough pea	*L. hirsutus*
Common vetch	*Vicia sativa*	Tangier pea	*L. tingitanus*
Hairy vetch	*V. villosa*	Flat pea	*L. sylvestris*
Horsebean	*V. faba*	Lentil	*Lens culinaris (esculenta)*
Narrowleaf vetch	*V. angustifolia*		
		COWPEAS	
Cowpea	*Vigna sinensis*	Guar	*Cyamopsis tetra-gonoloba*
Asparagus bean	*V. sesquipedalis*		
Common lespedeza	*Lespedeza striata*	Jackbean	*Canavalia ensi-formis*
Korean lespedeza	*L. stipulacea*		
Sericea lespedeza	*L. cuneata*	Peanut	*Arachis hypogaea*
Slender bushclover	*L. virginica*	Velvetbean	*Stizolobium deeringianum*
Striped crotalaria	*Crotalaria mucro-nata*		
		Lima bean	*Phaseolus lunatus (macrocarpus)*
Sunn crotalaria	*C. juncea*		
Winged crotalaria	*C. sagittalis*	Adzuki bean	*P. angularis*

TABLE 30–4 (cont.)

Common Name	Latin Name (Genus–Species)	Common Name	Latin Name (Genus–Species)
Florida beggar-weed	*Desmodium tor-tuosum*	Mat bean	*P. aconitifolius*
Tick trefoil	*D. illinoense*	Mung bean	*P. aureus*
Hoary tickclover	*D. canescens*	Tepary bean	*P. acutifolius* var. *latifolius*
Kudzu	*Pueraria thunber-giana*	Partridge pea	*Chamaecrista fasciculata*
Alyceclover	*Alysicarpus vagi-nalis*	Acacia	*Acacia linifolia*
		Kangaroo thorn	*A. armata*
(No common name)	*Erythrina indica*	Wild indigo	*Baptistia tinctoria*
		Hairy indigo	*Indigofera hirsuta*
Pigeonpea	*Cajanus cajan (indicus)*		

BEAN

Garden beans, kidney bean, Navy bean, pinto bean	*Phaseolus vulgaris*	Scarlet Runner bean	*P. coccineus (multiflorus)*

LUPINE

Blue lupine	*Lupinus angusti-folius*	Washington lupine	*L. polyphyllus*
		Sundial	*L. perennis*
Yellow lupine	*L. luteus*	Texas bluebonnet	*L. subcarnosus*
White lupine	*L. albus*	Serradella	*Ornithopus sativus*

SOYBEAN

All cultivars of soybeans　　*Glycine max*

SPECIFIC STRAINS

Birdsfoot trefoil	*Lotus corniculatus*	Sainfoin	*Onobrychis vul-garis (sativus)*
Big trefoil	*L. uliginosus*		
Foxtail dalea	*Dalea alopecu-roides*	Crownvetch	*Coronilla varia*
		Siberian pea-shrub	*Caragana arbor-escens*
Black locust	*Robinia pseudo-acacia*		
		Garbanzo	*Cicer arietinum*
Trailing wild bean	*Strophostyles hel-vola*	Lead plant	*Amorpha can-escens*
Hemp sesbania	*Sesbania exaltata*	Kura clover	*Trifolium am-biguum*

the legume to be established has not been grown previously on the land, or if the legume is widely different from one previously planted.

For practical purposes the common legumes are grouped according to their association with certain strains or species of rhizobia. These "cross-inoculation groups" consist of those legumes that are inoculated with the same kind of rhizobia. For example, the alfalfa rhizobia can function with sweetclover and with certain other legumes. However, the rhizobia that inoculate red clover belong to a different group and cannot function with alfalfa. While the demarcations between groups are less distinct than previously believed, knowledge of these groups remains valuable in rhizobia differentiation. The USDA listing of all the legumes constituting the different cross-inoculation groups is shown in Table 30–4.

Many seed disinfectants are toxic to legume bacteria, but some of the bacteria have a high degree of tolerance if certain precautions are taken. Seed treated to protect them against harmful bacteria and fungi can be inoculated by making a paste with the inoculant and mixing this thoroughly with the seed. If the seed are planted within two hours, successful nodulation may follow. If necessary to hold the seed longer, they should be reinoculated. Complete fertilizers should not be allowed to come in contact with inoculated legume seed. Phosphates are not as harmful as nitrogen or potassium. It is a common practice, however, to apply fertilizers at the same time that legumes are being seeded. If the concentration or amount of fertilizer drilled with the seed is not enough to affect germination or to injure the seedlings, it will not be harmful to the legume bacteria applied to the seed. Ground limestone is beneficial and sometimes is used to pellet the inoculated legume seed.

The success of inoculation can be determined by digging up a few plants from the field and evaluating nodulation. Numerous small nodules scattered throughout the secondary root system extremities is a sign of ineffective nodulation. Effectiveness is evidenced by the presence of large nodules located near the primary root or taproot. Effective nodules also are pink inside as a result of the presence of leghemoglobin, which participates in the nitrogen-fixation process.

Although the amount of nitrogen fixed by legumes varies with the plant stand, soil, and environmental conditions, averages for various crop species range from less than 50 pounds/acre (56 kg/ha) to almost 200 pounds/acre (224 kg/ha) (Table 30–5).

Establishment of Legume–Grass Seedings

LEGUME–GRASS MIXTURES USUALLY DESIRABLE. Usually grass–legume mixtures are preferable to either a grass or a legume alone. Some of the advantages are the following:

1. Yields of legume–grass mixtures tend to be higher than those of either component.
2. Legumes supply nitrogen to the grasses, so that they yield more than grasses grown alone and also have a higher protein content.
3. The soils in many fields are quite variable; mixed seeding may furnish crops adapted to each soil condition.
4. The presence of grass reduces the likelihood that legumes will be heaved out on tight soils with repeated freezes and thaws.
5. Mixtures usually resist the encroachment of weeds better than do pure stands of legumes.

TABLE 30–5. Average Amount of Nitrogen
Fixed Per Acre by Selected Legumes

Legume	Nitrogen Fixed Per Acre (Pounds)
Alfalfa	194
Ladino clover	179
Lupines	151
Sweetclover	119
Alsike clover	119
Red clover	114
White clover	103
Cowpeas	90
Annual lespedezas	85
Vetch	80
Peas	72
Soybeans	58
Peanuts	42
Beans	40

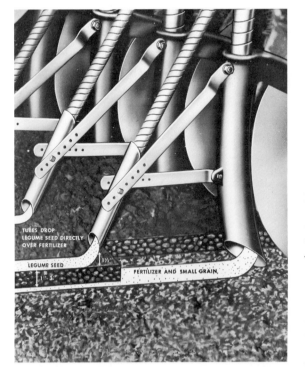

FIGURE 30–3. *"Band seeding" of grasses and small-seeded legumes places the seed just below the soil surface, directly above a band of fertilizer, placed there as a part of the seeding operation. Research has shown a decided advantage in some areas for the band-seeding method of establishing grasses and legumes.*

6. Grass reduces the danger of bloat when legumes are grazed.

7. The presence of grass roots results in increased aggregation of soil particles and thus aids in resistance to erosion losses.

8. Grasses reduce legume lodging and maintain quality by saving leaves; grass–legume hay is more easily cured than pure legume hay.

MAKING THE SEEDING. Seedbeds should be sufficiently firm to permit a uniform, shallow covering of the seed. If the ground is plowed just before seeding every effort should be made to firm it thoroughly. This ensures that seed will not be placed too deep and also improves seed-to-soil contact. Seeding rates usually have been much higher than necessary when good cultural practices are followed. Because of many hazards, a high seeding rate may be justified under some conditions. In humid areas sowing forages in small-grain companion crops is still practiced widely, although this procedure has declined somewhat in recent years. The primary reason is to obtain a return from the land while the forages are being established. The companion crop provides an erosion-resistant growth, but it also has been found to compete with the new forage seedlings. Herbicides for weed control may be substituted for companion crops in spring seedings. Companion crops are seldom used with late summer seedings because of a lessened weed problem.

FIGURE 30–4. *A commercial cultipacker seeder for grasses and small-seeded legumes. The first of two corrugated rollers firms the surface, the seed is dropped, and a second roller firms the seed in contact with the soil.* [SOURCE: Southern Illinois University.]

METHODS OF SEEDING. There are at least three commonly used techniques for seeding forages:

1. *Broadcast seeding.* The seed is broadcast on a well-prepared seedbed and may be covered lightly or firmed in contact with the soil by mechanical means. Late winter or early spring seedings may be successful without covering because of freezing and thawing action of the soil. Frost-seeding of red clover in winter grains is widely practiced.

2. *Cultipacker seeding.* The first of two corrugated rollers on a cultipacker seeder firms the soil surface, the seed is dropped, and a second roller firms the seed in contact with the soil. This method often is more reliable than is broadcast seeding.

3. *Band seeding.* Considerable research has shown a decided advantage in establishing grasses and legumes by placing the seed in drill rows directly above, but not in contact with, a band of fertilizer placed approximately 1.5 inches (3.8 cm) below the soil surface. Press wheels may be used to ensure coverage and good seed–soil contact.

REFERENCES AND SUGGESTED READINGS

1. ALLEN, G. C., E. F. HODGES, and M. DEVERS. *National and State Livestock–Feed Relationships,* USDA-ERS Stat. Bul. 446 (1972).

2. ALLEN, O. N. "Symbiosis: Rhizobia and Leguminous Plants," In *Forages* (Ames: Iowa State Univ. Press, 1973).

3. ALLEN, O. N. "The Inoculation of Legumes," In *Forages* (Ames: Iowa State Univ. Press, 1966).

4. BROWNING, G. M. "Forages and Soil Conservation," In *Forages* (Ames: Iowa State Univ. Press, 1973).

5. COWAN, J. R. Foreword to *Grasslands* (Ames: Iowa State Univ. Press, 1974). edited by H. B. Sprague.

6. DECKER, A. M., T. H. TAYLOR, and C. J. WILLARD. "Establishment of New Seedings," In *Forages* (Ames: Iowa State Univ. Press, 1973).

7. ERDMAN, L. W. *Legume Inoculation: What It Is—What It Does,* USDA Farmers' Bul. 2003 (1967).

8. GOULD, F. W. *Grass Systematics* (New York: McGraw-Hill, 1968).

9. HARRISON, C. M. (Ed.). *Forage Economics—Quality,* Spec. Publ. 13 (Madison, Wisc., American Society of Agronomy, 1968).

10. HEATH, M. E. "Grassland Agriculture," In *Forages* (Ames: Iowa State Univ. Press, 1973).

11. HODGSON, R. E. "The Place of Forage in Animal Production, Now and in Years Hence," *Forages of the Future,* Proc. Res. Ind. Conf. Sponsored by Am. Forage and Grassland Council, Chicago (1968).

12. HODGSON, H. J. "Forages, Their Present Importance and Future Potential," *Agr. Sci. Rev.,* 6:2, 1968.

13. INGALLS, J. J. "Bluegrass," *Kansas Magazine* (1872). Reprinted in full in the USDA *Forage Crops Gazette,* 3–4 (1939–1940); (also reprinted in part in *Grass, The 1948 Yearbook of Agriculture*).

14. JACOBS, V. E. "Forage Production Economics," In *Forages* (Ames: Iowa State Univ. Press, 1973).

15. KELLER, W., *et al.* (Eds.). *Forage Plant Physiology and Soil Range Relationships,* Spec. Publ. 5 (Madison, Wisc.: American Society of Agronomy, 1964).

16. RAYMOND, W. F. "The Nutritive Value of Forage Crops," In *Advances in Agronomy,* vol. 21 (New York: Academic Press, 1969).

17. ROHWEDER, D. A., W. D. SHRADER, and W. C. TEMPLETON, JR. "Legumes, What Is Their Place in Today's Agriculture?" *Crops and Soils,* 29(6):11 (1977).

18. SMITH, D. *Forage Management in the North* (Dubuque, Iowa: W. C. Brown, 1962).

19. USDA Agr. Stat. (1977).

20. USDA Agr. Stat. (1966).

21. USDA. Crop Production Annual Summary (1972).

22. U.S. Department of Commerce *1974 Census of Agriculture, United States Summary* (1977).

CHAPTER 31
PASTURES AND PASTURE IMPROVEMENT

> It is highly important to the welfare of this nation that every effort be made to maintain the consumption of animal products at a high level. Much of our land is of little use except in livestock production. . . . A livestock economy permits a great flexibility between the production of meat and that of grain, as conditions may require. Level grazing land can be out to the plow or returned to pasture with the ebb and flow of the need for grain. . . . Finally, pasture lands have tremendous potentialities in terms of storage of soil fertility that can be readily released by cultivation.
>
> FIRMAN E. BEAR

Grasslands of the World

Grazing lands were vital to primitive man long before cattle were domesticated. Forage culture is mainly a product of European and American civilization. Haymaking in Great Britian dates from 750 B.C. Columella, the Roman, described in great detail the growing of hay crops and the significance of proper curing about 50 A.D.

Between one-fifth and one-fourth of the earth's surface is grassland which, together with the livestock associated with it, occupies a key position in world food production. This vast area may be divided into (1) natural grasslands, containing native plant species; and (2) improved grasslands, usually seeded to mixtures of grasses and legumes that are predominantly introduced species. Grasslands are of value, in general, only as they are utilized by livestock in producing animal products for man's use. Efficiency of conversion from vegetable matter to animal products becomes more important as population increases.

Pasture and Grazing Lands of the United States

Just as is true for the world, pastures and grazing lands together make up by far the largest single item of land use for the United States. Of the land area considered in farms, all kinds of pastures, rangeland, and grazing areas total about 0.5 billion acres (0.2 billion ha) or about 54 percent of the farmland area.

Cropland used only for pasture occupies about 13 percent of the total pasture and range area; permanent grassland pasture and range, about 80 percent; and woodland grazed, 7 percent.

There has been a gradual improvement of grassland pastures by brush clearing, reseeding, fertilizing, and shifting of cropland to pasture. These yield-increasing activities have been accompanied by the removal of some areas of low productivity from grazing use. Although there is a greater emphasis on pasture improvement than previously, still about 90 percent of the grazing lands are listed as unimproved. This illustrates the need for a greater emphasis on pasture improvement.

Large areas are grazed as the only feasible agricultural use to be made of them. This land is important, however, in the overall agricultural production supplying about one-third the nutrients for all livestock. Studies in feed values indicate that on the basis of present land uses approximately 7 acres (2.8 ha) of pasture and grazing land will average the production of 1 acre (0.4 ha) of cropland; the range in carrying capacity is from two animal units or more per acre (0.4 ha) to only one animal unit per 45 or more acres (18 ha). The acreage of pasture and grazing land per animal unit of roughage-consuming livestock is 3–4 acres (1.2 to 1.6 ha) in the Northeast, Corn Belt, and Great Lakes states; 9 acres (3.6 ha) in the northern Great Plains states; 14 (5.7 ha) in the South, including the southern Great Plains states of Oklahoma and Texas; and more than 38 (15.4 ha) in the 11 Western states as a group.

Cropland pasture is the most productive type. This refers to grasses, legumes, or mixtures grown at varying intervals and durations on land suitable for row-cropping. A given acreage of cropland pasture may produce five or six times as much forage as does an acre (0.4 ha) of farm grassland pasture and 25 times as much feed as the same area of nonfarm pasture or range.

Natural Grasslands

The natural grasslands consist mainly of undeveloped land which, because of rough topography, poor or unsuited soil, insufficient precipitation, lack of irrigation water, or other reasons cannot be used successfully for crops or for improved pasture without considerable cost. It is suitable for grazing and capable of supporting uncultivated and unfertilized forage, primarily native grasses and other forage plants.

The principal native or unimproved grazing lands are now found in the West and lower South. Those of the West are predominantly grasslands, or desert shrublands, too dry for arable farming. An important part, however, is mountain woodland, which is moist enough for trees, but generally too rough for tillable farming. Those of the South are principally forested grazing lands in the Coastal Plains, together with some important areas of wet prairies and marshes.

Control and limitation of grazing in accordance with carrying capacity, allowance of growth for natural reseeding, artificial reseeding of open areas in abandoned fields, and removal of competing brush and other woody species are among the chief methods of restoring and improving the grazing lands in this area.

Studies in range management indicate that heavy continuous grazing results in reduced gains per acre, failure to carry cattle for the full season, and a gradual deterioration of range conditions. Some system of intermittent grazing, such as deferred or rest-rotation, may be both desirable and essential for maintaining the vegetative stand, depending on the individual range plants.

Improved Pastures

Pastures in this category are classified as improved, even though they may be producing far below their capacity. They usually occur on land that was plowed at some time and where the native species were replaced by introduced species.

It is almost universally agreed that the lowest cost in animal gains is obtained when animals are on pasture. Thus, the primary aim of a good pasture plan is to provide productive, nutritious pasture for just as long a grazing season as possible. To do this, more than one type of pasture on any given farm is needed. In the North, winter rye or wheat can be used to extend the normal pasture season, both in the fall and the spring. Nitrogen fertilization of permanent grass pastures will make them ready for grazing from 10 days to 2 weeks earlier than if not fertilized. Sudangrass or forage sorghums can be used to provide a palatable, nutritious pasture of high carrying capacity in midsummer, when most pasture grasses are least productive. Rotation seedings of grasses and legumes can be used for either pasture or hay production, depending on the need. Deferred grazing of cool-season perennials can provide grazing in winter months.

Northern Pastures

Most improved pasture is found in the Northeastern Corn Belt, Great Lakes, and northern Great Plains states. Pasture production is well suited to the Northeastern and Great Lakes states, considering the topography, rainfall, and nearness to population centers.

The four northern Great Plains states, with more than 70 million acres (28 million ha) in pasture and rangeland, have a much greater combined acreage than the adjoining Corn Belt and Great Lakes states. Here native short grasses, principally grama and buffalograss, are found. Tall grasses predominate toward the eastern margin of the area and on deep sandy soil. The Flint Hills area in eastern Kansas and the Sand Hills of northeast Nebraska are among the few extensive areas of native grazing land remaining in the humid, tall-grass prairie region.

In the Central and Lake States and Northeast, introduced pasture species dominate. Kentucky bluegrass is more common in permanent pastures than any other grass. Timothy also is grown extensively in these regions

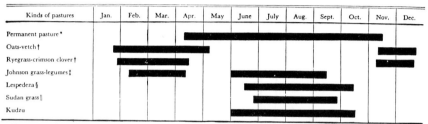

Kinds of pastures	Jan.	Feb.	Mar.	Apr.	May	June	July	Aug.	Sept.	Oct.	Nov.	Dec.
Permanent pasture*												
Oats-vetch†												
Ryegrass-crimson clover†												
Johnson grass-legumes‡												
Lespedeza§												
Sudan grass‖												
Kudzu												

* Kentucky bluegrass-white clover with soil treatment.
† Seeded to mixture as orchard-ladino or brome-alfalfa with soil treatment.
‡ If grain crop is taken, grazing season must be shortened at least six weeks.

Kinds of pastures	Jan.	Feb.	Mar.	Apr.	May	June	July	Aug.	Sept.	Oct.	Nov.	Dec.
Permanent pasture*												
Renovated pasture†												
Rotation pasture†												
Hay-meadows												
Small grain-lespedeza‡												
Sweet clover												
Sudan grass												

* Dallisgrass or carpetgrass on lowland, Bermuda-grass on upland with white, hop and Persian clovers and proper lime-fertilizer treatments.
† Bur clover, crimson clover or other adapted winter legumes may be used.
‡ Winter oats-rough peas, bur and crimson clovers can also be seeded where Johnsongrass is used.
§ May be seeded in small grain but not with winter growing clovers.
‖ Tift Sudan should be used in the humid southeastern U.S.

FIGURE 31–1. *With pastures recognized as making possible the lowest-cost animal products, the contribution of different species and mixtures to long-season, high-production grazing is indicated for the South (upper) and the North (lower).*

for pasture or hay. Orchardgrass, smooth bromegrass, and tall fescue are important in certain areas. White or Ladino clover, red clover, alfalfa, birdsfoot trefoil, and annual lespedezas are the principal pasture and hay legumes.

Pastures in the South

The 16 Southern states have about 40 million acres (16 million ha) of cropland used for pasture; 117 million acres (47 million ha) of permanent grassland pasture and range; and nearly 17 million acres (6.9 million ha) of forest land which is grazed. The wide variation in climate and soil permits the growth of many different pasture crops. The dominant pasture plants in this region are perennials, both cool-season or warm-season. Cool-season perennials adapted to some parts of the South are orchardgrass, tall fescue, Kentucky bluegrass, white clover, and red clover. Warm-season perennial forages grown in this area include bermudagrass, dallisgrass, carpetgrass, bahiagrass, johnsongrass, 'Pangola' digitgrass, napiergrass, sericea lespedeza, and alfalfa.

Annual species often are used to supplement perennial pastures for grazing or hay. Summer annuals used most widely are annual lespedezas, sorghum–sudangrass hybrids, and millets. Winter annuals include winter small grains, annual ryegrass, crimson clover, and arrowleaf clover. A combination of cool-season perennials or winter annuals with warm-season perennials or summer annuals can provide a long grazing season in the South.

The Western Range

The western half of the United States has long been known as the Western Range. The main body of the Western Range lies west of an irregular line running north and south through the Great Plains from North Dakota to Texas. Rangelands surround many important irrigated valleys in dry-farming areas.

Year-long grazing is widely practiced in the southern part of the Western Range. Seasonal grazing is practiced throughout most of the central and northern part of the Western Range. Grasslands and brush lands surrounding the mountain ranges are used for winter and spring–fall grazing; the mountains provide summer grazing.

The Northern Plains, Southern Plains, Mountain, and Pacific regions have well over three-fourths of the U.S. permanent grassland pasture and range acreage. Highly productive irrigated pastures on the best farmland is a relatively recent development in some regions. Alfalfa and other hays may be grown under irrigation as well as in dryland areas.

Improved Pastures

Improved grasslands are classed as (1) permanent, (2) rotation, or (3) supplemental, or temporary. There is a place for all three types on most farms, each making a definite contribution toward the full-season program. Improved permanent pastures, balanced with rotation and supplemental or temporary pastures, are needed on most farms to make high-quality pasturage continuously available through the grazing season.

Improvement of Permanent Pastures

Permanent pastures may occur on land that is too steep for or otherwise unsuited for use in a regular crop rotation. In most cases pastures in this category are low-producing because of low fertility, poor drainage, encroachment of weeds and brush, or a combination of these factors. They can be improved in several ways. The use of herbicides or mowing to control weeds and brush usually will result in a high return per dollar invested, but it may not increase return on investment greatly if soil fertility is limiting or if productive pasture species are not present.

The Renovation of Permanent Pastures

It often is possible to increase permanent pasture production by two to five times with renovation. Renovation refers to improvement of pastures by partial or complete destruction of a sod, with the application of lime and fertilizer and the use of proper weed-control procedures, for establish-

FIGURE 31-2. *An unimproved, weedy pasture in middle Tennessee. Large areas used for pasture give low returns. The returns from many of these acres can be doubled and trebled by renovation—liming and fertilizing, tearing up the old sod, and seeding pasture-type legumes.* [SOURCE: Southern Illinois University.]

ment of desirable forage species. The steps in renovation generally include (1) testing the soil; (2) applying lime, P, and K; (3) grazing closely or otherwise utilizing heavily before time to seed; (4) disturbing at least 50 percent of the sod, and possibly as much as 80 percent by mechanical means, such as disking, or with herbicides to kill or suppress the sod; and (5) planting seed of high-yielding legumes or grasses by broadcasting, with a drill or cultipacker seeder, or with one of the no-till pasture seeders.

MAINTAINING LEGUMES IN PASTURES. The importance of legumes in humid pastures has long been recognized. Grass sods become "sod bound" and unproductive rather quickly because of nitrogen deficiency if this element is not supplied through legumes or by application of nitrogen fertilizers. Legumes are important also because of their high feed value, which is superior to grasses in protein and calcium content. Mixtures of grasses and legumes often are more productive than either alone. Research at the University of Illinois Dixon Springs Station compared beef cow performance on tall fescue alone and on tall fescue–legume pastures. Legumes included with the grass increased beef calf weaning weight, grade of calves at weaning time, and conception rate of cows. The grasses in a forage

FIGURE 31–3. *Pasture renovation in many instances can be accomplished by minimum tillage methods. This no-tillage power drill is being used to introduce legumes into an unproductive, sod-bound grass.* [SOURCE: Deere and Company.]

mixture are important as well. They aid in holding the soil against erosion on sloping land and greatly reduce the danger of bloat.

The successful establishment of legumes is dependent upon the selection of adapted species and cultivars, with control of competition from the companion crop and from weeds. Control of annual weeds by mowing usually is necessary. Seedling vigor varies considerably between legume species; for example, where birdsfoot trefoil is recommended for the renovation of permanent pastures in the Midwest, stand failures usually can be traced to poor control of weeds or the smothering effect of the associated grass. Persistence is related to the perennial or reseeding habit of a legume.

Of first importance in keeping a pasture productive is to manage the grazing so as to favor the maintenance of the legume. This usually means regular applications of phosphate and potash. Also important is grazing sufficiently close in the early part of the season, when the grass is most productive and vigorous, so the grass will not compete excessively with the legume component.

Upright plants like alfalfa require periods of uninterrupted growth for renewal of root reserves. The survival of such plants through long periods in pastures is dependent on a system of alternate grazing. The statement "take care of the legumes and the grasses will take care of themselves" applies in a large part of the humid regions.

CONTROLLING UNDESIRABLE VEGETATION. Nearly all permanent pastures need weed and brush control. Time is well spent in maintaining a pasture so that it can be mowed regularly. Brush and small trees left to grow for a few years are expensive to remove; they should not be allowed to get started. Chemicals may be effectively used to control brush regrowth; the mower and sprayer together complete the job of killing undesirable plants in pastures.

LIME AND FERTILIZER USE. Liming acid pasture soils may improve production because it encourages legumes to come into the sward, but this is a slow process. Grasses may respond favorably to applications of phosphorus and potash, but the increase probably will not be economical if legumes are not present. Nitrogen fertilization of grass, on the other hand, can be expected to produce large increases on most soils. Whether the increases are profitable will depend on the cost of the nitrogen and the utilization of the increased growth. Increased grass growth from nitrogen fertilizer usually comes early in the season, at which time a surplus of succulent pasturage may already be available. Advantage can be taken of this early growth by getting animals into pasture a week or 10 days earlier than into grass not receiving nitrogen. The animals also benefit because the feed value is higher, especially in protein content.

When a grass–legume mixture is present, fertilizer management is considerably different than that of grasses alone. Kentucky researchers studied the effect of N, P, and K fertilizer on the maintenance of white clover with tall fescue. Nitrogen fertilizer consistently decreased the percent clover, regardless of the amount of P and K applied. Adquate P and K, but not supplemental N, was recommended for maintenance of the legume.

Grazing Management

Lack of proper grazing management can be disastrous to pastures. Either overgrazing or undergrazing can be detrimental to stands or can alter the original botanical composition. The most difficult problem to be solved in the management of grazing arises from the seasonal differences in the rate at which pasture plants grow. Saving part of the spring growth for summer and fall grazing is not the solution on improved grasslands because of poor quality of the mature grass. When possible, high-quality hay or silage should be made from surplus growth. When this cannot be done the number of animals should be adjusted so as to utilize the herbage during the early part of the season, with rotation and annual pastures used to supplement permanent pastures in the summer and fall.

Overgrazing can result in poor root growth and weed encroachment.

Grazing too late in the fall prevents the synthesis of sufficient carbohydrate reserves, and plants are more susceptible to winter injury. The height to which pastures should be grazed depends upon the species. Low-growing forages, such as Kentucky bluegrass and white clover, can be grazed relatively close without injury. On the other hand, it is known that hay-type grasses and legumes, such as smooth bromegrass, timothy, alfalfa, and birdsfoot trefoil, will not tolerate close grazing. They give greater production of quality feed when rotationally grazed and never grazed close.

Rotation Pastures

Not only are rotation pastures usually more productive than most permanent pastures, but the yields of intertilled crops that follow are increased. These pastures are on some of the best land and are regularly fertilized in rotation. The biggest problem seems to be that of proper utilization. Rotation pastures often contain upright-growing legumes, such as alfalfa, which cannot stand close, continuous grazing. Consequently, most rotation pastures are best utilized by alternate grazing. One type of alternate grazing, known as *rotational grazing,* consists of subdividing a pasture into a series of units, with the cattle regularly moved from one to another. A more intensive type of grazing known as *strip* or *ration grazing,* requires that the animals be moved daily. Also, field chopping the fresh forage daily and bunk feeding are becoming popular as a means of increasing carrying capacities. It has been shown that the acre carrying capacity may be considerably increased by this procedure, but that usually it is not economically practical for fewer than perhaps 25 animals.

Supplementary or Temporary Pasture

Temporary pasture crops produce relatively large acre yields in a short time after seeding. These annuals may supplement permanent or rotation pastures or they may serve only in emergencies. Unexpected shortages of pasture can be met most economically by providing additional hay or pasture from annual cool-weather crops or from annual warm-weather crops. Sudangrass or millet can provide supplemental grazing from July to mid-September in areas where adapted. Small grains are especially good sources of winter pasture. With proper management, a grain such as winter wheat can be grazed and then harvested as a cash crop. Overseeding annuals into stands of perennial grasses is another supplemental forage source common to the South and Southwest. Small grains and legumes, such as crimson clover and vetch, can provide fall and winter grazing after overseeding on warm-season perennial grass sods such as bermudagrass.

Stockpiling, accumulating growth by diverting part of established pastures for later use, can provide supplemental pasture at a low cost during a time when forage is deficient. It is a common procedure in some areas to stockpile fall growth of cool-season grasses for grazing during the winter months. Supplemental forage crops should be planned for regularly; there

FIGURE 31-4. *Small grains are good sources of supplementary winter pasture. With proper grazing management, a grain such as winter wheat can be grazed in the late fall and early spring without greatly affecting grain yield.* [SOURCE: University of Tennessee Agricultural Experiment Station.]

are periods in practically every year when they can be used to advantage. If not grazed they can be made into hay. They are adapted to growing under weather conditions unfavorable for high production of rotation and permanent pasture. The cost of producing these annual crops is somewhat higher per unit of feed than that of other types of pasture, but feed produced during these periods has higher feed value than does feed produced during spring and early summer.

Forage Nutrition Research

Much research is devoted to the nutritional value of forage crops. A review by W. F. Raymond considers the nutritive value of forage in terms of the factors that determine the level of nutrient intake by ruminant livestock. Nutrient intake is treated as a product of intake of feed times digestibility of feed times efficiency of utilization of digested feed. Because animals differ in nutritional requirements, the nutritive value of forages must be equated with nutrient needs. Estimates of forage digestibility have been measured in vivo and from chemical composition. There are several basic patterns of digestibility among different forage species, different plant fractions, and different stages of maturity. Several factors affect forage digestibility, including environment, climate, fertilizers, and feed supplements.

FIGURE 31–5. *The grazing season of warm-season perennial grasses, such as bermundagrass, can be extended by overseeding annual or perennial cool-season species in the fall.* [SOURCE: University of Tennessee Agricultural Experiment Station.]

Factors controlling feed intake include intrinsic (features inherent in the forage) and extrinsic (method of presentation of the forage). Intrinsic factors include a relationship between forage digestibility and intake, differences between species, and effect of fertilizers. The energy value of the digested nutrients made available to an animal may be expressed by several methods—digestible energy, metabolizable energy, and net energy. In ruminant metabolism volatile fatty acids and rumen acid patterns also play roles. The requirements of the animal for mineral nutrients in forages cannot be overlooked. They include phosphorus, sodium, calcium, potassium, cobalt, copper, molybdenum, lead, iodine, sulfur, selenium, and silica. Also producing an effect are estrogenic compounds, *Phalaris* alkaloids, bloat-producing compounds, and the nitrate content.

Methods of measuring the nutrient intake by grazing animals include estimating the fecal output of grazing animals, estimating the digestibility of grazed forage, and using fistulated techniques.

In forage improvement programs plant breeders strive to improve the nutritive value of the various species. Processing greatly affects the nutritive value of forages. Raymond concludes that many of the world's livestock are at a low level of productivity because they are underfed. He sees a need in the future for integrating new nutritional concepts into practical systems of utilizing, with high-producing animals, increasing levels of forage production as a result of agronomic research.

Drylot Feeding vs. Pasturing

Drylot feeding of beef animals is a common practice. In addition, with an increasing number of acres of pasture and rangeland being used for nonagricultural practices and with the inflated prices of real estate, drylot cow-calf production is being considered. Most dairy farmers have turned to drylot feeding because of high retail costs and the desire to maximize production from every acre of farmland.

In a Tennessee experiment with dairy cows involving pasturing versus harvest of a grass–legume mixture it was concluded that because a good grazing management program requires a source of supplemental feed from time to time, a combination of the two methods may be more desirable than either system alone in terms of daily milk production.

An Arizona study in drylot cow-calf production compared total confinement and partial confinement with grazing irrigated pastures when available. Advantages of total confinement include less land investment, adverse weather conditions eliminated, closer observation of the cattle possible, and production records more readily obtained. Disadvantages include increased labor and equipment costs. With cows in drylot for nine consecutive calf crops, annual feed costs per cow-calf unit were approximately $10 less than those maintained in partial confinement. In another experiment, the gross return to labor, taxes, and interest for cows and calves in partial confinement was considerably higher in comparison to cows and calves in total confinement. Much of the difference was in the feed cost for maintaining the cow in total confinement.

REFERENCES AND SUGGESTED READINGS

1. ACCORD, C. R. Irrigated Alfalfa Pastures Are Profitable," *Proc. Western Section, Amer. Soc. An. Sci.,* 21 (1970).
2. ANONYMOUS. "No-Till Pasture Renovation System Now a Reality," *Crops and Soils,* 29(8):23 (1977).
3. BAXTER, H. D., *et al. Pasturing vs. Harvesting of a Grass-Legume Mixture,* Tenn. Agr. Sta. Bul. 454 (1969).
4. BEAR, F. E. From Presidential Address, American Society of Agronomy, 1949.
5. BLASER, R. E., D. D. WOLF, and H. T. BRYANT. "Systems of Grazing Management," In *Forages* (Ames: Iowa State Univ. Press, 1973).
6. BROWN, C. S., and J. E. BAYLOR. "Hay and Pasture Seeding for the Northeast," In *Forages* (Ames: Iowa State Univ. Press, 1973).
7. CHAMBLEE, D. S., and A. E. SPOONER. "Hay and Pasture Seedings for the Humid South," In *Forages* (Ames: Iowa State Univ. Press, 1973).

8. DENNIS, R. E. *Irrigated Pastures,* Univ. Ariz. Coop. Ext. Ser. and Agr. Exp. Sta. Bul. A-49 (1969).

9. HEATH, M. E. "Forages in a Changing World," In *Forages* (Ames: Iowa State Univ. Press, 1973).

10. HEATH, M. E. "Hay and Pasture Seedings for the Central and Lake States," In *Forages* (Ames: Iowa State Univ. Press, 1973).

11. HERBEL, C. H., and A. A. BALTANSPERGER. "Ranges and Pastures of the Southern Great Plains and the Southwest," In *Forages* (Ames: Iowa State Univ. Press, 1973).

12. HINDS, F. C., G. F. CMARIK, and G. E. McKIBBEN. "Fescue for the Cow Herd in Southern Illinois," *Illinois Research,* 16(1):6 (1974).

13. KELLER, W., and L. J. KLEBESADEL. "Hay, Pasture, and Range Seedings for the Intermountain Area and Alaska," In *Forages* (Ames: Iowa State Univ. Press, 1973).

14. LOVE, R. M., and C. M. McKELL. "Range Pastures and Their Improvement," In *Forages* (Ames: Iowa State Univ. Press, 1973).

15. McGINTY, D. D., *et al. Production of Feeder Calves in Intensive Management Systems,* Univ. Ariz. Cattle Feeder's Day, P-16 and P-12 (1969 and 1970).

16. NEWELL, L. C., and R. A. MOORE. "Hay and Pasture Seedings for the Northern Great Plains," In *Forages* (Ames: Iowa State Univ. Press, 1973).

17. OLSEN, F. J., and D. M. ELKINS. "Renovation of Tall Fescue Pasture with Lime-Pelleted Legume Seed," *Agron. Jour.,* 69:871 (1977).

18. RAYMOND, W. F. "The Nutritive Value of Forage Crops," In *Advances in Agronomy,* Vol. 21 (New York: Academic Press, 1969). ed. N. C. Brady.

19. ROHWEDER, D. A., and W. C. THOMPSON. "Permanent Pastures," In *Forages* (Ames: Iowa State Univ. Press, 1973).

20. SMITH, D. *Forage Management* (DuBuque, Iowa: W. C. Brown, 1962).

21. TEMPLETON, W. C., and T. H. TAYLOR. "Some Effects of Nitrogen, Phosphorus, and Potassium Fertilization on Botanical Composition of a Tall-Fescue-White Clover Sward," *Agron. Jour.,* 58:569 (1966).

22. TURELLE, J. W., and W. W. AUSTIN. *Irrigated Pastures for Forage Production and Soil Conservation in the West,* USDA Farmer's Bul. 2230 (1967).

23. Virginia Polytechnic Institute. *Managing Forages for Animal Production,* VPI Res. Div. Bul. 45 (1969).

24. USDA *Agr. Stat.* (1977).

25. U. S. Dept. of Commerce. *1974 Census of Agriculture, United States Summary* (1977).

26. WEDIN, W. F., and A. G. MATCHES. "Cropland Pasture," In *Forages* (Ames: Iowa State Univ. Press, 1973).

CHAPTER 32

HAY AND HAYMAKING

HAY is produced by drying down a green forage to 15 percent moisture or less for storage without heating losses. Nutrients and energy in hay cost less per unit than in any other feed except pasture and corn silage. Hay is the most important winter feed in northern regions, both as to the amount fed and the nutrients it contains. Its chief purpose is to provide energy at a lower cost than concentrates, but hay also is a source of minerals, vitamins, and fiber. About 12 percent of the total feed consumed by all classes of livestock is hay; the amount for dairy cattle is about 25 percent. Since hay contains 40–70 percent complex carbohydrates that require fermentative digestion by bacteria, 96 percent of hay is consumed by ruminants.

Hay Quality

A hay grading system is used in some hay markets. This has as its basis characteristics such as color, odor, leafiness, stem texture, and amount of foreign matter. Unfortunately, grade is not always an accurate predictor of feeding value and animal performance. The principal factors affecting hay quality are:

1. *Forage species used for hay and time during season and growth stage at which harvested.* An early cutting generally has greater digestible dry matter and energy levels than more mature forage. For a combination of field stand maintenance, productivity, palatability, and nutritive value, most legumes are best cut in the late bud to early bloom stage; the grasses are best cut in boot to early head stage. Also, the aftermath growth may be finer-stemmed and freer from weeds than the first cutting.
2. *Leaf content.* Leafiness is a good index of palatability and thus influences intake.
3. *Extent to which it is damaged by weather and handling.* As much as 20–40 percent of the nutritive value may be lost from rain-damaged hay. In addition to weather, the handling procedures, which can influence factors such as leaf loss, are critical in maintaining quality.
4. *Physical form in which it is fed.* The leaf content, texture, and brittleness of hay affect intake and animal performance. The intake of

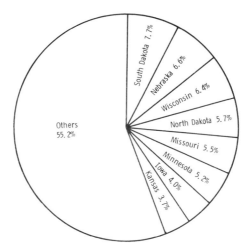

FIGURE 32–1. *Leading states in United States total hay area (acreage).*
Eight Midwestern and Great Plains states have about 45 percent of the
acreage. [SOURCE: Southern Illinois University.]

ground, pelleted hay may be as much as 30 percent greater than other
hay in other forms.

BASIC FORMS. In general, forages are harvested in five basic forms:
(1) loose hay; (2) baled or packaged hay; (3) chopped hay; (4) silage or
haylage; and (5) pellets, cubes, or wafers.

Loose hay. Loose hay, except large, compressed stacks, continues to decline in importance because of the difficulty in handling and storage. In
low-precipitation areas, loose hay may be stored in piles exposed to the
weather or under roofed structures having no walls. Difficulty of handling
loose hay restricts it to feeding near storage areas.

Baled hay. The standard package for hay is baled hay, the most common
form of forage entering marketing channels. About 90 percent of the total
hay crop is baled, with only about 5 percent loose, 3 percent chopped, and
2 percent field cubed.

Pellets, cubes, wafers. Hay may be compressed into pellets, cubes, or wafers. Advantages are: (1) handling can be completely mechanized; (2) high-
bulk density allows economical transportation and storage; (3) hay can
be self-fed; (4) feeding losses are reduced; and (5) leaf and stem portions
cannot be separated by the animals. Pellets are made from ground, dehydrated hay. Dehydration reduces feed losses from poor drying conditions.
Additional protein is realized as well as carotene, B-complex vitamins,
and minerals. Alfalfa is the main crop dehydrated. Ground dehydrated
forage is bulky, dusty, and difficult to handle. Pelleting can overcome these
deficiencies; 90 percent of dehydrated alfalfa is pelleted.

Of cubes and wafers, the cube is the most common and important. It
is about 1¼ inches (3.2 cm) square and 1–2 inches (2.5–5.0 cm) long. Com-

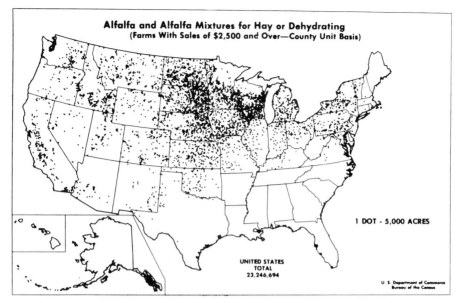

FIGURE 32–2. *Acreage of alfalfa and alfalfa mixtures for hay or dehydrating in the United States, shown geographically.* [SOURCE: U.S. Department of Commerce.]

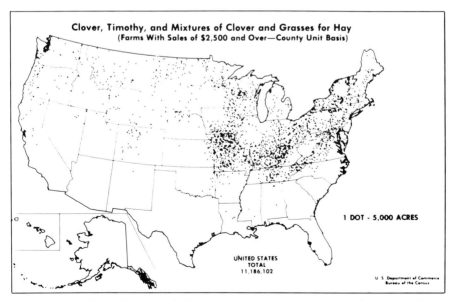

FIGURE 32–3. *Acreage of clover, timothy, and mixtures of clover and grasses for hay in the United States, shown geographically.* [SOURCE: U.S. Department of Commerce.]

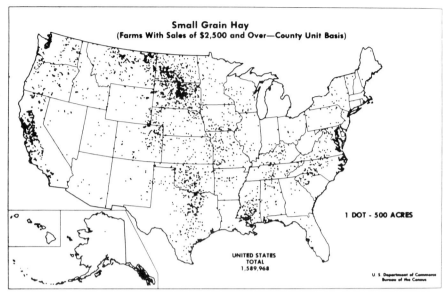

FIGURE 32-4. *Acreage of small grain hay in the United States, shown geographically.* [SOURCE: U.S. Department of Commerce.]

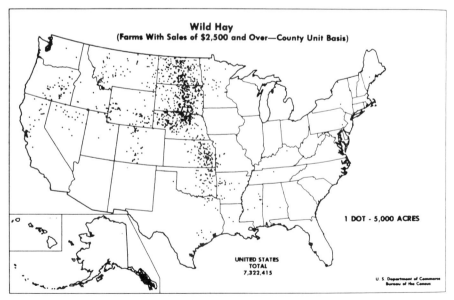

FIGURE 32-5. *Acreage of wild hay in the United States, shown geographically.* [SOURCE: U.S. Department of Commerce.]

TABLE 32–1. Comparison of Chopped Alfalfa Hay and Cubed Alfalfa for Growth of Hereford Heifer and Steer Calves

Form of Hay	Average Daily Gain	Average Daily Feed	Average Daily Feed Waste	Feed Efficiency
		Heifers		
Chopped hay	0.62	14.1	0.6	22.7
Cubes	1.40	17.4	0.7	12.4
		Steers		
Chopped hay	1.08	11.9	0.1	11.0
Cubes	1.84	17.2	0.4	9.3

Source: Butcher, J. E., and N. J. Stenquist, "Cubed Hay Gives Better Gains," Utah Science, 33(2):57 (1972).

mercial production of alfalfa cubes started in the Sacramento Valley of California in the early 1960s as an attempt to package hay in units small enough to be readily handled in bulk. Cubes provide less waste, higher hay consumption, reduced storage requirements, and better control of feeding. Practical cubing is limited to geographic areas where hay may be field cured to 12 percent moisture, primarily arid hay areas in New Mexico, Arizona, and California. Despite this limited geographic adaptation, this method is expected to increase drastically in areas of the West and Southwest. Cubing increases density to as much as 55 pounds/ft³. Machines for cubing may be mobile or stationary.

Chopped hay. Butcher and Stenquist at Utah State University evaluated heifer and steer calf performance with chopped alfalfa hay as compared to cubed hay of the same quality (Table 32–1). Cubed alfalfa hay produced significantly better gains, about three-fourths pound (0.34 kg) per head per day, than did chopped hay. A portion of this better animal performance could be attributed to a greater daily intake of cubes than chopped hay.

Hay Harvesting and Processing

The hazard in loss of palatability and nutritive value resulting from unfavorable weather is great. It is greater for legume forages than for the grasses. The leaves of legumes are particularly high in proteins and carotene, both of which are readily lost by leaching. This loss is greatly augmented when the leaves shatter in the field in the curing process.

Leaves comprise 50 percent of alfalfa plants, when in good condition for cutting. These leaves contain 70 percent of the protein and 90 percent of the carotene of the plant. The relatively thin, fine-textured leaves give up their moisture more readily than the stems and are more subject to leaching and shattering. If half the leaves shatter during the haymaking process, this loss can be equivalent to 1,000 pounds (454 kg) or more of protein supplement, depending on the yield level.

Hay has been much slower to mechanize than other field crops; however, rapid growth in mechanization of harvesting and handling has occurred in recent years. Hay harvesting and processing is complicated because of the following factors: (1) great bulk and weight and difficulty of handling and storage; (2) easy loss of palatability and nutritive value in handling and curing; (3) initial high moisture content which must be reduced to safe storage levels; (4) the short harvest period; and (5) sloping or rolling land conditions which may be unfavorable to mechanical operations. Hay quality and ease of making hay have been improved greatly by relatively new machines for cutting, conditioning, windrowing, baling, and handling hay.

Mowing may be done with a cutter-bar mower, but the trend is toward a greater use of equipment that can mow, condition, and windrow. Conditioners pass freshly cut hay between smooth or corrugated rollers under pressure to crush or crack the stem so that rate of stem-drying will be comparable to that of leaves. Thus, conditioning helps to preserve leaves. Direct windrowing increases field curing time as compared to partial curing in swath before raking into the windrow. However, conditioning reduces drying time and thus, the two methods are comparable. Mower-condi-

FIGURE 32–6. *The trend is toward equipment that can mow, condition, and windrow forage for hay in one operation. Conditioning, crushing or cracking the stem speeds stem-drying and helps to preserve leaves. Mower-conditioners and windrowers may be pull-type machines or self-propelled, as the one shown.* [SOURCE: Deere and Company.)

tioners and windrowers may be either pull-type machines or self-propelled. If the hay is not direct-windrowed, it must be raked into windrows. A reel-type side delivery rake is more common than wheel-type side delivery or dump-type rakes.

Baling

Bales may be rectangular packages of variable sizes, most commonly about 36 inches (91 cm) long and weighing 70–80 pounds (32–36 kg). Equipment is available to make small round bales 14–22 inches (36–56 cm) in diameter and 36 inches (91 cm) long, and large bales or stacks weighing a few hundred pounds up to several tons. The large round bales and stacks are a relatively recent development made possible by great strides in mechanization—fully mechanized "one-man handling systems." These systems are available also for small rectangular bales, by which one man can bale, handle, and transport hay to the storage area.

Small rectangular bales may be handled in several ways: (1) bales dropped to ground and later picked up by hand; (2) bales transported from baler down a chute where they are stacked by hand onto a trailer or wagon;

FIGURE 32–7. *Small rectangular bales may be handled in several ways. In this system, a thrower on the baler throws bales into a trailing wagon.* [SOURCE: International Harvester Company.]

FIGURE 32–8. *Another system for handling hay is the use of automatic bale wagons, which pick up, stack, and transport conventional bales. Bale wagons may be tractor-drawn or self-propelled, as the one shown.* [SOURCE: Sperry New Holland.]

(3) a thrower on the baler throws bales into a trailing wagon; (4) bales are accumulated in groups of 4 to 8 as discharged and handled by special attachments or equipment; and (5) bale loaders pick up bales from the ground and load a truck or trailer. Automatic bale wagons, for picking up and stacking conventional bales, may be tractor-drawn or self-propelled. Some have the capacity to stack and haul more than 100 bales in a load.

Small round bales or large round bales or stacks remain in the field where they fall from the baler, or may be moved and stored on the edge of a field or in the cattle feeding area.

Chopped hay may be stored at lower moisture than any other hay. Types of field choppers are (1) radial-knife or flywheel, (2) reel-type, and (3) flail-type. Chopped hay is blown into a forage wagon following, and then conveyed to storage with a blower.

Large Hay Packages

The trend in haymaking systems in parts of the Midwest is away from conventional bales and toward stacks or large, round bales. Large, round bales most often weigh 1,000–1,500 pounds (454–680 kg) but may be as much as 3,000 pounds (1361 kg). Most stacks weigh 1, 3, or 6 tons (0.9, 2.7, or 5.4 MT). Various mechanized equipment is available so that one person can transport and feed large packages. The best feeding system

A

FIGURE 32–9. *Large hay packaging. The trend in hay-making in some parts of the United States is toward stacks or large, round bales. (A) Equipment for making large, round bales. Most bales weigh 1000 to 1500 pounds (454 to 680 kg) but may weigh as much as 3000 pounds (1361 kg).* [SOURCE: International Harvester Company.] *(B) Equipment for making large stacks. Most stacks weigh 1, 3, or 6 tons (0.9, 2.7, or 5.4 MT).* [SOURCE: Southern Illinois University.]

B

for large bales involves placement on the ground in a row, with individual bales about one foot (30 cm) apart to allow water to drain away and allow good air movement on all sides. Some cattlemen use temporary electric fences around the group of bales and feeding racks for individual bales to prevent excess feed waste and trampling.

Studies have shown that large hay packages stored outside will keep well. Weathering losses may be as low as 5 percent, depending primarily on the density and shape of the hay package. Table 32–2 shows results of studies at Purdue University on weathering of various hay packages. Weathering losses were not too serious, being confined to the top 2–4 inches (5–10 cm) in large, round bales and the top 4–6 inches (10–15 cm) in looser stacks. About 20 percent of the hay in small, round bales was weather damaged, while about 11–13 percent was damaged in large bales or stacks. The TDN content of the weathered portion decreased about 17 percent for grass hay and 23 percent for alfalfa hay.

Wilson *et al.* at Penn State University measured average storage losses of 12 percent for large, round bales and 17 percent for large stacks.

Haying System Comparisons

With several haying system options available to producers, labor and cost comparisons become important. Petrtiz at Purdue University estimated the total costs, including annual fixed costs and operating expenses, per

TABLE 32–2. Weathering Losses of Grass and Legume Hay Stored Outside in Various Package Forms, 1972–1973

Type of Package	Portion of Each Package Weathered	Total Digestible Nutrients Unweathered Core	Weathered Outside	TDN Loss Due to Outside Storage
Grass hay (1972)				
Average of 3				
big packages	16.4	54.1	38.1	9.87
Small round bale	20.6	55.3	33.0	16.87
Grass hay (1973)				
Average of 3				
big packages	11.1	58.9	41.7	7.57
Small round bale	20.2	60.0	42.6	13.45
Alfalfa hay (1973)				
Average of 3				
big packages	12.8	56.7	33.3	11.42
Overall average				
big packages				
(1972, 1973)	13.47			9.62

Source: Smith, W. H., *et al.* *Suggestions for the Storage and Feeding of Big-Package Hay,* Purdue Univ. Coop. Ext. Serv. Pub. ID-97 (1974).

A

B

FIGURE 32–10. *(A) Cubes provide less waste, higher hay consumption, reduced storage requirements, and better control of feeding. The cube is about 1¼ inches (3.2 cm) square and 1 to 2 inches (2.5 to 5.0 cm) long.* [Photograph with permission of University of Arizona Agricultural Experiment Station.] *(B) Handling cubed hay, which can be completely mechanized.* [SOURCE: J. I. Case.]

TABLE 32–3. Estimated Total Costs/Ton for Varying Tonnages Harvested Per Year, for Specific Hay Harvesting Systems

Type of Harvesting System	Total Costs/Ton Harvested Per Year				
	50 T	100 T	200 T	300 T	500 T
Hydraulic stack wagon (Stack wagon, stack-mover, special hay feeder)	21.84	11.67	6.58	4.89	3.53
Large round bale system (Baler, loader, trailer, special feeder)	23.44	12.47	6.98	5.16	3.69
Standard bale system (Baler, wagons, elevator)	21.88	13.44	9.22	7.81	6.69
Bale accumulator system (Baler, bale accumulator, special loader and fork, wagon)	30.50	17.00	10.25	8.00	6.20
Bale thrower system (Baler, bale thrower, wagons, elevators)	23.92	13.71	8.61	6.90	5.54
Automatic bale wagon system (Baler, tractor-drawn automatic bale-loading wagon)	28.19	15.37	8.96	6.82	5.11

Source: Petrtiz, D. "Economics of Big-Package Haymaking," Paper presented at Purdue Univ. Farm Science Days Program, 1973.

TABLE 32–4. Estimated Harvesting and Feeding Costs/Ton for Three Systems of Handling Hay

Harvesting System	Cost/Ton Harvested Per Year			
	250 T	500 T	1,000 T	2,000 T
System 1 (Conventional baler, automatic bale wagon for hauling to storage, total hay costs)	$29.91	$23.68	$20.56	$19.01
System 2 (Large round baler, front-end loader and truck for hauling to storage, no feeding panels)	$23.56	$20.17	$18.48	$17.64
System 3 (Same as System 2, except panels around bales to control hay waste)	$25.53	$22.14	$20.45	$19.62

Source: Smith, L. A., *et al. Hay in Round and Conventional Bale Systems,* Auburn Univ. Agr. Exp. Sta. Circ. 216 (1975).

TABLE 32–5. Calculated Labor and Energy Distribution for Haymaking System Producing 15–35 Tons/Day

Haymaking System	Average Time (Min/Ton)	Average Fuel Equiv. (Gal/Ton)	Time (%)		
			Mow/Rake	Baler	Wagons
Conventional bales (transported to farmstead and unloaded)	60	0.41	14	15	52
Large round bales (dropped, road-sided with tractor lift)	18	0.55	44	22	34
One-ton stacks (road-sided with tractor-mounted mover)	20	0.41	30	30	40

Source: Wilson, L. L., and W. L. Kjelgaard. "Large Hay Packages Compared with Bales," *Penn. State Univ. Sci. Agric.,* 24(4):2 (1977).

ton of hay for various hay harvesting systems (Table 32–3). At low tonnage, all systems had high costs per ton. At 50 tons (45 MT) per year, the hydraulic stack system and the standard bale system had the lowest costs per ton. At large tonnages of 300 and 500 tons (272 and 454 MT) of hay harvested per year, the large package systems had the lowest costs per ton, while the standard bale system had the highest cost.

L. A. Smith *et al.* at Auburn University estimated the harvesting and feeding costs per ton of hay with three handling systems (Table 32–4). Costs were less for a large, round bale system without feeding panels than for the conventional system, which had the highest costs per ton in these tests. These studies did not consider losses from feeding or by spoilage, or hay utilization.

Researchers at the Pennsylvania Station calculated labor and energy distribution for three haymaking systems involving 15–35 tons (14–32 MT) per day (Table 32–5). They concluded that neither round bales nor stack systems were likely to reduce fuel needs in haymaking. The initial time needed for baling and transporting was less for large bales or stacks, but when all elements of handling were considered, the time needs may be similar to those of conventional bale systems. However, time-consuming operations with large bale operations are spread out over a longer period, which helps distribute labor.

Marketing

Much of the hay is fed on the farm where produced or is fed nearby. Farmers tend to adjust their livestock numbers to expected production or to nearby supplies. This leaves them vulnearble to drought, unusually

wet weather, or insect invasions. Under these conditions costs soar, livestock production suffers, and some herds are sacrificed. Some areas, however, are usually self-sufficient; others export or import hay. The market
for hay is not as well defined as are markets for other commodities, such
as wheat, cotton, and tobacco, which are generally regarded as cash crops.
Areas are limited where farmers regularly grow hay for sale.

Truckers from the South who transport materials to the Midwest often
return with a load of hay from Wisconsin or other states rather than return
with an empty trailer. In some communities hay auction systems have
been organized. Farmers bring salable hay to a central location where
an auction is held periodically. This gives hay producers additional marketing possibilities, and allows buyers to inspect and compare hay before a
transaction. Some hay producers offer contracts for a portion of their expected hay production at a set or guaranteed price for the year, and take
orders on the remaining hay throughout the year.

The palatability and nutritive value of hay vary more than for any other
field crop harvested and marketed. This fact makes grades and descriptions
of hay bought and sold, with much of it never seen by purchasers, of particular importance. The grading of hay requires very careful and uniform
attention to many details, procedures, and requirements, all of which are
fully described in a USDA handbook on hay and straw standards.

Unfortunately, hay grades do not always correspond directly with the
value of hay as measured by feeding value and animal performance. Perhaps the grading standards should be revised to include consideration of
protein content and other nutritive or chemical factors.

Developments and Future

Developments that have facilitated haymaking and that have been discussed previously include an increase in mechanization, such as development of "one-man haying systems," and the development of equipment
for making and handling large round bales or stacks. Other developments
of great importance are the following:

1. *High-moisture storage.* This reduces the elapsed time and lessens the
   period of exposure to weather hazards. This has been made possible
   by the use of chemical preservatives. Preservatives such as propionic
   acid can be applied to wet hay just prior to baling by means of a
   spray attachment on the baler. Such chemicals inhibit molding and
   reduce dry matter losses during storage.
2. *Artificial drying.* This refers to forced circulation of heated or unheated air through loose, baled, or chopped hay in storage. Hay may
   be partially field-cured, to 35–45 percent moisture, moved to storage,
   and dried. This has the advantage of maintaining quality through
   greater leaf and color retention. A limited number of producers are
   using solar-heated air systems to dry either small rectangular bales

or large round bales which may have an initial moisture content of 40–45 percent.

The future hay bale may be a denser bale with reduced bulk for more economic handling, storing, and feeding, and to make shipping or transport easier. In order to maximize the potential for more economical handling, storing, and feeding, it will be necessary to use high-yielding crops, harvested at the optimum quality, consolidated into high-density units adaptable to automated mechanical handling. At present, the use of field-going cube machines in the more arid areas is the most promising. For the more humid areas it appears that partial field curing, followed by drying and consolidating in stationary units, offers the best possibility.

REFERENCES AND SUGGESTED READINGS

1. Alfalfa Cubing and Wafering Conference. Proceedings Calif. Grain and Feed Assn. and Univ. of Calif., Davis, Calif. (June 1966).
2. ANGUS, R. C. *Arizona Hay, Price-Quality Relationships,* Ariz. Agr. Exp. Sta. Tech. Bul. 157 (1963).
3. ANONYMOUS. "Hay Cubes Come of Age," *Farm Quarterly,* p. 48 (1971).
4. BARNES, K. K. "Mechanization of Forage Harvesting and Storage," In *Forages* (Ames: Iowa State Univ. Press, 1973).
5. BOWERS, W., and A. R. RIDER. "Hay Handling and Harvesting," *Agric. Engineering,* 55(8):12 (1974).
6. BROOKS, O. L., *et al. Pelleted Coastal Bermudagrass—Comprehensive Investigations,* Ga. Agr. Exp. Sta. Res. Bul. 27 (1968).
7. BUTCHER, J. E., and N. J. STENQUIST. "Cubed Hay Gives Better Gains," *Utah Science,* 33(2):57 (1972).
8. BUTLER, J. L. "Economic Concepts in Reducing Bulk of Forage for Improved Handling, Storage, and Feeding," In *Forage for the Future,* Proc. Res. Ind. Conf. Sponsored by Amer. Forages and Grassland Council, Chicago (1968).
9. CHRISMAN, J., G. O. KOHLER, and E. M. BICKOFF. "Dehydration of Forage Crops," In *Forages* (Ames: Iowa State Univ. Press, 1973).
10. DOBIE, J. B., and R. S. CURLEY. *Hay Cube Storage and Feeding,* Calif. Agr. Exp. Sta. Ext. Serv. Circ. 550 (1969).
11. *Handbook of Official Hay and Straw Standards,* USDA Production and Marketing Administration.
12. KELLOGG, D. W. *Effects of Cubed Alfalfa Hay on Milk-Fat Percentage and Milk Production,* N.M. State Univ. Agr. Exp. Sta. Res. Rep. 169 (1970).
13. KNAPP, W. R., D. A. HOLT, and V. L. LECHTENBERG. "Propionic Acid as a Hay Preservative," *Agron. Jour.,* 68:120 (1976).
14. LECHTENBERG, V. L., *et al.* "Big Package Hay Care," *Crops and Soils,* 28(5):12 (1976).
15. PARSONS, S. D. *Big-Package Haymaking, Packaging and Handling Equipment Alternatives,* Purdue Univ. Coop. Ext. Serv. Publ. AE-85 (1973).
16. PETRTIZ, D. "Economics of Big-Package Haymaking," Paper presented at Purdue Farm Science Days Program, 1973.
17. REID, J. T. "Quality Hay," In *Forages* (Ames: Iowa State Univ. Press, 1973).

18. RENOLL, E. S., *et al. Stack and Bale Systems for Hay Handling and Feeding,* Auburn Univ. Agr. Exp. Sta. Bul. 455 (1974).

19. SMITH, L. A., *et al. Hay in Round and Conventional Bale Systems,* Auburn Univ. Agr. Exp. Sta. Circ. 216 (1975).

20. SMITH, W. H., *et al. Suggestions for the Storage and Feeding of Big-Package Hay,* Purdue Univ. Coop. Ext. Serv. Publ. ID-97 (1974).

21. USDA. Agr Stat. (1977).

22. USDA-ERS, *Alfalfa Dehydration, Separation, and Storage,* Mktg. Res. Rep. 881, (1970).

23. WILSON, L. L., and W. L. KJELGAARD. "Large Hay Packages Compared With Bales," Penn. State Univ. *Science Agric.,* 24(4):2 (1977).

CHAPTER 33

SILAGE, GREEN CHOP, AND SUCCULAGE

Silage

SILAGE is forage with a fairly high moisture content that has been preserved, usually by fermentation in the absence of free air. *Ensile*, technically speaking, is a verb meaning "to prepare for and place in a silo." A silo is the container in which silage is stored. The terms *ensilage* and *silage* have been applied rather loosely and in common usage have the same meaning.

Crops Used for Silage

Corn and sorghum are the crops that have been used more extensively for silage than any other. Silage that contains considerable grain is known as high-energy silage. Grasses and legumes, used for low-energy silage, have become increasingly important in recent years.

Any crop that provides high-quality forage for livestock feeding in the fresh, green, succulent state can be stored satisfactorily in the silo. Converting forage into silage does not change its nutritive value. Low-grade forage is just as low grade in the silo as out of it. Ensiling may make coarse, woody plants and parts of plants more palatable, however, with the result that a greater percentage will be consumed (as, for example, cornstalks).

The most significant silage developments in recent years have been (1) an increase in the use of oxygen-free or air-tight silos to minimize losses; and (2) the increased use of haylage, low-moisture silage made from forage crops.

High-Energy Silage

CORN SILAGE. Corn has been used much more extensively in the making of silage than has any other crop. In many ways it is an ideal silage crop. It is palatable and nutritious, produces high tonnage, and has the right moisture and carbohydrate content at harvest for satisfactory results. Of the total U.S. corn crop, about 13 percent, or 11 million acres (5 million ha), goes into the silo annually. In some of the more northern areas, especially where dairying is important, more than 50 percent of the corn crop is stored as silage.

Corn surpasses most other forages in acre production of dry matter and digestible nutrients in areas to which it is adapted. Although the U.S. average yield of corn silage is just over 10 tons/acre (22 MT/ha), under the

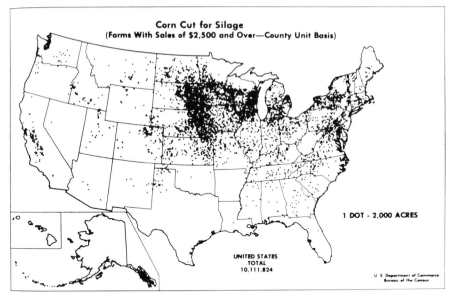

FIGURE 33–1. *Acreage of corn cut for silage in the United States, shown geographically.* [SOURCE: U.S. Department of Commerce.]

most favorable conditions corn will produce 25–30 tons or more of green forage per acre (56–67 MT/ha). Larger yields of total dry matter and digestible nutrients sometimes are obtained when the crop is planted at a higher population than when grown for grain.

Studies at the Ohio Station indicate that the best corn silage is obtained when harvested between the dent and glaze stages of kernel maturity. At this point the forage contains about 28–34 percent moisture and the ears about 50 percent moisture. Advantages of harvesting at this stage include greater yield, maximum dry matter digestibility, high crude protein digestibility, and higher intake.

Earlier cutting sacrifices acre production and may result in a poor quality of silage. Later cutting may give a somewhat greater acre production, but if the crop is so dry that it does not pack sufficiently well to exclude air, molding and loss are likely to occur. Corn that has passed the best stage for silage is improved by adding water as the crop goes into the silo.

Corn silage is low in crude protein, and the silage itself or the animal's diet must be supplemented to provide a balanced ration. Urea, a nonprotein nitrogen source, is used widely because it provides nitrogen at a lower cost. Sometimes ground limestone is added at ensiling time to increase the content of organic acids such as lactic acid.

The general feeding of corn silage has materially lowered the cost of producing both milk and beef over a large part of the country. It gives especially good results when used with legume hay. Well-matured corn silage averages about 35 percent as much digestible nutrients as good-qual-

FIGURE 33–2. *Cutting corn silage. Self-propelled forage chopper and a forage wagon.* [SOURCE: Sperry New Holland.]

ity hay. For fattening cattle and sheep such silage usually has been found to be worth half as much per ton as good legume or mixed hay.

SORGHUM SILAGE. Sorghums are second in importance to corn as a silage crop. Sorghum is similar to corn in its suitability for silage making. Prior to 1945 more than half the sorghum acreage was harvested as forage or was put into the silo. Since 1945, with an increased proportion of the crop planted to combine-type varieties, the acreage harvested as grain has increased significantly, with a proportionate decrease in the forage acreage and some decrease in the acreage stored as silage. Presently just over 4 percent of the total sorghum acreage, or about 0.8 million acres (0.3 million ha), goes into silos.

Yields of sorghum silage, both in dry matter tonnage and total digestible energy, are less than corn silage. Moisture stress conditions in some areas of the United States, however, may favor sorghum silage. The feeding value usually is similar to forage sorghums but somewhat less than that of well-eared corn silage, because sorghum silage is not as rich in grain. Moreover, more of the sorghum grain may pass through cattle undigested. The silage appears to be slightly less palatable than corn silage.

Grain sorghums are best harvested for silage when the seeds are in the

FIGURE 33–3. *Dairy cattle on a Maryland farm being fed silage from a tractor-pulled power-driver feed wagon. Silage is the backbone of the feeding program on many dairy farms.* [SOURCE: USDA.]

late milk to medium dough stage and the oldest seeds are hard dough, with 35–45 percent moisture. The best stage for forage sorghum is medium dough, with 25–30 percent moisture.

Interplanting soybeans with corn or sorghum, and interplanting sorghum with corn for silage has been suggested as a means of improving production and improving quality of silage. Results with interplanting have not often been favorable.

Grass-Legume Silage

The use of grasses and legumes for making silage did not gain much acceptance until the 1930s. Since that time the tonnage of grasses and legumes ensiled has increased steadily to the present level of about 10 million tons (9 million MT). The main contributing factor to its increasing use probably has been an increase in gastight silos, which has enabled suitable storage of haylage, a low-moisture (40–50 percent) chopped hay. Haylage is often advantageous over making hay because with proper procedures more nutrients are preserved and less field and storage losses occur. In addition to the low-moisture silage, forage grasses or legumes may be prepared for ensiling by (1) direct cutting, or (2) wilting.

The wilting procedure for silage making was developed in the 1940s. When wilting crops for silage, the moisture is reduced to 60–70 percent by partial drying in the swath or windrow for 1–4 hours. Direct cut crops are taken directly to the silo upon cutting. Advantages of wilting include

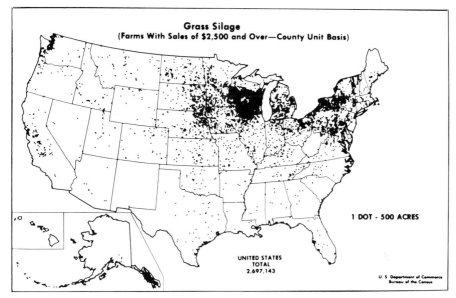

FIGURE 33–4. *Acreage of grass cut for silage in the United States, shown geographically.* [SOURCE: U.S. Department of Commerce.]

less weight to store and feed, faster harvesting, reduced seepage losses, no expense of added preservatives, pleasant odor, and high dry-matter intake by animals. Disadvantages include possibility of weather damage is increased, more exacting management is required and an additional field

FIGURE 33–5. *The use of grasses and legumes for making silage has increased steadily since the 1930s. Haylage, a low-moisture chopped hay, has contributed much to this increase.* [SOURCE: Deere and Company.]

operation is required. Direct-cut advantages include one operation from field to silo, less chance of field losses, and easier packing.

Disadvantages are expense of preservatives, objectionable odors, additional water to be handled and fed, increased seepage, and low dry-matter intake by animals.

The best stages to harvest forages for silage are orchardgrass, tall fescue, and small grains—boot stage just before heads emerge; smooth bromegrass, timothy, and reed canarygrass—early to medium head; alfalfa and sweet clover—late bud to early bloom; and birdsfoot trefoil, red clover, and alsike clover—one-fourth to one-half bloom.

The Ensiling Process

The cells of fresh-cut forage continue to respire, using up the oxygen in the air and giving off carbon dioxide. This is the aerobic phase of silage fermentation. The oxygen is used up within a few hours. This prevents the development of microorganisms responsible for molds which do not grow in the absence of oxygen. Under anaerobic conditions, acid-forming bacteria multiply rapidly. They depend upon readily available carbohydrates in the crop being ensiled, and thus break down the sugars, producing organic acids such as lactic acid (the acid of sour milk) and some acetic, propionic, formic, and succinic acids. The presence of these acids prevents the growth of undesirable bacteria which cause rotting or putrefaction.

Concentration of lactic acid in the silage may reach 8–9 percent of the dry matter under favorable conditions. When this degree of acidity has developed, a pH of 4.5 or below, the bacteria themselves are killed, checking further fermentation and "pickling" the crop. If additional oxygen-carrying air does not get into the mass of silage, it will keep for long periods with little or no change.

The respiration and fermentation processes result in a considerable increase in temperature, but if the material is well compacted so that little air is present, the temperatures seldom get as high as 100°F (38°C).

SILAGE LOSSES. In the ensiling process a certain part of the nutrients are oxidized into carbon dioxide and water and are therefore lost. Some losses occur in the field, depending on the degree of drying. Losses that may occur in the silo are (1) surface spoilage, (2) seepage, and (3) gaseous loss. In a well-built conventional tower silo the silo losses may be less than 15 percent, depending on initial moisture content. In trench and stack silos the loss is considerably greater because of the greater surface exposure to oxygen-carrying air. Gas-tight tower silos will reduce surface spoilage to zero and often will decrease total silo losses to less than 10 percent.

Immature forages with a high water content will suffer some loss from seepage. These losses usually increase as moisture content of the ensiled crop increases. Seepage losses are minimal when the moisture content of silage is less than 70 percent. Gaseous loss of dry matter may result from such factors as excessive air in the silo or failure of pH to decline rapidly.

Estimates by Shepherd and his co-workers of the loss from grass silage storaged in different types of silos and with different moisture contents are shown in Table 33–1.

ESSENTIALS FOR GOOD SILAGE. High-quality silage can result if certain practices are observed: (1) Harvest a crop of good quality at the proper stage of growth; (2) field dry to 65–70 percent moisture before fine-chopping; (3) exclude oxygen-carrying air by filling the silo rapidly and compacting thoroughly; and (4) ensure that the crop being ensiled has adequate carbohydrates for the activity of acid-forming bacteria, the acids preventing the growth of undesirable bacteria that cause spoilage.

Good-quality silage will have these characteristics: (1) low pH (4.2–4.5); high content of lactic acid but little if any butyric acid; (2) freedom from molds, mustiness, or other objectionable odors; (3) green color, not brown or black; and (4) firm texture, no sliminess. These characteristics usually are associated with high acceptance by livestock.

Types of Silos

Silos are of two types, upright or horizontal. High-quality silage can be produced in either type if air is properly excluded.

UPRIGHT OR TOWER SILOS. Upright silos are constructed from a variety of materials, including wood, concrete, tile, steel, and glass-coated steel. Many are lined with plastic sheeting or asphalt paper to give added protec-

TABLE 33–1. Estimates of Minimum Dry-Matter Losses of Grass Silage in Different Types of Silos and at Different Moisture Levels

| Kind of Silo | Moisture as Stored (%) | Dry-Matter Losses (%) | | | | | |
		Surface Spoilage	Fermenta- tion	Seepage	Total Silo	Field	From Cutting to Feeding
Conventional tower	80	3	9	7	19	2	21
	70	4	7	1	12	2	14
	60	4	9	0	13	6	19
Gas-tight tower	80	0	9	7	16	2	18
	70	0	7	1	8	2	10
	60	0	5	0	5	6	11
Trench	80	6	10	7	23	2	25
	70	10	10	1	21	2	23
Stack	80	12	11	7	30	2	32
	70	20	12	1	33	2	35

tion against air leaks and are designed or reinforced to withstand high pressures.

Conventional tower silos are the most popular type for silage storage in the United States. They do not always exclude all oxygen. Gas-tight silos with bottom unloaders have some advantages over the conventional tower silo. Among these are no top spoilage and lower overall dry matter losses, greater flexibility and ease in refilling and emptying, and possibility of ensiling lower-moisture material without excessive spoilage. Feeding from upright silos can be made completely automatic by the use of mechanical unloaders, conveyors, and bunks.

Snow fence with a lining of reinforced waterproof paper or of plastic or heavy galvanized wire fence with such a lining makes a useful temporary silo. Additional labor and care are required to construct and pack this temporary stack properly. Height of the silo should not exceed the diameter by more than 4 feet (1.2 m).

HORIZONTAL SILOS. Trenches, bunkers, and stacks are inexpensive but require greater skill in filling and sealing than upright silos because of the large area of forage susceptible to damage by surface exposure. Thorough packing is essential; use of plastic covers can help to exclude air and water and thus reduce losses in the silo. Horizontal silos can be used for self-feeding by installing a portable feed gate at one end. Sidewalls of trench silos may be lined with concrete, wood, or masonry blocks. Bunker

FIGURE 33–6. *One of the most significant silage developments in recent years has been an increase in the use of oxygen-free or airtight silos to minimize losses. Airtight silos, such as the ones shown here, have enabled suitable storage of haylage, low-moisture silage made from forage crops.* [SOURCE: Southern Illinois University.]

silos are built with most of the floor at or above ground level; the sidewalls, concrete or wood, require support. Stack silos that have no sidewalls or walls require no silo construction at all. This is the simplest type of silage storage, but also is the least efficient because of the large surface area exposed. Sealing the silage in a stack depends entirely on the surface sealing material after the stack is built.

Silage Additives

Silage preservatives may be needed if a crop with over 70 percent moisture is ensiled. Normally they are not needed with material with less than 70 percent moisture or with crops such as corn or sorghum that have high levels of readily available carbohydrates. Additives may be either (1) grains or feeds, such as ground corn or wheat, or molasses; or (2) chemical additives, such as sodium metabisulfite, lactic acid, or formic acid. Preservatives should not be confused with nutrient additives that improve the feed value.

GROUND DRY CONCENTRATES. Ground dry concentrates, including such conditioners as corn grain and cob meal, ground corn, oats, or barley, improve silage fermentation by (1) absorbing excess moisture, (2) modifying the energy content of the mixture, and (3) providing available carbohydrates for desirable bacterial growth. From 15 to 20 percent of the nutritive value of the concentrates may be lost in the fermentation process. Dried-pulp products, as beet or citrus pulp, may be substituted for ground grains. The feedstuffs or dried-pulp products are added at the rate of 100–300 pounds/ton (41–123 kg/MT) of fresh storage.

MOLASSES. Molasses as a silage conditioner provides readily available sugar for the growth of bacteria and makes the silage slightly more palatable but it increases seepage losses. From 25 to 50 percent or more of the nutritive value of molasses may be lost through fermentation and seepage. Molasses is added at the rate of 40–100 pounds/ton (17–41 kg/MT) of fresh forage.

CHEMICAL PRESERVATIVES. Use of many of the chemical preservatives has declined in recent years. They are used to produce desirable fermentation by lowering pH or by decreasing growth of bacteria that produce undesirable changes. Sodium metabisulfite is typical of this type of preservative. Lactic acid and formic acid have been added to forage being ensiled with inconsistent and variable results.

CORN SILAGE ADDITIVES. Common non-protein nitrogen sources that may be added to a crop being ensiled include urea and biuret. Anhydrous ammonia may be added when mixed with other materials such as molasses and minerals. When added to corn at the time of ensiling at a rate of 10 pounds/ton (4 kg/MT), urea increases the crude protein equivalent of the silage at a very low cost. It also increases concentrations of lactic and acetic acids. Urea can substitute for oilseed proteins to a large degree and can provide most of the supplemental crude protein needs of cattle being fed corn silage.

Ground limestone has been added to corn silage at a rate of 10–20 pounds/ton (4–8 kg/MT) to increase the content of organic acids, particularly lactic acid. Addition of this material has not resulted in consistently better cattle gains or feed efficiency.

Green Chop

Forages cut green and fed to livestock while in the fresh condition are known as green chop, soiling, or zero grazing crops. In many dairy regions of the United States, green chop is the major feed source for dairy cows during at least five months of the year. Silage or hay take the place of green chop when pastures are not producing. Green chop is used to some extent for sheep but they generally utilize it less effectively than cattle.

Where cattle are allowed to graze selectively, the forage grazed from a field is often of higher quality than green chop. The results are influenced, however, by factors such as plant maturity and grazing vs. chopping height. Milk production per cow per day is generally about the same with green chop, rotational grazing, and strip grazing. Researchers in Wisconsin found that milk per dairy cow did not differ significantly for green chop as compared to strip grazing and hay. Green chop requires more labor and machinery investment than grazing systems, but this may be cancelled by the larger amount of energy expended in grazing.

High-producing grass–legume crops, such as alfalfa and brome, are excellent for green chop. Summer annual species, such as sudangrass, forage sorghums, and millets, also work well in this utilization system.

Succulage

In parts of Europe root crops have been grown extensively in the past to provide livestock with succulence. Such crops have never been used widely in the United States because of the large amount of labor involved as compared with the requirements for silage feeding. Moreover, only limited areas in the United States are well suited to the growing of root crops. Only in areas where the summers are too cool or the season too short for corn or on farms where the animals are too few in number to use silage economically does the growing of root crops appear to be justified.

RAPE. A different type of succulage is provided by rape, which is especially valuable when pastured by hogs. It may be seeded alone or in a mixture with oats for grazing. Rape can be seeded in early spring in the North or during the fall in the South at the rate of 6–7 pounds/acre (6.7–7.8 kg/ha). It is ready to graze 8–10 weeks later. Its nutritional value compares favorably to that of alfalfa and other legumes. 'Dwarf Essex' is the most common rape cultivar.

REFERENCES AND SUGGESTED READINGS

1. AGRI-FIELDMAN. *Research Report on Haylage,* Arlington Heights, Ill. (1975).
2. Alabama Coop. Ext. Serv. *Silage Production, Storage, and Utilization Meeting* February 1969).
3. ANONYMOUS. Hay-Crop Silage," *Successful Farming,* p. 47 (July 1969).
4. BLASER, R. E., D. D. WOLF, and H. T. BRYANT. "Systems of Grazing Management," *Forages* (Ames: Iowa State Univ. Press, 1973).
5. BRYANT, H. T., *et al. The Value of Alfalfa—Orchardgrass Silage with and Without Sodium Metabisulfite for Milk Production,* Va. Agr. Exp. Sta. Bul. 534 (1962).
6. CUMMINS, D. G. *Interplanting of Corn, Sorghum, and Soybeans for Silage,* Ga. Res. Bul. 150 (1973).
7. FOSTER, J. R. "Forages for Swine and Poultry," In *Forages* (Ames: Iowa State Univ. Press, 1973).
8. HILDEBRAND, S. C., *et al. Corn Silage,* Mich. State Univ. Ext. Bul. E-665 (1969).
9. HOGLUND, C. R. *Comparative Losses and Feeding Values of Alfalfa and Corn Silage Crops When Harvested at Different Moisture Levels and Stored in Gastight and Conventional Tower Silos: An Appraisal of Research Results,* Mich. State Univ. Agr. Econ. Publ. 947 (1964).
10. JOHNSON, R. R., and K. E. MCCLURE. "Corn Plant Maturity IV. Effects on Digestibility of Corn Silage in Sheep," *Jour. Animal Sci.,* 27:535 (1968).
11. JOHNSON, R. R., *et al.* "Corn Plant Maturity I. Changes in Dry Matter and Protein Distribution in Corn Plants," *Agron. Jour.,* 58:151 (1966).
12. LARSENS, H. J., and R. F. JOHANNES. *Summer Forage, Stored Feeding, Green Feeding, and Strip Grazing,* Wisc. Agr. Exp. Sta. Bul. 257 (1965).
13. LINDAHL, I. L., *et al.* "Effect of Management Systems on the Growth of Lambs and Development of Internal Parasitism. III. Field Trials with Lambs on Soilage and Pasture Involving Medication with N. F. and Purified Grades of Phenothiazine," *Jour. Parasitol.,* 56:991 (1970).
14. MORRISON, F. B. *Feeds and Feeding,* 22nd ed. (Clinton, Ia.: Morrison, 1959).
15. REID, J. T. "Forages for Dairy Cattle," In *Forages* (Ames: Iowa State Univ. Press, 1973).
16. SHEPHERD, J. B., *et al. Developments and Problems in Making Grass Silage,* USDA-BDI-Inf. 149 (1953).
17. TERRILL, C. E., I. L. LINDAHL, and D. A. PRICE. "Sheep: Efficient Users of Forage," In *Forages* (Ames: Iowa State Univ. Press, 1973).
18. THOMPSON, W., *et al. Corn Silage.* Ky. Coop Ext. Serv. Misc. Bul. 336 (1968).
19. USDA. Agr. Stat. (1977).

CHAPTER 34
ALFALFA

THERE are about 20 relatively common legumes in the *Medicago* genus but of these, only three are widely distributed in the United States. Of these, alfalfa has outstanding value in many parts of the world. Others commonly found here are burclover and black medic.

History and Distribution

Alfalfa, *Medicago sativa,* is believed to have originated in southwest Asia. Wild forms and related species are found in much of central Asia and even in Siberia. Pliny, one of the early Roman writers (23–79 A.D.), states that alfalfa was introduced into Greece as early as 490 B.C. by the invading Medes and Persians as feed for their chariot horses and other animals. As early as 60 A.D. Columella, another early Roman writer, promoted the virtues of alfalfa for soil improvement and as a remedy for "sick beasts." Later alfalfa spread into Italy and other European countries. Early Spanish explorers took it to Central and South America.

This legume was grown by both George Washington and Thomas Jefferson about 1790. However, alfalfa had its real beginnings in this country as an extensively grown commercial crop in 1851, when it was brought into California from Chile. It was found ideally suited to the fertile limestone soils of the West and gradually spread eastward.

An especially winter-hardy alfalfa was needed if the crop was to succeed in the more northerly states. Such a cultivar became available as a result of seed brought from Germany to Minnesota in 1857 by Wendelin Grimm. Persistent effort to maintain the crop and to expand its acreage, by sowing seed from such plants as survived, produced a cultivar with unusual resistance to winter injury. The cultivar 'Grimm' was the result. Through many years the name 'Grimm' was synonymous with winter-hardy alfalfa.

Alfalfa is a crop with worldwide distribution. In many European countries it is still called *lucerne.*

In 1900 the United States had fewer than 3 million acres (1.2 million ha) of alfalfa; by 1950 it had more than 15 million acres (6.1 million ha). And today alfalfa is harvested on more than 25 million acres (10.1 million ha). In earlier years more than 80 percent of the acreage was in the western half of the country. However, now the higher percentage is in the Corn Belt states and eastward, with over 3 million acres (1.2 million ha) in Wis-

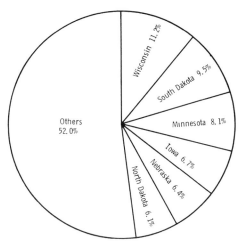

FIGURE 34–1. *Leading states in United States alfalfa and alfalfa mixtures hay area (acreage). Six Midwestern and Great Plains states have almost one-half of the total acreage.* [SOURCE: Southern Illinois University.]

consin and 2 million acres (0.8 million ha) each in Minnesota and South Dakota (See Figure 34–1.)

The recognized value of alfalfa as a forage crop and the availability of winter-hardy cultivars, better cultural practices, and greater use of lime, phosphorus, and potash have been responsible for the expanded acreage.

Adaptation

Alfalfa is adapted to a wide range of climatic conditions. The most favorable soils are deep loams with good drainage characteristics. In addition to good drainage, a soil well supplied with P, K, and lime is essential to good production. It fails on acid soils. It is tolerant of alkaline but not highly alkaline conditions.

This legume has been recognized for many years as well suited to dry climates. It is highly drought resistant but becomes somewhat dormant during dry periods. It is grown extensively in moisture-deficient areas under irrigation, with high acre production. Fourteen or 15 tons/acre (31–34 MT/ha) yields are not uncommon under these conditions.

Some cultivars can be grown in Southern states and others are notably winter-hardy. Proper cultivar selection makes possible its wide distribution and extensive use.

Plant Description

The alfalfa plant is a perennial which usually grows to a height of 2–3 feet (0.6–0.9 m). A plant may have from 5–25 or more stems (tillers) arising from a woody crown. This crown is the overwintering structure. Some types have short rhizomes and are considered broad-crown cultivars. New

FIGURE 34–2. *Alfalfa is "Queen of the Forages." Among the legumes it is unexcelled for a combination of high feed value, productivity, and palatability.*

stems arise from buds near the crown area as older ones mature or are harvested. Alfalfa stands may last from a few to many years, depending primarily on the disease susceptibility of a given cultivar.

Leaves are pinnately trifoliolate, with about one-third of each leaflet tip serrated. The plant is characterized by a strong taproot which may penetrate to a depth of 25 feet (7.6 m) under favorable conditions, though the usual depth is more nearly 3–8 feet (0.9–2.4 m). The characteristic flower color is purple but may be yellow or variable in certain cultivars and species. The flowers are borne in loose racemes, with 5–20 flowers per raceme. Seed pods are twisted in spirals and contain one to several seeds per pod.

Importance and Use

Alfalfa is one of the most important forages in the United States. Among the legumes it is unexcelled for a combination of high feed value, productivity, and palatability. Although some of the clovers have a higher plant protein percentage, alfalfa produces more protein per acre than any forage legume. The acre production of digestible protein has averaged over twice as much for alfalfa hay as for clover and over four times that of an acre of corn silage. It is high in minerals and vitamins which contributes to its feeding value and is a particularly good source of vitamin A. This forage plant is known for its versatility, being valuable for livestock feed in the form of hay, silage, pasture, and pellets or meal. Its greatest value is for hay, but in a mixture with a grass it is used extensively for rotation pasture. It is an excellent hog pasture, and despite the possible bloat problems it is being used more extensively for cattle and sheep grazing. The increasingly common procedure of including grass with alfalfa results in a reduced bloat hazard and greater erosion control. If diseases or unfavorable soil or climatic conditions reduce the alfalfa stand, the grass assures a sod cover and continued livestock feed. Alfalfa dehydration is an important industry in such states as Nebraska, Kansas, and California. Some dehydrated alfalfa is fed to livestock as meal, but most now is pelleted, which eliminates the dust problem.

Species and Types

There are more than 50 named cultivars of alfalfa grown in the United States. No one cultivar is adapted throughout the country; each is adapted to a limited area. Alfalfa cultivars grown in the United States were developed principally from two species, *Medicago sativa* and *Medicago falcata.*

1. *Medicago sativa*—These are purple-flowered types with erect stems and narrow crowns.
    a. *Common*—These alfalfas originated from Chilean introductions into California about 1850. They consist of regional strains adapted to different climatic regions and are identified by the state of origin such as 'Kansas Common.' All are susceptible to bacterial wilt, and superior cultivars developed by breeding generally are available. Several new cultivars of common have been developed that are recognized as markedly superior for certain climatic areas of adaptation. The most striking of these is 'Buffalo,' developed from 'Kansas Common.' It is recognized for its high yield, is resistant to bacterial wilt, and is moderately winter-hardy. 'Cody,' a selection from 'Buffalo,' is highly resistant to spotted alfalfa aphid; otherwise it is similar to 'Buffalo.'
    b. *Flemish*—These are purple-flowered cultivars which are only moderately winter-hardy and susceptible to bacterial wilt. Such cultivars often are used for rotations or short-term stands. Plants recover quickly after cutting and often give high yields over a short lifespan.
    c. *Turkestan*—Plant types resistant to bacterial wilt, these alfalfa are used widely as breeding stock for incorporating wilt resistance.
2. *Medicago falcata*—This species is native to Siberia and very cold-hardy. These alfalfas are yellow-flowered and have a decumbent growth habit and widely branching roots. Some of the creeping-rooted cultivars, such as 'Rambler' and 'Travois,' originated from *M. falcata.*
3. *Medicago media*—This species includes variegated types which originated from a cross between *M. sativa* and *M. falcata.* They may have

flower colors ranging from purple to blue, yellow, or white. Most cold-resistant cultivars grown successfully in the northern United States and Canada are variegated types. 'Vernal' is a variegated cultivar.

CREEPING ALFALFAS. In general, alfalfa has been recognized as a plant having a strong taproot, having many shoots developing from a crown at the ground surface, but forming no rhizomes. There is some interest in

A

B

FIGURE 34–3. (A) Spaced nursery with creeping-rooted alfalfa clones on the left and noncreeping clones on the right. (B) Root systems of a creeping-rooted alfalfa plant. These creepers originated from intercrossing M. sativa and M. falcata. [SOURCE: D. H. Heinrichs, Canada Department of Agriculture.]

creeping alfalfas for use in pastures and possibly for greater tolerance to frost heaving.

There are two types of root systems in alfalfa which enable the plant to spread horizontally. They generally are referred to as rhizomatous and creeping-rooted. In the first type the spread is attributed to lateral expansion of the low-set crown by short horizontal stems; in the second the spread is brought about by horizontal roots from which shoots may arise at irregular intervals. 'Rhizoma,' 'Nomad,' 'Rambler,' 'Victoria,' and 'Spredor' are examples of creeping cultivars.

Cultural and Management Practices

Alfalfa is not adapted to a soil with an acid reaction; thus, most areas planted to alfalfa east of the Mississippi River require lime applications. Adequate P and K are also necessary for establishment and maintenance. In recent years it has been recognized that K fertility is particularly critical for high yields and good stand maintenance, and subsequently higher rates of K have been recommended than previously. In some areas of the United States, particularly the East and Southwest, boron applications for establishment and maintenance have proven beneficial.

In the southern parts of the alfalfa-growing region, late summer seeding is practiced most often. Such a procedure lessens the weed problems encountered with spring seedings. However, seeding date should be early enough to allow adequate seedling development before cold weather so the plants will not winter kill. In other areas of the country, seedings are made with small grain companion crops, such as oats or winter wheat. While the companion crop will afford some competition during establishment, it will help to control weeds and will give a return during the seeding year. A spring seeding date coupled with the use of a herbicide has become popular in some areas where a hay yield during the year of seeding is desired.

Alfalfa should be seeded in deep, well-drained soil on a fine, well-prepared seedbed. While much alfalfa is broadcast-seeded, band-seeding often gives best stands. Seeding rate recommendations range from 10 to 15 pounds/acre (11–17 kg/ha) in the Lake and Central States and Western states with irrigation to 5–8 pounds/acre (6–9 kg/ha) in drier areas. Seed should be inoculated with the proper *Rhizobium* strain immediately prior to seeding if alfalfa has not been grown on the area in recent years. A small percentage of the alfalfa acreage has been planted to blends, but these generally have not proved superior to improved single cultivars.

Utilization

HAY. With a high-quality crop such as alfalfa, which offers the possibility of several hay crops per season if cut at a somewhat immature stage

FIGURE 34-4. *Alfalfa is not adapted to a soil with an acid reaction. Most areas planted to alfalfa east of the Mississippi River require lime applications to bring the pH up to about 6.5.* [Photograph with permission of University of Tennessee Agricultural Experiment Station.]

of growth, the importance of cutting at the right time is evident. If harvesting is delayed unnecessarily, palatability and nutritive value are lowered. Previous recommendations were to harvest at early flower, about one-tenth to one-fourth bloom. More recent recommendations are to cut in prebloom (bud) to first flower stages, with the increased quality being worth some yield sacrifice. Alfalfa should not be cut during a period of at least 4 weeks before the last killing frost to allow the roots to accumulate carbohydrate reserves for cold hardiness and spring regrowth. Removing a last cutting in the fall after this carbohydrate accumulation often is not injurious. The number of annual hay cuttings ranges from one in drier areas to six or more in irrigated regions. In many of the Lake and Central States, three to five cuttings are common.

In some areas of the lower Midwest characterized by alternate freezing and thawing during the winter, the additional management problem of frost-heaving occurs. Frost heaving results in alfalfa plants being gradually pushed upward until a portion of the crown area is exposed, resulting in lower yields or stand losses. Seedlings can be completely uprooted in severe frost-heaving periods. This problem can be lessened somewhat by alfalfa cultivar selection, soil site selection, and planting alfalfa in a mixture with a grass.

Seed Production

More than 0.3 million acres (0.12 million ha) of alfalfa seed are harvested annually in the United States. The highest acreages are found in South

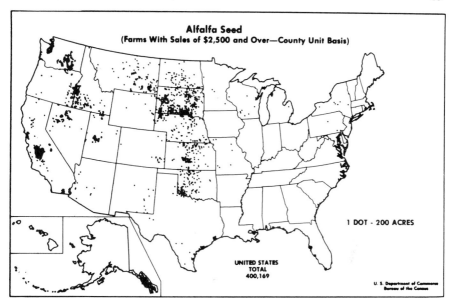

Alfalfa Seed
(Farms With Sales of $2,500 and Over—County Unit Basis)

1 DOT - 200 ACRES

UNITED STATES
TOTAL
400,169

U. S. Department of Commerce
Bureau of the Census

FIGURE 34–5. *Leading states in alfalfa seed production, shown geographically.* [SOURCE: U.S. Department of Commerce.]

Dakota, Kansas, California, Idaho, Washington, Montana, and Nebraska. (See Figure 34–5.) Although the average U.S. yield is about 225 pounds/ acre (252 kg/ha) some states more than double that average.

The more important requirements for a good seed crop are a normal growth of an adapted cultivar; the presence of an abundance of pollinating insects; absence of harmful insects; and bright, sunny weather during bloom and harvest. These conditions can best be obtained under irrigation. The development of self-pollinating alfalfa strains, such as 'Ellerslie-1,' may rejuvenate what is a sagging seed industry in some areas. Nearly all of the alfalfa seed is harvested by combining directly.

Diseases

About 30 diseases caused by fungi, bacteria, viruses, and nematodes damage alfalfa. Diseases cause estimated annual losses of about 25 percent of alfalfa forage and about 10 percent of the seed yield. These losses cost alfalfa producers about $400 million each year.

Bacterial wilt is generally recognized as the most destructive alfalfa disease in the United States. It is found in practically all parts of the country. The bacterial growth clogs the fibrovascular bundles, with the result that the plant becomes stunted and yellow and dies after a few months. The only known control is to grow wilt-resistant cultivars, and a long-term stand is dependent on selection of such a resistant type.

Other diseases affecting alfalfa include crown and root rots, anthracnose,

FIGURE 34–6. *A bacterial wilt-resistant alfalfa cultivar (center) as com-pared to wilt-susceptible cultivars on either side. Bacterial wilt is the most destructive alfalfa disease. A long-term stand is dependent upon selection of a resistant cultivar.* [SOURCE: Wisconsin Agricultural Experiment Station.]

and various foliar diseases. Of these the most destructive may be *Phyto-phthora* root rot, which is a universally occurring disease in wet soils. Cultivars such as 'Agate' have been developed for use in the upper Midwest where this disease is a problem.

Insects

Some 100 insects are injurious to alfalfa. These include insects which damage alfalfa by feeding on foliage, sucking sap, and feeding on roots, as well as those that cause heavy losses to the seed crop. The worst injurious insects include the alfalfa weevil, the spotted alfalfa aphid, the pea aphid, the potato leafhopper, the meadow spittlebug, and the lygus bug. Of about $300 million annual losses from insects, approximately $50 million is attrib-uted to the alfalfa weevil alone. Cultivars such as 'Team,' 'Weevlchek,' and 'Arc' have some tolerance of the weevil. Insecticide sprays are neces-sary in areas south of Southern Illinois to prevent damage on the first cutting of hay.

Alfalfa Improvement

HYBRID ALFALFA. There has been much research in an effort to produce hybrid alfalfa seed on a commercial basis. The isolation of cytoplasmic male-sterile plants was a big step toward this accomplishment. There are two basic types of male sterility—genetic and cytoplasmic. Male-sterile plants appear only sporadically in populations of self- and cross-pollinated species, presumably as a result of mutation. It has been reported that in 1963, in a single study, more than 50,000 plants from 28 cultivars of alfalfa were examined, with the result that 10 nearly sterile plants were found. Four of the 10 exhibited the male-sterile trait.

Alfalfa was first hybridized on a large scale in the late 1960s. There are now several hybrid alfalfas available. However, because of the cost of producing the seed, the hybrids must be priced somewhat higher than the standard cultivars. It remains to be seen whether these alfalfa hybrids will be sufficiently superior to the standard cultivars to justify the increased seed cost. In the mid- and late 1970s there appeared to be some declining activity in breeding and release of new hybrids.

A brief review of reports on alfalfa improvement in the 1970s indicates that many of the breeding efforts are being made and most likely will continue in these areas:

1. Increased pest resistance with the development of cultivars resistant to such insects as alfalfa weevil and such diseases as *Phytophthora* root rot.
2. Increased nitrogen fixation potential, which would be useful in agricultural systems for improving pasture and grain production.
3. Development of nonbloating cultivars, which could be grazed without fear of livestock bloat losses.

Other *Medicago* Species

Other *Medicago* species that have some importance as forages in the United States are the burclovers and black medic.

Burclover

The common name burclover is used for most of the annual species of *Medicago*. Native to the Mediterranean region, these species came into this country as escapees.

There are two species of considerable importance—California burclover, *M. hispida,* which has wide distribution and use in the Western states, and the spotted burclover, *M. arabica,* which is grown considerably in the Southern states. These have been in the South and West for many years.

The acreage of burclover grown in pure stands is small, but it occurs in mixtures on millions of acres of pasture and rangeland throughout the Southern and Pacific Coast states. California burclover is an important constituent of the range pastures of the foothills throughout California and Arizona. In California the volunteer crop is utilized as a winter cover to decrease soil erosion and add organic matter and fertility to the soil.

All burclovers are cool-weather plants that lack winter hardiness and cannot endure summer heat. In the South burclovers start growth in the fall, continue through the winter, and mature in the early spring. They require a moderately fertile soil. A satisfactory growth is obtained on moderately acid as well as on lightly saline soils.

In the Western states burclover is seldom seeded; it volunteers on range and orchard lands. In the humid region of the Southeast, seeding is required for original establishment, but once established can be maintained almost indefinitely if well managed.

Black Medic

Native to Europe and Asia, black medic, *M. lupulina,* appears to have brought into this country with clover seed. It gets its name from the black pods or husks in which the seeds develop.

It is an annual with a decumbent growth habit. The stems, which make a growth of 1–2 feet (0.3–0.6 m) may ascend when the stand is thick. The stems are rather fine and leafy. Most of the acreage throughout the United States has volunteered in pastures or waste areas.

REFERENCES AND SUGGESTED READINGS

1. ANDERSON, L. D., *et al. Pest and Disease Control Program for Alfalfa Seed,* Calif. Agr. Exp. Sta.–Ext. Serv. (January 1969).
2. BEATTY, E. R., *et al. Alfalfa for Georgia,* Univ. of Ga. Agr. Exp. Sta. Res. Bul. 211 (1978).
3. BOLTON, J. L. *Alfalfa—Botany, Cultivation, and Utilization* (New York: Interscience, 1962).
4. DENNIS, R. E., *et al. Alfalfa in Kansas,* Ariz. Coop. Ext. Serv. and Agr. Exp. Sta. Bul. A-16 (1966).
5. HANSON, C. H. (Ed.). *Alfalfa Science and Technology* (Madison, Wisc: American Society of Agronomy, 1972).
6. HANSON, C. H., and D. K. BARNES. "Alfalfa," In *Forages* (Ames: Iowa State Univ. Press, 1973).
7. HEINRICHS, D. H. "Creeping Alfalfas," In *Advances in Agronomy,* vol. 15 (New York: Academic Press, 1963). ed. A. G. Norman.
8. JENSEN, E. H., *et al. Environmental Effects on Growth and Quality of Alfalfa,* Western Reg. Res. Pub. Univ. Nev. Agr. Exp. Sta. (1967).
9. LUESCHEN, W. E., *et al.* "Field Performance of Alfalfa Cultivars Resistant and Susceptible to Phytophthora Root Rot," *Agron. Jour.,* 68:281 (1976).
10. PULLI, S. K., and M. B. TESAR. "Phytophthora Root Rot in Seeding-Year Alfalfa as Affected by Management Practices Including Stress," *Crop Science,* 15:861 (1975).
11. ROMINGER, R. S., D. SMITH, and L. A. PETERSON. "Yield and Chemical Composition of Alfalfa as Influenced by High Rates of K Topdressed as KCl and K_2SO_4," *Agron. Jour.,* 68:573 (1976).

12. SEETIN, M. W., and D. K. BARNES. "Variation Among Alfalfa Genotypes for Rate of Acteylene Reduction," *Crop Science,* 17:783 (1977).
13. USDA *Agr. Stat.* (Washington, D.C.: Government Printing Office, 1977).
14. USDA-ARS. *Losses in Agriculture,* Agr. Handbook 291 (1965).
15. USDA-ARS. *Report of the Twenty-Fifth Alfalfa Improvement Conference* (1976).

Chapter 35
THE TRUE CLOVERS

THE legumes in the *Trifolium* genus usually are recognized as the true clovers. A number of other small-seeded legumes generally are known as clovers but do not belong to this group. Of those most extensively grown, the sweetclovers and burclovers are examples.

Red Clover

Red clover, *Trifolium pratense,* is the most widely grown of the true clovers. It is suitable to use for hay, pasture, silage, and soiling. Alone, or with grasses, red clover continues to be one of the most extensively grown legume hay crops in the Northeastern United States and the Lake and Central states.

Originating in southwestern Asia, red clover came into use much later than alfalfa. It came to North America with the early English colonists. Although the climatic conditions appeared favorable, red clover did not succeed well following its introduction. Persistence in saving seeds from plants that did survive resulted ultimately in a strain differing significantly from that generally grown in Europe.

Distribution and Adaptation

Red clover is an important forage in northern and western Europe, west central Asia, the United States, Canada, New Zealand, and Australia. There are several types of red clover, differing significantly in growth characteristics. The wild red clover found in England is not found anywhere in North America. Most American cultivars of the early-flowering type are referred to as medium red clover. This type generally gives two hay crops per year. Most plants die at the end of the second season of growth.

The American mammoth red clover is the principal late-flowering type grown in this country. It is generally regarded as a cultivar of red clover, though by some as a subspecies. It makes a coarser growth and produces only one main crop. In the seeding year its growth is confined to a rosette of leaves at the crown. It is not extensively grown.

Red clover is best adapted where summer temperatures are moderately cool to warm and where adequate moisture is available throughout the growing season. The red clover region extends from the Atlantic Coast to the eastern part of North and South Dakota, Nebraska, and Kansas and south into Tennessee and North Carolina. Red clover also is an important crop in Washington, Oregon, Montana, and other Western states, where it is grown under irrigation except at high elevations. It is used to some extent as a winter annual in the southeastern states.

Plant Description

Numerous rebranching, leafy stems develop from the crown. Red clover does not have stolons or rhizomes. Leaves and stems ordinarily are pubescent or hairy although some European strains are relatively free of hairs. The leaf is divided into three oblong leaflets, each of which has a light-colored marking or variegation in the center. Leaves are palmately trifoliolate. The flowers are borne in heads at the tips of branches, with up to 125 flowers per head. The rose-purple flowers resemble the well-known pea flower but are much smaller and more elongated. The mitten-shaped seed are borne one per pod and vary in color from yellow to dark purple. The root system is a taproot with many rebranching secondary branches.

Importance and Use

It is estimated that 12–14 million acres (4.9–5.7 million ha) of red clover are grown annually in the United States. Almost 0.4 million acres (0.16 million ha) are harvested for seed in the United States. The crop fits better than any other small-seeded legume into the more commonly used rotations of its area of adaptation. It is outstanding in value for hay, pasture, and

FIGURE 35–1. *Palmately trifoliolate leaves of red clover. Leaves and stems ordinarily are pubescent or hairy and have a light-colored marking (variegation) on each leaflet.* [SOURCE: Southern Illinois University.]

soil improvement. In addition it most often is seeded with timothy, orchard-grass, or tall fescue and may be a part of a broad forage mixture with grasses and other legumes. Although a perennial, red clover usually functions as a biennial, with most of the plants dying after the second year, primarily of crown and root-rot diseases.

Cultivars and Strains

Red clover is largely self-sterile, and must be cross-fertilized. The result is that red clover is exceedingly variable in the characteristics of individual plants. Such a condition favors the development of regional strains and cultivars. There are naturalized red clovers in many parts of the world. Strains and cultivars grown in North America differ significantly from those of Europe. American strains and cultivars are heavily pubescent, or hairy. There is now little doubt that this character developed by the survival of such plants, when plants less pubescent were weakened and many of them killed as a result of leafhopper damage. In Canada, both single-cut and double cultivars are grown, whereas in the United States, the double-cut types predominate.

In North America the two principal sources of strains are (1) old, well-established farm strains produced locally, and (2) improved cultivars developed through breeding programs. Among the better naturalized cultivars are 'Chesapeake' from eastern Maryland and 'Pennscott' from Lancaster County, Pa. More important cultivars developed through breeding and selection include 'Lakeland,' a synthetic cultivar developed at the Wisconsin station, 'Ottawa,' developed at Canada's experimental farm, 'Kenland' and 'Kenstar,' developed by the Kentucky station and the USDA. Cultivars such as 'Kenland' and 'Kenstar' were developed for greater persistence because of resistance to diseases such as southern anthracnose. 'Kenstar' often lasts for three full years in contrast to the two-year lifespan of most cultivars.

Cultural and Management Practices

Relatively fertile, well-drained soils of high moisture-holding capacity are best for red clover. Loams and even fairly heavy-textured soils are preferred to light, sandy soil. Red clover will grow on moderately acid soils, but for maximum yields soils should contain adequate amounts of calcium. There are three common seeding methods practiced in the United States:

1. Broadcasting seed on winter small grains in late winter or early spring. This is most common in areas where winter wheat is grown extensively.
2. Seeding with a spring small grain companion crop. The small grain can be allowed to mature and be harvested for grain, or it can be harvested earlier for silage or hay.
3. Seeding alone or with a grass and/or other legumes either in the spring, late summer, or early fall. South of Tennessee, where red clover

FIGURE 35–2. *Band-seeded red clover. This legume may be seeded alone or with a grass in the spring, late summer, or early fall or with a spring small grain companion crop. Red clover often is broadcast on winter small grains in late winter or early spring.* [SOURCE: Southern Illinois University.]

is grown as a winter annual, red clover may be seeded in the late fall or early winter.

Most hay management recommendations call for cutting established red clover stands at early bloom. If the clover is grazed, a rotational grazing system is preferred. Graze or harvest for hay only until 45 days before frost to allow carbohydrate reserves to accumulate.

The most serious insect and disease problems are grasshoppers, leafhoppers, clover leaf weevil, crown and root rots, anthracnose, leaf spots, and rusts.

In most of the areas where red clover is grown, root and crown rots virtually destroy stands by midsummer of the second year. This is why red clover behaves as a biennial rather than as a perennial. Most of the improvement efforts are centered on better persistence through greater insect and disease resistance.

A seed crop in humid areas is uncertain; yields are likely to be relatively low. However, the principal seed production areas are the Central and North Central states even though yields often average less than 100 pounds/acre (112 kg/ha). (See Figure 35–3.) An increasing amount of seed is being produced in the Northwest, particularly in Oregon, Washington, and Idaho, where yields often have averaged more than 300 pounds/acre (336 kg/ha). For all practical purposes, red clover is self-sterile and seed producers must depend primarily on bees for cross-pollination. It is a common practice in the Central states to harvest the first crop for hay, and the regrowth for seed.

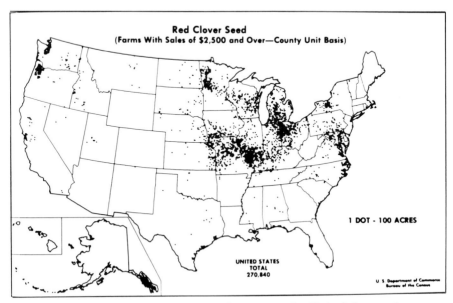

FIGURE 35–3. *Red clover seed production in the United States, shown geographically.* [SOURCE: U.S. Department of Commerce.]

The White Clovers

The exact center of origin of white clover, *Trifolium repens,* is not known but it may have been the Near East. It apparently was cultivated in the Netherlands during the last half of the sixteenth century, and in England soon after 1700. White clover undoubtedly came to our Atlantic Coast with the British colonists who brought seed with them. The fact that it was well adapted and spread rapidly is indicated by the name the Indians gave it, "white man's foot grass." It was common in Ohio and Kentucky as early as 1750. The large type of white clover, ladino, was not successfully produced in the United States until the early 1900s. Its popularity grew rapidly during the 1940s and 1950s.

Distribution and Adaptation

White clover has an almost universal occurrence. Its geographic distribution seems to be limited only by the cold of the Arctic, the drought of the deserts, and the heat and plant competition of tropical jungles. This widespread occurrence of white clover presents an intriguing and perplexing problem of seed dispersal and production. Cases are known where fair stands of white clover appeared without seeding following the removal of second-growth timber and the application of mineral fertilizer. Volunteer stands also often develop in fields where the clover is not known to have been seeded. White clover makes its best growth on clay and silt soils in humid states and on irrigated soils in the West at a pH of 6–7. It is not considered tolerant of highly acid or highly alkaline soil conditions.

Plant Description

White clover plants have a short, primary stem with a number of internodes. Axillary buds on the primary stem give rise to long, creeping stolons which stretch out in all directions. Leaves arise from the nodes of a stolon, one trifoliolate leaf per node on a long petiole. Leaflets are almost sessile, elliptical to heart-shaped, glabrous, and variegated. The flowering heads are produced on peduncles somewhat longer than leaf petioles. The flowers are normally white with a pinkish hue, with from 20–40 or more per head. The yellow to reddish seed is largely heart-shaped and small, much smaller than either red clover or alsike clover. The plant produces a primary taproot, but is considered to be relatively shallow-rooted. Secondary roots arise from nodes of stolons. The plant usually behaves as a perennial. In the Deep South, however, white clover usually is grown as a winter annual. Even in the more northern states white clover often disappears under drought and heat conditions. When this happens, re-establishment usually occurs from seeds which may remain viable in the soil through several years.

FIGURE 35–4. *Palmately trifoliolate leaves of Ladino clover are borne on long petioles. Leaves are glabrous (without hairs) and variegated.* [SOURCE: Southern Illinois University.]

Importance and Use

White clover ranks at the very top in palatability and nutritive value. Crude protein content ranges from about 20 percent at full bloom to over 30 percent at prebloom stages. It is higher in crude protein and essential amino acids than alfalfa, red clover, or birdsfoot trefoil. This high quality is a result of the fact that grazing animals generally harvest only leaves and petioles.

This legume is used widely in the United States for pasture, hay, silage, and to a lesser extent as a cover crop and in lawn mixtures. It is almost always seeded for pasture in association with grasses but may be seeded alone for swine or poultry pastures. Because of its succulence and high palatability there is considerable bloat danger when grazed by ruminants.

In most permanent pastures in the more humid northern states the indigenous white clover volunteers when climatic conditions are favorable. This is not true for the large type of white clover, ladino, which must be seeded. In the lower South, white clover is not known to volunteer—it is necessary to seed for original establishment. It thereafter volunteers in the fall, producing a maximum growth in the late winter and early spring months.

FIGURE 35–5. *A characteristic growth and use of Ladino clover.*

Cultivars and Strains

There are three general types or forms of white clover. They may be classified as the (1) large, (2) intermediate, and (3) low-growing forms. Half the white clover acreage in the United States is seeded to the large types. The most widely grown variety of the large type is ladino clover. Others are 'Merit' from the Iowa station, 'Regal' from Auburn University, and 'Tillman' from the South Carolina station and USDA. Of the intermediate types one of the best known is the 'Louisiana S1,' adapted to the Gulf States. There are many regional low-growing strains in the more northern states, developed in pastures closely grazed through long periods.

LADINO. There is no information on the origin of ladino clover other than that it is found growing in the Po Valley, Italy. Several introductions of seed lots had been made into the United States by 1910, but it was not until after 1930 that ladino came to be used extensively and its value widely recognized.

Ladino often is referred to as *giant white clover.* It grows from two to four times as large as the indigenous types of the North. In the Northern states ladino is more subject to winter injury than the common intermediate and low-growing types, winter loss of stands not being uncommon. But it is still the most productive white clover under favorable soil and climatic conditions. Neither ladino nor the northern common types are adapted to the lower South, but improved large and intermediate types are available for those areas.

Culture and Management Practices

Ladino clover often is seeded in a mixture with at least one other legume and a grass, except when to be used for hog or poultry pasture, or as a cover crop in orchards. The rate of seeding varies from 0.5 to 2 pounds/acre (0.6–2.2 kg/ha). Early seeding with good moisture conditions is important. The best soil pH is above 6.0. Under irrigation in the West it is a common practice to seed ladino alone and to graze the early growth, following later with the harvest of a seed crop. It often is difficult to maintain ladino clover when it is part of a grass–legume grazing mixture. Proper fertilization practices, with no application of N fertilizer to stimulate the grass excessively coupled with adequate P and K for the clover, plus good grazing management, are essential to clover maintenance. If the ladino clover component disappears, this legume is well suited for pasture renovation. Many of the same diseases affecting red clover also affect ladino clover. In addition, virus infection frequently builds up in ladino so that 5 percent or more of the plants in some fields are infested by the third year.

White clover is particularly attractive to all kinds of bees. There are some seed production centers in Louisiana and Mississippi, but practically all ladino seed is produced in the irrigated valleys of the West. California, with over 11,000 acres (4,452 ha) harvested for seed, is by far the leading producer of ladino seed. Under certain conditions in the South the application of 0.5 pound of boron per acre (0.6 kg/ha) has more than doubled the seed production of white clover, the increase resulting from more flowers per head being pollinated. Boron is known to be necessary for normal pollen functioning in several plant species and may influence bee activity, increasing the quantity and quality of nectar.

Alsike Clover

Alsike clover, *Trifolium hybridum,* is said to have been introduced into England about 1830 and into the United States ten years later. It is grown rather widely in northern Europe and eastern Canada. In the United States it does best in the Great Lakes states and eastward into New England. It is more tolerant than red clover of acid soil and poor drainage.

This legume tillers extensively from the crown. The stems of alsike clover are notably more slender than those of red clover. The crop is likely to lodge badly, and for that reason is usually grown with a grass such as timothy. The stems, petioles, and leaves lack the pubescence of red clover, and the flower heads are smaller and pink or pinkish-white. Seeds are about one-third the size of red clover and are dark green to greenish-yellow.

The time and method of seeding are the same as for red clover. The crop is seldom grown alone; however, 2–3 pounds/acre (2.2–3.4 kg/ha) often is added to a red clover–timothy mixture. On soils inclined to be decidedly wet, alsike often is seeded with redtop, both of these crops being well suited to such an environment. Unlike red clover, alsike usually produces only one crop of hay per season.

The same diseases that affect red clover are likely to damage alsike. However, alsike is considered resistant to two red clover diseases, northern and southern anthracnose.

Crimson Clover

Crimson clover, *Trifolium incarnatum,* apparently was well known throughout much of southeast Europe before its introduction into the United States from Italy in 1819. Its spread was rapid throughout the southeastern states prior to 1900.

Distribution and Adaptation

Crimson clover, which is used as a winter annual, especially for pasture, is best adapted to the Southeast. Its geographic adaptation extends from the Gulf Coast as far north as Maryland, southern Ohio, southern Indiana, and southern Illinois, and westward to eastern Texas.

Crimson clover is tolerant of medium soil acidity and is adapted to growing on both sandy and heavy clay soils. After becoming well established, it makes more growth at lower temperatures than most of the other clovers.

Plant Description

Crimson clover seedlings establish rapidly, forming a dense crown or rosette type of leaves, from which a central taproot develops with many fibrous branches. The trifoliolate leaves, with broadly oval leaflets that are densely hairy, arise from equally hairy stems at the many nodes. The plants grow to a height of 1 to 3 feet (0.3 to 0.9 m) at which time growth is terminated by the formation of rather long, narrow flower heads with 75 to 125 florets each. In comparison with the flower heads of other clovers, they are impressively pointed, conical, and bright crimson. Even though crimson is self-fertile, insects are necessary to bring about proper pollination and seed setting. Seeds are reddish yellow to yellow and noticeably larger and more rounded than seeds of other clovers. The plants die after seed formation in the summer.

Importance and Use

Crimson clover generally is regarded as the most important winter annual legume throughout many of the Southeastern states. It provides good

FIGURE 35-6. *A characteristic growth in early spring of a reseeding culti-var of crimson clover.*

winter grazing and produces large yields of easily harvested hay or seed. It fits well into the grazing sequences with other crops. Crimson clover is responsible in large part for the early expansion of winter grazing in the South.

It also is recognized as a very valuable green manure crop. Its growth not only adds large amounts of nitrogen and organic matter to the soil, but the dense cover it provides through the winter and early spring months reduces soil erosion to a minimum. It is used in pecan and other orchards for this purpose.

There has been a marked increase in the use of crimson clover since 1942, because of development of reseeding and volunteering cultivars, recognition of the relatively high fertilizer requirements for rapid stand establishment and vigorous growth, and the need for thorough inoculation and an appreciation of its value for winter grazing.

Cultivars and Stains

Of great importance in the practical utilization of crimson clover is the availability of reseeding cultivars. Until about 1940 the crimson clover generally planted throughout the South was not designated as of any particular cultivar, with more than half the seed imported from Europe. In 1935 three strains were found that gave volunteer stands in the fall from seed shattered the preceding spring. These were combined to form the 'Dixie' cultivar. The more or less impervious seed coat of this cultivar, which gives a high percentage of hard seed, makes it possible for the seed to remain alive in the soil throughout the summer period, germinating gradually during the fall months. Several other strains with the reseeding character have been found. Among the other commonly used cultivars are 'Autauga,' 'Auburn,' 'Talladega,' 'Chief,' 'Kentucky,' and 'Tibbee.'

Culture and Management Practices

Crimson clover is seeded from mid-July to November, depending on the location and intended use. If fall grazing is desired, seeding must be early enough to get establishment before cold weather. When the crop is planted alone, the seeding rate is normally 20–30 pounds/acre (22–34 kg/ha). Seeding rate is reduced when planted with winter grains, annual ryegrass, or other winter legumes. Seed inoculation with the proper *Rhizobium* strain is important when the crop is grown for the first time.

Crimson clover gives excellent results when seeded in established stands of bermuda, dallis, johnson, and broad-leafed bahiagrass. Seedings have not been so successful in other Southern grasses, however. A Coastal bermudagrass–crimson clover mixture is a productive combination for the lower South.

With good stands and an abundance of honeybees, seed production may be as much as 1,000 pounds/acre or more (1,120 kg/ha). The average harvested yield is about 300 pounds/acre (334 kg/ha). Seed crops generally are combined directly from plants allowed to stand until the seed crop is fully mature, and some harvest losses occur. The leading state in seed production is Oregon.

Other True Clovers

In addition to the clovers that have wide distribution and use, there are others of more limited importance.

HOP CLOVER. Indigenous to Europe, it appears likely that hop clover was brought to the United States in white clover seed. There are three species: small hop clover, *Trifolium dubium,* large hop clover, *T. campestre,* and field hop clover, *T. agrarium.* All three are rather small-growing plants, seldom exceeding 15 inches (38 cm) in height. Their small heads of yellow flowers turn brown upon maturity and resemble the flowers of the hop plant. They are adapted to infertile, eroded soils in the South. Small hop clover appears spontaneously throughout most of the southern half of the United States and in both the Atlantic and Pacific Coast states, and farther north where moisture and other conditions are favorable. The large hop clover is not so generally found but has about the same distribution. Field hop clover is found growing sparsely in much of the northern and east–central part of the United States.

SUBTERRANEAN CLOVER. Native to the Mediterranean area, subterranean clover or subclover, *T. subterraneum,* has proved to be of great value in Australia and New Zealand, where it is grown extensively for pasture. It is also used as a rangeland legume in the Pacific Northwest. Naturalized ecotypes have been found in the Southeast, where subclover has good potential as a reseeding legume in permanent pastures. It is a decumbent winter annual whose rather inconspicuous white to pinkish flowering heads contain only 3–7 florets. Subclover is particularly well adapted to areas with warm moist winters and dry summers. Its reseeding mechanism results from peduncle elongation toward the ground after fertilization, with stiff forked bristles of the seed bur allowing the seed to bury itself in the soil.

Arrowleaf Clover

T. vesiculosum is a relatively new winter annual introduced from Italy which has become a major pasture legume in some parts of the Southeast. Its geographic adaptation extends from east Texas to South Carolina and north from Arkansas and Tennessee to the Gulf. The acreage of arrowleaf clover is expanding rapidly and is expected to increase farther on droughty upland soils of the lower South. It gives more late spring production than crimson clover and persists better that white clover on droughty soils. This legume thrives on well-drained soils and is not tolerant of poor drainage, extreme acidity, or high pH conditions.

Early growth is from a leafy rosette. The stems are thick and hollow and often purple when mature. Leaflets are arrow-shaped, and some have a white V-shaped mark. Long, pointed, white stipules are located at the base of each leaf petiole. The conical flower heads are white initially but later turn pink to purple. Forage quality is high. It extends the grazing season of small grains, such as wheat or rye. Arrowleaf clover can be over-seeded in the autumn on warm-season perennial grass sods such as bermudagrass to furnish winter and spring grazing.

Ball Clover

T. nigrescens, adapted to loam and clay soils in the lower Southeast, is closely related to white clover. It tolerates heavy grazing, reseeds well under grazing pressure, and gives more later spring production than crimson clover in its area of adaptation.

PERSIAN CLOVER. *T. resupinatum* is a winter annual with weak, hollow stems that attain a length of 2 feet (0.6 m) or more. It has value in mixtures for hay and pasture on low, moist soil in the extreme South.

STRAWBERRY CLOVER. *T. fragiferum* is a stoloniferous perennial that resembles white clover in general appearance and habit of growth. The pink to white flower heads somewhat resemble strawberries. It is especially adapted to extremely wet land and to alkaline conditions. It probably was introduced accidentally in clover seed.

BERSEEM CLOVER. *T. alexandrinum* is an upright-growing winter annual somewhat resembling alfalfa. It has come into limited production in the Imperial Valley of southern California and in the Yuma and Salt River valleys of Arizona. Its use is restricted because of poor winter hardiness.

ROSE CLOVER. *T. hirtum* is a winter annual that is widely adapted on rangelands in California. It is also suited to the Southeast but has not been used extensively in that region. This species is highly palatable and can reseed itself under proper grazing management.

KURA CLOVER. *T. ambiguum*, found in southern Russia, is of some interest in the United States, but there is no known acreage planted to it. This perennial legume has rhizomes, good pest resistance, and persists well, but it does not nodulate effectively in the field. It is similar to red clover in general appearance, but the leaves and flower heads are much larger.

CLUSTER CLOVER. *T. glomeratum* is an introduced small winter annual that became established in southern Mississippi as an escapee.

LAPPA CLOVER. *T. lappaceum* is another small winter annual that has become naturalized in the black lands of Alabama and Mississippi.

ZIGZAG CLOVER. *T. medium* closely resembles medium red clover in general appearance of leaves and flowers. It is a perennial, with strong, creeping rhizomes. It is winter-hardy and persistent but produces seed only sparingly.

REFERENCES AND SUGGESTED READINGS

1. BARNETT, O. W., and P. B. GIBSON. "Identification and Prevalence of White Clover Viruses and the Resistance of *Trifolium* Species to These Viruses," *Crop Science*, 15:32 (1975).
2. DONNELLY, E. D., and J. T. COPE, JR. *Crimson Clover in Alabama*, Auburn Univ. Agr. Exp. Sta. Bul. 335 (1961).
3. ENSMINGER, L. E., and E. M. EVANS. *Establishment and Maintenance of White Clover–Grass Pastures*, Auburn Univ. Agr. Exp. Sta. Bul. 327 (1960).
4. GIBSON, P. B. *White Clover*, USDA Agr. Handbook 314 (1966).
5. GORZ, H. J., G. R. MANGLITZ, and F. A. HASKINS. "Resistance of Red Clover to the Clover Leaf Weevil," *Crop Science*, 15:279 (1975).
6. HERMANN, F. J. *A Botanical Synopsis of the Cultivated Clovers (Trifolium)*, USDA Agr. Monograph 22 (1953).
7. HOVELAND, C. S. *Arrowleaf Clover*, Auburn Univ. Agr. Exp. Sta. Leaflet 67 (1962).
8. JUSTIN, J. R., *et al. Red Clover in Minnesota*, Univ. of Minnesota Ext. Bul. 343 (1967).
9. KNIGHT, W. E., and C. S. HOVELAND. "Crimson Clover and Arrowleaf Clover," In *Forages* (Ames: Iowa State Univ. Press, 1973).
10. LEFFEL, R. C. "Other Legumes," In *Forages* (Ames: Iowa State Univ. Press, 1973).
11. LEFFEL, R. C., and P. B. GIBSON. "White Clover," In *Forages* (Ames: Iowa State Univ. Press, 1973).

12. MORELY, F. H. W. "Subterranean Clover," In *Advances in Agronomy*, Vol. 13 (New York: Academic Press, 1961). ed. A. G. Norman.

13. REYNOLDS, J. H., and C. D. PLESS. "Forage Yield of Red Clover Treated with Furadan," *Tennessee Farm and Home Science Progress Report* 104 (1977).

14. SMITH, D., and R. R. SMITH. "Response of Red Clover to Increasing Rates of Topdressed Potassium Fertilizer," *Agron. Jour.*, 69:45 (1977).

15. TAYLOR, N. L. "Red Clover and Alsike Clover," In *Forages* (Ames: Iowa State Univ. Press, 1973).

16. TAYLOR, N. L., J. K. EVANS, and G. LACEFIELD. *Growing Red Clover in Kentucky,* Univ. of Ky. Leaflet AGR-33 (1975).

17. USDA Agr. Stat. (1977).

18. USDA. *Persian Clover—A Legume for the South,* Leaflet 484 (1960).

19. USDA. *Strawberry Clover—A Legume for the West,* Leaflet 464 (1960).

20. USDA. *White Clover for the South,* Leaflet 498 (1961).

CHAPTER 36
BIRDSFOOT TREFOIL, LESPEDEZA, AND OTHER FORAGE LEGUMES

Birdsfoot Trefoil

THERE are several species of birdsfoot trefoil. Only two are of importance in this country, and one of these is more generally grown than the other. The broadleaf birdsfoot trefoil, *Lotus corniculatus,* is much more productive and more widely adapted than the narrowleaf, *L. tenuis.* We are concerned here with the broadleaf birdsfoot trefoil.

Origin and History

In Europe, birdsfoot trefoil appears to go back toward the beginning of recorded agricultural history. It was present in many grazing regions throughout Europe, Asia, and northern Africa. The Mediterranean Basin likely is the center of origin. Although its high nutritive value and palatability were generally recognized, the legume appears to have been seeded only to a limited extent and under the more unfavorable soil conditions.

It is not known how or when birdsfoot trefoil first came to the United States. In New York, western Oregon, and northern California, it was found to have become naturalized and to have spread along roadsides and into pastures and meadows. It first came to the attention of the New York station in 1934. Two forms were found to occur: the broadleaf on the upland soils and the narrowleaf type on the heavier, more moist soils.

Distribution and Adaptation

Birdsfoot trefoil is grown on an estimated 2 million acres (0.9 million ha) in the United States. The known range of adaptation in the United States east of the Great Plains is from the eastern part of Kansas, Nebraska, and Dakotas to Vermont, and south to the Ohio River. The region of greatest adaptation and most extensive use is in New York, Pennsylvania, and Vermont. On the Pacific Coast there is a considerable acreage in California, Oregon, and Washington. It is also used as a permanent pasture legume in several north central states.

Birdsfoot trefoil is a winter-hardy perennial. It has developed a reputation for maintaining itself under unfavorable soil conditions. Like other crops, however, it makes higher yields on the more productive soils or when properly fertilized. Even where it persists on the less fertile soils, it is not until lime and phosphate have been applied in adequate amounts that birdsfoot trefoil contributes significantly to the prosperity of the area. It is more tolerant of poor drainage, low fertility, and an acid soil reaction than most other small-seeded legumes.

Plant Description

Birdsfoot trefoil resembles a fine-stemmed alfalfa. A single plant long established in a thin stand may have as many as a hundred or more stems from a single crown. These stems usually are from 20–40 inches (51–102 cm) in length. Each leaf consists of three leaflets attached at the tip of the petiole with two additional leaflets at the base, giving a total of five leaflets.

The flowers, which are lemon yellow to orange-yellow, are shaped like the sweetpea flower. They are much larger than those of alfalfa, red clover, or sweetclover. The arrangement is a typical umbel with flowers attached in the center at their pedicel bases in groups of four to eight. The ovary of the fertilized flower develops into the seed pod, usually about an inch in length. The seed are olive green to dark brown and a little smaller

FIGURE 36–1. *Stem and leaf characteristics of broadleaf birdsfoot trefoil. This fine-stemmed legume has leaves with five leaflets.*

FIGURE 36–2. *Eight pods of birdsfoot trefoil (one split open and twisted) attached at a single point. Each pod is about an inch in length, and dark brown to black in color. The attachment and spread of the pods give the appearance of a bird's foot; hence the name.*

than red clover seed. Because several seed pods are attached at a single point, they are forced to spread apart, giving the appearance of the several toes of a bird's foot; hence the name. The root system is best described as intermediate between that of alfalfa and red clover. The tap root has more lateral branches than alfalfa. No stolons or rhizomes are present. Fine-stemmed and leafy, the forage is palatable either in the green or cured condition, with a protein content similar to that of alfalfa.

Importance and Use

Birdsfoot trefoil is particularly well suited for use as a permanent pasture legume. With the wide soil tolerances, persistence under grazing by cattle and sheep, and high palatability and nutritive value, it is a better choice than red and ladino clovers for long-term pastures or meadows in parts of the Northeast. Its strongest feature probably is the ability to produce on poorly drained, shallow soils. It is very important in the Northeast and North Central states on areas too wet to grow alfalfa. An unusual and important characteristic of birdsfoot trefoil is that animals have not been known to bloat on this legume. Apparently this is the result of an accumulation of tannins which precipitate soluble protein to reduce bloat potential.

This legume works well for pasture renovation. Some cultivars will reseed well under grazing which contributes to its persistence.

Birdsfoot trefoil is valuable for hay as well as for pasture. Because it is somewhat slower in establishment than other well-known hay legumes such as alfalfa and red clover, it is often used for hay only under conditions not suitable for the more generally grown hay legumes. Because of its susceptibility to lodging, slow establishment and high seed cost, it is not well suited for hay in a short rotation. It differs from other legumes in staying succulent, leafy, and of high nutritive value long after the seeds have matured. Stands will not persist well under frequent cutting intervals, low cutting height, and late fall harvests.

Cultivars and Strains

There are two distinct types of broadleaf birdsfoot trefoil, the American and the European. European types make a more erect growth, begin to bloom ten days to two weeks earlier, and recover more quickly after cutting. The characteristics of the European type would seem to be favorable when the crop is to be used for hay or for rotational grazing. The American strain, with its more spreading, decumbent habit of growth, can better withstand the close grazing to which many permanent pastures are subjected.

In New York, seed that traces back to the naturalized stands first observed in the vicinity of Preston Hollow are certified under the name 'Empire.' Other pasture type cultivars are 'Dawn,' which was developed in Missouri for greater resistance to root rots; 'Leo,' developed in Canada for spring vigor and good winter hardiness; and 'Carroll,' developed in Iowa for greater winterhardiness when grown in the north central states.

The New York station has released a cultivar of the European type under the name 'Viking,' considered to be a hay type. 'Mansfield' is another European type sometimes recommended for hay.

Cultural and Management Practices

In establishing birdsfoot trefoil, of first importance is the fact that the seedlings start off slowly and are not able to endure much competition. It is generally advised that birdsfoot trefoil not be seeded with other legumes and that if seeded with a small grain companion crop, the rate of seeding the small grain be much reduced. Because of the slow seedling growth of this legume, companion crops have often been found to be detrimental to establishment. Better results often have been obtained by using a herbicide such as EPTC for weed control in a spring seeding. The usual rate of seeding is 4–6 pounds/acre (4.5–6.7 kg/ha). Inoculation of the trefoil seed with a specific *Rhizobium* culture is essential on most soils. Because the trefoil lodges badly, it is advisable to seed a grass with it. Kentucky bluegrass, timothy, and orchardgrass have been most used. Because of the small seed size and slow establishment, a smooth, firm seedbed is essential.

Mowing, grazing lightly, or applying selective herbicides such as 2,4,-DB help reduce weed competition in established trefoil. Sometimes a combination of methods is needed.

Birdsfoot trefoil has been especially recommended in the renovation of unproductive bluegrass pastures. The usual procedure is to tear up or shallow-plow the sod in the late fall or in the early spring or to apply herbicides to kill most of the grass. Lime, phosphate, and potash are applied according to the needs indicated by soil tests, and the legume is seeded in the early spring. Various types of planters or sod drills, many of which use the principle of minimum tillage, are often useful in renovation procedures.

Pollination and good seed set is dependent on insects such as bees. The birdsfoot trefoil pods mature unevenly and if humidity is relatively low are likely to dehisce, shattering the seed to the ground shortly after it becomes mature. The possibility of 100–300 pounds of seed per acre (112–336 kg/ha) appears not to be unusual, but the seed actually saved in harvesting is usually only a fraction of this.

The first crop of trefoil usually is harvested for seed. Seed yields are lower when the first crop is taken for hay and the second for seed. Two seed crops may be possible in some areas of the West. Trefoil is harvested when a large percentage of the pods are light brown to brown and before much shattering takes place. Three methods are used: (1) mowing before combining, (2) direct combining, and (3) direct combining after defoliation. The most common method is mowing, windrowing, and then combining from a partially dried windrow.

Annual losses of birdsfoot from diseases may be as high as 20 percent. Damp, humid, warm weather favors root diseases which may reduce the stand. Attacks by crown and root rot fungi have largely eliminated stands in the southernmost area of adaptation. *Stemphylium* leaf spot also is widespread and may cause extensive defoliation in some years. The leaves and stems of birdsfoot are glabrous so the plant is susceptible to injury by the potato leafhopper.

Other Species

In addition to the broadleaf birdsfoot trefoil there are two other species of some interest and value in this country—the narrowleaf birdsfoot, *L. tenuis,* and big trefoil, *L. pedunculatus.* The narrowleaf trefoil is adapted to low wet soils in areas of California, Oregon, and New York. It makes a considerably smaller growth and is more shallow-rooted than the broadleaf trefoil; in addition, it is less heat and drought tolerant. In California the narrowleaf has been shown to be strikingly salt tolerant. Big trefoil has become naturalized along the northern Pacific Coast. It is a long-lived perennial, with its area of adaptation limited to regions of mild winter temperature. This trefoil spreads by means of rhizomes, as well as from seed. It is especially well adapted to wet, poorly drained soils and is tolerant to brackish overflow water. Improved cultivars of big trefoil such as 'Beaver,' 'Columbia,' and 'Marshfield,' are available for use in Oregon and Washington.

The Lespedezas

There are more than 100 species of lespedeza. All are perennial except two, which are annual. The two annuals, striate lespedeza, *Lespedeza striata,* and Korean lespedeza, *L. stipulacea,* have great agricultural value in this country. One perennial species, commonly known as sericea, *L. cuneata,* also is grown on a considerable acreage in the South.

The striate lespedeza, also known as common, came from the Orient. The first record of it is in Georgia in 1846. A later introduction, under the name *Kobe,* was made in 1920. Korean lespedeza came in 1919 as a seed sample sent by a missionary in Korea. The perennial lespedeza, sericea, as it is known today, is a selection from seed introduced from Japan in 1924. There are several shrubby lespedezas, such as *L. bicolor,* which are useful as ornamentals or for wildlife cover and feed.

Plant Description

The annual lespedezas grow to a height of from a few inches to perhaps 2 feet (0.6 m), depending on the environment. The seedlings develop two, opposite, unifoliolate leaves in comparison to the one unifoliolate leaf of the true clovers. Striate lespedeza can be distinguished from Korean by the more narrow leaflets. Korean has larger, more prominent white stipules at the base of the leaf petiole. Both have prominent, white "skeleton-like" leaf veins. The plants are fine-stemmed and leafy, but they do not have the succulence of most small-seeded legumes. These annuals have shallow taproot systems.

Light pink to purple flowers appear in the late summer to fall. The flowers of striate lespedeza occur in leaf axils along the entire length of the stem, while those of Korean appear in clusters at the tip of branches. Lespedezas form both small inconspicuous flowers—cleistogamous—which do not open and are self-pollinating, and normal, showy-petaled flowers—chasmogamous—which are largely cross-pollinated. The seed shatter rather readily at maturity; the crop normally reseeds well.

The perennial lespedeza sericea is an erect plant with coarse stems. If left uncut, sericea may grow to a height of 3 feet (0.9 m) or more and the stems can become quite woody and unpalatable. The leaflets of sericea are long, narrow, and square on the ends. Although a perennial, the top growth is killed back each year and new growth arises from crown buds the following spring.

Distribution and Adaptation

Collectively, the area of practical use is from the Atlantic Coast states westward into eastern Texas, Oklahoma, and Kansas, and northward into

FIGURE 36-3. *The perennial sericea lespedeza is an erect plant with coarse stems which can become woody if left uncut. Leaflets are long and narrow.* [SOURCE: Southern Illinois University.]

southern Illinois, Iowa, and Indiana. Sericea is well adapted to this entire region, while Korean is most adapted to the upper two-thirds of the region, and striate in the southern part of the region. One of the most valuable characteristics of the lespedezas is their ability to establish, grow and mature seed on soils so low in fertility that other crops and most weeds fail. They are heat and drought resistant and tolerant of acid soils. Since their peak production is during the summer months, they can provide grazing at a time when most species in a cool-season perennial pasture in the northern area of their adaptation are furnishing little livestock feed.

Importance and Use

Lespedezas are used widely for pasture, hay, and soil conservation. The annual lespedezas have their greatest use as pasture. Sericea can be grown for hay or can be used for grazing either alone or in combination with a grass. It is an excellent soil-building legume. It will grow on depleted, acid soils of low fertility. A good growth of lespedezas for pasture does not occur until late spring or early summer. For the early part of the grazing season other grasses or legumes in the mixture must provide pasturage. Lespedezas often develop abundant growth about the time the carrying capacity and nutritive value of permanent grass pastures have declined. From early summer to early fall lespedeza has good carrying capacity.

While annual lespedezas can be harvested for hay, yields are commonly quite low. However, good annual lespedeza hay is nearly equal to alfalfa hay in feeding value. Higher hay yields can be expected from sericea, but the feeding value is somewhat lower than comparable annual lespedeza hay.

Cultivars

Korean, striate, and sericea are species of lespedeza, not different cultivars. Among the improved cultivars of Korean are 'Iowa 6,' 'Rowan,' 'Yadkin,' and 'Summit.' These cultivars generally are more resistant than common Korean lespedeza to diseases and pests such as bacterial wilt, powdery mildew, tar spot, and root knot nematodes.

Several improved sericea cultivars have been developed and released. 'Arlington,' from the USDA, was the first improved cultivar in general use. 'Serala,' and more recently 'Serala 76,' developed primarily at the Alabama station, are finer-stemmed, more acceptable cultivars than the common sericea. 'Interstate' and 'Interstate 76' are low-growing cultivars re-

FIGURE 36–4. *A field of 'Serala' sericea lespedeza for grazing and hay near Cherokee, Alabama.* [SOURCE: Southern Illinois University.]

leased from the Alabama station in cooperation with the Georgia station and the USDA, primarily for highway roadbank stabilization. These can be used for grazing or hay as well as for conservation purposes. Plant breeders are interested in low-tannin sericea lines for improved palatability and animal performance.

Cultural and Management Practices

Annual lespedezas usually are seeded in the early spring at a rate of 20–30 pounds/acre (22 to 34 kg/ha). Broadcasting seed on the surface without covering often can be successful. Good land use requires that lespedeza be grown and managed in association with other crops, such as small grains. A common practice is to seed in fall-sown small grain or in spring-sown oats, utilizing the small grain for pasture instead of harvesting for grain. Usually by the time the small grain crop has been fully grazed the lespedeza has developed to furnish good pasture until frost. After it is established lespedeza may be managed so as to maintain a continuous stand in the small grain in successive years through volunteer reseeding.

Sericea has a high percentage of hard seed and should be scarified before planting. Recommended seeding rate is often 30 pounds/acre (34 kg/ha) or more, but use of a herbicide to reduce weed competition in the spring may reduce the seeding rate to as little as 10 pounds/acre (11 kg/ha). The key to good-quality sericea pasture or hay is to graze or cut early; older plants are too coarse and woody. Sericea will not tolerate close grazing or clipping. Two cuttings of hay per year can be expected.

The annual lespedezas are heavy seed producers. Yields of 600 pounds/acre (672 kg/ha) have been obtained but yields of 200–300 pounds/acre (224–336 kg/ha) are more common. Seed ordinarily is combined directly as soon as the leaves are dry and the seedpods brown. Sericea seed yields range from about 300–900 pounds/acre (336–1,008 kg/ha). Sometimes the first growth is removed for hay and seed harvested from the second growth.

The Sweetclovers

There are two principal sweetclovers of agricultural importance, *Melilotus alba* and *M. officinalis*. Reported to have been found growing in Virginia as early as 1739, sweetclover is native to temperate Europe and Asia. Its value as a soil improving crop was recognized by 1900, and its use gradually spread through the Corn Belt and Great Plains thereafter.

Distribution and Adaptation

Thriving under a wide range of soil and climatic conditions, all the sweet-clovers have one important restriction in that they do not tolerate an acid soil condition. They are deep-rooted and drought resistant. Recognized as particularly valuable for soil improvement and in mixtures for pasture, their use has been limited in the East because of the tendency of soils in

FIGURE 36–5. *Pinnately trifoliolate leaves of sweet clover are similar to those of alfalfa. Leaflets of sweet clover are serrated completely along the margins, while those of alfalfa are only about one-third serrated.* [SOURCE: Southern Illinois University.]

that area to an acid reaction. The geographic regions of maximum adaptation include the northern, central, and southern Great Plains, and much of the Corn Belt.

Plant Description

The sweetclovers of agricultural importance are typically biennial. The first season's growth of the biennials consists of one central, much-branched stem. Toward the end of the first year several buds form at the crown, which is slightly below the soil surface. In the early spring of the second year, coarse, rapid-growing stems push up, usually several from each crown. The trifoliolate leaves are similar to alfalfa, except that leaflets are serrated completely along the margins. Leaves have a high content of coumarin, which gives the forage its characteristic bitter taste and its sweet-smelling aroma when cut. The sweetclover flowers are either white or yellow, depending on the species, and are borne in long, loose racemes. The seeds shatter readily as they mature.

Importance and Use

This legume has no equal as a soil-improving crop. The strong, fleshy taproots of sweetclover penetrate the soil deeply and improve aeration. Roots break down and decay rapidly at maturity. Plants may grow to a height of 5–6 feet (1.5–1.8 m), giving a high tonnage of dry matter per acre. In addition to its use for soil improvement, such as green manure, sometimes it is used for pasture, hay, or silage. In the Dakotas it is utilized

FIGURE 36-6. *Sweetclover flowers are either white or yellow, depending on the species and are borne in long, loose racemes.* [SOURCE: Southern Illinois University.]

for hay and silage, while in the southern Great Plains it is more important for pasture. It is also used in crop rotations and for honey production.

Species and Cultivars

The two most extensively used species are the yellow-flowered biennial, *M. officinalis,* and the white-flowered biennial, *M. alba.*

In general, the biennial yellow matures 10–15 days earlier, is finer stemmed, and gives a better quality but smaller yield of hay. The yellow is credited with being more tolerant to adverse conditions such as drought and competition with a companion crop. The biennial white is more productive, either for green manure or for pasture.

Improved cultivars of *M. officinalis* are 'Madrid,' which originated from seed introduced from Spain; 'Goldtop,' developed in Wisconsin; and 'Yukon,' derived from 'Madrid' by natural selection under Canadian conditions. Cultivars of *M. alba* include 'Denta' from the Wisconsin station, and 'Arctic' and 'Polara' from Canada. While the biennial *M. alba* is most important agriculturally, several annual cultivars of this species are used to some extent. Among these are 'Hubam' from Iowa, 'Floranna' from Florida, and 'Israel' introduced from Israel and released by the Texas station.

An unusual yellow-flowered species, usually known as sourclover, *M. indica,* is used as a green manure in the Southwest and in the lower Pacific Coast area.

Coumarin

The high content of coumarin gives all sweetclovers a rather bitter taste. When stored without having been well cured, the ensuing heat and mold may result in the formation of a toxic substance, a breakdown from the coumarin. This toxic substance, dicoumarol, reduces the clotting power of the blood when the forage is eaten by animals, with the result that the animals may bleed to death from slight wounds or from internal hemorrhages. This is known as the "bleeding" disease. Cultivars such as 'Arctic,' which are low in coumarin, now are available.

Culture and Management

Sweetclover has a high percentage of hard seeds and should be scarified before planting. Seeding rate is 10–15 pounds/acre (11–17 kg/ha). If the soil pH is below 6.0, the soil reaction should be modified by liming. In the Corn Belt seedings are usually made in the spring, in either spring or winter small grains as a companion crop.

For grazing the first's years growth, which consists of single main stems, sweetclover should be managed carefully. Regrowth must come from buds along the stem and not the crown area. Growth in the second year may be grazed heavily up to midsummer.

High quality hay may be made from the first year's growth, but the second year's growth is less suitable. If cut at the bud stage and cured properly, feeding value compares favorably to that of alfalfa hay.

When used for green manure, it may be plowed under in the fall of the seeding year, but turning under the following spring when it is about 6 inches (15 cm) tall is preferable.

Diseases do not cause major losses to sweetclover in the Great Plains, but are more severe in the Corn Belt. Root rots and black stem have occurred frequently in humid areas. The adult sweetclover weevil is the major insect pest, destroying stands by defoliating new seedlings.

Most of the sweetclover seed is produced in the prairie area of western Canada, in northwest Minnesota, and in the eastern parts of the Dakotas, Nebraska, Kansas, and Oklahoma, and Texas.

The Vetches

The vetches, in the genus *Vicia,* are widely distributed over the world and include about 150 species. Some 15 are native to the United States. The species in commercial use, however, are all native to Europe or western

Asia. The species of most economic importance in the United States are hairy vetch, *Vicia villosa,* common vetch, *V. sativa,* and purple vetch, *V. benghalensis.* Hairy vetch is by far the most important, with well over three-fourths of the vetch acreage planted to this species. All are semiviny winter annuals with pinnate leaves. Most are decidedly hairy except for the common vetch, which is relatively smooth. The predominant flower color is purple.

Importance, Adaptation, and Use

The most extensive and important uses of the vetches are as cover and green manure crops. Grown as a winter annuals in the South, they serve as valuable soil-improving crops at a time when the land is not occupied by a cash crop such as cotton, peanuts, or corn. Common vetch, purple vetch, and hairy vetch also have been used extensively as orchard cover crops in the Pacific Coast states.

Hairy vetch is the most winter-hardy of the vetches and is adapted to light, sandy soils as well as to heavier soils. It is used most commonly as a winter cover crop in Oklahoma, Arkansas, Texas, and Louisiana. Hairy vetch is cold tolerant enough to be a winter annual in northern regions as well as in the South and West and thus is fall-seeded. In the South and in other mild climates, the best time for seeding is October 1 to November 15. It may be seeded with small grains for pasture, hay, or silage. 'Madison' is an improved cultivar.

FIGURE 36–7. *Hairy vetch is a semi-viny winter annual with pinnate leaves. Note the terminal tendrils for clasping or climbing.* [SOURCE: Southern Illinois University.]

FIGURE 36–8. *A good growth of hairy vetch in winter rye, ready to be plowed down for green manure. This legume is a valuable soil-improving crop in the South and Southwest.*

The species *V. villosa* contains the subspecies *varia*, or winter vetch, which is more heat tolerant but less cold tolerant than hairy vetch. 'Auburn,' 'Oregon,' and 'Lana' are improved cultivars of winter vetch, formerly called woollypod vetch.

Common vetch is less winter hardy than hairy vetch. The better cultivars are 'Williamette' and 'Warrior.' Purple vetch is the least winter-hardy of those in commercial use. It is similar in appearance to hairy vetch, but is distinguished by wine-colored flowers and hairy pods.

Crownvetch

Crownvetch, *Coronilla varia,* is a hardy, herbaceous perennial which spreads by creeping roots. The plant has some characteristics similar to hairy and common vetches, but it is not a true vetch. For many years crownvetch was used primarily as an ornamental plant. In more recent years the value of crownvetch for erosion control has been widely accepted, and it is used on highway embankments, stripmine spoil banks, and similar areas. It is gaining in importance for forage purposes, being used for pasture and hay. The major shortcomings appear to be slow germination,

FIGURE 36-9. *The pinnate leaves of crownvetch are similar to those of hairy vetch, but crownvetch is not a true vetch. A terminal leaflet is present on each leaf rather than a tendril.* [SOURCE: Southern Illinois University.]

weak seedlings, slow growth and poor competition with weeds during establishment. Crownvetch is high in protein, about the same as alfalfa, but yields are generally lower than those of alfalfa. On shallow, droughty soils it may outyield most other legumes. It cannot stand close and frequent clipping and grazing, and does not recover quickly after cutting. Normally only two hay cuttings a year may be obtained.

Crownvetch is adapted north of the 35th parallel. Good persistent stands have been obtained on poor, acid soils, but to establish quickly, soils must be limed and fertilized. This legume will not tolerate poorly drained soil conditions. Seed should be scarified and inoculated before planting. Weeds must be controlled with herbicides or by proper management until crownvetch becomes established.

Seed harvest is complicated by uneven maturity caused by an indeterminate flowering habit and the tendency of the seedpod to shatter readily when disturbed. Field losses from machine harvest of seed often exceed 50 percent.

The three available cultivars are 'Emerald,' 'Chemung,' and 'Penngift.' Crownvetch breeding and improvement efforts are centered around greater seedling vigor and improved forage quality.

Lupine

There are more than 250 species of the genus *Lupinus*. Most of these are native to the Americas, but the species used agriculturally are all native

FIGURE 36–10. *Crownvetch flowers and seedpods. The inflorescence is an umbel, with individual flowers lavender and white in color. Flowers and the resulting pods are attached at a single point.* [SOURCE: Southern Illinois University.]

to Europe. The three cultivated lupines in the United States are blue lupine, *L. augustifolius;* yellow lupine, *L. luteus;* and white lupine, *L. albus.* Known and used to some extent before the time of Christ, they were of comparatively little importance until about 200 years ago, when they came into common use in central Europe. In the United States they have been in commercial use only since about 1950. The blue lupine is the one most grown in this country and is used as a soil-improving crop in the Southeastern states.

The lupine makes a coarse, upright growth to a height of 2–3 feet (0.6–0.9 m). Some species normally contain an alkaloid that is poisonous. Lupine improvement began in 1928 when alkaloid-free lines, nonpoisonous to livestock, were discovered.

Use in the United States is confined to an area south of central South Carolina, Georgia, and Mississippi to Austin, Texas. The lupines are grown as winter annuals, to be plowed down for the benefit of crops to follow, and a lesser extent for winter and early spring grazing and for silage.

Miscellaneous Forage Legumes

Cowpea

The cowpea, *Vigna sinensis,* is believed to have been grown for several centuries as a food plant over wide areas of Africa, Europe, and Asia. Introduced into the South early in the 1700s, it was for many years the most extensively grown legume, being used both as a food and as a hay crop. It was an important crop as far north as southern Missouri, Illinois, and Indiana. Acreage of cowpeas in the United States has been declining since the 1940s. Most of the cowpea acreage in the South has now given way to soybeans, clovers, and other crops.

This warm-season annual crop can be used for pasture, hay, silage, soil improvement, or human food. Cowpeas can add nitrogen and organic matter when used in corn and cotton rotations in the South. It is a viny to semiviny plant with weak stems, large leaves, and many curved pods. This legume is adapted to a wide range of soil conditions, but tolerates acid

FIGURE 36–11. *Cowpeas should be cut for hay when the pods are fully mature and begin to yellow, and before leaf drop. This legume may be interplanted with corn or sorghum and cut for silage.* [SOURCE: Southern Illinois University.]

soils better than alkaline, and well-drained soils better than poorly drained.

Cowpeas should be cut for hay when the pods are fully mature and begin to yellow, and before leaf drop. Hay is difficult to cure, but properly cured cowpea hay is equal in nutritive value to red clover hay. This crop also may be interplanted with corn or sorghum and cut for silage. Good forage varieties are vigorous, upright, disease-resistant, and retain their leaves until late in the season.

Some cowpeas, such as Crowder and Blackeye types, are harvested for table use.

Kudzu

Kudzu, *Pueraria thunbergiana,* is a native of Japan and China and was introduced as an ornamental into this country about 1875. In the early 1900s interest developed in kudzu for use as a forage or for soil conservation in the humid South.

It is a long-lived, coarse-growing, stoloniferous, viny perennial legume. The running vines often extend 40–50 feet (12–15 m) in a single season, with roots and new plants established at each node. It is of particular value for erosion control and as a soil-building and pasture crop. Kudzu

FIGURE 36–12. *Kudzu used for soil stabilization on a highway roadbank in Mississippi. This rapid-growing legume can serve as a good forage and soil conserving crop, but may become a pest, covering trees and weakening or killing them by excluding light.* [SOURCE: Southern Illinois University.]

is adapted to the same general region as lespedeza, that part of the United States south of the Ohio River and east of the Great Plains.

Kudzu plantings are established by setting one- or two-year-old plants in rows, with the rows up to 30 feet (9 m) apart and the plants spaced 5 feet (1.5 m) in the row. Digging and planting kudzu crowns is done while the plants are dormant.

If allowed to grow unhampered, kudzu can become a pest, covering trees and weakening or killing them by excluding light.

Velvetbean

Velvetbean, *Stizolobium deeringianum*, is a vigorous-growing, summer annual used for fall and winter pastures in southern Georgia, northern Florida, and other limited areas of the Southeast. Popularity of this crop for forage has declined, starting in the 1940s.

Most cultivars are viny; they attain a stem length of 40 feet (12 m) or more if conditions are favorable. The trifoliolate leaves are large and numerous; flowers are borne in pendant clusters. The pods are covered with hairs which sting like nettles. Velvetbeans are well adapted to the less fertile and sandier soils of the Southeastern states, where they are used as a soil-improving crop. They are often planted with corn, used as support for the viny stems.

Because velvetbeans are hardy, yield well, and decay readily, they make a good green manure crop. Sometimes they are grown for soil improvement on newly cleared land. Interest in velvetbean seed has increased because of the content of L-DOPA, used in treatment of Parkinson's disease.

Rough Pea

The rough pea, *Lathyrus hirsutus*, is a winter annual with weak stems and a decumbent growth. Other names are wild winter pea, caley pea, and singletary pea. It is native to the Mediterranean region. The principal uses are for winter cover and early spring pasture in the lower South. Once seeded, the crop volunteers almost indefinitely, because of the hard seed that continue viable in the soil for a number of years.

Hairy Indigo

Hairy indigo, *Indigofera hirsuta*, is an upright-branching summer annual legume adapted to the Gulf Coast. It has moderately coarse stems that become woody with age, and the plants attain a height of 4–7 feet (1.2–2.1 m). The leaves resemble those of vetch. The plant is native to tropical Asia, Australia, and Africa. It may be used for pasture, hay, or green manure. A small acreage is reported in Florida.

Guar

Guar, *Cyamopsis tetragonoloba,* is an upright, coarse-growing, summer annual legume that is drought resistant. It has been used to some extent in California, Arizona, and Texas as a green manure crop. The seed has commercial use, yielding an industrial gum. Guar was introduced from India in 1903.

Florida Beggarweed

Florida beggarweed, *Desmodium purpureum,* is native to tropical and subtropical America and has spread as far north as Florida and adjoining states. It is an upright, branching plant attaining a height of 4–7 feet (1.2–2.1 m). In the United States it behaves as a summer annual and is used for pasture and soil improvement.

REFERENCES AND SUGGESTED READINGS

1. AL-TIKRITY, W., *et al.* "Seed Yield of *Coronilla varia* L., *Agron. Jour.,* 66:467 (1974).
2. BALDRIDGE, J. D., *et al. Using Birdsfoot Trefoil in Missouri,* Univ. Mo. Science and Technology Guide (1966).
3. CLARK, N. A. *Crownvetch,* Univ. of Md., The Agronomist (1977).
4. COPE, W. A., and J. C. BURNS. "Components of Forage Quality of Sericea Lespedeza in Relationship to Strain, Season, and Cutting Treatment," *Agron. Jour.,* 66:389 (1974).
5. DECKER, A. M., *et al.* "Permanent Pastures Improved with Sod-Seeding and Fertilization," *Agron. Jour.,* 61:243 (1969).
6. DONNELLY, E. D. "Registration of Interstate Sericea Lespedeza," *Crop Science.,* 11:601 (1971).
7. DONNELLY, E. D., W. B. ANTHONY, and J. W. LANGFORD. "Nutritive Relationships in Low- and High-Tannin Sericea Lespedeza Under Grazing," *Agron. Jour.,* 63:749 (1971).
8. GORZ, H. J., and W. K. SMITH. "Sweetclover," In *Forages* (Ames: Iowa State Univ. Press, 1973).
9. GUTEK, L. H., *et al.* "Variation of Soluble Protein in Alfalfa, Sainfoin and Birdsfoot Trefoil," *Crop Science,* 14:495 (1974).
10. HART, R. H., A. J. THOMPSON III, and W. E. HUNGERFORD. "Crownvetch–Grass Mixtures under Frequent Cutting: Yields and Nitrogen Equivalent Values of Crownvetch Cultivars," *Agron. Jour.,* 62:287 (1977).
11. HEATH, M. E. *Crownvetch,* Purdue Univ. Agron. Guide AY-177 (1969).
12. HENSON, P. R., and W. A. COPE. *Annual Lespedezas Culture and Use,* USDA Farmers' Bul. 2113 (1969).
13. HENSON, P. R., and H. A. SCHOTH. *Vetch Culture and Uses,* USDA Farmers' Bul. 1740 (1968).
14. HERMANN, F. J. *Vetches in the U.S., Native, Naturalized, and Cultivated,* USDA Agr. Handbook 168 (1960).

15. HOVELAND, C. S., and W. B. ANTHONY. "Cutting Management of Sericea Lespedeza for Forage and Seed," *Agron. Jour.,* 66:189 (1974).

16. HOVELAND, C. S., and E. D. DONNELLY. "A Comeback for Sericea," *Crops and Soils,* 21:2 (November 1968).

17. HOVELAND, C. S., *et al. Management of Sericea for Forage and Seed,* Auburn Univ. Agr. Exp. Sta. Circ. 222 (1975).

18. HOVELAND, C. S., *et al. Sericea–Grass Mixtures,* Auburn Univ. Agr. Exp. Sta. Circ. 221 (1975).

19. LEFFEL, R. C. "Other Legumes," In *Forages* (Ames: Iowa State Univ. Press, 1973).

20. MCKEE, G. W., R. A. PFEIFFER, and N. N. MOHSENIN. "Seedcoat Structure in *Coronilla varia* L. and Its Relation to Hard Seed," *Agron. Jour.,* 69:53 (1977).

21. MOORER, M. M., and E. D. DONNELLY. *Vetches for Seed Production, Green Manure and Winter Grazing,* Auburn Univ. Ext. Serv. Circ. 613 (1962).

22. OFFUTT, M. S., and J. D. BALDRIDGE. "The Lespedezas," In *Forages* (Ames: Iowa State Univ. Press, 1973).

23. Phillips Petroleum Co. *Introduced Grasses and Legumes* (Bartlesville, Oklahoma, 1960).

24. ROBINSON, R. G. "The Shortcomings of Crownvetch," *Crops and Soils,* 21:18 (1969).

25. SEANEY, R. R. "Birdsfoot Trefoil," In *Forages* (Ames: Iowa State Univ. Press, 1973).

26. SMITH, D. and R. M. SOBERALSKE. "Comparison of Growth Responses of Spring and Summer Plants of Alfalfa, Red Clover, and Birdsfoot Trefoil," *Crop Science,* 15:519 (1975).

27. SMITH, W. A. "Sweetclover Improvement," In *Advances in Agronomy,* Vol. 17 (New York: Academic Press, 1965). ed. A. G. Norman.

28. USDA. *Trefoil Production for Pasture and Hay,* Farmers' Bul. 2191 (1967).

29. WHEATON, H. N. *Crownvetch,* Univ. Mo. Science and Technology Guide (1968).

Chapter 37

COOL-SEASON PERENNIAL GRASSES

THERE are thousands of species of grass, and practically all those extensively used throughout the humid sections of the North and Midwest are introductions from Europe.

Grasses with wide adaptation extensively grown in this area are the grasses to be considered here: Kentucky bluegrass, bromegrass, timothy, orchardgrass, tall fescue, redtop, and reed canarygrass.

Kentucky Bluegrass

Kentucky Bluegrass, *Poa pratensis,* is believed to have been brought from Europe by Early English colonists before 1700. As settlers moved westward, it spread rapidly in that direction.

There are many species of *Poa* native to North America, however. Kentucky bluegrass is the most important of the 69 species of *Poa* found in North America. Kentucky bluegrass is called "junegrass" or "smooth-stalked meadow grass" in Europe. It got the name "Kentucky" bluegrass from the early importance in Kentucky and as a recognition of the excellent pastures there.

Well known in Europe and in parts of western Asia, it is found throughout North America, except in arid regions and at alpine altitudes. In the United States it is the most extensively grown of all grasses in the humid Northern states but is not common in the Gulf states. It is best adapted to well-drained, highly productive soils of limestone origin. On soils inclined toward an acid reaction or low in fertility this grass gives way to such grasses as redtop and Canada bluegrass.

Plant Description

Kentucky bluegrass is an aggressive rhizomatous plant which forms a dense, low-growing sod. Leaves are mostly basal and blades are flat or folded with boat-shaped tips. The leafy, erect, unbranched culms grow to a height of 12–30 inches (30–76 cm). The inflorescence is an open, pyramidal-shaped panicle. Spikelets are awnless, but lemmas have a heavy midnerve like the keel of a boat. Each of the seeds has a mat of cobwebby hairs at its base.

It begins growth very early in the spring, producing a succulent, nutritious, and palatable forage. As heads begin to form in June, however, the

FIGURE 37–1. *Kentucky bluegrass is rhizomatous and forms a dense, low-growing sod. Leaf blades are narrow, often folded, with boat-shaped tips.* [SOURCE: Southern Illinois University.]

grass becomes less palatable and nutritious, with little or no production during the heat of midsummer. Kentucky bluegrass is largely apomictic, producing seeds without fertilization. Those seeds that are reproduced sexually are highly variable and do not breed true.

Importance and Use

Kentucky bluegrass is considered by some as the most important pasture grass in North America. It contributes largely to the grazing obtained from 50–100 million acres (20–40 million ha) of humid pastureland. It is given a low rating by some, however, because of its relatively low summer yield and its high fertility requirements. The seedlings develop slowly, and as a result, it is recommended only for permanent pastures or for pastures in long rotations. Probably 90 percent of the Kentucky bluegrass in America developed spontaneously. Kentucky bluegrass usually is sown in mixtures that include a more rapidly establishing grass, such as timothy.

Seed Production

Seed production for Kentucky bluegrass centers in Kentucky, as well as in a north central region, and a northwest region of the United States. Some seed is harvested from pasture fields managed for good seed yields, and the remainder is harvested from fields grown only for seed. Bluegrass seed grown in Kentucky and some in the north central region are harvested with strippers, self-cleaning metal combs. Some strippers are essentially rotating, spike-studded cylinders that brush the seed into a box which is part of the machine. All other seed in the north central district and that in the northwest district are harvested with self-propelled combines. Most crops are swathed into windrows first, but some bluegrass is combined standing.

Other Species

There are about 200 species of *Poa* distributed throughout the world. After Kentucky bluegrass, the most important *Poa* is Canada bluegrass, *Poa compressa,* also brought from Europe. This grass dominates only on soils that are too acid, droughty, or infertile for Kentucky bluegrass. Canada bluegrass has a distinct, easily recognized blue-green color. It produces few basal leaves, in marked contrast to Kentucky bluegrass. It is inferior to Kentucky bluegrass for grazing.

Three other *Poa* species grown somewhat extensively in certain areas are mutton bluegrass, *P. fendleriana,* Texas bluegrass, *P. arachnifera,* and big bluegrass, *P. ampla.* Roughstalk bluegrass, *P. trivalis,* a native of Europe, where it is used much in pastures, is used for shady lawn seeding in the United States.

Superior cultivars of Kentucky bluegrass have been developed for lawn purposes. Generally different cultivars are selected for lawn and pasture uses.

Smooth Bromegrass

Smooth brome, *Bromus inermis,* is the most widely used of the cultivated bromegrasses. There are about 60 species of *Bromus,* with 36 listed as occurring in the United States. A southern type was introduced from Hungary and a northern type from Siberia in the late 1800s. The Southern type is most important in the northern and midwestern regions. It came into great prominence and gained popularity over a wide area because of its outstanding value for pasture, hay, and erosion control.

Adaptation and Distribution

Smooth brome is adapted to most temperate climates. For North America its range of adaptation is roughly the Corn Belt, in the Great Plains states under irrigation, and the Pacific Northwest. It is resistant to drought and to extremes in temperature. During droughty summer periods it becomes more or less dormant. It has a higher fertility requirement in general than other common grasses. Also, overgrazing of smooth brome can be disastrous to stands, and careful grazing management must be practiced.

FIGURE 37–2. *Smooth bromegrass culms usually grow to a height of 3 to 4 feet (0.9 to 1.2 m). Some leaves are borne on the culms. The inflorescence is a large open panicle whose branches are whorled at the node.* [SOURCE: Southern Illinois University.]

Plant Description

Smooth brome is a long-lived perennial grass with strong creeping rhizomes allowing it to form a coarse sod. The culms usually grow to a height of 3–4 feet (0.9–1.2 m). Soft, palatable leaves are glabrous and broad with a constriction in the shape of a W located about midway on the leaf blade. Leaves are basal as well as culm-borne. The flowering head is a large open panicle whose branches are whorled at the node. Spikelets are numerous and each spikelet contains 5–10 florets.

Utilization

Smooth brome is a highly palatable and nutritious grass for pasture or hay. It came into prominence during drought years when its performance was outstanding in comparison with other grasses more extensively grown at that time. When seeded in a mixture with other grasses and legumes, the bromegrass usually is the dominant grass. For hay, a bromegrass–alfalfa combination is most favored over a large area, with ladino added for use also as pasture.

Cultivars

Seed from the Dakotas and Canada, tracing back to Siberian origin, usually is designated as Northern Commercial. This type sods more slowly and produces a more open sod than strains that became established farther south. The more aggressive, sod-forming type, of central European origin, has been found to be better adapted to the central Corn Belt and west into the eastern Great Plains. This Southern type exhibits superior seedling vigor, so that stands are established with greater certainty. Southern strains that have come to be generally recognized and used are the 'Barton,' 'Blair,' 'Baylor,' 'Beacon,' 'Lincoln,' 'Sac,' and 'Southland.' These are closely related cultivars derived from old fields originating from some of the earliest introduction.

Seed Production

Most of the Southern-type seed is produced in Nebraska and Kansas. Considerable quantities of the Northern type come in from Canada, with some production also in North and South Dakota. Heavy applications of nitrogen fertilizer are required for maximum seed production. It is important that seed fields be kept from weeds, and especially from weedy bromes.

Other Bromes

There are many different brome species in North America, both introduced and native. These vary from early-maturing winter annuals to the slower-maturing, more productive perennials. For the most part the annuals are weedy grasses of little or no value. There are a number of native perennial brome species important in the mountain region of the western United States.

Timothy

Timothy, *Phleum pratense,* is native to northern Europe and eastward through Siberia. It has been known as "Herd" grass and "cat's tail," but obtained the name "timothy" from Timothy Hanson who played an important role promoting this grass in Maryland in the 1700s. It is said that timothy has been cultivated through a longer period than any other grass except perennial ryegrass.

Timothy is adapted to cool and humid climates such as in the Northeast. In the United States, for a long period of years it was the leading hay grass east of the Missouri and south of Kentucky.

Plant Description

Timothy is a perennial bunch grass with erect culms, 20–40 inches (51–102 cm) tall, arising from a bulbous stem base called a corm, where large amounts of carbohydrates are stored. It forms large clumps or bunches through tillering, and results in an open sod. There are no stolons or rhizomes. It bears a dense cylindric spikelike panicle inflorescence with one-flowered spikelets. The culms bear many leaves, which are long, broad, and pliable. The roots are relatively shallow and fibrous. Although individual timothy shoots are typically biennial, new shoots develop vegetatively each year from the older ones and the plant thereby maintains itself as a perennial.

Timothy is grown primarily for hay, but may be included also in mixtures for pasture. It is highly palatable. Its productivity compares favorably with other cool-season grasses early but the yield often declines rapidly after one or two growing seasons. Census reports show that more than 80 percent of the entire timothy acreage is east of the Missouri River and north of the Ohio River. The acreage of timothy has decreased steadily since the early 1900s. Bromegrass and orchardgrass are replacing it to a considerable degree.

In the past the timothy seed crop was harvested with a grain binder, with the bundles cured in shocks. Now timothy seed is harvested from the standing crop with a combine. Important seed-producing states are Minnesota, Missouri, Ohio, and Illinois.

Orchardgrass

Orchardgrass, *Dactylis glomerata* L., is native to western and central Europe. It was introduced into the United States in Colonial times, probably accidentally, as were most of the cultivated grasses. It was mentioned as early as 1785 by Thomas Jefferson as being grown in Virginia. It apparently was grown in several parts of New England in the early 1800s. The grass owes its common name to its shade tolerance and to its consequent occurrence in orchards. In England and in other parts of the world it is commonly known as "cocksfoot" because of the shape of the spikes of the panicle inflorescence.

Distribution and Adaptation

Orchardgrass is found throughout much of the temperate zone of the Northern Hemisphere. In North America it is found in the eastern Canadian provinces and southward to the northern parts of the Gulf states. It also occurs in the high-rainfall areas of the mountain states and in irrigated areas throughout much of the West. It is of greatest agricultural importance in the southern half of the timothy-bluegrass belt, from southern New York to southern Virginia and westward through Kentucky, Tennessee, and southern Missouri. It is an important grass for both hay and pasture in much of this area. It is less winter-hardy than smooth brome, timothy, or Kentucky bluegrass and is better adapted farther South than these species. It is deep-rooted and more drought resistant than timothy and Kentucky bluegrass but less so than smooth brome or tall fescue. Orchardgrass does best on well-drained, deep soil, but it will grow on thin and infertile soils more so than smooth brome or timothy, but less than tall fescue. It responds well to high levels of soil fertility, however. It does not do well on very poorly drained sites.

Plant Description

Orchardgrass is a long-lived perennial. It has no stolons or rhizomes and makes a distinctly bunch-type of growth. This open sod makes it a good companion for legumes. The plant bears mostly basal leaves which are characteristically folded and form a V in cross section. The root system is intermediate between the deep-rooted tall fescue and shallow-rooted timothy and Kentucky bluegrass. The flowering culms are 2–4 feet (0.6–1.2 m) tall and bear panicles with digitate spikelets.

Importance and Use

It is estimated that there may be 1 million acres (0.4 million ha) of orchardgrass in its area of adaptation. In parts of Virginia, Kentucky, and Tennessee it is the major grass used for forage. Here it is found in most permanent pastures, as well as being used extensively for hay.

Because of its productivity and vigor of growth, orchardgrass is an important constituent in high-producing, intensively managed pastures through-

FIGURE 37–3. *Orchardgrass, a bunch grass, bears mostly basal leaves which are characteristically folded and form a "V" in cross section.* [SOURCE: Southern Illinois University.]

FIGURE 37–4. *The flowering culms of orchardgrass are 2 to 4 feet (0.6 to 1.2 m) tall and bear panicles with digitate spikelets.* [SOURCE: Southern Illinois University.]

out a rather large area, from Maryland and New Jersey on the east to and including Kentucky and Tennessee on the west.

The grass starts growth early in the spring and grows rapidly, maturing about three weeks earlier than timothy. It recovers rapidly after grazing or mowing and produces relatively well throughout the growing season.

Management Practices

Because of its aggressive growth habits orchardgrass is likely to crowd out legumes seeded with it unless the rate of seeding the grass is kept relatively low.

It is important that seedings with orchardgrass be kept vegetative by mowing or grazing. It is highly palatable and nutritious early but both palatability and nutritive value decline rapidly as plants mature, more so than most cool-season grasses.

Virginia, Kentucky, and Missouri are the leading orchardgrass seed producing states.

FIGURE 37–5. *Orchardgrass clones in a space-planted nursery at the Tennessee Station. Evaluation of dozens of clones could result in an improved orchardgrass cultivar for the Upper South.* [Photograph with permission of the University of Tennessee Agricultural Experiment Station.]

Tall Fescue

Tall fescue, *Festuca arundinacea,* was introduced from Europe more than a century ago, but did not receive much attention in the United States until the 1930s. Oregon released an ecotype selection as the cultivar 'Alta' in 1923, and Kentucky released a natural selection found on a Kentucky farm in 1931 as the cultivar 'Kentucky 31.' These two cultivars still occupy a large part of the tall fescue acreage.

Adaptation and Distribution

Tall fescue is probably the most widely adapted cool season grass, being able to grow under a wide range of soil drainage and pH conditions. It has grown at a pH as high as 9.5 and as low as 4.7. It grows well on heavy clay soils and also on poorly drained soils. It can tolerate cold weather but will not grow under hot conditions during the summer. Tall fescue is recommended for long-lived permanent pastures and is used largely for beef cattle. It is well suited in an accumulated-forage system for autumn and winter grazing.

The U.S. acreage of tall fescue is estimated to be more than 3 million acres (1.2 million ha). Kentucky and Tennessee dominate in acreage with approximately 1 million acres (0.4 million ha) each.

Figure 37-6. *Tall fescue forms an open panicle inflorescence.* [Source: Southern Illinois University.]

Plant Description

Tall fescue is a long-lived perennial with a very deep root system which gives good drought tolerance. It has no stolons or rhizomes but tillers extensively to form bunches. The leaves are broad and flat with a short ligule and small auricles. The upper surface of the leaf is rough and prominently ribbed, while the under surface is smooth and glossy. The grass forms an open panicle inflorescence.

Utilization and Management

Generally tall fescue pastures are used for beef cattle. Seedings are made with white or Ladino clover or alsike clover for either wet or upland pastures or with red clover or annual lespedeza for upland pastures. This grass is aggressive which makes it difficult to maintain legumes in a mixture. It gets tough and unpalatable in the late spring as it approaches maturity. Under certain conditions it can result in fescue poisoning and grass tetany disorders in livestock. Its palatability generally is considered to be lower than orchardgrass, smooth bromegrass, timothy, or Kentucky bluegrass.

Cultivars

'Kentucky 31' and 'Alta' cultivars are still recommended in many states. Newer cultivars are 'Kenwell,' 'Fawn,' and 'Aronde' which are reported to have greater palatability or digestibility. 'Kenhy' is a new tall fescue–ryegrass hybrid released by the USDA and the University of Kentucky and is considered a more palatable and digestible grass. The new cultivar 'Missouri-96' is reported to give higher daily cattle gains than 'Kentucky 31.'

Redtop

Redtop, *Agrostis alba,* was brought from Europe by the early colonists. There are 32 *Agrostis* species recognized in the United States. All of these go under the common name of *bent,* except redtop. The bentgrasses are recognized for their value in lawn and golf course seeding.

Distribution and Adaptation

Redtop probably has a wider range of adaptation than almost any other cultivated grass. It succeeds over most of the United States except in the dry regions and in the extreme South. It is recognized as having considerable drought resistance yet is one of the grasses best adapted to wet land. It grows well on acid and wet soils and on soils of low fertility. It is aggressive and often spreads into areas where it is not wanted.

The Plant

On fertile soil, redtop grows to a height of about 3 feet (0.9 m). It is sod-forming by means of creeping rhizomes. The plant makes a leafy growth, with rather long and wide leaves, medium in texture. The inflorescence is in the form of an erect panicle that takes on a bright reddish color, from which the grass gets its name. Redtop seeds are very small, averaging about 5 million per pound (11 million per kg).

Culture and Use

In the past, redtop was used extensively in pasture and hay mixtures, especially on the more acid, low-fertility soils and on poorly drained soils. It is now recognized as considerably less palatable and nutritious than other productive gasses; therefore, it is not so generally recommended.

Reed Canarygrass

Reed canarygrass, *Phalaris arundinacea,* is one of the few grasses reported to be indigenous to the temperate portions of all five continents. Its use as a cultivated grass in Europe goes back to the 1800s. The first known date of cultivation in the United States was 1885 in Oregon. It came into rather extensive use in Minnesota about 1900 and after.

Distribution and Adaptation

Reed canarygrass is adapted and found growing throughout much of the northern half of the United States and the southern half of Canada. The greatest acreage is on the Pacific Coast in Oregon, Washington, and northern California. When seeded in most cases it has been on poorly drained soil in areas subject to flooding and silting. For such environments it is markedly superior to other grasses. Its natural habitat is poorly drained, wet areas, but in recent years it has been recognized also as one of the

Figure 37–7. *Reed canarygrass is a tall-growing, coarse, sod-former. Leaves are wide, relatively long, medium-textured and have prominent, papery ligules.* [Source: Southern Illinois University.]

most drought-tolerant of the cool-season grasses when grown on upland soil.

Plant Description

Reed canarygrass is a tall, coarse, sod-forming, cool-season grass. It normally makes a growth of 4–7 feet (1.2–2.1 m). The closed panicle inflorescence develops at the ends of culms in cylindric clusters 3–6 inches (7.6–15.2 cm) long. The seed shatters promptly when ripe. The culms bear many wide, relatively long, medium-textured leaves which have prominent and papery ligules. A very tough, dense sod is formed by means of vigorous, thick rhizomes which push out from the crown.

Uses and Culture

Reed canarygrass has been considered primarily as a pasture grass on poorly drained soils. Because of its wetland adaptation, it can produce green succulent feed after upland pastures have been dry and dormant. This grass is reported to have palatability problems because of its content of

FIGURE 37-8. *The culms of reed canarygrass are 4 to 7 feet (1.2 to 2.1 m) long and bear closed panicle inflorescences in cylindric clusters.* [SOURCE: Southern Illinois University.]

several basic alkaloids. Breeding work is underway to develop low alkaloid lines.

Reed canarygrass is unchallenged in its value to heal and control gulleys. Small pieces of sod embedded at intervals in the bottom and across gulleys when the soil is moist either in early spring or late summer spread and sod quickly. A lower-growing dwarf reed canarygrass has been developed as an improved plant for erosion control purposes.

In general, seeding a legume with this grass has not been successful because of the smothering effect of the grass. Seed is often low in germination and this should be checked with care. Reed canarygrass must be well fertilized, especially with nitrogen.

Improved cultivars that are available include 'Rise,' 'Vantage,' 'Ioreed,' 'Superior,' and 'Frontier.'

Seed Production

There is a period of only two to three days between the ripening of the first seed and the time when the seeds have fallen to the ground. Much

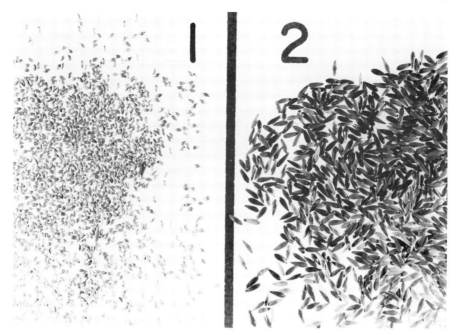

FIGURE 37–9. *Relative seed size and shape of timothy (1) and reed canary-grass (2).* [SOURCE: Southern Illinois University.]

care is necessary in handling such seeds to prevent heating and a resulting low germination.

REFERENCES AND SUGGESTED READINGS

1. ANONYMOUS. "New Fescue Variety, Missouri-96, Released," *Crops and Soils,* 30(5):20 (1978).

2. ANDERSON, J. M., and L. M. SAFLEY. *Orchardgrass-Ladino Clover and Fescue-Ladino Clover Pasture for Beef Cows and Calves With and Without Creep Feeding.* Univ. of Tennessee Agriculture Exp. Sta. Bul. 422 (1967).

3. ARCHER, K. A., and A. M. DECKER. "Autumn-Accumulated Tall Fescue and Orchardgrass. I. Growth and Quality as Influenced by Nitrogen and Soil Temperature," *Agron. Jour.,* 69:601 (1977).

4. ARCHER, K. A., and A. M. DECKER. "Autumn-Accumulated Tall Fescue and Orchardgrass. II. Effects of Leaf Death on Fiber Components and Quality Parameters," *Agron. Jour.,* 69:605 (1977).

5. BALASKO, J. A. "Effects of N, P, and K Fertilization on Yield and Quality of Tall Fescue Forage in Winter," *Agron. Jour.,* 69:425 (1977).

6. BALDRIDGE, D. E., and C. W. ROATH. *A Comparison of Potomac, Pennlate, and Other Orchardgrass Varieties in Montana.* Montana State Univ. Agr. Exp. Sta. Bul. 633 (1970).

7. BAXTER, H. D., *et al. Comparison of Alfalfa-Orchardgrass Mixture With Or-*

chardgrass as Hay for Lactating Dairy Cattle. University of Tennessee Agr. Exp. Sta. Bul. 547 (1975).

8. BROWN, C. S., *et al. Management and Productivity of Perennial Grasses in the Northeast. IV. Timothy.* West Virginia Agriculture Experiment Station Bulletin 570T (1968).

9. BUCKNER, R. C., and J. R. COWAN. "The Fescues," In *Forages* (Ames: Iowa State Univ. Press, 1973).

10. DECKER, A. M., *et al. Management and Productivity of Perennial Grasses in the Northeast. I. Reed Canarygrass.* West Virginia Agr. Exp. Sta. Bul. 550T (1967).

11. DOBSON, J. W., C. D. FISHER, and E. R. BEATY. "Yield and Persistence of Several Legumes Growing in Tall Fescue," *Agron. Jour.,* 68:123 (1976).

12. FERGUS, E. N., and R. C. BUCKNER. "The Bluegrasses and Redtop," In *Forages* (Ames: Iowa State Univ. Press, 1973).

13. FRIBOURG, H. A., and J. H. REYNOLDS. *Yield and Stand Responses to Orchardgrass Subjected to Different Management Treatments and Nitrogen Fertilizer Levels,* Univ. of Tennessee Agr. Exp. Sta. Bul. 451 (1968).

14. GIST, G. R. *Grasses and Legumes for Forage and Conservation.* Ohio State Univ. Ext. Bul. 478 (1966).

15. HARRISON, C. M., and J. F. DAVIS. *Reed Canarygrass for Wet Lowland Areas of Michigan.* Michigan State Univ. Ext. Bul. E-517 (1966).

16. HIGH, T. W., JR., *et al. Producing Yearling Steers on Irrigated Bluegrass-Clover.* Univ. of Tennessee Agr. Exp. Sta. Bul. 387 (1965).

17. JUNG, G. A., and B. S. BAKER. "Orchardgrass," In *Forages* (Ames: Iowa State Univ. Press, 1973).

18. KROTH, E., R. MATTAS, L. MEINKE, and A. MATCHES. "Maximizing Production Potential of Tall Fescue," *Agron. Jour.,* 69:319 (1977).

19. MARTEN, G. C., and M. E. HEATH. "Reed Canarygrass," In *Forages* (Ames: Iowa State Univ. Press, 1973).

20. MARTEN, G. C., R. M. JORDAN, and A. W. HOVIN. "Biological Significance of Reed Canarygrass Alkaloids and Associated Palatability Variation to Grazing Sheep and Cattle," *Agron. Jour.,* 68:909 (1976).

21. NEWELL, L. C. "Smooth bromegrass," *Forages* (Ames: Iowa State Univ. Press, 1973).

22. OCUMPAUGH, W. R., and A. G. MATCHES. "Autumn-Winter Yield and Quality of Tall Fescue," *Agron. Jour.,* 69:639 (1977).

23. POWELL, J. B., and A. A. HANSON. "Timothy," In *Forages* (Ames: Iowa State Univ. Press, 1973).

24. TAYLOR, T. H., and W. C. TEMPLETON, JR. "Stockpiling Kentucky Bluegrass and Tall Fescue Forage for Winter Pasturage," *Agron. Jour.,* 68:235 (1976).

25. TEMPLETON, W. C., JR., T. H. TAYLOR, and J. R. TODD. *Comparative Ecological and Agronomic Behavior of Orchardgrass and Tall Fescue.* Univ. of Kentucky Agr. Exp. Sta. Bul. 699 (1965).

26. USDA. *Grass Waterways in Soil Conservation,* Leaflet 477 (1960).

27. WASHKO, J. B., *et al. Management and Productivity of Perennial Grasses in the Northeast. III. Orchardgrass.* West Virginia Agr. Exp. Sta. Bul. 557T (1967).

28. WRIGHT, M. J., *et al. Management and Productivity of Perennial Grasses in the Northeast. II. Smooth Bromegrass.* West Virginia Agr. Exp. Sta. Bul. 554T (1967).

CHAPTER 38
WARM-SEASON GRASSES

THERE are many grasses used in the South; some have been grown for years, others are relatively new. Grasses most commonly grown in this area include bermudagrass, dallisgrass, bahiagrass, carpetgrass, johnsongrass, and the ryegrasses.

Bermudagrass

Bermudagrass, *Cynodon dactylon,* is a warm-season grass found throughout the tropics and subtropics. It is believed to have originated in India. It was introduced to Georgia about 1750 and by 1800 was recognized to be one of the most important grasses for the South. Because of its aggressiveness, farmers interested only in cotton or other row crops condemned it as one of the worst weeds of the South.

Adaptation and Use

Bermudagrass is best adapted to the area south of the line connecting the southern boundaries of Virginia and Kansas. It will grow on any moderately well-drained soil, provided it has an adequate supply of moisture and plant nutrients. It grows better on heavy soils than on light, sandy soil. Its nutritive value is greatly influenced by the stage of growth and the available soil nutrients. In Georgia the protein content has been doubled at almost any stage of growth by heavy applications of nitrogen fertilizer.

The bermudagrasses are widely used in the Southeastern United States for turf purposes with several new cultivars of fine-leafed turf-type bermudagrasses having been released.

Plant Description

Bermudagrass is a creeping, perennial, sod-forming grass that propagates by stolons, rhizomes, and in some instances seed. The leaves are short, flat, and narrow, two-ranked, and characterized by a ligule, which is a fringe of white hairs at the leaf base. Slender, spike flowering heads develop in clusters of three to eight from flattened culms. This grass may spread as much as 15–20 feet (4.6–6.1 m) under favorable conditions. A fibrous root system can develop from each node of a stolon or rhizome.

FIGURE 38–1. *Common bermudagrass is a creeping, perennial, sod-form-ing grass. Leaves are short, flat, narrow, and characterized by a ligule which is a fringe of hairs at the leaf base.* [SOURCE: Southern Illinois University.]

Improved Strains

Probably few, if any, other cultivated grasses have been so improved by breeding and selection as has bermudagrass. The best known of the improved cultivars is 'Coastal,' a cultivar well adapted throughout most of the bermudagrass Belt. This F_1 hybrid was developed at the Coastal Plain Station in Georgia by crossing 'Tift' bermudagrass with an introduction from South Africa. It tolerates more frost, makes more growth in the fall, and remains green much later than common bermuda. It has made over twice the production of common bermuda and has been outstanding in its ability to grow and produce in the late summer and fall in comparison with other summer-growing grasses. 'Coastal' bermuda grows tall enough to be cut for hay on almost any soil, whereas the growth of common bermudagrass often is too short to cut. Four to five cuttings per year for 'Coastal' bermuda is not uncommon.

Other improved types superior to common bermuda for pasture or hay are the following:

1. *'Suwanee'*—A tall-growing hybrid developed at Tifton, Georgia, it is better adapted than 'Coastal' to light, sandy, infertile soils.

2. *'Midland'*—An F_1 hybrid of Coastal and a cold-hardy common from Indiana, this was released by the Oklahoma station for use in areas where 'Coastal' winter-kills.

FIGURE 38–2. *A shade tolerance experiment with bermudagrass cultivars and variable degrees of shading at the Mississippi Station.* [SOURCE: Mississippi Agriculture and Forestry Experiment Station.]

3. *'Coastcross-1'*—A sterile F_1 hybrid of 'Coastal' and a plant introduction from Kenya, it grows taller, has broader and softer leaves, and produces stolons that spread more rapidly than those of 'Coastal.'

Propagation

Common bermuda may be propagated by planting either seed or sprigs. Improved hybrids such as 'Coastal' and 'Suwanee' must be propagated by means of sprigs. Commercial planters are available for planting sprigs, or they can be broadcast and covered with a disk-harrow followed by a cultipacker.

Cultural and Management Practices

Planting a legume with the grass when it is to be used for either hay or pasture is recommended. Almost any legume will make an excellent growth in association with bermudagrass, provided soil moisture and plant nutrients are available and the planting is properly managed. White, crimson, and arrowleaf clovers are the legumes most commonly grown with bermudagrass in the South.

Best results are obtained for pasture from common bermuda when it is continuously grazed and not allowed to get over 4–6 inches (10–15 cm) in height. With types like 'Coastal' best results are obtained when allowed to get to a height up to 12 inches (30 cm) and then grazed rotationally. When a legume is grown with the grass, grazing and fertilization practices should favor the legume.

FIGURE 38-3. *A bermudagrass-dallisgrass pasture in Mississippi typical of many Southeastern pastures for beef cattle herds.* [Photograph with permission of Mississippi Agriculture and Forestry Experiment Station.]

Adequate P and K fertilizer, but N only from the legume, should maintain the same botanical composition. Bermudagrass grown alone will respond to high rates of N fertilizer.

Dallisgrass

Dallisgrass, *Paspalum dilatatum,* native to South America, was introduced sometime in the late 1800s. The grass spread from the New Orleans area throughout the whole Gulf Coast region.

Distribution and Adaptation

Dallisgrass is found from New Jersey to Tennessee and Florida and west to Texas. It is adapted to practically every condition throughout the Cotton Belt where the rainfall is as much as 30 inches (76 cm). It makes its best growth on bottomland because of the greater available fertility and moisture. More tolerant to excessively wet soils, it is at the same time more drought resistant than bermuda or carpetgrass.

Plant Description

Dallisgrass is a rather stout perennial and grows in clumps to a height of 2–4 feet (0.6–1.2 m). It produces many basal leaves and bears many leafy culms. After cutting or grazing it makes rapid recovery. It spreads by short fleshy rhizomes but grows mainly in clumps, and reproduces by seed. The seed are borne in spreading racemes and are arranged in two rows. They are fringed with long, white, silky hairs.

Use and Management

Dallisgrass is one of the most valuable pasture plants for the humid South. It is the first grass to begin growth in the spring and makes continuous growth throughout the whole season; it is not injured by moderate frost and is the last grass to become dormant in the fall. Its bunch habit of growth makes it adapted for growing in association with several legumes. It should be grazed or clipped regularly for best production.

Bahiagrass

Bahiagrass, *Paspalum notatum,* is a native to the West Indies and South America. It was first introduced into Florida in the early 1900s.

Distribution and Adaptation

Bahiagrass is adapted principally to the coastal areas of the southern United States. This grass will do well from east Texas to the Carolinas, and as far north as northern Arkansas and central Tennessee. It grows best on sandy soils.

Plant Description

This deep-rooted warm season perennial grows to a height of 6–24 inches (15–61 cm), with culms arising from short, stout, woody rhizomes. The dense sod formation allows few other plants to encroach. The numerous basal leaves are flat and generally hairy. The inflorescence is a panicle with a two or three forked seedhead with flat, shiny spikelets.

Use and Management

In its area of adaptation its productivity usually is higher than most other grasses. It has become a popular pasture grass in the lower South and Southwest because it tolerates a wide range of soil conditions while giving moderate yields on infertile soils. It is established by seed, and it withstands close grazing. On sandy soils it is better than dallisgrass. It is used primarily for spring and summer pasture for beef cattle, and is less adapted for hay production. Higher yields and longer seasons of production can be obtained by close clipping and heavy grazing. In some areas bahiagrass is planted for erosion control such as on highway roadbanks.

Carpetgrass

Carpetgrass, *Axonopus affinis,* native to Central America and the West Indies, was introduced sometime prior to 1852. It is well adapted to sandy or sandy loam soils, particularly where moisture is near the surface most of the year. Adapted to soils of low fertility, it produces seed abundantly and tends to become established naturally in the Gulf Coast area.

Plant Description and Use

A low-growing sod-forming perennial, carpetgrass spreads both by stolons and by seed. The dense sod and aggressive growth make it difficult to maintain legumes in pastures where this grass predominates. Carpet-grass is the most common permanent grass in unimproved pastures throughout the Gulf Coast area. Its prolific seeding has caused it to become established naturally in vast areas of cut-over timber.

In general, it is not recommended for use in improved, high-producing pastures and is rarely seeded today because of the better production and nutrition of other grasses, such as dallisgrass and bahiagrass.

Culture and Management

This grass can withstand more severe grazing than most of the grasses used for permanent pasture in the Deep South. Where grown in mixtures the preference of animals for other grasses often results in the carpetgrass crowding out these grasses much sooner than if the area were not grazed. Heavy utilization is needed to prevent excessive seedhead development which lowers animal acceptability and quality.

Johnsongrass

Johnsongrass, *Sorghum halepense,* was first introduced as a forage plant but is now considered a serious weed pest in some areas of the United States. However, it is still regarded as an important hay and pasture grass in some parts of the Southeast. Because of its aggressiveness and extensive system of underground stems or rootstocks, it often is very troublesome as a weed in cultivated crops such as cotton and corn.

Introduced from the Mediterranean area in the early 1800s, johnsongrass spread rapidly over the entire Cotton Belt. It is adapted to all sections where cotton is grown. It does best on heavy clay soils of relatively high fertility and high water-holding capacity.

Plant Description

Johnsongrass is a warm season perennial that grows to a height of 3–6 feet (0.9–1.8 m). It spreads by seed and an extensive creeping scaley rhizome system. The leaves are broad and long and of medium texture and have thickened, light-colored midveins. The inflorescence is an open panicle resembling sudangrass.

Use and Management

Johnsongrass hay compares favorably with other grass hays. It also is used extensively for pasture in the black prairie belt of Alabama, Missis-sippi, and Texas, and has a high carrying capacity. Any cutting treatment before maturity will reduce the production of rootstocks and the yield of hay.

The Ryegrasses

There are two important cultivated species of ryegrass: Italian ryegrass, *Lolium multiflorum,* and perennial ryegrass, *Lolium perenne.* These grasses are believed to have been introduced into this country in Colonial days.

Distribution and Adaptation

The ryegrasses are used extensively throughout the humid South—especially the Italian ryegrass—where it is often sold under the name *common* or *domestic.* They are also grown west of the Sierra Nevadas and the Cascade Range. Practically all the ryegrass seed used in the South is produced in western Oregon. Ryegrasses have a wide range of soil adaptation, but generally, they are not adapted to regions of extremely high or low temperatures, droughty conditions, or to infertile soils.

Plant Description

Italian ryegrass is considered an annual bunchgrass, but under some conditions it performs as a biennial. It grows 2–3 feet (0.6–0.9 m) in height and is leafy, succulent, and palatable. The seedheads, or spikes, are slender and usually weak. Perennial ryegrass functions as a short-lived perennial. It makes a growth of from 1–2 feet (0.3–0.6 m), with leaves shorter and more erect than those of Italian ryegrass and quite stiff. The seed spikes make a stiffer, more erect growth, producing awnless seed. The stems are slightly flattened while those of Italian ryegrass are round. Seeds of the Italian ryegrass have awns of varying lengths. The leaves of perennial ryegrass are folded in the bud, while Italian is rolled in the bud.

Use and Management

The most important and extensive use of Italian ryegrass is as a winter and early spring pasture in the South and for pasture and hay in the upper Pacific Coast area. It can be overseeded in summer perennial pastures such as bermudagrass to furnish extra winter grazing days. In the South it is used extensively for fall seeding on permanent warm-season lawns, giving a pleasing appearance during the winter months. Seeding in the South is in the late summer or early fall. When seeded with small grains for temporary pastures in the North, the seeding is as early in the spring as possible. Perennial ryegrass can be used as a component for pasture mixtures, but it will not persist for as long as many other pasture species.

REFERENCES AND SUGGESTED READINGS

1. BENNETT, H. W. "Johnsongrass, Dallisgrass, and Other Grasses for the Humid South," In *Forages* (Ames: Iowa State Univ. Press, 1973).

2. BURTON, G. W. "Bermudagrass," In *Forages* (Ames: Iowa State Univ. Press, 1973).

3. EVANS, E. M., L. E. ENSMINGER, B. D. DOSS, and O. L. BENNETT. *Nitrogen and Moisture Requirements of Coastal Bermuda and Pensacola Bahia,* Auburn Univ. Agr. Exp. Sta. Bull. 337 (1961).

4. FRAKER, R. V. "The Ryegrasses," In *Forages* (Ames: Iowa State Univ. Press, 1973).

5. FRIBOURG, H. A., *et al.* "Adaptation and Productivity of Some New Bermudagrasses," *Tennessee Farm and Home Science, Progress Rep.* 104:9 (1977).

6. HOVELAND, C. S. *Bermudagrass for Forage in Alabama,* Auburn Univ. Agr. Exp. Sta. Bul. 328 (1960).

7. ROLLINS, G. H., C. S. HOVELAND, and K. M. AUTREY. *Coastal Bermuda Patures Compared with Other Forages for Dairy Cows,* Auburn Univ. Agr. Exp. Sta. Bul. 347 (1963).

8. WARD, C. Y., and V. H. WATSON. "Bahiagrass and Carpetgrass," In *Forages* (Ames: Iowa State Univ. Press, 1973).

CHAPTER 39

NATIVE AND RELATED INTRODUCED GRASSES

NATIVE grasses and closely related introduced species are used chiefly for reseeding abandoned farmlands, severely eroded sites, droughty soils, rangelands, and other areas where the more commonly cultivated grasses often fail. In addition, there is increasing interest in warm-season perennial native grass pastures to supply summer feed in selected geographic areas when the predominant cool-season species are largely dormant.

Seed harvest of most of the better native grasses is difficult. Much of the native grass seed is collected directly from wild stands. Harvested seed often is of poor quality. Some species, like little bluestem, are difficult to clean. Others, such as buffalograss, have low seed viability. A good many others, such as Canada wildrye have awns or appendages which interfere with seeding. It has been possible to propagate some species only vegetatively.

Of the many species, those that will be considered here as probably the most important are (1) the wheatgrasses, (2) the bluestems, (3) the gramagrasses, (4) buffalograss, (5) switchgrass, and (6) the lovegrasses. A few others are mentioned briefly.

The Wheatgrasses

Some 150 species of the genus *Agropyron* are widely distributed in the temperate regions of the world. Of about 30 species found in North America, six of the more important species should be considered. These are well adapted to the northern Great Plains, the Intermountain region, and the higher altitudes of the Rocky Mountain states. The wheatgrasses are widely used in these areas because of their hardiness, drought resistance, wide climatic and soil adaptation, early season forage production and wind and water erosion control properties. Of the wheatgrass species being discussed, crested, intermediate, and tall wheatgrasses and quackgrass were introduced into the United States, and western and slender wheatgrasses are considered to be native to North America. (See Table 39–1.)

Plant Characteristics

The wheatgrasses are mostly perennials, mostly cool-season, and either with or without creeping rhizomes. Most of the wheatgrasses produce an abundance of seeds which germinate readily even at relatively low tem-

FIGURE 39–1. *A desert range in the West showing typical growth of bunch grasses and small forbs.* [SOURCE: USDA, Soil Conservation Service.]

peratures. This characteristic gives the young seedlings a better opportunity to become established in competition with other grasses and with weeds. Because of seedling vigor and the ease with which stands can be established, wheatgrasses have been used extensively for revegetating lands abandoned from crop production and for grazing lands on which native grass cover has been depleted.

Crested Wheatgrass

Crested wheatgrass is the most important introduced wheatgrass and has been more extensively seeded in the northern Great Plains and in the prairie provinces of Canada than any other wheatgrass. Introduced from Siberia just before 1900, it attracted little attention for 20 years.

The drought resistance, extreme winter-hardiness, long-lived character-

TABLE 39–1. Common and Latin Names and Bunch- or Sod-Forming Characters of Six Prominent Species of Wheatgrass

Name		Native or Introduced	Bunch- or Sod-Forming
Common	Latin		
Crested wheatgrass	*Agropyron cristatum*	Introduced	Bunch
Intermediate wheatgrass	*Agropyron intermedium*	Introduced	Sod
Quackgrass	*Agropyron repens*	Introduced	Sod
Tall wheatgrass	*Agropyron elongatum*	Introduced	Bunch
Western wheatgrass	*Agropyron smithii*	Native	Sod
Slender wheatgrass	*Agropyron pauciflorum*	Native	Bunch

istic, and ability to withstand intense grazing, combined with heavy yield of nutritious forage in early spring, make crested wheatgrass one of the most valuable forage grasses available to farmers and ranchers of the northern Great Plains. It has low tolerance to alkali soils or flooded conditions, however.

This grass is ready for grazing earlier than any other native or introduced species. Growth practically stops during the dry, hot months of the summer and resumes only when cool, moist weather comes in the early fall.

Crested wheatgrass is extremely valuable to supplement the grazing from native grass pastures. This is because of the extra volume of succulent, high-protein forage available in the early spring, before the normal growth of native grasses is ready for grazing.

Intermediate Wheatgrass

Intermediate wheatgrass, a perennial, sod-forming grass, was introduced from Russia in the 1930s. It appears to have considerable value in the northern and central Great Plains and in the Pacific Northwest. It is well suited for use both as pasture and as hay. It produces an abundance of leafy forage palatable to all classes of livestock, especially during the early spring. Like crested wheatgrass, growth almost ceases during the hot dry summer period. It appears to be somewhat less winter-hardy and shorter-lived than crested wheatgrass and also requires better grazing management.

Quackgrass

A native of Europe, quackgrass is notorious throughout the northern part of the humid East and Central states as a noxious weed. It spreads both from seed and by creeping rootstocks. Under many conditions it can be used as valuable hay and pasture. Quackgrass possesses nearly all the qualities desired in a hay plant, being palatable, nutritious, prolific, hardy, leafy, easily harvested, and adapted to wide variations of soil and climate. When harvested for hay, quackgrass should be cut not later than the early bloom stage to reduce the danger of scattering viable seed. It can also be useful as a vegetative cover to prevent soil erosion on steep slopes or embankments.

Tall Wheatgrass

Tall wheatgrass, *Agropyron elongatum,* makes a bunch-type growth. It is rapidly gaining in importance in the northern Great Plains, despite its recent introduction. The outstanding advantage of this grass over some of the other wheatgrasses is its tolerance of wet, alkaline conditions. It shows promise of producing pasture or hay on large areas of land out of production because of irrigation seepage, high water table, or alkaline condition.

FIGURE 39-2. *Tall wheatgrass space-planted breeding nursery at the New Mexico Station.* [Photograph with permission of New Mexico State Agricultural Experiment Station.]

Western Wheatgrass

Western wheatgrass is a native perennial, dense, sod-forming grass with rather general distribution and adaptation, except in the Southeast. It is best known and most widely valued in the central and northern parts of the Great Plains, where it often is the dominant grass species in native plant associations. It starts growth early in the spring, before buffalograss and blue grama. It is one of the first grasses to become re-established on abandoned cropland. Few other grasses are as tolerant of alkali soils.

Like other wheatgrasses, the early growth is much more palatable than that produced later in the season. Its growth characteristics—including drought resistance, winterhardiness, wide adaptation to soil and climatic conditions, and ability to spread rapidly by means of running rootstocks, or rhizomes, and by seed—make it particularly valuable for revegatation and control of soil erosion.

Slender Wheatgrass

Slender wheatgrass is a perennial bunchgrass native to the Northern states and Canada. It is valuable in the Rocky Mountain States and the northern Great Plains. It begins its growth fairly early in the spring and produces an abundance of forage that is readily grazed. It is easy to establish because of high germination and seedling vigor, and is highly tolerant of alkali soils. However, it has less drought tolerance than western or crested wheatgrass and is relatively short-lived. If allowed to mature, the forage furnishes nutritious winter grazing. A good-quality hay is obtained if cut before the plants become woody.

The Bluestems

There are a large number of different species of the bluestems, but only three native and two introduced species are important forage plants in this country. These are all warm-season grasses; they start late in the spring and make their growth during the summer months.

Big Bluestem

Big bluestem, *Andropogon gerardi,* is a vigorous native perennial that often grows to a height of 6 feet (1.8 m) when moisture and soil nutrients are in adequate supply. Its strong deep roots and short underground stems produce a sod highly resistant to erosion. The seedheads often branch into three parts resembling a turkey's foot, and the lower leaves are usually covered with silky hairs. Its major range of distribution is moist, well-drained loams of relatively high fertility in the Central states and on the eastern edge of the Great Plains. It was the dominant native grass of many such areas. The leafy forage is palatable to all classes of livestock. Few of the prairie grasses can equal big bluestem in production or quality of forage. It makes a good-quality hay if cut before it becomes stemmy and seedheads form. Seed processing is moderately difficult.

FIGURE 39–3. *Kaw big bluestem. Few prairie grasses can equal big bluestem in production or quality of forage. Its major range of distribution is moist, well-drained loams of relatively high fertility in the Central States and on the eastern edge of the Great Plains.* [SOURCE: USDA, Soil Conservation Service.]

Little Bluestem

Little bluestem, *A. scoparius*, grows to heights of 2–4 feet (0.6–1.2 m). Compared to big bluestems, its major distribution is in more westerly and drier areas of the Great Plains because of its greater drought tolerance. It is often found on gravelly soils on ridges and other exposed locations. One of the areas of native growth is the flint hills section of east central Kansas and Oklahoma. It can be identified by flat, bluish-colored basal shoots and folded leaf blades. Under most growing conditions it is less palatable than big bluestem. Seed processing is difficult.

Sand Bluestem

The sand bluestem, *A. hallii,* is a native grass resembling big bluestem but is easily distinguished from it by the fact that the heads are conspicuously hairy. It is a perennial with creeping underground stems. It occurs on deep, sandy soils. Its major distribution is in western Nebraska, Kansas, Oklahoma, and parts of Texas and New Mexico.

The two introduced bluestems which have some importance as forage plants are yellow bluestem, *Bothriochloa ischaemum,* a perennial semi-prostrate bunchgrass from India and old world bluestem, *B. caucasica,* a perennial bunchgrass from Russia.

FIGURE 39–4. *The gramas are warm-season grasses of major importance in the Great Plains.* [SOURCE: USDA, Soil Conservation Service.]

The Gramagrasses

The gramas are warm-season grasses of major importance in the Great Plains. Two species have proved of outstanding value to their environment.

Sideoats Grama

Sideoats grama, a perennial bunch-type grass, *Bouteloua curtipendula,* grows to a height of 2–3 feet (0.6–0.9 m). It is easily recognized by its long flower stalk, with short dangling purplish spikes. Leaf blades are flat with hairs and bumps along the edges. It is found on the more favorable sites of the Great Plains; in the drier sections it is replaced by blue grama. Sideoats grama is usually found growing in association with the bluestems and is reasonably palatable in its range of adaptation. It seeds freely, with the seed maturing in late summer. Seed yield is good, and the seed crop can be combined easily.

Blue Grama

Blue grama, *B. gracilis,* makes a smaller, finer growth than sideoats grama and is considerably more drought resistant. It is found growing in association with buffalograss. Although it has no stolons it produces a dense

FIGURE 39–5. *A black grama breeding nursery at the New Mexico Station.* [Photograph with permission of New Mexico State University Agricultural Experiment Station.]

mass of fine roots and forms a sod with its low, basal type of growth. The fine, curling basal leaves have a distinctive gray-green color. Plant height at maturity is 6–12 inches (15–30 cm). It will withstand more drought and more alkali than most other grasses. Highly palatable, it retains its feeding value into the winter months. There is a wide range of types from north to south in the Great Plains, each well adapted locally.

Several other species, such as slender grama, hairy grama, black grama, and Rothrock grama, are of some importance on the native range and have shown promise in localized areas or in isolated trials.

Buffalograss

Buffalograss, *Buchloe dactyloides,* is a low-growing perennial that spreads by stolons. The plant height usually is 2–6 inches (5–15 cm). The foliage is grayish-green until maturity, when it becomes a light straw color. Buffalograss is of primary importance in the central and southern Great Plains. It is noted for its drought resistance and tolerance to alkali.

Not only is it highly palatable and nutritious in the green, summer stage, but it maintains its feeding value in the dry, cured stage during winter months. It tolerates heavy grazing. The seeds are enclosed in a hard bur, with one or several seeds per bur. Seed harvest is difficult. Male and female flowers are borne on different plants. It can be proagated either by seed or sod pieces.

Switchgrass

Switchgrass, *Panicum virgatum,* is a tall-growing, native perennial with short rhizomes which allow it to form a sod. It is coarse-stemmed, broad-leaved plant which makes a growth of 3–5 feet (0.9–1.5 m). It can be identified by a cluster of hairs where the blade attaches to the sheath. It occurs naturally in much of the United States east of the Rocky Mountains. Its chief importance is in the Great Plains region. The excellent seed yields, high seedling vigor, and good forage yields give switchgrass some value for hay, pasture, and erosion control, but its palatability and feeding value are low. It occurs naturally on fertile soils with adequate moisture, yet it will produce better growth and cover on droughty, infertile, eroded soils than many prairie grasses. Improved cultivars are available.

Two other *Panicum* species are of some importance. Vine mesquite, *P. obtusum,* is native to the Southwest, where it is used to some extent in seeded pastures and for range reseeding. Because of its long creeping stems, which root at the nodes, it is a valuable erosion-control plant. Blue panicgrass, *P. antidotale,* is an introduction which shows promise in the Southwest. It is a tall-growing, coarse, vigorous plant like switchgrass. It roots at the nodes of the culms and is definitely adapted to hot summers and mild winters.

Figure 39-6. *Switch-grass occurs naturally east of the Rocky Mountains and is important chiefly in the Great Plains region. Its good seed and forage yields give it some value for hay, pasture, and erosion control but palatability and feeding value are low.* [Source: USDA, Soil Conservation Service.]

The Lovegrasses

The lovegrasses are best known for their ability to grow on low-fertility soils and on sandy soils. In general, they produce an abundance of seeds which germinate readily and establish easily. One native species and three introduced species are worthy of consideration.

Sand Lovegrass

Sand lovegrass, *Eragrostis trichodes,* is native to the central and southern Great Plains. It is an erect perennial bunchgrass and usually makes a growth of 2–3 feet (0.6–0.9 m). It is productive from April to late October and generally is classed as palatable, more so than weeping lovegrass. It also has some value as a winter grass. Sand lovegrass seeds are small and usually seeded at the rate of 0.5 pound/acre (0.6 kg/ha) in mixtures. Seed is easily combined and processed.

Weeping Lovegrass

Weeping lovegrass, *E. curvula,* is a perennial bunchgrass with an extensive but shallow root system. It makes a growth of 2–4 feet (0.6–1.2 m). Introduced from East Africa, it has been used considerably in Oklahoma for erosion control and for forage on low-fertility soil. North of the Oklahoma–Kansas line it may winter-kill. Palatability is not high except during the season of lush spring growth.

Other Lovegrasses

Boer lovegrass, *E. chloromelas,* and Lehmann lovegrass, *E. lehmanniana,* are introduced perennials which have shown some promise as forage plants. Boer is a long-lived, erect grass but less winter-hardy than weeping lovegrass. Lehmann, which has prostrate stems which root at the nodes and form new plants, is the least winter hardy of the three introduced lovegrasses. It is drought-resistant and has been proved effective in erosion control.

Other Dryland Grasses

The Dropseed Grasses

The dropseed grasses of the genus *Sporobolus* are "invader" grasses which grow under unfavorable conditions. Their presence on the range often indicates overgrazing, drought, or unfavorable soil conditions. The forage is of relatively low value. Among the most common species are sand dropseed, giant dropseed, and tall dropseed.

Wildryes

The wildrye grasses are closely related to the wheatgrasses. There are several native and introduced species of some agricultural value. Russian wildrye, *Elymus junceus,* is an introduced, drought-resistant, bunch-type grass that starts early in the spring, the leaves of which tend to remain green during the summer. Canada wildrye, *E. canadensis,* is a large, coarse, short-lived perennial bunchgrass, widely distributed in Canada and in the United States. It is productive but only fair in palatability. There are a number of other wildryes with limited usefulness, including the blue, Virginia, basin, beardless, and dune.

Needlegrasses

Many needlegrasses are found in the arid rangelands of the West. The most important species is green needlegrass, *Stripa viridula.* Needle-and-Thread, *S. comata,* is one of the most commonly occurring species. Green needlegrass is a perennial with awned seed, but not of the type injurious to livestock, as are some needlegrasses, such as needle-and-thread. Green needlegrass is found as a secondary constituent of the native prairies of the northern Great Plains. There are some 100 different species of the needlegrasses in the United States.

REFERENCES AND SUGGESTED READINGS

1. American Society of Agronomy. *Forage Plant Physiology and Soil Range Relationships,* Spec. Publ. 5 (Madison, Wisc.: American Society of Agronomy, (1964).

2. GAY, C. W., and D. D. DWYER. *Poisonous Range Plants,* New Mexico Coop. Ext. Serv. Circ. 391 (1967).

3. HITCHCOCK, A. S. *Manual of the Grasses of the United States,* USDA Misc. Publ. 200 (revised 1951).

4. HUMPHREY, R. R. *Arizona Range Grasses* (Tucson: Univ. of Arizona Press, 1970).

5. HYDE, R. *Cool Season Grasses in Kansas,* Kansas State Univ. Ext. Bul. C-257 (1968).

6. JORDAN, G. L. *Pelleted Seeds for Seeding Arizona Rangelands,* Univ. of Arizona Agr. Exp. Sta. Tech. Bul. 183 (1967).

7. Phillips Petroleum Co. *Pasture and Range Plants* (Bartlesville, Oklahoma, 1963).

8. ROGLER, G. A. "The Wheatgrasses," In *Forages* (Ames: Iowa State Univ. (Press, 1973). eds. M. E. Heath, D. S. Metcalfe, and R. E. Barnes.

9. SCHWENDIMAN, J. L., and V. B. HAWK. "Other Grasses for the North and West," *In Forages* (Ames: Iowa State Univ. Press, 1973).

10. USDA. *Grass Seed Production and Harvest in the Great Plains.* Farmer's Bull. 2226 (1967).

11. USDA. *Irrigated Pastures for Forage Production and Soil Conservation in the West.* Farmer's Bull. 2230 (1967).

12. WRIGHT, L. N. *Blue Panicgrass,* Univ. of Arizona Agr. Exp. Sta. Tech. Bul. 173 (1966).

Part VII
Crop Research
and Improvement

CHAPTER 40
SCIENCE AND AGRICULTURAL RESEARCH

Few occupations . . . involve such varied knowledge and skill as farming. It looks simple to drop some seed into the ground, let it grow up under the blessed influence of sun and rain, and then harvest the crop. But at least as far back as the days of the Romans men realized that the business was not so simple after all. . . . Today a comparative handful of able or successful farmers are required to feed all the rest of the population, and this vast undertaking would be impossible without the aid of science. Indeed, it can hardly be said any longer that science aids agriculture; rather, agriculture under modern conditions is in itself a science, and one with many complicated and indispensable divisions. Whether he knows this consciously or not, the modern farmer constantly uses the results of research in genetics, soils science, the science of nutrition, physiology, bacteriology, parasitology, entomology, plant pathology, engineering, weather science, and many others. They all have an intensely practical bearing on his everyday work with soils, crops, and herds. Moreover, the farmer . . . must know how to gear his operations into a market affected in a hundred ways by the complications of modern industry, commerce, and government. He is forced to farm and to act upon his judgments about . . . economic and political events in his own country and in other countries, because these things, too, have a direct practical bearing on his affairs.

GOVE HAMBRIDGE

Science, in any field of interest, consists essentially of asking carefully structured questions and then answering these questions by precise methods of study.

Agricultural research has been defined as a systematic method of gaining and applying new knowledge effectively (1) in biological, physical, and economic phases of producing, processing, and distributing farm and forest products; (2) in consumer health and nutrition; and (3) in the socio-economic relations of rural living. As a result of such research we come to an appreciation of the cause-and-effect relationships between relevant variables. This enables us to predict results and to develop decisions and policies on the basis of factual information.

The scientific method may best be represented as an attitude and philosophy on the part of those actively concerned: a question is posed, pertinent evidence is collected, an explanatory hypothesis is formulated, its implications are deduced, the evidence is tested experimentally, and the hypothesis then is accepted, rejected, or modified accordingly.

Need of Research Recognized

Knowledge of crop production first developed by the accumulation of experiences. Until the last 100 years or so most of the knowledge about crop production came directly from farmer observations. By way of illustration, the 1789 annual report of the Bath and West Agricultural Society, England, contains the following:

The benefits of agriculture, and indeed the absolute necessity of its improvement, are felt more powerfully than ever, the truth of which no other argument than the increasing demand for the necessities of life is required to confirm. That this has been an age of speculative as well as practical experiment will be allowed. It is not in human wisdom to devise at once the means of reaping from all bountiful nature the fullest products of her power. It now remains to invite gentlemen who are in the habit of agricultural experiment to communicate whatever facts they may ascertain, that are either new in themselves or which they may deem of a useful tendency in the advancement of knowledge.

Farmers, seeing the advantage of applying science to practice, began gradually to establish experimental farms to make discoveries for their own immediate use. In Germany, in 1852, a company of Saxon Farmers formed the Leipzig Agricultural Society. This was the first station where farmers themselves brought science to their own farms to aid them in their farming operation. Some 87 stations, supported largely by farmers, were established in Germany in the next 24 years. The government, impressed with the value of such stations and research, came to their support and also organized additional ones. Other European countries soon followed the example set by Germany. A number of stations soon were established at Benbloux in France. In Belgium the first and central station was established in 1872. In Italy the first such station was organized during the next 15 years.

In the early American Colonial period a considerable number of groups formed to study and promote better agricultural methods. Among the best known of these was the Albermarle Agricultural Society of Virginia, organized in 1817. Among its 30 original members were Thomas Jefferson, two later governors of Virginia, a future senator and justice of the Supreme Court, and many statesmen, physicians, and farmers. James Madison was its first president. The society obtained many lots of seed of new crops and cultivars from abroad for trial plantings.

President Washington, in his inaugural address to the Congress, urged the establishment of a National Board of Agriculture, but it was not until 1862 that the United States Department of Agriculture was authorized.

Benjamin Franklin, as a result of his many years in Europe as a representative of the United States, was an American member of the Royal Society of England, chartered by King Charles II in 1662. In 1727 Franklin sponsored a similar American organization, the American Philosophic Society. This was followed in 1780 by the American Academy of Arts and in 1848 by the American Association for the Advancement of Science. The AAAS has come to have more than 300 affiliated societies, making it the world's largest scientific organization, with a combined membership of more than 2 million.

Congress chartered the National Academy of Sciences in 1863, with the recognized function to "investigate, experiment and report whenever called upon by any department of the government." The number of new members admitted yearly is limited. They are chosen for their outstanding achievements in basic research.

Crop Science vs. Good Crop Practices

It is evident that there is a difference between crop science and good crop practices. The establishment of scientific principles requires the replication of exact conditions. This seldom is possible for the average farmer, no matter how keen an observer he may be. Crop science is based on carefully planned and replicated plantings and treatments, from which detailed records are obtained, usually through a period of years, before any deductions or conclusions are advanced. Farmers generally have come to recognize the value of the results reported from the agricultural experiment stations and from other agricultural research agencies.

In the British Empire there was early recognition that agricultural research, both in theory and in practice, is of first importance to the Empire. By the early 1950s the British Agricultural Council listed some 700 agricultural research stations in the British Commonwealth nations. Agricultural research in other developed countries throughout the world has increased in much the same way that it has in the British Empire and the United States.

The Development of Crop Science

The concern of mankind for raw materials for food, clothing, and housing resulted gradually in better methods, with greater returns for the effort expended. With this production has come a large body of supporting knowledge about crops and crop production, and about the soil and its management. This increase in knowledge has enabled farmers in the developed nations to produce with ever-increasing efficiency.

Recognizing the importance of crop science, governments have established extensive agricultural research centers, with laboratories of many kinds, and test stations, all staffed with experts. The last century has seen tremendous developments and a rapid accumulation of knowledge. This stream of knowledge has been continuous since the beginning of history. We think of chemistry, physics, and astronomy as ancient subjects, but the crude beginnings of the study of crop production far antedates any of these. An appreciation of crop science as we have it today can be acquired only by a study of the length, breadth, depth, and characteristics of this stream of knowledge.

Science has many branches. *Crop science* is that branch of thought concerned with the observation and classification of knowledge concerning crops, especially with the establishment of verifiable principles. Many of these must be arrived at by hypothesis and deduction. Scientific thinking is its motive power; the problem is its unit of advance. Proficiency in crop science can be gained only by learning to think out and apply principles in the solution of problems. The term *research* is used to describe the systematic application of thinking geared to the discovery of new facts and new laws from which certain generalizations can be made.

INTERRELATED SCIENCES. In addition to the direct development of crop science, there are large areas of interrelated and overlapping sciences and fields of knowledge which have become a part of crop science. A particular science has been defined as, "A concentration of knowledge about a particular subject, surrounded by interrelated areas of other sciences." Some of the scientific fields and areas of knowledge closely related to crop science are indicated by the following:

Soil science is so closely related to *crop science* that the two commonly are included under the term *agronomy.*

Agronomy is the usual unit name for the department or section administering the closely related crops and soils work in colleges, universities, and other organizations and agencies.

Botany is fundamental to an understanding of plant processes, relationships, and diseases.

Economic entomology attempts to solve the problems of insect control.

Genetics develops many of the principles used by the plant breeder.

Chemistry is a basic science, knowledge of which is necessary for an understanding of such plant life processes as photosynthesis, transpiration, selective absorption, and nutrition.

Bacteriology is necessary for an understanding of soil microorganisms and their relationship to availability and supply of plant nutrients in the soil and such processes as decomposition of organic residues.

The Methods of Crop Science

The methods of crop science are the fundamental methods of all the sciences. Four general steps may be said to characterize the scientific method:

THE PROBLEM IS DEFINED. The problem may start with an idea that some change is needed, that some improvement should be made, that improvement is possible. The problem is to find how to bring about the desired end.

A HYPOTHESIS IS PROPOSED. The hypothesis, or inference, is based on past experience or information already known. It serves as the guide for the investigation. The inference is a scientist's supposition relating to the solution of the problem. In formal research it is the hypothesis.

A SEARCH FOR EVIDENCE IS MADE. Beginners in crop science, of necessity, will be limited to a study of journal articles, bulletins, books, and other printed material already available. Advanced workers not only survey thoroughly the related literature, but also devise tests, methods, and experiments for obtaining new evidence.

A CONCLUSION IS ESTABLISHED. The conclusion arrived at or accepted may be a revision of the original inference with which the investigation began. The conclusion is tested according to the standards of scientific thinking and may be arrived at by considering these questions: Is the evidence sufficient? Have all the differences been accounted for? Have bias and prejudice been eliminated? Have all the important facts bearing on the problem been considered? Does the conclusion conflict with any important facts? Is the conclusion strongly supported by the available evidence and judged to be the best solution of the problem?

Why Agricultural Research?

It has been said that the miracle of United States agricultural production had its beginning in 1862 with the establishment of the Land-Grant colleges and the United States Department of Agriculture by Congressional action. The Hatch Act of 1888 provided the first state agricultural experiment stations. The Smith-Lever Act established the Cooperative Extension Service on a national basis in 1914 as a war measure. This legislation made it possible to get the results of agricultural research applied quickly to farm production programs, at the same time acquainting the scientist–research staffs with the farm problems most pressing for solution.

Among the many reasons that have been stated for agricultural research are the following:

1. It provides the producer with the information needed to produce efficiently.
2. It aids in ensuring an abundance of agricultural products of better quality produced at a lower cost per unit.
3. It means a lower cost for agricultural products to the consuming public as a result of the lower cost of production.
4. It is possible for the American producer to compete better in both foreign and home markets because of the decreased cost of production.
5. It enables the United States to share its agricultural abundance with the people of less fortunate nations and to aid them in applying improved technological methods to their own production efforts.
6. It makes possible the maintenance and improvement of soil and water resources, while these resources are at the same time utilized effectively in production practices.

Importance Of Agriculture And Research

Agriculture is the nation's largest industry. The total agricultural and forestry complex accounts for more than one fourth of the gross national product. Agricultural exports return a great percentage of the dollars that the United States spends overseas for defense, foreign aid, tourism, and investments. This helps level the balance of payments.

U.S. farms spend billions of dollars each year on the goods and services necessary for production. They use more petroleum than any other industry and one-third as much steel as the automobile industry. Out of every 100 jobs in private employment, about one-third are directly or closely related to agriculture and forestry.

Although production is the basic segment, agriculture also includes suppliers of feed, fertilizer, chemicals, seeds, machinery, tools, buildings, petroleum products, electric service, and other necessary items. These businesses provide employment and supply services and farm products. In addition, agriculture supplies workers with raw products to transport, process, manufacture, and market.

Goals Of Agricultural Research

The original goals and objectives of agricultural research were well stated in the Congressional acts establishing the United States Department of Agriculture and Land-Grant colleges and agricultural experiment stations.

The act establishing the United States Department of Agriculture says in part, "There shall be at the seat of government a Department of Agriculture, the general design and duties of which shall be to acquire and to

FIGURE 40–1. *An experiment at a state agricultural experiment station on soil, water, and nutrient losses under different cropping systems. Farmers generally have come to recognize the value of results from agricultural experiment stations and other agricultural research agencies.*

diffuse among the people of the United States useful information on sub-jects connected with agriculture, in the most general and comprehensive sense of the word. . . ."

With regard to the purpose and activities of the state agricultural experi-ment stations, the Hatch Act states:

It shall be the object and the duty of the state agricultural experiment stations . . . to conduct original and other researches, investigations, and experiments bear-ing directly on and contributing to the establishment and maintenance of a perma-nent and effective agricultural industry in the United States, including researches basic to the problems of agriculture in its broadest aspects, such investigations as have for their purpose the development and improvement of the rural home and rural life, and the maximum contribution by agriculture to the welfare of the con-sumer, as may be deemed advisable, having due regard to the varying conditions and needs of the respective states.

STATEMENT OF OBJECTIVES. The present and future functions and ob-jectives of agricultural research were set out in some detail in a study sponsored jointly by the United States Department of Agriculture and the National Association of State Universities and Land-Grant Colleges. This report states objectives as follows:

1. Insure a stable and productive agriculture for the future, through wise manage-ment of natural resources.
2. Protect forests, crops, and livestock from insects, diseases, and other hazards.
3. Produce an adequate supply of farm and forest products at decreasing real pro-duction costs.
4. Expand the demand for farm and forest exports by developing new and improved products and processes and enhancing product quality.
5. Improve efficiency in the marketing system.
6. Expand export markets and assist developing nations.
7. Protect consumer health, and improve nutrition and well-being of the American people.
8. Assist the rural Americans in improving their level of living.
9. Promote community improvement, including development of beauty, recreation, environment, economic opportunity and public services.
10. Enhance the national capacity to develop and assimilate new knowledge, and new or improved methodology for solving present problems, or new problems that will arise in the future.

Proved Worth of Agricultural Research

The efficiency of American agriculture is recognized as "one of the mira-cles of the century." As a result of the improvements in the technology of production, American consumers are the best fed, clothed, and housed people in the world, and at the lowest real cost.

Perhaps the strongest proof of the effective value of agricultural research, as it has been conducted through these past years, is the fact that whereas late in the past century one American farm worker could meet the food

TABLE 40-1. Agriculture's Productivity Since 1930

Year	U.S. Population (July 1) (Millions)	Total Farm Output (1967 = 100)	Persons Supplied per Farm Worker	Output per Man-hour (1967 = 100)	Cropland Harvested (Millions of Acres)
1930	123.1	52	10	16	360
1940	132.1	60	11	20	331
1950	151.7	74	16	34	336
1955	165.3	82	20	44	333
1960	180.7*	91	26	65	317
1965	194.6*	98	37	89	292
1968	201.2*	102	43	106	297
1977	217.5*	121	56	173	337

* Includes 50 states.

and fiber needs of himself and about five others, today he provides these needs for himself and 55 others. This increased efficiency of agricultural production has greatly increased the nation's available manpower, releasing millions who formerly did farm labor to other productive activities (See Table 40-1.) This increased production efficiency can be illustrated also by noting U.S. farm production and output index since 1959 (Figure 40-2). Nearly 25 percent of the U.S. population lived on farms in 1937, but the number fell to 15 percent in 1950, and by 1978 it was less than 4 percent.

Food costs in the United States are now lower relative to income than ever before and are lower than anywhere else in the world. In 1900 consumers paid 40 percent of their income for food, as compared with 20 in 1960, 17.7 percent in 1968, and 16.8 percent in 1977.

Who Does the United States Agricultural Research?

The agricultural research in the United States is done by the United States Department of Agriculture, the state agricultural experiment stations, and industry, including private foundations.

The interrelated and cooperative programs of the USDA and the SAES cover research at more than 500 locations in all 50 states and in Puerto Rico.

Much of the work of the USDA is in cooperation with the different state agricultural experiment stations and with other research and educational agencies and institutions. However, many laboratory and field investigations of several of the sections are conducted at the Beltsville Research Center, Beltsville, Maryland, near Washington, D.C. This research center is probably the world's largest and most comprehensive institution devoted to the scientific solution of agricultural problems. The Agricultural Re-

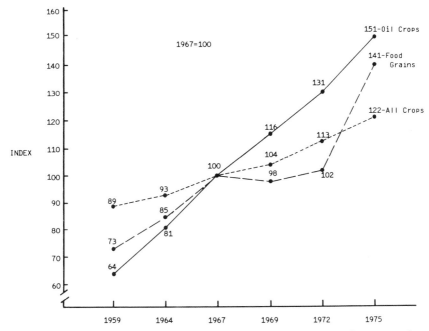

FIGURE 40–2. *United States farm production and output index. Agricultural research has contributed to crop production and output index increases of 22 to 51 percent since 1967.* [SOURCE: Southern Illinois University.]

search Administration also directs research at several regional laboratories and has other laboratories doing extensive research on industrial utilization of farm products.

Each state operates an agricultural experiment station, and branch stations also are maintained in most states. These state agricultural experiment stations receive financial support from state legislative appropriations, the federal government, produce sales, and industry and other nonfederal sources.

A large share of the expenditures of the United States Department of Agriculture is in support of the work of the USDA research scientists located at the different state agricultural experiment stations. Results of research at these stations are shared by the two agencies.

Industry and private foundations support substantial research and development programs promoting technological improvements in agriculture. This work emphasizes improved seed, plants, animal breeding stock, pesticides, fertilizers, feeds, produce development, consumer testing, and equipment and improved facilities for production, processing, and marketing. Most of these expenditures are devoted to the development of products manufactured and distributed by large industrial firms.

Why Agricultural Research Is Government Financed

Public funds are used in the support of agricultural research in view of the fact that the nation as a whole is directly benefited. When agricultural materials are produced as efficiently as possible the cost of each unit produced is decreased, with the result that the cost to the consuming public is decreased.

Much of the U.S. agricultural research is devoted to current production needs and the capacity to meet these needs in the future, not only to produce enough to meet our needs of farm and forest products, but to do so at decreasing real cost. Marketing includes the development of new products, the improvement of product quality, and the improvement of marketing efficiency.

It is impossible for a million individual farm producers to organize and conduct an effective research program, no matter how evident the need for such research. Agricultural research makes it possible to produce with increased efficiency, with greater assurance of the necessary production from year to year. Agricultural research also is necessary to maintain and improve soil and water resources and at the same time use them in production with maximum efficiency.

Justification for Continued Research

Several facts should be considered in justifying continuation of agricultural research. The lag between the time of new discoveries and their general application often is considerable. Moreover, eliminating production research would have little effect on production for a decade or more and should not be related to production deficits or surpluses at the present time. Another consideration is the fact the research is not an activ-

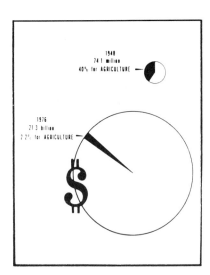

FIGURE 40–3. *Agriculture's share of federal funds for research and development has declined from 40 percent of the total in 1940 to only 2.2 percent in 1976. Returns from agricultural research of the past are recognized as outstanding and of great benefit to the public welfare.*

FIGURE 40-4. *A major alternative energy source for agriculture could be solar energy, already used on a limited scale for drying grain, heating pig nurseries, and warming greenhouses. Additional solar research is needed in order to tap the sun's maximum potential.* [Phototgraph with permission from *Illinois Research,* Illinois Agricultural Experiment Station.]

ity that can be turned on and off at will. Technological research in agriculture that might be discontinued now might be re-established at some later time, but it would then take years to restore, and at greatly increased costs.

The returns from investments in agricultural research are a contribution to economic growth through the discovery and development of new and superior resources which produce a higher rate of return in relation to costs than is true of resources previously employed. The returns from investments in agricultural research of the past are recognized as outstanding and of great benefit to the public welfare in general. That agricultural research has been an important factor in increasing agricultural productivity is undisputed.

REFERENCES AND SUGGESTED READINGS

1. *Climate and Man: Yearbook of Agriculture* (Washington, D.C.: USDA, 1941).
2. DUROST, D. D., and E. T. BLACK. *Changes in Farm Production and Efficiency, 1977,* USDA-ESCS Stat. Bul. 612 (1978).
3. FREEMAN, O. L. "Scientific Agriculture: Keystone of Abundance," In *Science for Better Living: Yearbook of Agriculture* (Washington, D.C.: USDA, 1968).
4. KELLOGG, C. E., and D. C. KNAPP. *The College of Agriculture: Science in the Public Service* (New York: McGraw-Hill, 1966).
5. KLOSE, N. *America's Crop Heritage* (Ames: Iowa State Univ. Press, 1950).
6. KREBS, H. A. "The Making of a Scientist," *Agr. Science Review,* 6:2 (1968).

7. LANDSBERG, H. H. *Natural Resources for U.S. Growth* (Baltimore: Johns Hopkins Univ. Press, 1964).

8. MAYER, L. V., and E. O. HEADY. *Projected State and Regional Research Requirements for Agriculture in the U.S. in 1980,* Iowa State Univ. Res. Bul. 568 (1969).

9. National Association of State Universities and Land-Grant Colleges and the USDA. *A National Program of Research for Agriculture.* Washington, D.C. (October 1966).

10. NELSON, W. L. (Ed.). *Research with a Mission,* Spec. Publ. 14 (Madison, Wisc.: American Society of Agronomy, 1969).

11. REAGON, M. D. "Basic and Applied Research: A Meaningful Distinction," *Science,* 155:1383–1386 (1967).

12. RONNINGEN, T. S. "Systems Research in Agriculture," *Agr. Science Review,* 6:4 (1968).

13. THOMAS, G. W. *Progress and Change in the Agricultural Industry* (Dubuque: W. C. Brown, 1969).

14. USDA. *Agric. Stat.* (1977).

15. USDA-ESCS. *Food Consumption Prices Expenditures,* Suppl. for 1976 to Agric. Econ. Rep. 138 (1978).

16. USDA-ESCS. *National Food Review,* Publ. NFR-1 (1978).

17. USDA. *Fact Book of U.S. Agriculture,* Misc. Pub. 1063 (1976).

18. USDA. *1978 Food and Agricultural Outlook.* Food and Agriculture Outlook Conference (Washington, D.C., 1977).

19. USDA. *1977 Handbook of Agricultural Charts,* Agric. Handbook 524 (1977).

20. WHITNEY, R. S. (Ed.). *Challenge to Agronomy for the Future,* Spec. Pub. 10 (Madison, Wisc: American Society of Agronomy, 1967).

Chapter 41

CROP BREEDING AND IMPROVEMENT

Not since the dawn of civilization has man ceased in his effort to improve the plants that provide his food. The length of time required to domesticate such crop plants as wheat, barley, sorghum, millet, corn, cotton, and many others is not known.

It is not surprising that the improvement of field crops has occupied a prominent place in agriculture. Many thousands of different cultivars have been developed. Obtaining foreign cultivars and types through world exploration and plant introduction was an important part of the early work of the USDA.

Plant breeding is the art and science of changing and improving the heredity of plants. Before breeders possessed the scientific knowledge now available, they relied largely on their skill and judgment in selecting superior types. They were able to recognize variations between plants of the same species which could be used as the basis for establishing new cultivars. With increased knowledge of genetics and the related plant sciences, plant breeding has become less of an art and more of a science. Plant breeders have been successful in influencing such characteristics as yield, early maturity, drought resistance, winter hardiness, disease resistance, insect resistance, and quality. Plant breeding depends on the sciences working together, not only genetics and cytogenetics, but also plant physiology, plant pathology, entomology, plant biochemistry, agronomy, botany, statistics, and computer science. The most successful accomplishments in plant breeding are the result of the team approach between specialists.

Basic Hereditary Principles

All scientific progress is the result of discoveries by many men; when combined, these discoveries become basic principles upon which theories and laws are formulated. The discovery by Camerarius in the latter part of the seventeenth century that plants possess sex and therefore may be cross-bred, like animals, was an important contribution to the science of plant breeding. Darwin's theory of evolution gave an added stimulus to scientific research. The classification of plants by Linnaeus provided an orderly method of grouping plants on the basis of their characteristics. Studies by Weissman established the basic concept that the germplasm of plants was passed on from one generation to the next. The Dutch botanist

De Vries showed that mutations occur in plants, thus providing an explanation for sudden inherent changes. Vilmarin, a French breeder, outlined the principles of the "progeny test," which holds that the best way to determine the breeding behavior of a plant is to grow it and to study its progeny. Johannsen, a Danish plant breeder, developed the pure-line concept and explained why part of the variation among plants is due to noninheritable differences.

Mendelian Theory

The discovery that had the greatest influence on plant breeding, however, was made by Gregor Mendel, an Austrian monk. Mendel in 1866 published the results of his work on the inheritance of characters in peas. Many previous workers had made crosses between plants and had observed the segregation of characters in later generations, but Mendel was the first to count the number of kinds of segregates obtained by crossing true-breeding parents. Thus, he discovered that an orderly ratio of types was repeatedly obtained from the same crosses. From these facts he formulated the first laws of heredity, from which the present science of genetics has developed. Much that had been observed in the past then had new meaning. The implications of his theory opened up a new area of crop improvement.

A vast amount of later work has been done in the fields of genetics, cytology, and cytogenetics to add new basic knowledge. Proof by Morgan that visible characters are due to the action of genes located on chromosomes provided an explanation of the physical basis for transmission of characters from parents to their progeny. Much work has been done since Mendel's time to study characters dependent for their expression on many, rather than a few, genes. Characters such as yielding ability, lodging resistance, quality of grain or forage, and many others are complex in their inheritance.

Physical Basis for Inheritance

All living organisms are formed from cells. Examination of a cell under the microscope shows that it has several distinct parts: the cell wall, which gives the cell its unity; the cytoplasm, with its many structures; and most important from the standpoint of heredity, the nucleus. The nucleus in a resting cell is made up of a large number of granules known as *chromatin*.

Growth is made possible by the multiplication and subsequent enlargement of cells. Cell division is known as *mitosis*. In the first stage of mitosis threads are formed from the chromatin, which become shorter and thicker by forming a spiral. These structures are called *chromosomes*. The number of chromosomes is relatively constant for each plant species. For example, cells in a corn plant each have 20 chromosomes; common oats, 42; barley, 14; and smooth bromegrass, 56. If we closely examine the chromosomes within a cell we find that some of them differ in length and in other characteristics.

In the second stage of mitosis the chromosomes become aligned in the central part of the cell; spindle fibers are formed and are attached to a structure of the chromosome called the *centromere.* Each chromosome consists of two halves, called *chromatids,* lying side by side. In the third and final stage the centromere divides and the two chromatids of each chromosome are dragged by the spindle fibers to opposite ends of the cell. Two daughter nuclei are then formed, each containing one of the two chromatids of each chromosome. A new cell wall is formed to separate the two nuclei and mitosis has been completed. Because the two new cells each contain one half of each chromosome, from end to end, they have the same genetic complement as the mother cell from which they were derived.

The change from vegetative to reproductive growth is a dynamic event in the life of a plant; from a genetic standpoint the changes that occur in the formation of the reproductive cells are of even greater significance.

In a study of the parts of a perfect flower—whether it is a "showy" flower like that of birdsfoot trefoil, alfalfa, or red clover or the less attractive flower of wheat, oats, barley, or bromegrass—each has the same reproductive parts, the *pistillate,* or female part, and the *staminate,* or male part. These two parts of the flower produce female and male sex cells essential to the formation of seed for the next generation.

In the progenitor cells, which ultimately produce the male and female gametes, a special kind of cell division occurs called *meiosis.* In this process the chromosomes arrange themselves in pairs; each member of the pair has an affinity for each other because they are *homologous* in structure. One member of each pair came from the previous male and female parent. Thus, a corn plant with 20 chromosomes in reality has 10 *pairs* of chromosomes. In the first cell division of meiosis one member of each pair of chromosomes is drawn by the spindle fiber to opposite ends of the cell, and when the cell wall is formed to separate the two resulting nuclei each of the two daughter cells has one-half the original chromosome number of the mother cell. This first division is followed by a second division in which each chromosome, as is the case in mitosis, splits longitudinally to form two more cells. The final result of meiosis is the production of four cells, each with half the chromosome number of the progenitor cell from which they were derived. By further nuclear divisions, without forming cell walls, a male gamete with two nuclei (vegetative and generative) is formed and from the generative nucleus two sperm cells are produced. A female gamete with eight nuclei is formed. Of the eight nuclei in the female gamete, one is the egg cell, two are called polar nuclei, two are called synergids, and three are called antipodals. The role of these male and female gametes is given in the following section.

Fertilization in Plants

When pollen from the anther is applied to the stigma of the pistillate flower, the pollen grain germinates and a pollen tube grows down through the stylar tissue. Growth of the pollen tube is under the control of the vegetative nucleus in the male gamete. The two sperm nuclei migrate down the pollen tube as it progresses in the stylar tissue. In corn the stylar tissue of the pistil is the silk. Thus, the pollen tube must grow several inches before it reaches the ovary. In other plants, like alfalfa and clover, the pistil may be only a few millimeters long. The pollen tube grows into the ovary and penetrates the ovary sac (the female gamete cell), where the two sperm cells are released. One of the sperm cells fuses with the egg nucleus to form a *zygote,* which after subsequent growth becomes the *embryo.* The other sperm cell fuses with the two polar nuclei, ultimately to form the *endosperm.*

Although varying somewhat in detail, the process of reduction division to form gametes and the union of gametes to restore chromosome numbers to the original level in the zygote is a universal process called *alternation of generations.* In some lower plant forms the *gametophytic generation* (1N) may be the more conspicuous part of the plant, but in seed plants the *spermatophytic generation* (2N) is the larger.

Relation of Reduction Division to Inheritance

To illustrate the relationships between reduction division, fertilization, and inheritance of characters, let us assume that a plant breeder has two cultivars of barley, one with rough awns (barbed) and the other with smooth awns. These two plant characters are designated RR in the rough-awned parent and rr in the smooth-awned parent. By reduction division all gametes of the rough-awned parent would have the single gene R on one of its seven chromosomes, and the smooth-awned parent would have the single gene r on one of its chromosomes. If these two parents were crossed, using the rough-awned cultivar as the male parent, its sperm nucleus carrying the gene R would unite with the egg nucleus carrying the gene r. The resulting zygote would be designated Rr. The F_1 generation plants would have rough awns because the rough-awn character is dominant over the smooth-awn character.

In reduction division of the F_1 generation plants, two kinds of male and female gametes would be produced, one with gene R and the other with gene r in a ratio of 1 to 1. If this F_1 plant were allowed to self-pollinate, the following kinds of zygotes would be formed:

Female gamete R with the male gamete R	$= 1\ RR$
Female gamete R with the male gamete r	$= 1\ Rr$
Female gamete r with the male gamete R	$= 1\ Rr$
Female gamete r with the male gamete r	$= 1\ rr$

Because rough awn is dominant there would be three rough-awned plants to one smooth-awned plant in the F_2 generation. This basic principle of independent recombination illustrates one of the laws formulated by Mendel. This orderly segregation and recombination could not occur unless there were an orderly process of reduction division and the union of the gametes in fertilization to form the next generation of plants.

Applications of Genetics to Plant Breeding

The major objective of the plant breeder is to combine parental cultivars in crosses to obtain new recombinations of characters not found in any of the original parents. To illustrate this point let us assume a second character in barley, such as resistance and susceptibility to stem rust. This is another character determined by a single factor pair (*AA* and *aa*), with resistance dominant. Genes for rust reaction are on different chromosomes than those for awn type; that is, the two characters are *independently* inherited. If the rough-awned parent were rust resistant, its genotype would be *RRAA,* and if the smooth-awned parent were rust susceptible, its genotype would be *rraa.* When these parent cultivars are crossed, the genotype of the F_1 would be *RrAa*. In the formation of gametes from this plant, four different kinds would be obtained, namely, *RA, Ra, rA,* and *ra.* In the random union of these gametes from self-pollination, 16 combinations are possible, as illustrated in the following:

Female Gametes	Male Gametes			
	RA	*Ra*	*rA*	*ra*
RA	*RRAA*	*RRAa*	*RrAA*	*RrAa*
Ra	*RRAa*	*RRaa*	*RrAa*	*Rraa*
rA	*RrAA*	*RrAa*	*rrAA*	*rrAa*
ra	*RrAa*	*Rraa*	*rrAa*	*rraa*

Some of these zygotes are alike in genotype, and if assembled by kinds they would be in the following ratio:

Number	Genotype	Expression in F_2
1	*RRAA*	Rough-awned, rust resistant
2	*RrAA*	Rough-awned, rust resistant
2	*RRAa*	Rough-awned, rust resistant
4	*RrAa*	Rough-awned, rust resistant
1	*RRaa*	Rough-awned, rust susceptible
2	*Rraa*	Rough-awned, rust susceptible
1	*rrAA*	Smooth-awned, rust resistant
2	*rrAa*	Smooth-awned, rust resistant
1	*rraa*	Smooth-awned, rust susceptible
16		

Plants in the F_2 that are homozygous—such as *RRAA, RRaa, rrAA,* and *rraa*—will breed true for these two characters in the next (F_3) generation, but plants that are heterozygous for one pair or two pairs of genes will segregate in the next generation. An important fact in this example is that some plants, those with the genotype *rrAA,* have a *new combination* not found in either of the original parents. These plants would be smooth-awned and rust resistant. Another important fact from this example is that the breeder must grow his material beyond the F_2 generation to determine if they are true breeding lines.

In the preceding example of a cross between two parent cultivars that differed by two factor pairs, only a small number of F_2 plants would be required to obtain plants with a new combination of characters. In crop breeding, problems are not as simple as this illustration. Most important agronomic characters—such as yielding ability, lodging resistance, time of maturity, grain or forage quality, and resistance to drought or to low winter temperatures—are determined by the action of many factor pairs. Instead of independent inheritance, genes are often linked; that is, two or more genes are on the same chromosome. Sometimes genes for a desired character are on the same chromosome as the genes for an undesirable character. As a consequence of these complexities the breeder must grow a very large population of plants in the F_2 generation, often 10,000 or more, to ensure a reasonable probability that desired recommendations will occur. From this large number he selects those which appear to have desired traits and grows their progenies in the next generation to determine the effectiveness of selection. This process of selecting and testing progenies is repeated until true breeding lines are obtained. Although this type of research requires a great deal of time and effort it has unusual appeal because it is creative. The production of a new and superior cultivar, when widely grown in its area of adaptation, may give millions of dollars of added farm income, an exceptionally high return on investment in research.

Methods of Field Crop Pollination

Before attempting to develop a breeding plan for any crop it is necessary to know its method of pollination. Most field crops can be divided into three general categories: (1) those naturally self-pollinated, (2) those often cross-pollinated, and (3) those naturally cross-pollinated.

1. Naturally Self-Pollinated Crops

Examples: Wheat, oats, barley, flax, rice, soybeans, peas, beans. In these crops both the staminate and pistillate parts are enclosed within the glumes or petals of a single flower. Although the glumes during anthesis are forced slightly open by the pressure exerted at their base by the enlarging lodicules, the quantity of pollen shed outside the glumes is very small in comparison with that which falls directly on the stigma. Only a small percentage of natural crossing, normally less than 1 percent, occurs between nearby plants.

2. Often Cross-Pollinated Crops

Examples: Cotton, sorghum, sudangrass. Crops in this group may vary considerably in the extent to which cross-pollination occurs; accurate classification of crops in this group is difficult. Species in this group have perfect flowers, but their mode of pollination is such that receptive stigmas may be exposed to both self- and cross-pollination, either by wind-blown pollen, such as that of the sorghums, or by pollen carried by insects, as is the case with cotton.

3. Naturally Cross-Pollinated Crops

Examples: Corn, rye, smooth bromegrass, crested wheatgrass, orchardgrass, timothy, alfalfa, red clover, white clover, sweetclover, birdsfoot trefoil, and a large number of other forage grasses and legumes. Species of crop plants classified as naturally cross-pollinated may be divided into three groups: (a) those with perfect flowers, (b) those with monoecious flowers, and (c) those with dioecious flowers.

A. THOSE WITH PERFECT FLOWERS. Crops in this group, such as rye, bromegrass, and red clover, are structurally capable of natural self-pollination, but normal self-pollination is greatly reduced because of self-sterility. In grasses the glumes open widely at anthesis, exposing the stigma, thus providing maximum opportunity for cross-pollination. In addition, the anthers generally are extruded to the outside of the lemma and palea before

FIGURE 41–1. *A bromegrass spikelet within one minute after anthesis, showing the feathery stigmas and the anthers dangling at the end of their long filaments. Bromegrass is a naturally cross-pollinated crop.*

anthesis occurs. In legumes, in addition to varying degrees of self-sterility, flowers are generally attractive to pollinating insects.

B. THOSE WITH MONOECIOUS FLOWERS. Crops in the monoecious group, such as corn, have two types of flowers. The staminate flowers in corn are in the tassel, whereas the pistillate flowers form on the cob. Although both male and female flower parts are on the same plant, there is only limited opportunity for natural self-pollination.

C. DIOECIOUS CROPS. Crops in the dioecious group, such as hemp, cannot be self-pollinated, because an individual plant produces either all staminate or all pistillate flowers. Complete separation of sexes in these plants presents problems in breeding somewhat comparable to those in the improvement of domestic animals.

Methods of Breeding Self-Pollinated Crops

Some of the earliest work in crop improvement was with the small grains. The development of improved cultivars may be accomplished in several ways, depending on the specific needs and available facilities. Only a very brief description can be given of the following procedures.

Improvement Through Plant Introductions

In the earlier periods of crop improvement, extensive programs of plant introduction brought in cultivars developed in other parts of the world. One of the most striking examples of such an introduction is the Turkey winter wheat, brought to Kansas by a small group of Mennonites from southern Russia, and Red Fife spring wheat, obtained indirectly from Poland through Scotland and Canada. Soybean cultivars first grown in the United States were introductions from the Orient. Sorghum types came to this country as introductions from Africa.

Plants flowed into the United States in a haphazard and uncontrolled manner until 1898, when a special controlling office was formed in the USDA. New, introduced plants are held in quarantine at a Plant Introduction Station for a few months up to five years. These plants are observed in the greenhouse for unwanted microorganisms, insects, insect eggs, and diseases. After the period of quarantine, they can be distributed. Regional or Federal Plant Introduction Stations are located at Geneva, New York; Glenn Dale, Maryland; Experiment, Georgia; Savannah, Georgia; Coconut Grove, Florida; Sturgeon Bay, Wisconsin; Ames, Iowa; Pullman, Washington; and Chico, California. Some Stations are specialty Centers, such as the Interregional Potato Introduction Station at Sturgeon Bay.

Currently an estimated 8,000 new plant introductions are brought into the United States each year, most at the request of plant breeders. In recent years plant introductions are primarily used as source material for specific characters. In oats, for example, genes for crown-rust resistance were ob-

FIGURE 41–2. *Improvement through plant introduction: Geneticist R. L. Bernard of the USDA examines a vinelike ancestor of modern soybeans. Grown from seed collected in the Orient by Dr. Bernard, these ancient plants may provide new cultivars with resistance to disease, insects, and nematodes.* [SOURCE: USDA.]

tained from 'Landhafer,' a cultivar introduced from South America. Examples of the current emphasis in introductions are:

1. *Corn*—Hundreds of races are being introduced from Central and South America to broaden our gene base and give added disease resistance.
2. *Cotton*—Many plants are introduced from Mexico and Central America and tested for resistance to insects such as pink bollworm and lygus bug, and diseases such as verticillium wilt.
3. *Soybeans*—Types from the Oriental countries are being tested for resistance to the soybean cyst nematode, particularly race 4.

Improvement by Selection

Two methods of selection are practiced in breeding new cultivars of self-pollinated crops: mass selection and pure-line selection. In mass selection elimination of undesirable types may be achieved by roguing them as the cultivar grows in the field. Another approach is to tag desirable

A

B

C

FIGURE 41–3. *Small-scale wheat hybridization. (A) The seed parent of wheat must be emasculated before any pollen is shed in this self-pollinated crop. Forceps are used to remove the anthers. Glumes have been cut back to facilitate emasculation and pollination. (B) After emasculation, head bags are slipped over the spike to prevent natural cross-pollination. (C) The stigma of the seed parent is pollinated with pollen collected from the pollen parent. Tags are attached to the plants to show the parents and the date of the cross.* [SOURCE: Pioneer Hi-Bred International, Inc.]

types as they grow and harvest only tagged plants at maturity. In each instance the best plants in the cultivar are identified and bulked. Although it has been an effective breeding method, mass selection often is too slow. A refinement of mass selection is to harvest the best plants separately and grow them out as pure lines for comparison.

Pure-line selection was used in the early breeding for wilt resistance in flax by Bolley at the North Dakota station. When flax was grown on wilt-infested soil, Bolley saved seed from the few surviving plants; from these the leading cultivars were developed through more than 25 years. The pure-line method consisted of selecting a large number of single heads and observing their progenies when grown in short rows. The most desirable progeny rows were saved and compared with existing cultivars. Lack of continued progress, together with the greater opportunities for improvement by hybridization, led to the gradual decline of this method of breeding.

Improvement by Hybridization

Use of hybridization in plant breeding was necessary when it became apparent that pure-line selection could not solve problems confronting the plant breeder. This change was a gradual one. A better understanding of genetic principles was first required, and information was needed on sources of parental material and on the inheritance of characters to be recombined. During the past two decades nearly all breeding programs with self-pollinated crops have been based on crosses between parents carefully chosen to provide genetic variability for subsequent selection.

Different patterns of breeding by hybridization have come into general use, each adapted to the solution of specific problems. Regardless of the method used, certain basic concepts are followed. These include (1) a clear definition of the needs for improvement, based on the deficiencies of the existing cultivars; (2) careful choice of parents whose combined characteristics would meet these needs; and (3) adequate facilities to grow and evaluate segregates obtained after crosses have been made. Different systems followed in hybridization may be stated briefly as follows.

THE PEDIGREE METHOD. In the pedigree system individual plants are selected in the F_2 generation following the cross. Plants saved in the F_2 generation are harvested and threshed individually and their seed is space-planted the next year (F_3 generation) in short progeny rows to permit further study of individual plants. The best progeny rows are saved and the most desirable plants are again harvested individually and threshed. Their seeds are planted for the F_4 generation and handled in the same manner as that described for the F_3. This process is continued to the F_5 or F_6 generation. At this stage of breeding many of the lines will be uniform because successive generations of self-pollination results in homozygosity. Plants in uniform lines are then harvested on a row basis and tested for three to five years in replicated trials to eliminate those not superior to existing cultivars.

FIGURE 41–4. *Large-scale production of hybrid wheat can be accomplished by planting alternate strips of male sterile seed parent and male fertile pollen parent. Hybrid wheat is produced in the male sterile strips and is harvested and sold as hybrid seed wheat. Pollen parent strips are harvested and fed or marketed at local elevators.* [SOURCE: DeKalb Ag-Research, Inc.]

FIGURE 41–5. *A grain-sorghum breeding nursery, with selected plants bagged. Grain sorghum is an often cross-pollinated crop. Commercial production of hybrids is possible through the utilization of male sterility.* [SOURCE: Southern Illinois University.]

A   B

FIGURE 41–6. *Cotton breeding: (A) Cotton flower ready for emasculation and crossing. (B) Anthers are collected from the pollen parent flower (right) with a short section of a soda straw and the straw containing the anthers is slipped over the protruding stigma of the emasculated flower. The straw remains on the flower to protect it from foreign pollen, and the plant is tagged to indicate the parents and date of the cross.* [Photographs with permission of the Mississippi Agriculture and Forestry Experiment Station.]

THE BULK METHOD. The bulk system of breeding differs from the pedigree method in that single plants are not selected in each generation following the cross. Instead plants are harvested as bulk populations. Individual plants are first selected from the F_5 or F_6 generation. The selected lines are evaluated in rod-row trials.

THE BACKCROSS METHOD. The backcross system of breeding self-pollinated crops is different from either the pedigree or the bulk method. It is primarily used when the breeder has a desirable cultivar that is deficient in only one or two simply inherited characters. In the backcross system the desirable cultivar (called the recurrent parent) is crossed to another cultivar (called the donor parent) which has the gene or genes that the recurrent parent lacks. Rather than permit the F_1 plants to self-pollinate, as in the pedigree and bulk methods, they are crossed to the recurrent parent, hence the designation *backcross*. Progenies from the first backcross, designated BC_1, that have the gene or genes from the donor parent are again backcrossed to the recurrent parent. This procedure of selecting plants with the desired trait and then backcrossing is continued for several generations, usually to BC_5 or BC_6.

At the end of backcrossing, a generation of self-pollination is necessary to fix the gene from the donor parent in a homozygous condition. If proper selection has been made the "recovered" strains should resemble closely the original recurrent parent and differ from it only by the desired gene added from the donor parent.

Methods of Breeding Cross-Pollinated Crops

Perennial Forages

If one examines individual plants in cross-pollinated species—such as smooth bromegrass, orchardgrass, birdsfoot trefoil, and alfalfa—a striking feature is the wide variation in plant type which is the result of crossing among plants, a continuous source of new genetic recombinations. When these variable populations are subjected to different environmental forces, natural selection brings about gradual changes to produce strains differing in important characteristics. For example, when strains of red clover were collected from different sources in the United States and Europe, those from areas in which winter temperatures were not severe proved to be nonhardy. Much of the early work in the improvement of perennial cross-pollinated forages consisted of testing strains found in different areas. Natural selection resulted in the development of superior strains. Well-known examples are 'Grimm' alfalfa and the types of smooth bromegrass represented by the 'Lincoln,' 'Fischer,' and 'Achenbach' strains. These bromes are believed to have been originally introduced from southern Europe and are distinctly different from the northern type from Russia.

The objective in breeding work is largely to search for and isolate superior individual plants and to use them in the production of new strains and cultivars. The first step is to establish a nursery from different sources

FIGURE 41–7. *Cages for controlling pollination by bees in alfalfa breeding work at the New Mexico Station. Alfalfa is a cross-pollinated perennial forage legume.* [Photograph with permission of New Mexico State University Agricultural Experiment Station.]

or strains of the species. Plants are evaluated in the field nursery, perhaps consisting of 5,000 or more plants, for one or more years for vigor, leafiness, resistance to diseases, cold and drought tolerance, and so on. The next step is to evaluate general combining ability, as measured by the performance of outcrossed progeny. Three kinds of outcrossed progenies can be obtained: (1) those from open-pollinated seeds produced on the plant growing in the nursery, (2) those from a special isolated nursery planted with clone members (vegetative reproduction) of only selected plants, and (3) those from a special nursery planted with clone members of selected plants, arranged in alternate rows with a commercial strain of the same species.

Outcrossed seeds are planted in a replicated plot trial, preferably in broadcast seedings, and are evaluated for important characters for which improvement is sought. On the basis of these tests a few of the most promising clones, usually 8–10, are finally chosen and permitted to intercross in an isolated plot to produce the first generation of a new synthetic cultivar. Because the first generation of the synthetic cultivar may yield higher than the second generation, seeds from the second increase are used to compare the improved strain with existing strains and cultivars.

No mention has been made of the need for inbreeding, as is done in corn. Many species of forage crops exhibit a high degree of self-sterility; consequently, it is not even possible to produce inbred lines. When inbreeding is possible, observation of first-year "selfed" progenies may provide additional information on the potential value of a plant for inclusion in a synthetic cultivar.

Because breeding cross-pollinated perennial forage species is more time-consuming than breeding for such crops as corn, progress made to date has not been as striking. Notable advances, however, have been made in certain cases, particularly with alfalfa.

Hybrid Corn

The development of hybrid corn is one of the most important agricultural advances made in all the years since man first began cultivating food-bearing plants. Before the planting of hybrid corn, yields of 30–50 bushels/acre (1,882–3,136 kg/ha) were considered quite satisfactory. The acre yield more than doubled in a 30-year period after hybrids began to be planted, and the current average yield in the United States is more than 85 bushels/acre (5,331 kg/ha). Some states regularly average more than 100 bushels/acre (6,272 kg/ha); for such statewide average yields, many farmers must produce regularly 150 bushels/acre (9,408 kg/ha). Although many factors enter into the achievement of these high yields, the availability of hybrids has played a vital role.

No hybrid corn seed was commercially available prior to 1927, at which time it was offered first for sale on a continuing commercial basis by the Pioneer Hi-Bred Corn Company, organized by Henry A. Wallace. Through

FIGURE 41–8. *The inflorescences of a corn plant. Corn is classified as monoecious and is a cross-pollinated crop.*

many centuries corn growers saved their seed for the next season's crop each year as the crop was harvested. The revolutionary swing to the purchase each year of hybrid corn seed was made almost as rapidly as seed could be made available. There were three good reasons for this rapid changeover. First, it had been shown that the hybrids were markedly more productive than even the best open-pollinated cultivars. In addition, they were much more disease resistant and had significantly better stalk strength. At present the planting of hybrid seed corn is almost universal throughout the United States.

The Concept of Hybrid Corn Breeding

The idea of inbreeding corn appears to have originated with Dr. George Shull of Princeton University, although Dr. E. N. East at the Connecticut Agricultural Experiment Station worked on the same genetic problems at the same time. Both published in 1908. The primary interest of both appears to have been in basic genetic research rather than in corn breeding. Publications by East and Hayes in 1912 and by East and Jones in 1919 gave the most complete early information on the concept of utilizing hybrid vigor as a means of increasing corn productivity. Shull, in a paper in 1914, was the first to suggest the term *heterosis* for hybrid vigor. As a practical means of utilizing the heterosis of hybridization, Jones suggested in 1919 using seed of a cross between two single crosses to produce the seed to be used by the commercial corn grower.

How Hybrid Corn Seed Is Produced

The term *hybrid corn* applies to first-generation seed obtained by crossing single cross hybrids or selected inbred lines, lines selfed through a number of years.

Not all inbred lines are useful. The hybrid corn breeders must work with a tremendous number of plants in order to find the very infrequent combination of inbreds that has outstanding superiority. Although several methods of predicting the performance of inbreds, and of hybrids in combination, have been worked out, no means has been found to evaluate accurately either an inbred or a hybrid except by comparing them with other inbreds and hybrids under the same environmental conditions.

Hybrid seed corn commercially available may be a single cross, three-way cross, or four-way (double) cross. A single cross involves crossing two inbred lines, such as A × B. Single crosses did not become popular until the mid-1960s because the seed yield is low and thus the seed is more expensive. The three-way cross is produced by crossing a single cross hybrid with an inbred line, such (A × B) × C. The four way or double cross involves two single cross hybrids, such as (A × B) × (C × D). Alternate plantings of these two single crosses are made in a plot well isolated from any other corn. Usually, six rows of the ear-producing single cross are alternated with two rows of the pollen producing single cross. The tassels of the plants in the ear-producing single cross are pulled out (detasseling)

FIGURE 41–9. *In this corn breeding plot, ear shoots are bagged to prevent fertilization by stray pollen, and tassels are covered to collect pollen for selfing or for controlling crossing. Corn plants are selfed for several generations to produce inbred lines for use in producing hybrids.* [SOURCE: Southern Illinois University.]

A

B

FIGURE 41–10. *Field production of hybrid corn. (A) Detasseling machines move through hybrid seed production fields removing tassels from female rows so they can be pollinated by male rows. (B) Hand detasseling by hired workers follows mechanical detasseling machines to make sure the job is complete. The plants with tassels serve as the male parent to pollinate the detasseled female (ear) parent plants. Seed from the female rows is sold to farmers by seed companies as hybrid corn seed.* [SOURCE: DeKalb AgResearch, Inc.]

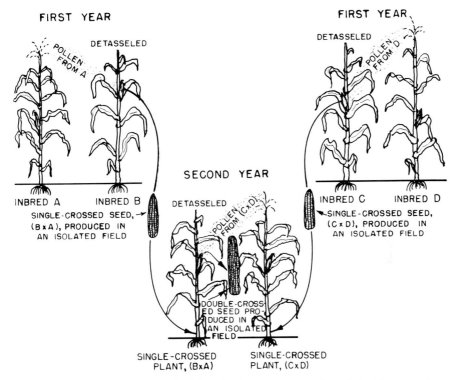

FIGURE 41-11. *The method by which single-crossed and double-crossed seed are produced is illustrated.*

before any pollen has been shed, or male sterile lines are used as ear parents, in this way ensuring the controlled cross-pollination desired. In producing the single-crosses the same general procedure is followed except that, because of low pollen yield, ordinarily only two rows of the ear-producing self are alternated with one row of the pollen-producing self. Three-way crosses are produced by using an inbred line as the pollen parent and a single cross hybrid as the ear parent, thereby obtaining greater vigor and higher yields of seed to be sold to farmers as hybrid corn. Performance of single crosses and three-way crosses has been sufficiently superior to double crosses to make their use profitable, in spite of the higher cost of seed. The use of double cross seed likely will become obsolete.

A tremendous amount of painstaking work through a period of years is necessary in order to find inbred lines which when combined in crosses with certain other selected and proved inbreds give the desired results. To obtain such results four inbred lines that have been found to yield well in all six possible single-cross combinations are used. These lines also must have been proved superior for lodging resistance, for other desirable characteristics such as disease and insect resistance, and for adaptation

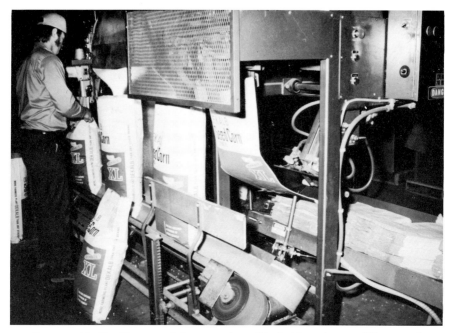

FIGURE 41–12. *Commercial processing of hybrid seed corn. When seed corn is harvested, it is taken to production buildings where it is husked, shelled, cleaned, sorted, dried, separated by size, and then bagged. Automatic bagging machines such as the one shown are capable of bagging, tagging, and sewing 16 bags per minute.* [SOURCE: DeKalb AgResearch, Inc.]

to the climatic conditions for which the particular hybrid is being produced.

It is evident that it is not practically possible for corn growers to produce their own hybrid seed. Such seed must be produced by organizations and individuals that make the production of hybrid corn seed their special business.

Male Sterility

The necessity of detasseling all the ear-producing parent plants in producing the single, three-way, and double-cross hybrid seed has been estimated to cost $20–$50 per acre. This necessitates the availability at a critical time of a rather larger labor supply; such availability is not always certain. Male sterility has been used, making most of this detasseling unnecessary. In such cases, the male reproductive organs, the anthers, are maldeveloped or aborted, so that no viable pollen is formed. This male sterility may result from genetic or cytoplasmic causes.

Several different sources of germplasm carrying the cytoplasmic sterility factor have become available, as have the necessary fertility restorers. Texas male sterile cytoplasm was used extensively in producing hybrid corn prior to 1970, when such corn was devastated by the southern corn leaf blight organism. This caused hybrid seed producers to return to the use of normal cytoplasm, which necessitated detasseling, and to search for other types of male sterility.

Polyploidy

Polyploidy is of value to the plant breeder because it adds genetic diversity. When species have more than two chromosome sets in their somatic cells they are referred to as polyploids. In contrast to the normal diploid ($2n$), they may be triploid ($3n$), tetraploid ($4n$), and so on. The basic number of chromosomes is referred to as a genome. Multiples of the same genome are autoploids. It is estimated that one third of the angiosperm species and over two thirds of the grass species are polyploid. Many cultivated crop species have evolved in nature as polyploids, including common wheat, cultivated oats, American upland cotton, tobacco, and sorghum.

Polyploid plants may arise by duplication of the chromosome sets for a single species or by combining chromosome sets from two or more species, *alloploidy.* In nature alloploidy is the more common. An alloploid in which the total chromosome complement of two species is combined to form a fertile species hybrid is called an *amphidiploid.* An amphidiploid receiving much attention is the rye–wheat hybrid, triticale.

Generally, autoploids are stockier and more vigorous than the diploids from which they are derived. They tend to have larger leaves, darker color, larger flowers, and larger seed. Autoploids of soybeans, barley, and other crops tend to have more vegetative growth and less seed. Tetraploid corn has more Vitamin A, and tetraploid tobacco contains more nicotine.

The use of colchicine, an akalaloid, has enabled the plant breeder to produce a vast number of polyploids. Although other chemicals have been used, colchicine has been the most successful. Colchicine interferes with spindle formation and prevents migration of the divided chromosomes to the poles during meiosis. The restitution nucleus has twice the normal number of chromosomes.

Improvement by Mutation Methods

Many plant breeders have explored the use of mutagens, agents which induce mutations, on crop plants. Mutagens are used primarily as a means of developing variability and have been most effective where specific objectives were desired. Mutagens may affect plants or plant parts by including changes in genes and structural changes in chromosomes. These effects are not mutually exclusive. Breaks in chromosomes involving structural changes may cause gene mutations, interpreted as gene changes.

With the advent of the atomic age, several ionizing mutagenic agents in addition to X rays are being used, including gamma rays, beta particles, fast electrons or cathode rays, alpha particles, protons, and neutrons. In addition, there has been an increasing interest in the use of chemical mutagens, including alkylating agents (mustards, ethylene imines, methane sulfonates, beta-propiolactone), urethane, the alkaloids, the peroxides, formaldehydes, and substances related to nucleic nitrous acids. Ultraviolet radiation also has been used with some degree of success.

Examples of new cultivars developed through the use of mutagenic agents are 'Luther' barley, with the chemical diethyl sulfate, and 'Interstate' sericea lespedeza, by means of irradiation.

Genetic Vulnerability

The National Research Council of the National Academy of Sciences, in a 1972 report entitled "Genetic Vulnerability of Major Crops," warned that most major crops are uniform and impressively vulnerable. Most cultivars used have a narrow genetic base. Many of the technological advances in crop production depend on a small number of genes. The dwarf wheats and rices of the Green Revolution are prominent examples. Genetic vulnerability was partly responsible for the southern corn leaf blight outbreak in 1970, which destroyed 500 million bushels of corn. The corn that was susceptible was virtually the only type grown that year.

Entire regions of the United States are devoted to near-identical crops. Cultivars grown in an area are frequently similar in appearance and heredity. Such narrow genetic bases leads to large areas that are vulnerable to disease or insect pests.

The National Research Council of the National Academy of Sciences reported that large acreages in the United States are dominated by only limited cultivars, as indicated by the following list:

Crop	Percent of Acreage	Number of Cultivars
Corn	71	6
Wheat	50	9
Soybeans	56	6
Rice	65	4
Cotton	53	2
Peanuts	95	9
Sugarbeets	42	2
Dry beans	60	2
Potatoes	72	4
Sweetpotatoes	69	1
Millet	100	3
Snap beans	76	3
Peas	96	2

Farmers can avoid risks if they plant several different hybrids or cultivars each year. Other possible solutions to lessen the risk are the use of synthetic cultivars, made up of a number of base parent lines; the use of blends of two or more strains; and the use of multiline cultivars, a controlled blending of closely related lines, each with differences in resistance to pests.

Trends in Crop Improvement

Some relatively new developments in crop improvement offer considerable promise in speeding the process of developing and releasing new cultivars, and possibly in revolutionizing hybridization techniques. These developments include *cloning,* or *micropropagation,* and the *cellular fusion* technique.

CLONING OR MICROPROPAGATION. Cloning can involve propagation of a plant either from meristematic cells or tissues or from somatic cells. Other propagation efforts involve producing haploid plants directly from pollen grains or from sex cells.

Cells or tissue is transferred to an artificial medium, where a cluster of cells called callus is formed. With the appropriate growing medium,

FIGURE 41–13. *A USDA researcher is shown culturing soybean tissues in solutions containing radioactive compounds. He follows the pathways these compounds take in the growing tissue to study how the plant produces protein and oil in a quest for the development of new cultivars.* [SOURCE: USDA.]

and possibly the addition of hormones, callus tissue can reproduce and differentiate into organs and eventually produce entire plants almost identical to the parent. Carlson at Michigan State has succeeded in regenerating whole plants of tobacco, potatoes, wheat, and barley from single somatic cells or tissue. Regeneration of plants from cells can be done with more than 50 plant species, and is used commercially by horticulturists. The major advantages of such a technique are (1) the resulting material is almost always pathogen-free, (2) the time to develop new cultivars is reduced greatly, and (3) a collection of plant material can be maintained in a small space.

Researchers at the University of Guelph in Canada have succeeded in producing haploid barley plants directly from sex cells. Chromosomes are doubled to the diploid number with the chemical colchicine, which results in a homozygous plant identical to the parent. Commercial companies presently are using this technique as a major shortcut to reduce hand crossing, greenhouse growing, and selection procedures. This accelerated breeding system can allow development of new cultivars ready for field testing in about 1½ years, as compared to 5–10 years required by conventional methods.

Iowa State University researchers have produced tobacco and several other plants from pollen grains. Such haploid plants are induced to double chromosomes, thereby producing true-breeding genetic stock plants. This bypasses the several generations of inbreeding required with some species to produce true breeding lines. Chinese scientists have reported success with the pollen technique in breeding rice and wheat.

CELLULAR FUSION. The most dramatic result of cell or tissue culture could be cross-species hybridization by means of cellular fusion, which bypasses incompatibility barriers and possibly could result in man-made or synthetic species. Before fusion can occur the rigid cell walls must be removed. Chemicals such as polyethylene glycol have been found to induce cell fusion at a high rate. Among the world leaders in this technique is the National Research Council's Prairie Regional Laboratory in Canada. This group has succeeded in producing hybrid cells of widely diverse plants such as barley–soybeans, soybeans–corn, soybeans–pea, and pea–carrot. But so far fused hybrid cells have not been induced to divide and grow into new plants.

It is exciting to consider the possibilities of a corn–grain sorghum cross with greater drought tolerance than corn, a grain crop with nitrogen-fixation ability, and even more exotic possibilities. The Canadian group does not expect or propose the development of such highly exotic, "superplants." They do expect to be able to make improvements in plant characteristics such as yield, protein content and disease resistance.

REFERENCES AND SUGGESTED READINGS

1. ALLARD, R. W. *Principles of Plant Breeding* (New York: Wiley, 1960).
2. ANONYMOUS. "Somatic Hybridization—A Key to New Plant Breeding," Mich. Agr. Exp. Sta. Spec. Res. Rep. 319 (Dec., 1976).
3. BRIGGS, F. N., and P. F. KNOWLES. *Introduction to Plant Breeding* (New York: Reinhold, 1967).
4. BROWN, W. L. *A Talk: Current Trends in Corn Breeding,* Iowa Seed Dealers Assn., Des Moines (1968).
5. CRABB, R. A. *Hybrid Corn Makers* (New Brunswick, N.J.: Rutgers Univ. Press, 1947).
6. DOBZHANSKY, T. *Genetics and the Origin of Species* (New York: Columbia Univ. Press, 1951).
7. DULL, S. "Cellular Fusion," In *The Furrow,* pp. 24–25 (1975).
8. ELLIOT, F. C. *Plant Breeding and Cytogenetics* (New York: McGraw-Hill, 1958).
9. FREY, K. J. (Ed.). *Plant Breeding* (Ames: Iowa State Univ. Press, 1966).
10. HARVEY, P. H. "How Vulnerable Are Our Crops?" *Crops and Soils,* 27(2):15–17 (1974).
11. HAYES, H. K. *A Professor's Story of Hybrid Corn* (Minneapolis: Burgess, 1963).
12. HEYNE, E. G., and G. S. SMITH. "Wheat Breeding," In *Wheat and Wheat Improvement,* ASA Monograph 13, edited by K. S. Quisenberry and L. P. Reitz (Madison: American Soc. of Agronomy, 1967).
13. JUGENHEIMER, R. W. "Germplasm for Present and Future Use," In *Corn, Improvement, Seed Production, and Uses* (New York: Wiley, 1976).
14. KLOSE, N. *America's Crop Heritage* (Ames: Iowa State Univ. Press, 1950).
15. LERNER, I. M. *The Genetic Basis of Selection* (New York: Wiley, 1958).
16. MANGELSDORF, P. D. *Genetics in the 20th Century* (New York: Macmillan, 1951).
17. MARTIN, J. H., W. H. LEONARD, and D. L. STAMP. *Principles of Field Crop Production* (New York: Macmillan, 1976).
18. MOSEMAN, A. H. "International Needs in Plant Breeding Research," *Proc. Plant Breed. Symposium* (Ames: Iowa State Univ. Press, 1966), pp. 409–419.
19. MYERS, W. M. Eng. (Ed.). *Recent Plant Breeding Research.* Svalof, 1946–1961 (New York: Wiley, 1963).
20. National Research Council, National Academy of Science. *Genetic Vulnerability of Major Crops* (1972).
21. POEHLMAN, J. M. *Breeding Field Crops* (Westport, Conn.: AVI, 1977).
22. SCHERY, R. W. *Plants for Man* (Englewood Cliffs, N.J.: Prentice-Hall, 1952).
23. SPRAGUE, G. F. *Corn and Corn Improvement* (New York: Academic Press, 1955).

CROP TERMS

Achene: A small, dry indehiscent, 1-seed fruit having a thin pericarp that is free from the seed, typical of buckwheat.

Acid soil: A soil with a reaction below a pH of 7.0 (usually less than pH 6.6). More technically, a soil having a preponderance of hydrogen ions.

Adventitious: Out of the ordinary place (as for buds or roots).

Aerobic bacteria: Bacteria that need free oxygen to live and function.

Aftermath: The second growth of meadow plants after the first crop has been cut.

Agronomy: The science of crop production and soil management.

Aleurone: The outer layer of cells, consisting of protein, of the seed.

Alkali soil: A soil containing alkali salts, usually sodium carbonate (with a pH value of 8.5 and higher).

Alkaline soil: A soil with a pH above 7.0 (usually above pH 7.3).

Alloploid or **Allopolyploid:** An organism with more than two sets of chromosomes in its body cells, each set derived from a different species.

Alternate: One after another singly at the nodes; often refers to leaf arrangement.

Amphiploid or **Amphidiploid:** An individual originating by hybridization between species and possessing the total chromosome complement of the parent species. Generally produced by doubling the chromosome number of the F₁ hybrid plant.

Anaerobic: Able to live in the absence of free oxygen.

Annual: Of one year's duration.

Anther: The pollen-bearing part of the stamen.

Anthesis: The time of the opening of the flower.

Apomixis: A type of asexual production of seeds (without gametic union), as in Kentucky bluegrass.

Arable: Land capable, without further substantial improvement, of producing crops requiring tillage.

Arid climate: A climate having annual precipitation of usually less than 10 inches (25 cm).

Asexual: Reproduction not involving the germ or sexual cells.

Auricle: Clawlike appendage projecting from the collar of the leaf.

Autoploid or **Autopolyploid:** An organism with more than two sets of chromosomes in its body cells, both sets derived from the same species.

Awn: A bristle-like structure, usually attached to the glume or lemma.

Backcross: Cross of a hybrid with one of the parental types.

Beard: The long awn or bristlelike hair of grasses borne at the top or on the backs of the lemma.

Biennial: A plant that produces leaves and roots the first year and in the second year flowers and develops seed, then dies.

Blade: The flat portion of a grass leaf above the sheath.

Biometry: The science dealing with the application of statistical methods to biological problems.

Biotype: A population in which all individuals have an identical genotype.

Bloat: Excessive accumulation of gases in rumen of an animal.

Boll: The subspherical or ovoid fruit of flax or cotton.

Boot: The upper leaf sheath of a grass; the stage at which the inflorescence expands the boot.

Brush: The pistillate end of the grass caryopsis, such as in wheat and oats.

Cambium: The growing layer of the stem.

Capsule: A dry, dehiscent fruit containing two or more carpels.

Carbohydrates: The chief constituents of plants, including sugars, starches, and cellulose (the ratio of molecules of hydrogen to oxygen is 2:1).

Carbon–nitrogen ratio: The relative proportion by weight of organic carbon to nitrogen in a soil.

Carotene: A yellow pigment in green leaves and other plant parts. Certain of the carotenoid pigments are a source of Vitamin A.

Caryopsis: The seed (grain) or fruit of grasses.

Cereal: A grass grown primarily for its edible grain and used for food or feed.

Character: The expression of a gene as revealed in the phenotype.

Chlorophyll: The green vegetable pigment of plants which is essential to the process of photosynthesis.

Chlorosis: Yellowing or blanching of leaves and other parts of chlorophyll-bearing plants.

Chromosomes: Dark-staining rodlike or threadlike bodies visible under the microscope in the nucleus of the cell at the time of cell division. The chromosomes carry the genes which control the development of the different plant characters.

Clay: Small mineral soil particles less than 0.002 mm in diameter (formerly less than 0.005 mm.).

Clones: Individual plants propagated vegetatively by rooting portions from a single original plant.

Coleoptile: A modified leaf sheath which surrounds the growing point and foliage leaves of seedlings of some plants, e.g., grasses.

Combine: A machine for harvesting and threshing seed in one operation.

Combining ability, specific: The performance of specific combinations of genetic strains in crosses in relation to the average performance of all combinations.

Companion crop: A crop sown with another crop, which is harvested separately. Used particularly for small grains in which forage crops are grown.

Convergent improvement: A more or less definite system of crossing, back-pollinating and selfing or sibbing, all accompanied by selection, in an effort to improve inbred lines of corn without interfering with their behavior in hybrid combinations.

Corm: A nutrient storage enlargement at the base of the stem, as in timothy.

Cotyledon: Seed-leaf; the primary leaf or leaves in the embryo. In some plants the cotyledon always remains in the seed; in others (as bean) it emerges on germination.

Coumarin: The odor and flavor substance of sweet clover; also found in lesser amounts in other plants.

Cover crop: A soil-protecting crop planted and left on the land to prevent leaching and erosion by wind or water (usually turned under for soil improvement).

Crop: Normally refers to plant products harvested by man.

Crop residue: A name given to cornstalks, legume chaff, and small grain straw that is returned to the land in grain farming.

Cross-fertilization or **cross-pollination:** Fertilization of the flower by pollen from another plant.

Crossing over: An interchange of segments between the chromatids of two homologus chromosomes at meiosis.

Cross inoculation: The inoculation by symbiotic bacteria of one legume species by those of another legume species.

Crown: The base of the stems where roots arise.

Crude fiber: The coarse, fibrous portion of plants, relatively low in digestibility and nutritive value.

Culm: The jointed stem of grasses.

Cultivar: A group of cultivated plants, distinguished by any significant character and which when reproduced sexually or asexually retain their distinguishing features; a variety.

Cytoplasm: The protoplasm of a cell, exclusive of the nucleus.

Dehiscence: The method or process of opening of a seed pod, or an anther.

Denitrification: The reduction of nitrates to nitrites, ammonia, and free nitrogen by soil organisms.

Dicotyledonous: Denoting a plant

which produces two seed leaves in each seed.

Dioecious: Denoting plants which bear the male and female flowers on separate plants.

Diploid: Having two sets of chromosomes. Body tissues of higher plants and animals ordinarily are diploid.

Disease: A disturbance of, or a deviation from, either the normal structure or the normal function of a plant, or both.

Dominant: (1) That one of two contrasted parental characters which appears in the individuals of the first hybrid generation to the exclusion of the alternative "recessive" character. (2) An individual possessing a dominant character in contrast to those individuals which lack that character and which are called "recessive."

Dormancy: Resting stage through which ripe seeds usually pass and in which nearly all manifestations of life come to an almost complete standstill.

Double cross: The progeny resulting from crossing two single crosses.

Double fertilization: Fusion of one of the sperm (male nuclei) from the pollen grain with the egg (female nucleus) to produce the embryo, and fusion of the other sperm with the polar nuclei to produce the endosperm.

Drill: A machine for sowing seeds in furrow.

Ear-to-row selection: Selection based on the performance of rows, each of which was planted with seed from an individual ear of corn.

Embryo: The rudimentary plantlet within seed; the germ.

Endosperm: The substance that surrounds the embryo in many seeds, as the starchy part of a kernel of wheat or corn.

Epicotyl: That portion of the axis of a seedling plant above the cotyledonary node.

Epigeal: Cotyledons emerging aboveground, as in soybean.

Erosion (land): The wearing away of the land surface by running water, wind, and other geological agents.

Exotic plant: An introduced plant not fully naturalized or acclimated.

F_1: The first generation offspring of a given mating.

F_2: The second generation; the first hybrid generation in which segregation occurs.

Fallow: Cropland left idle, usually for one growing season, while the soil is being cultivated to control weeds and conserve moisture.

Fertility (plant): The ability to produce sexually; the opposite of sterility.

Fertility (soil): The ability of a soil to provide the nutrients necessary for plant growth.

Fertilization (plant): The union of the male (pollen) nucleus with the female (egg) cell.

Fibrous root: A slender, much-branched, threadlike root, as in grasses.

Floret: A small flower from an inflorescence, as in a grass panicle or a composite head.

Fodder: Corn, sorghum, or other coarse grasses harvested whole and cured in an erect position.

Forage: Generally, such livestock feeds as pasturage, browse, hay, straw, and silage.

Forage crops: Plants grazed by animals or harvested for soiling, silage, or hay.

Fruit: The entire structure developed from the ovary after fertilization.

Fumigant: A toxic gas or volatile substance that is used to disinfest certain areas from various pests.

Fungicide: A chemical substance used to kill fungi.

Fungus: An undifferentiated plant lacking chlorophyll and conductive tissues.

Gamete: A mature male or female sex cell.

Gene: The unit of inheritance transmitted in the germ cell.

Genetics: The science which seeks to account for the resemblances and the differences that are exhibited among organisms related by descent.

Genotype: (1) The genetic makeup of an organism; the sum total of its genes, both dominant and recessive. (2) A group of organisms with the same genetic makeup.

Germ cells: Cells specialized in sexual reproduction; the ova and spermatozoa in animals, the egg cells and pollen grains in plants.

Germination: Resumption of growth of the embryo after it has been dormant; sprouting.

Glabrous: Smooth; not hairy.

Glaucous: Covered with whitish waxy bloom.

Glumes: The pair of bracts at the base of a spikelet.

Gluten: The protein in wheat flour, which enables dough to rise.

Grain: (1) A caryopsis. (2) A collective term for the cereals. (3) Cereal seed in bulk.

Grass: Botanically, any plant of the family *Gramineae.* Generally, in grassland agriculture, grass includes the forage species of *Gramineae* grown either alone or with legumes.

Grazing capacity: Number of animals a given pasture will support at a given time, or for a given period of time.

Green manure: Any crop or plant grown and plowed under to improve the soil.

Growth inhibitor: A natural substance that inhibits the growth of a plant.

Growth regulator: A natural substance that regulates the enlargement, division, or activation of plant cells.

Haploid: Single; referring to the reduced number of chromosomes in the mature germ cells of bisexual organisms.

Hay: Harvested forage of the finer stemmed crops that has been cured by drying.

Head: An indeterminate type of inflorescence (flowers in dense cluster).

Height classes in sorghum: For popular usage, three classes are recognized: *standard, dwarf,* and *combine* types, as representative of varieties carrying one, two, and three pairs of recessive genes for height, respectively. For more technical use, in describing genetic height classes, in which there may be zero, one, two, three or four pairs of recessive genes for height, the designations *tall, single dwarf, double dwarf, triple dwarf,* and *quadruple dwarf,* respectively, are proposed.

Herbaceous: Denoting plant growth relatively free of woody tissue, characterized by the absence of a persistent stem above the soil surface; dying down each year.

Herbage: The leaves, stems, and other succulent parts of plants upon which animals feed.

Herbicide: Any chemical substance used to kill herbaceous plants.

Heredity: The transmission of genetic characters from parents to progeny; the genetic characters transmitted to an individual by its parents.

Heterosis: Increased growth stimulus, often exhibited in the F_1 generation of a cross.

Heterozygous: Containing two unlike genes of an allelomorphic pair in the corresponding loci (positions) of a pair of chromosomes; the progeny does not breed true.

Hilum: The scar of the seed; its place of attachment.

Hirsute: Hairy.

Homozygous: Containing two like genes of an allelomorphic pair in the corresponding loci of a pair of chromosomes. The progeny of a homozygous plant breed true.

Hormone: A chemical growth-regulating substance that can be or is produced by a living organism.

Humus: The well-decomposed, more or less stable part of the organic matter of the soil.

Hybrid: The first generation offspring of a cross between two individuals differing in one or more genes.

Hybrid vigor: The increase in vigor beyond that of the parents of the hybrid.

Hybridization: Cross-fertilization of plants belonging to different genotypes.

Hydrocyanic acid: A poison, also called prussic acid, developed under certain conditions by cyanogenetic species.

Hydroponics: The growing of plants in aqueous chemical solutions.

Hypocotyl: The stem of the embryo or young seedling below the cotyledons.

Hypogeal: Cotyledons remaining underground, as in pea and the grasses.

Imperfect: Referring to a flower lacking either stamens or pistils.

Inbred line: (1) A pure line usually originating by self-pollination and selection. (2) The products of inbreeding.

Inbreeding: The mating of related individuals.

Indeterminate inflorescence: A plant that blooms over a long period (characteristic of plants such as buckwheat and alsike clover).

Indigenous: Produced or living naturally in a specific environment.

Inflorescence: The flowering part of a plant.

Inoculation: Introduction of *Rhizobium* (legume bacteria) on seed or into the soil; i.e., symbiotic bacteria for the benefit of legumes.

Insecticide: A chemical used to kill insects.

Integument: Walls of an ovule; future seed coat.

Internode: Region of the stem between two successive nodes.

Ion: An electrically charged element, group of elements, or particle.

Kafir: A type of grain sorghum, usually with straight heads; stout, juicy stalks; and small starchy seeds. Should be restricted to varieties of sorghum tracing to the original kafir type to which it was first applied or to more recent introductions with similar plant characteristics.

Keel: (1) A ridge resembling the keel of a boat on a plant part, as on the glume of durum wheat. (2) The pair of united petals in a legume flower.

Kernel: The matured body of an ovule, as a "kernel" of corn.

Leaching: Removal of nutrient materials in solution from the soil (usually in gravitational water).

Leaflet: One part of a compound leaf; secondary leaf.

Legume: (1) Any plant of the family *Leguminosae*. (2) The pod of a leguminous plant dehiscing on both sutures.

Lemma: The lower of the two bracts enclosing each floret in the grass spikelet.

Ley: In American terms, interpreted as a biennial or perennial hay or pasture portion of a rotation including cultivated crops.

Ligule: A membranous projection on the inner side of a leaf at the top of the sheath of many grasses.

Lime: As commonly used in agricultural terminology, calcium carbonate, $CaCO_3$, or calcium hydroxide, $Ca(OH)_2$. Agricultural lime refers to any of these compounds, with or without magnesium.

Limestone: Rocks composed essentially of $CaCO_3$.

Line: A group of individuals from a common ancestry. Genetically, a more narrowly defined group than a strain or variety.

Linkage: The type of inheritance in which the factors tend to remain together in the general process of segregation because located in the same chromosome.

Lister: An implement for furrowing land, often with also a planter attachment.

Loam: A soil composed of a mixture of two or more of the separates, clay, silt, sand, and gravel.

Lodicule: The organs at the base of the ovary of a grass floret which swell and force open the lemma and palea during anthesis.

Male sterility: A condition in which pollen is absent or nonfunctional in flowering plants.

Mass selection: A system of breeding in which seed from individuals selected on the basis of phenotype is composited and used to grow the next generation.

Meadow: An area covered with fine-stemmed forage plants grown for hay.

Meiosis: Two successive nuclear divisions, in the course of which the diploid chromosome number is reduced to the haploid.

Mendel's law: A theory of heredity as put forth by Mendel.

Micropyle: The opening through which the pollen tube had grown during fertilization.

Milo: A type of grain sorghum, usually with a compact elliptical or oval head borne on an erect peduncle, occasionally recurved or goose-necked, with slender pithy stalks and large starchy seeds. Should be restricted to varieties of sorghum tracing directly to the original milo type to which it was first applied, or to more recent introductions with similar plant characteristics.

Mitosis: A process of nuclear division in which the chromosomes are duplicated longitudinally, forming two daughter nuclei, each having a chromosome complement equal to that of the original nucleus.

Monocotyledon: Plant having one cotyledon, as in the grasses.

Monoecious: Male and female parts in separate flowers on the same plant.

Morphological: Pertaining to the form and structure of plants.

Mosaic: Symptom of certain viral diseases of plants characterized by intermingled patches of normal and light green or yellowish color.

Mutation: A sudden variation, resulting from changes in a gene or genes, that is later passed on through inheritance.

Mutant: An organism which has acquired a heritable variation as a result of mutation.

Native pastures: Native vegetation in untilled areas. (The term *resident* or *unimproved* should be used instead of *native* for pastures which include mainly species which are not native.)

Nematocide: A chemical compound or physical agent that kills or inhibits nematodes.

Neutral soil: A soil neither acid nor alkaline, with a pH of about 7.0, or between 6.6 and 7.3.

Nitrification: Formation of nitrates from ammonia (as in soils by soil organisms).

Nitrogen fixation: A conversion of atmospheric (free) nitrogen to a nitrogen compound chemically, or by soil organisms, or by symbiotic nitrogen-fixing bacteria living in the nodules of legume roots.

Nodule (legumes): A tubercle, particularly one formed on legume roots by the symbiotic nitrogen-fixing bacteria of the genus *Rhizobium*. (In some instances nodules are formed without active fixation or nitrogen resulting).

Nucleic acid: An acidic substance containing pentose, phosphorus, and pyrimidine and purine bases. Nucleic acids determine the genetic properties of organisms.

Nucleus: A deeply staining body of specialized protoplasm containing the chromosomes within a cell.

Nurse crop: *See* Companion crop.

Nutrients (plant): Materials which furnish energy for growth; also elements essential for plant growth.

Outcross: A cross to an individual not closely related.

Ovary: Ovule-bearing part of a pistil.

Ovule: The female sex cell with the immediate surrounding parts; the future seed.

Palea: The upper of the two bracts enclosing each floret in the grass spikelet.

Panicle: An inflorescence with a main axis and subdivided branches, as in oats.

Parasitic: Living in or on another living organism.

Pasture renovation: The process of subduing a pasture sod by cultivation, followed by fertilizing and reseeding for an improved pasture without subjecting the land to a row crop.

Pathogen: An organism capable of inciting a disease.

Pedicel: The stem which supports a single spikelet.

Peduncle: The stem which supports an inflorescence.

Perennial: A plant which normally continues three or more seasons.

Perfect: Denoting a flower having both pistil and stamens.

Pericarp: The modified and matured

ovary wall; as the bran layers of a cereal grain.

Permanent pasture: A pasture of perennial or self-seeding annual plants kept for several years grazing.

Petiole: The stalk of a leaf.

pH: The pH scale is the chemists' measure of acidity and alkalinity. pH 7 is neutral; pH's above this represent alkalinity; below it pH's represent acidity.

Phenotype: (1) Physical or external appearance of an organism as contrasted with its genetic constitution (genotype). (2) A group of organisms with similar physical or external makeup.

Phosphoric acid: The available phosphate, P_2O_5, of fertilizers.

Photoperiodic response: The response that a plant makes to length of day and night (light and dark), particularly in respect to floral initiation.

Photosynthesis: The process by which carbohydrates are manufactured by the chloroplasts, or chlorophyll-bearing cell granules, from CO_2 and water by means of the energy of sunlight.

Pistil: The seed-bearing organ of a flower composed of the ovary, style, and stigma.

Pistillate: Having pistils and no stamens; female.

Plumule: The apical bud of the plant.

Pod: A dry, many-seeded, dehiscent fruit; legume.

Pollen: The male sex cells produced in the anthers of flowering plants.

Pollination: The transfer of pollen from the anther to the stigma.

Polycross: An isolated group of plants or clones arranged in some fashion to facilitate random interpollination.

Polycross progeny: Progency of a selection, clone or line naturally outcrossed to other selections growing in the same isolated nursery, with all selections so arranged as to facilitate random pollination.

Polyploid: Plant having three or more basic sets of chromosomes.

Prairie: A level or rolling area of treeless land covered with grass (fertile soils usually have developed).

Profile: A vertical section of the soil through all its horizons and extending into the parental material.

Prostrate: Lying flat on the ground.

Protoplasm: The contents of a living cell.

Pubescent: Covered with fine, soft, short hairs; downy.

Pure line: A strain of organisms that is genetically pure (homozygous) because of continued inbreeding or self-fertilization, or through other means.

Race: A group of individuals having certain characteristics in common because of a common ancestry.

Raceme: An inflorescence in which the pediceled flowers are arranged on a rachis or axis.

Rachilla (little rachis): The axis of a spikelet in grasses.

Rachis: The axis of a spike or raceme.

Radicle: That part of the seed which following germination becomes the root.

Range: An extensive area of natural pastureland.

Recessive: Denoting a transmissible character which does not appear in the first-generation hybrid because it is masked by the dominant member of that factorpair from the other parent.

Reciprocal cross: Hybrids, the sexes of whose respective parents are reversed.

Recombination: Union of parental factors in individuals of the second or later generations after a cross.

Recurrent selection: A breeding system involving repeated cycles of selection and recombination with the objective of increasing the frequency of favorable genes for yield or other characteristics.

Regional strain: A strain developed under given environmental conditions as a result of the survival of the fittest through many generations.

Replication: System of repetition of an experiment to overcome accidental variations.

Reseeding variety: A variety that perpetuates itself by volunteering from seed (usually made possible because of a high percentage of "hard seed," or seed with high dormancy).

Respiration: The taking in of oxygen and giving off of carbon dioxide, for the purpose of releasing energy in plants.

Rhizome: An elongated underground stem, usually horizontal, capable of producing new shoots and roots at the node.

Rogue: (1) A variation from the type of a variety or standard, usually inferior. (2) To eliminate such inferior individuals.

Rootstock: Portion of a true root, or root with crown, capable of producing stems, or of having stems successfully grafted on them.

Rosette: A cluster of spreading or radiating basal leaves.

Rotation: A system of growing grain, legume, grass, and cultivated crops on a given area of land in such order and succession and in such a way as to keep the soil in a high state of productivity.

Roughage: Feed stuffs relatively high in crude fiber and low in total digestible nutrients, such as pasturage, hay, straw, and silage.

Runner: A creeping branch or stolon.

Saline soil: A soil containing an excess of soluble salts but not excessively alkaline (pH less than 8.5).

Sand: Small rock or mineral fragments having diameters ranging from 1 to 0.05 mm.

Scarification: The process of scratching the seed coat of "hard" or impervious seed to make germination possible.

Scutellum: That portion of the embryo which is next to the endosperm or starchy portion of the kernel; the single cotyledon of a monocot.

Seed: (1) The ripened ovule, enclosing a rudimentary plant and the food necessary for its germination. (2) To sow.

Seeding: Refers primarily to the distribution of small-seeded grasses and legumes on the surface or in the soil.

Seedling: The juvenile stage of a plant grown from seed.

Self-fertilization: The process by which egg cells of a plant are fertilized by the sperm cells of the same plant.

Selfed (self-pollinated): Pollination of an individual floret with its own pollen.

Selfing: The process of self-pollinating the plant.

Self-incompatability: Inability of a plant to set seed as a result of fertilization by its own pollen.

Self-sterile: *See* Self-incompatability.

Semiarid: A climate in which scattered short grass or shrubs prevail (usually an annual precipitation between 10 and 20 inches).

Seminal root: A root arising from the base of the hypocotyl.

Sessile: Without a pedicel or stalk.

Sheath (boot): The lower part of the leaf that encloses the stem.

Shoot: A newly developed stem with its leaves.

Short-grass prairie: A prairie covered with relatively low-growing native grasses (generally applied to the western part of the Great Plains).

Sib-mating: Mating between brother and sister.

Silage: Forage preserved in a succulent condition by partial fermentation.

Silo: A tight-walled structure for making and preserving silage.

Single cross: The progeny obtained by crossing two inbred lines.

Slip: A cutting, shoot, or leaf to be rooted for vegetative propagation.

Sod: The top few inches of soil permeated and held together with grass roots, or grass–legume roots.

Sod-bound: An unproductive nitrogen-deficient, or nitrogen-starved, grass sod.

Soiling crop: A crop grown to be cut and fed in a succulent condition.

Sowing: Refers to a uniform distribution of seeds on or in the soil by broadcasting or drilling.

Spike: An unbranched inflorescence in which the spikelets are sessile on the rachis, as in wheat and barley.

Spikelet: The unit of inflorescence in grasses, consisting of two glumes and one or more florets.

Spontaneous combustion: Self-ignition in a substance by the chemical action of its constituents, as in high-moisture hay.

Sprig: Stolon (runner) or rhizome of turf-grass with little or no adhering soil.

Square: An unopened flower bud of cotton with its subtending bracts.

Staminate: Having stamens but no pistils.

Stand: Refers to the number of plants per unit area or density of population of agronomic plants.

Sterile: Incapable of sexual reproduction.

Stigma: That part of the pistil that receives the pollen.

Stipule: One of a pair of lateral leaflike outgrowths at the base of the petiole; a part of the leaf.

Stolon (runner): A creeping stem above the soil surface (roots usually form at the nodes).

Stomata: Pores or openings in the surface of the leaves through which the gases and water vapor pass.

Stool: The aggregate of a stem and its attached tillers, i.e., a clump of stems of a single plant.

Stover: The mature, cured stalks of corn or sorghum from which the grain has been removed.

Strain: A somewhat distinct group of plants within a variety of an open-pollinated species which differ consistently from other plants of the same variety (often used synonymously with *variety*).

Stubble mulch: Crop residue left on the surface of the land that is being cropped as a means of conserving water and soil.

Style: The more or less elongated part of the pistil between the ovary and stigma.

Subhumid climate: A climate with sufficient precipitation to support a moderate to dense growth of tall and short grasses (usually 20–30 inches (51 to 76 cm) of rainfall or more).

Subspecies: A taxonomic rank immediately below that of a species.

Succotash: Grains grown in mixture, e.g., oats and barley.

Sucker: (1) A tiller. A shoot produced from a crown or rhizome from axillary buds. (2) To produce suckers. (3) To remove suckers.

Supplemental pasture: A crop used to provide grazing as a supplemental use (*examples:* Aftermath of forages grown for seed; sudangrass to supplement other pastures, etc.).

Symbiotic nitrogen fixation: The fixation of atmospheric nitrogen by bacteria which multiply in nodules on the roots of legumes.

Synthetic variety: A variety obtained by combining selected cross-pollinated lines or plants.

Tall-grass prairie: A prairie covered with tall native grasses (generally refers to that area in the eastern part of the Great Plains and in the Corn Belt originally covered with tall native grasses).

Tap-rooted: Primary root system predominant (as in alfalfa).

Tassel: (1) The staminate inflorescence of corn, composed of panicled spikes. (2) To produce tassels.

Temporary grasses: Those grasses, used as a companion crop or for quick temporary cover, which will not form a permanent turf.

Tendril: Slender, coiled organ used in climbing, as in vetch.

Testa: Seed coat.

Tetraploid: A plant with four times the primary chromosome number.

Tiller: An erect, secondary stem that grows from the crown buds of the grasses.

Tilth: The physical condition of a soil in respect to its fitness for the growth of crop plants.

Topcross: A cross between an inbred line and an open-pollinated strain of corn.

Topdressing: A selected or prepared mixture of soil which may contain physical conditioning materials, nutrients, and pesticides, and which is spread over turfgrass areas for the purpose of improving the surface, adding to the nutrient-supplying ability of the soil, or applying pesticides (used for leveling, covering stolons or sprigs in vegetative planting, and as an aid in controlling thatch and maintaining biological balance).

Total digestible nutrients (T.D.N.): The total of all the digestible organic nutrients in a feed, including protein, carbohydrates, and ether extract.

Transpiration: The evaporation of moisture through the leaves.

Tuber: A short, thickened subterranean branch, as the potato.

Turf: The upper stratum of soil filled with the roots and stems of low-growing, living plants, especially grasses.

Umbel: An indeterminate inflorescence in which peduncles of a cluster arise from the same point.

Unisexual: Flower containing either stamens or pistils, but not both.

Variegation: The barring (water mark) on leaves, as in red clover.

Variety: A group of individuals within a species that differ in some minor respect from the rest of the specie.

Venation (veination): The arrangement of veins.

Vernation: Arrangement of leaves in the bud.

Viability: State of being alive; capable of germinating.

Vigor: Strength in growth; power.

Vitamin: An organic compound, occurring in minute amounts in natural foods and feeds, which must be available to the animal in order that a specific metabolic function or reaction may proceed normally.

Warm-season grass: Species of the *Gramineae* family that make their major growth during the warmer part of the year.

Water table: The upper limit of the part of the soil or underlying material wholly saturated with water.

Weed: A plant that in its location is more harmful than beneficial; a plant out of place.

Winter annual: A plant which germinates in the fall, lives over winter, and produces its seed the following spring, after which it dies.

Xenia: The immediate visual effect of pollen on the resulting seed.

Zygote: The product of united or fused gametes.

Adapted from (1) H. D. Hughes, M. E. Heath, and D. S. Metcalfe, "Terminology," *Forages* (Ames: Iowa State Univ., 1962); (2) "A Glossary of Special Terms," *Soils and Men: Yearbook of Agriculture* (Washington, D.C.: USDA, 1938); (3) "A Glossary of Genetic Terms," *Yearbook of Agriculture* (Washington, D.C.: USDA, 1936); (4) *Manual of the Grasses of the U.S.,* USDA Misc. Publ. 200; (5) *Periodic Reports of the Committees on Terminology,* American Society of Agronomy, Madison, Wisc.; (6) A. W. Burger, *Laboratory Studies in Field Crop Science;* (7) L. F. Graber and J. M. Lund, *Laboratory Manual for Students of Agronomy;* (8) W. H. Leonard, R. M. Love, and M. E. Heath, "Crop Terminology Today," *Crop Science,* 8:257–261 (1968); and (9) several botanical glossaries and texts.

INDEX

Agricultural changes, 74–94
  future, 92–94
  present-day, 74–92
Agricultural land uses, 33–41
Agricultural research, 725–731
  goals, 726–727
  government financed, 730
  importance of, 726
  justification for, 730–731
  objectives, 727
  proved worth, 727
  reasons for, 725
  who does, 728–729
Alfalfa, 636–645
  adaptation, 637
  creeping, 640
  culture and management, 641
  diseases, 643–644
  history and distribution, 636–637
  hybrid, 644–645
  importance and use, 638–639
  improvement, 644
  insects, 644
  plant, 637–638
  seed production, 642–643
  species and types, 639–641
  utilization, 641–643
Alsike clover, 656–657
American Indian, contribution, 10–11
Arrowleaf clover, 660

Bahiagrass, 703
  plant, 703
  use and management, 703
  distribution and adaptation, 703
Ball clover, 660
Barley, 426–431
  adaptation and distribution, 426–428
  classification, 429
  insects and diseases, 431
  origin and history, 426
  plant, 428
  production practices, 431
  types, 428
  uses, 430–431
Bermudagrass, 699–702
  adaptation, 699
  culture and management, 701–702
  plant, 699
  propagation, 701
  strains, 700
  use, 699
Berseem clover, 660–661
Birdsfoot trefoil, 663–668
  adaptation, 663–664

Birdsfoot trefoil [cont.]
  big, 667–668
  cultivars, 666
  culture and management, 666–667
  distribution, 663
  importance and use, 665–666
  narrowleaf, 667
  origin and history, 663
  plant, 664
Black medic, 646
Bluestem, 711–712
  big, 711
  little, 712
  sand, 712
Buckwheat, 441–442
Buffalograss, 714
Burclover, 645–646

Carpetgrass, 703–704
  culture and management, 704
  plant, 703
  use, 704
Castorbean, 500–501
Cellular fusion, 756
Centers of production, world, 15–18
Climate and weather, 99–116
Cloning, 755–756
Clovers, 648–662
  alsike, 656–657
  arrowleaf, 660
  ball, 660
  berseem, 660–661
  cluster, 661
  crimson, 657–659
  hop, 659
  kura, 661
  lappa, 661
  Persian, 660
  red, 648–652
  rose, 661
  strawberry, 660
  subterranean, 660
  white, 653–656
  zigzag, 661
Cluster clover, 661
Contour farming, 214–215
Corn or maize, 333–365
  adaptation, 340
  diseases, 354–355
  distribution, 341–342
  fertilizers, 346–348
  for forage, 351
  harvesting, 355–356
  history, 335–336
  improvement, 358–359

Corn or maize [cont.]
  in rotations, 344–345
  insects, 353–354
  origin and development, 333–335
  plant, 336–337
  planting methods, 348–351
  production practices, 343–344
  seed selection, 348
  storage and drying, 357–358
  tillage, 345–346
  types, 337–339
  uses, 343–344
  weed control, 352–353
Cotton, 468–491
  adaptation, 469–470
  cultivation, 480–481
  diseases, 488
  distribution, 470–471
  fertilization, 478
  fiber, 475–476
  ginning, 484–487
  harvest-aid chemicals, 481
  harvesting, 481
  history, 468
  improvement, 488–489
  insect pests, 487
  mechanical harvesters, 483–484
  origin and species, 468–469
  plant, 474–476
  planting, 478–480
  production, 476–481
  seed, 475, 486–487
  seedbed, 478
  world production, 471–473
Cover crops, 216–217
Cowpea, 679–680
Crambe, 504–505
Crimson clover, 657–659
  cultivars and strains, 658–659
  culture and management, 659
  distribution and adaptation, 657
  importance and use, 657–658
  plant, 657
Crop classification, 167–175
  agronomic use, 173–174
  botanical, 167–173
  special purpose, 174
Crop improvement, 740–745
  cellular fusion, 756
  cloning, 755–756
  hybridization, 743–745
  micropropagation, 755–756
  mutation methods, 753–754
  plant introductions, 740–741
  selection, 741–743
  trends, 755–756
Crop science, 723–725
  development, 723–724
  methods, 724–725
  vs. good crop practices, 723
Cropping systems, 229–233
Crownvetch, 676–677
Cultivation, 272–277
  implements, 273–274
  methods and yield, 274–277

Dallisgrass, 702–703
  distribution and adaptation, 702
  plant, 702
  use and management, 702–703
Day length, 107
Diseases, 311–318
  control, 315–318
  infectious, 312–314
  noninfectious, 314–315
Double cropping, 229–231
Drainage, 292–296
  extent of, 293–294
  legal aspects, 295–296
  tile, 294–295
Dropseed grasses, 716
Dry edible beans, 572–574
  types, 573–574
Dry edible peas, 574–575
Dryland farming, 296–301
  and rotations, 299–301
  and tillage, 299–301
  crops, 299
Dust-mulch theory, 296–297

Exports, world agricultural, 19–26

Fertilization, in plants, 736
Fertilizers, commercial, 160–161
Fiber flax, 508
Flax, 492–494
Florida beggarweed, 682
Food crops, in worship, 5
Food production, encouraging world, 61–66
Forages, 581–596
  complementary benefits, 585–587
  economic aspects, 587
  establishment, 591–594
  legume inoculation, 587–591
  livestock use, 583–584
  nutrition research, 605–607
  relation to other feeds, 582–583

Genetic vulnerability, 754–755
Genetics, 737–738
Gramagrass, 713–714
  blue, 713–714
  sideoats, 713
Grass waterways, 213
Grasses, 684–717
  cool-season perennials, 684–698
  native and related, 707–717
  warm-season, 699–706
Grasslands, 596–598
  improved, 598
  natural, 597–598
  United States, 596–597
  world, 596
Green chop, 634
Guar, 506, 682

Hairy indigo, 681
Hay, 609–624
  baling, 615–616
  developments and future, 622–623

Hay [*cont.*]
  forms, 610–613
  harvesting, 613–615
  large packages, 616–618
  marketing, 621–622
  quality, 609–610
  system comparisons, 618–621
Hemp, 508
Henequen, 508
Herbicides, 325–328
History of Crop Production, 3–11
Hop clover, 659
Hybrid corn, 747–753
  breeding concept, 748
  how produced, 749
  male sterility, 752–753
Hybridization, 743–753
  backcross method, 745
  bulk method, 745
  improvement by, 743

Inheritance, 734–735
Insects, 303–311
  control methods, 305–311
Intercropping, 232–233
Irrigation, 279–292
  and crops, 285–287
  desalination, 288
  extent of, 280–281
  quality of water, 287
  salt-tolerant crops, 288
  sprinkler, 289–290
  subsurface, 291–292
  surface, 290–291
  United States, 281–282

Johnsongrass, 704
  plant, 704
  use and management, 704
Jute, 510

Kenaf, 510
Kentucky bluegrass, 684–686
  importance and use, 685
  plant, 684–685
  seed production, 685–686
Kudzu, 680–681
Kura clover, 661

Land, suited to production, 13
Lappa clover, 661
Legume, inoculation, 587–591
Lespedeza, 668–671
  adaptation, 669
  cultivars, 670
  culture and management, 671
  distribution, 668–669
  importance and use, 669–670
  Korean, 668
  plant, 668
  sericea, 668–671
  striate, 668
Liebig, and soil productivity, 12, 52
Lovegrass, 715–716
  sand, 715
  weeping, 715
Lupine, 677–678

Malthusian theory, 12, 51–52
Manila hemp, 508–509
Manure, 158–160
  animal, 158
  green, 159–160
Medicago species, 636–647
Mendelian theory, 734
Micropropagation, 755–756
Millets, 384–386
Moisture, 104–107
Multiple cropping, 229–233
Mustard, 502–504

Needlegrasses, 716
No-tillage, 265–272
Nutrients removed by crops, 157–158

Oats, 421–426
  adaptation and distribution, 421
  fertilization, 423
  harvesting, 424
  improvement, 424–426
  insects and diseases, 424
  origin and classification, 420
  plant, 420
  rotations, 423
  seeding, 423–424
  uses, 421–422
Orchardgrass, 689–692
  distribution and adaptation, 689–690
  importance and use, 690–691
  management, 691
  plant, 690

Pastures, 598–607
  grazing management, 603–604
  improved, 598
  northern, 598–599
  renovation, 600–603
  rotation, 604
  southern, 599
  supplementary, 604–605
  vs. drylot feeding, 607
  western, 600
Peanuts, 563–571
  cultural practices, 568–569
  diseases and insects, 571
  distribution and adaptation, 565–568
  harvesting, 570–571
  history, 563
  importance, 565
  plant, 563–565
  planting, 569–570
  rotations, 568
  utilization, 571
Peppermint, 506–507
Perilla, 508
Persian clover, 660
Plant growth, 182–186
  differentiation, 185
  diffusion, 183–184
  factors, 185–186
  movement of solutes, 184–185
  nutrients, 185
  photosynthesis, 182–183

Plant growth [cont.]
  respiration, 183
  transpiration, 183
Plant structure, 176–181
  cells, 181
  flowers, fruits, and seed, 179–181
  roots, 176
  stems and leaves, 176–179
Plowing, 255–261
  depth, 259
  time, 261
Pollination, 738–740
Pollution, 117–126
  agricultural, 118
  air, 118–119
  crop injury, 117–118
  factors in control, 125–126
  nutrients in water sources, 122–
    123
  pesticides, 120–122
  salinity in waters, 123
  sediment, 123–124
  soil, 124–125
  water, 120
Polyploidy, 753
Popcorn, 359–361
Population, 49–67
  and food supply, 49–67
  growth and control, 57
  world increase, 56–57
Potatoes, 546–562
  chemical weed control, 555
  cultivars, 555–556
  cultivation, 555
  culture, 550–558
  fertilizer and lime, 551–552
  growth requirements, 550
  harvesting, 556–557
  importance, 547
  insects and diseases, 558
  plant, 548–550
  planting, 554–555
  rotations, 550–551
  seed, 552
  storage, 557–558
  sweet, 558–561
  white, 546–558
Production adjustments, 70–95
Production factors, 41–47

Quackgrass, 709

Ramie, 509–510
Rapeseed, 502
Red clover, 648–652
  cultivars and strains, 650
  culture and management, 650–652
  distribution and adaptation, 648–
    649
  importance and use, 649–650
  plant, 649
Redtop, 694
  adaptation, 694
  culture and use, 694

Redtop [cont.]
  distribution, 694
  plant, 694
Reduction division, 736–737
Reed canarygrass, 694–697
  adaptation, 694–695
  distribution, 694
  plant, 695
  seed production, 696–697
  uses and culture, 695–696
Remote sensing, 328–329
Research, 721
Reservoirs, 217–218
Rice, 431–441
  cultivars and types, 434–437
  milling and uses, 438–440
  pests, 440–441
  plant, 432–434
  production practices, 437
Roots, 188–200
  factors in development, 191–192
  functions, 190
  secretions, 190–191
  selected crops, 192–196
  structure, 190
  top-root ratio, 196–200
  types, 188–190
Rose clover, 661
Rotations, 220–228
  advantages, 220
  and yield, 221–222
  crop sequence, 223–224
  for pest control, 222–223
  legumes in, 228–229
  sod crop, 225–227
Rough pea, 681
Rye, 409–413
  adaptation-distribution, 410
  cultivars, 412–413
  insects and diseases, 413
  origin, 410
  plant, 410–411
  production practices, 413
  uses, 412
Ryegrass, 705
  distribution and adaptation, 705
  plant, 705
  use and management, 705

Safflower, 499–500
Science, 721
Scientific method, 721–722
Seed, 235–253
  breeder, 251
  buried, 244
  certified, 249–252
  coatings, 241
  dormancy, 236–237
  federal legislation, 246–249
  foundation, 251
  germination, 238–241
  hard, 237–238
  longevity in storage, 241–244
  registered, 251–252
  scarification, 238

Seed [cont.]
  size, 244–245
  testing, 245–246
Sesame, 498–499
Silage, 625–634
  additives, 633–634
  corn, 625–627
  crops used for, 625
  ensiling process, 630–631
  grass-legume, 628–630
  high-energy, 625–628
  losses, 630–631
  sorghum, 627–628
Silos, 631–633
  horizontal, 632–633
  upright, 631–632
Sisal, 509
Small grains, 416–419
  new cultivars, 416–417
  production practices, 417–418
  weather and pests, 418–419
Smooth bromegrass, 686–688
  adaptation and distribution, 686
  cultivars, 688
  plant, 687
  seed production, 688
  utilization, 687
Soil, 143–163
  air, 146
  conservation districts, 210
  conservation movement, 209–212
  conservation practices, 212–218
  erosion, 205–208
  general characteristics, 143–144
  land-use classes, 212
  organic matter, 147–148
  organisms, 147–149
  profile, 145
  solution, 146
  survey, 212
  temperature, 146
  water movement, 145
Soils, 144–157
  acidity, 154–156
  classification, 144–145
  halomorphic, 156–157
  micronutrients, 152–154
  nitrogen, 149–150
  phosphorus, 150–151
  potassium, 151–152
Sorghum, 366–384
  adaptation, 367
  botany and cytology, 369–371
  breeding, 373
  broomcorn, 371–372
  distribution, 367–368
  fertilization, 376
  grain, 371
  harvesting, 378–379
  heat and drought resistance, 369
  in rotations, 375
  irrigation, 378
  origin and history, 366
  pests, 380–381

Sorghum [cont.]
  plant, 368–369
  planting methods, 377
  planting rates, 377–378
  production practices, 375–381
  prussic acid poisoning, 369
  seedbed preparation, 376
  storage, 379–380
  time of planting, 376–377
  types, 370–371
  uses, 372–373
  weed control, 378
Soybeans, 444–465
  fertilizers and lime, 452–453
  harvesting and storing, 457–458
  improvement, 462–465
  in rotations, 450–452
  inoculation, 453
  pests, 458–549
  plant, 449–450
  planting methods, 453–456
  processing, 459–461
  production practices, 450–459
  United States production, 445–448
  utilization, 459–462
  weed control, 456–457
  world production, 444–445
Spearmint, 507
Strawberry clover, 660
Strip cropping, 215–216
Stubble-mulch farming, 217
Subsoiling, 260–261
Subterranean clover, 660
Succulage, 634
Sudangrass, 381–384
  adaptation, 382
  cultural practices, 383–384
  for hay and silage, 384
  for pasture, 384
  plant, 382–383
Sugar, 511–528
  legislation, 511–512
Sugarbeets, 512–523
  climate and soil, 516–517
  contract production, 517
  cultivars, 522
  fertilization, 519
  harvesting, 520–522
  history, 512–515
  improvement, 522
  insects and diseases, 522
  irrigation, 517–518
  plant, 515–516
  planting, 519
  processing, 522
  rotations, 517
  seedbed, 518
  thinning, 520
  weed control, 520
Sugarcane, 523–528
  harvesting, 525–526
  history, 524–525
  pests, 526–527
  plant, 525

Sugarcane [cont.]
  planting, 525
  processing, 526
Sunflower, 494–498
Sweetclover, 671–674
  coumarin, 674
  cultivars, 673–674
  culture and management, 674
  distribution and adaptation, 671–672
  importance and use, 672–673
  plant, 672
  species, 673–674
Sweetcorn, 361–363
Sweetpotatoes, 558–561
  cultivars, 559–560
  curing and storing, 561
  diseases and insects, 561
  fertilizers, 560
  harvesting, 560–561
  plants and cuttings, 560
  rotation, 560
  season and soil, 559
  seedbed, 560
  transplanting, 560
Switchgrass, 714

Tall fescue, 692–693
  adaptation, 692
  cultivars, 693
  distribution, 692
  plant, 693
  utilization and management, 693
Temperature, 99–104
Terracing, 216
Tillage, 254–272
  history, 254–255
  listing, 257
  minimum, 263–272
  plow-plant, 267
  primary, 256–261
  secondary, 261–262
  strip, 267–268
  wheeltrack planting, 266–267
Timothy, 688–689
  plant, 688–689
Tobacco, 529–545
  classification and types, 532
  climate and soil, 534–536
  cultivars, 534
  curing, 540–542
  fertilizers and manure, 536–537
  field planting, 538
  grading and marketing, 543–544
  harvesting, 539–541
  insects and diseases, 544–545
  plant, 530–531
  rotations, 536
  seedbeds, 537–538
  stripping and packing, 542–543
  suckering, 538–539
  topping, 538
Triticale, 413–414

Velvetbean, 681
Vetches, 674–676
  adaptation, 675
  importance, 675
  use, 675

Water, 129–141
  absorption and storage, 137–139
  movement in soil, 139–141
  source of, 129–130
Water requirements of plants, 130–137
Weather, 107–115
  crop yield relationships, 112–115
  forecasting, 107–109
  modifications, 111–112
  National Agricultural Weather Service, 109–111
Weeds, 319–329
  biological control, 328
  chemical control, 325–328
  control, 323–328
  losses, 321–322
Wheat, 388–409
  classes, 395–400
  classification, 394–395
  durum, 396
  fertilizers, 401
  hard red spring, 396
  hard red winter, 396–400
  harvesting, 405
  history and origin, 393–394
  improvement, 408–409
  pasturing, 404
  pests, 405–407
  plant, 392–393
  production practices, 400–407
  seeding, 401–404
  soft red winter, 400
  United States distribution, 391–392
  utilization, 407–408
  white, 400
  world distribution, 388–391
Wheatgrass, 707–711
  crested, 708–709
  intermediate, 709
  plant, 707–708
  slender, 710–711
  tall, 709
  western, 710
White clover, 653–657
  cultivars and strains, 655–656
  culture and management, 656
  distribution and adaptation, 653
  importance and use, 654
  plant, 653
Wildryes, 716
World food problem, 11–12, 49–67
World trade, 26–29
Wormseed, 505
Wormwood, 505

Zigzag clover, 661